Biologically Active Natural Products

ACS SYMPOSIUM SERIES 380

Biologically Active Natural Products

Potential Use in Agriculture

Horace G. Cutler, EDITOR

Agricultural Research Service,

U.S. Department of Agriculture

Developed from a symposium sponsored
by the Division of Agrochemicals
at the 194th Meeting
of the American Chemical Society,
New Orleans, Louisiana,
August 30–September 4, 1987

American Chemical Society, Washington, DC 1988

Library of Congress Cataloging-in-Publication Data

Biologically active natural products.

(ACS Symposium Series, ISSN 0097-6156; 380).
"The distillation of a six-day American Chemical Society symposium from 30 August–4 September 1987"—Pref.

Includes bibliographies and index.

1. Natural products in agriculture—Congresses.

I. Cutler, Horace G., 1932– . II. American Chemical Society. Division of Agrochemicals. III. American Chemical Society. Meeting (194th: 1987: New Orleans, La.). IV. Series.

S587.45.B56 1988 631 88-7859
ISBN 0-8412-1556-1

CIP

PRINTED IN THE UNITED STATES OF AMERICA

ACS Symposium Series

M. Joan Comstock, *Series Editor*

Foreword

The ACS SYMPOSIUM SERIES was founded in 1974 to provide a medium for publishing symposia quickly in book form. The format of the Series parallels that of the continuing ADVANCES IN CHEMISTRY SERIES except that, in order to save time, the papers are not typeset but are reproduced as they are submitted by the authors in camera-ready form. Papers are reviewed under the supervision of the Editors with the assistance of the Series Advisory Board and are selected to maintain the integrity of the symposia; however, verbatim reproductions of previously published papers are not accepted. Both reviews and reports of research are acceptable, because symposia may embrace both types of presentation.

Contents

INDEXES

Preface

BIOLOGICALLY ACTIVE NATURAL PRODUCTS have been studied by investigators from many disciplines. The symposium on which this book was based brought together plant physiologists, plant pathologists, marine biologists, entomologists, plant taxonomists, plant ecologists, medicinal chemists, mycologists, pharmacologists, and others whose scientific curiosity about natural products extends into many areas. They believe and have presented evidence to show that natural products may indeed have a place in agriculture. More important, however, is their assertion that although these diverse structures have high specific activity, the residual effects in the environment may be minimal.

I am grateful to my colleagues and friends who gave refreshing insight to their research, as well as to the chairpersons who diligently sought speakers and topics for the sessions. The section on products from microorganisms was organized by Basil A. Burke of the Plant Cell Research Institute, Inc., Dublin, CA. Co-chairs of the section on products from higher plants were Richard G. Powell, Northern Regional Research Center, Agricultural Research Service, U.S. Department of Agriculture, Peoria, IL, and David E. Gianassi, Department of Botany, University of Georgia, Athens, GA. Coordinators of the section on products from insects or affecting insects were Murray S. Blum of the Department of Entomology, University of Georgia, Athens, GA, and Kevin C. Spencer of the Department of Medicinal Chemistry and Pharmacognosy, University of Illinois, Chicago, IL. I also acknowledge the generous support of the Division of Agrochemicals of the American Chemical Society.

HORACE G. CUTLER
Richard B. Russell Research Center
Agricultural Research Service
U.S. Department of Agriculture
Athens, GA 30613

May 13, 1988

Chapter 1

Natural Products and Their Potential in Agriculture

A Personal Overview

Horace G. Cutler

Richard B. Russell Research Center, Agricultural Research Service, U.S. Department of Agriculture, Athens, GA 30613

Biologically active natural products are derived from three major sources: the fermentation of microorganisms, higher plants, and insects. However, compounds derived from these sources may act within each or all of these domains. That is, compounds derived from microorganisms and higher plants may affect insects and vice versa. Often, natural products are obtained in limited quantities and small yields do not lend themselves to extensive testing. During the past three years there has been increased synthesis of natural product templates and their analogs for evaluation in biological systems. Relative to these approaches, assorted natural products from microorganisms, including oligopeptides, acyclic polyketols and some relatively simple molecules are examined. The brassinosteroids and photodynamic herbicides, from higher plants, are discussed. Finally, compounds that are produced by insects, or which affect insects, are reviewed.

In common with my primal ancestors, whose life began in a garden, my first recollections were not so much of people but of trees, flowers, the sun, rain, clouds, blue sky, and the sound of the cuckoo. Especially imprinted on my senses was the peppery smell of lupin at the early age of two, and there followed the scent of roses and English lavender: dire warnings about foxglove and deadly nightshade were issued as I wandered about gardens. An introduction to the world of secondary metabolites had started early in life and subsequently led to my first scientific job, in the mid 1950's, at the Boyce Thompson Institute for Plant Research when it was located in Yonkers, New York. While at the Institute, I came under the intellectual guidance of Lawrence J. King, who was something of a genius, and that led me into the area of plant growth regulators, especially the natural product of both

Published 1988 American Chemical Society

animal and plant origin, indole-3-acetic acid. There followed another period, at the same location, with A. J. "Chuck" Vlitos who was investigating flowering in plants, indole chemistry, and the growth promoting effects of long chain fatty alcohols in collaboration with Donald G. Crosby (a topic that was quite revolutionary at the time). This pioneering atmosphere was further enlivened by I.D.J. "Dai" Phillips who was an exchange pre-doctoral student from P.F. Wareing's laboratory in Aberystwyth, Wales. His exuberance for life, and natural products of growth regulatory ilk, was contagious. He was hot on the trail of the then structurally unknown substance from _Acer pseudoplatanus_ (and cotton bolls), abscisic acid. Contemporaneously, the team of P.W. Zimmerman and A.A. Hitchcock had synthesized the herbicide 2,4-dichlorophenoxyacetic acid at Boyce Thompson and, as they stated, that synthesis was the result of their examining the model of indole-3-acetic acid. The former being a substituted phenyl, the latter a phenyl-pyrrole. I also married one of their technicians and, I suppose, my thoughts concerning natural products for synthetic templates became subconsciously fixed as I came to know them better.

A three year stint in the West Indies (Trinidad) led me into the world of bush medicine (the use of tropical plants for medicinals) and took me down that curious path of ethnobotany. The range of useful plants was extraordinary. There were plants in the genus _Verbena_ that induced increased lactation in nursing mothers, grasses that broke fevers, and the bark of a tree which, when steeped in boiling water, gave rise to a tea that caused erections of some duration in males. The trees yielding this compound were obvious even to the untrained eye because near the local villages the bark was generally stripped from ground level to the height which an adult could reach standing on the seat of a bicycle. There are several yarns surrounding the sexual efficacy of the bark, some apocryphal, but all amusing.

The third part of the triology, insofar as the science is concerned, involved my training in the biochemistry of nematodes with Lorin Krusberg, at the University of Maryland. It happened that one of the nematodes under scrutiny, _Ditylenchus triformis_, could be cultured on the fungus _Pyrenochaeta terrestris_. Three flasks of that fungal substrate yielded more indole-3-acetic acid than I had seen from the extraction of one-quarter acre of sugarcane, and convinced me that fungi would be excellent sources of biologically active compounds. My psychological compass was, once again, fixed on a natural product course. Hence, my interests and training cut across several fields and at some point the thought crystallized that apparently divergent disciplines were inter-related.

There are several temptations that confront the author of a chapter of this nature. One such is the propensity to intellectually bludgeon anyone who dares to insinuate that natural products have no real use in agriculture either as agrochemicals, or as templates for the further synthesis of biodegradable agrochemicals, by quoting a series of products that have been used successfully. Among them would be indole butyric acid, a homolog of the original compound, indole-3-acetic acid, that first aroused the curiosity of plant physiologists because it was shown to be responsible for the phototropic response of plants. And indole-3-butyric acid has been used by amateur gardeners and horticulturalists to induce rooting in plant cuttings.

Another well known natural product is gibberellic acid (GA_3) which, again, has high specific activity and limited use in grapes and celery where the responses are quite dramatic. We tend to forget that thirty-five years ago it was a common horticultural practice to girdle grapes by cutting into the bark of the vines just above ground level in order to increase yields. This was a delicate procedure and often resulted in the death of the mature vines. With the advent of GA_3 the art of girdling disappeared and was replaced by spraying vines at flowering time with 100 ppm GA_3 solutions to increase yields 250%! The natural product ethylene, found in fungi and higher plants, has been used in both the pure form and as a derivatized chemical (2-chloroethylphosphonic acid)to ripen a variety of crops from bananas to cherries, to oranges.

The chemistry of the insecticide pyrethrin, the natural product of the Pyrethrum daisy consisting of two viscous liquid esters $C_{21}H_{28}O_3$, or $C_{22}H_{28}O_5$, is also well known and it was precisely these molecules that laid the foundation for the synthesis of the highly successful pyrethroids. Various natural lures (pheromones) have also played a decisive part in controlling insects.

Some of the most interesting insecticidal and antiparasitic natural products to enter the field have been the avermectins. Their history is unique because of their complexity, the fact that the fermentation products go through a synthetic sequence to arrive at the final products and their duality as both agrochemicals and pharmaceuticals. Certainly, the average Board of Directors would likely be initially negative to developing such elaborate molecules.

Another temptation would be to attempt to convince the reader that in the three areas covered by the symposium -- natural products from microorganisms, secondary metabolites from higher plants, and natural products from insects or that affect insects -- there are myriad examples of secondary metabolites that have been isolated that have biological activity and considerable potential as agrochemicals. However, space does not allow for that discussion and further elaboration would also detract from the contributions made by the other authors in this volume. Instead, I have chosen a different approach, which is to discuss compounds that have been isolated in each of these three categories, have been shown to possess unique biological activity, and have then been synthesized. The reason for this approach is that in these days the burden of proof for a structure no longer necessarily requires that a compound be synthesized. The advent of sophisticated NMR techniques and X-ray crystallography have virtually eliminated that necessity. Generally, the need for synthesis arises because, in many cases, there is insufficient quantity of the metabolite to make extensive examination of its properties in biological systems. With respect to products from microbes, for example, the speed with which a fermentation system can be cranked up to produce large quantities of a derived metabolite is directly proportional to the number of substrates on which the organism can be grown, the number of people available to do the work, and the access of a high yielding strain. For those who have forgotten, the history of the discovery and production of penicillin bears recalling. And the recent synthesis of avermectin A_{1a} by Danishefsky et al., (1) lends credence to this point of view even though, at present, the avermectins are produced by fermentation of *Streptomyces avermitilis*. Finally, doing chemistry for the sake of doing chemistry has been relegated to limited situations. Funding for such ventures is not really feasible or available in today's management climate so that when funds are forthcoming they are made so for very specific reasons, and commercial ones at that.

The examples given are neither all that exist, nor do they necessarily represent those that will eventually find their way into the marketplace. They are, however, diverse structures with specific activity and, in some instances, their synthesis has been improved upon. Most of the examples represent work that has taken place during the past three years, a time during which the synthesis of natural products seems to be going through an exponential stage. Some have a certain dichotomy or polychotomy in that they are active in more than one system. That appears to be the character of biologically active natural products and, consequently, the reader may not be in total agreement as to their categorization.

Synthesis of compounds based on templates from microorganisms

Earlier, we reported that the cyclic oligopeptides were compounds that had attracted much attention from synthetic chemists because of their high specific activity and their assortment of unnatural amino acids, both L and D (2,3,4). Their usefulness as agrochemicals ranges from direct to indirect. For example, the iturins, which are cyclic octapeptides, may be used to control soft rot, *Monilinia fructicola*, in stored peaches (5); the fragments of tentoxins, a cyclic tetrapeptide, show both plant growth promotory and herbicidal properties (see Edwards, et al.; Lax, et al., in this volume). The AM toxins I, II, III from *Alternaria mali*, a pathogen of apple trees, of which AM toxins I and III are particularly potent and produce interveinal necrosis in the susceptible cultivar "Indo" within 18 hours following treatments with 0.1 ppb (6,7), are particularly useful tools for determining the mechanisms whereby these compounds act. Hence, the question might be posed: why are apple varieties like "Indo" highly susceptible to *A. mali* toxins while resistant varieties like "Jonathan" require 1 ppm of AM toxin I and 10 ppm of AM toxin III to induce necrosis? Once the site, or sites, of activity are known in the susceptible cultivars there exists the possibility of protecting these vulnerable loci from the action of the toxin(s). With regard to the selectivity of these peptide toxins, much structure-activity work has been done with AM toxin I (Figure 1). First, it has been determined that the toxin response is directly dependent upon the backbone conformation of the amino acids, which are all L, and that the presence of L-α-hydroxyisovaleric acid, α-aminoacrylic acid, L-α-amino-δ-(p-methoxyphenyl)-valeric acid are important, while the presence of L-alanine is not critical (8,9,10,11,12). Furthermore, synthesis of the retroenantio - AM toxin I in which the peptide sequence was reversed and, thereby, the configuration of each residue, revealed, upon bioassay with "Indo" apple leaves, that all biological activity was deleted. In an attempt to ascertain whether there was a specific receptor in apple leaves the enantio - AM toxin I, an antipode of that toxin, was synthesized. That is, the amino acid sequence was identical to the original toxin but they were all D amino acids. It was postulated that if the enantio-toxin was active in the bioassay the interaction between AM toxin I and the membrane of the apple leaf cell is not a biological one, but rather a physicochemical one between, say, a peptide and a lipid. However, no biological responses were noted with concentrations up to 100 ppm. This is several orders of magnitude above the threshold amount necessary to produce a biological response with AM toxin I. Therefore, it appears that the enantiotoxin did not interact with the receptor site and it suggests that the recptor recognizes the chirality of the molecule and that the receptor may be a protein. No doubt the characterization of the active site and the subsequent manipulations that must follow will have a major impact on the apple growing industry.

Certainly, the oligopeptides obtained from microorganisms offer a wealth of active products both in their original and degraded states. There is a wide range of synthetic permutations and the offering of templates on which to base the production of novel herbicides. Two examples are proffered. The first is the well known compound glyphosate (N-[phosphonomethyl] glycine), an eminently successful herbicide. The second involves some research carried out by the Tanabe Seiyaku Co., Ltd., Japan, which involves both their Applied Biochemistry and Microbiological Research Laboratories. They initially found that amino acid analogs, the α-isocyanoacetic acid derivatives, were potent seed germination inhibitors (13,14), and had similar properties in assays as 2,4-dicholophenoxyacetic acid. It is significant that the authors state, "We had great interest in peptide compounds containing the isocyano group". But of the six homologs of the N-(α-isocyanoacetic) amino acid methyl esters synthesized, none were as active as N-isocyanoacetyl-L-valine methyl ester (Figure 2) which inhibited the stems and roots of germinating cucumber seed 80-100% at 100 and 10 ppm, and inhibited the shoot germination of rice, but not the roots, at 100 ppm. There was no effect on radish germination, indicating selectivity. It should be noted that D or L-valine is an amino acid that is often found in peptidic, phytoactive natural products.

Small molecules isolated from natural sources nearly always appeal to the synthetic chemist. The pyrenochaetic acids A, B, and C (Figure 3) are in this class and have been shown to be phytotoxins. They originate in the pathogen *Pyrenochaeta terrestris* which is responsible for onion pink root and are readily produced by fermentation. The most phytotoxic of this trio is pyrenochaetic acid A which completely inhibited the root growth of onion, at 250 ppm, and lettuce at 500 ppm. Because of the relatively greater phytotoxicity, pyrenochaetic acid A was synthesized and proved to be as active as the natural product (15). In addition, the regio- isomer of pyrenochaetic acid A was produced (Figure 4) and it inhibited lettuce root growth 80% at 500 ppm. No data were given for the effects on onion root growth and comparison of the data for biological activity versus structure were tantalizingly brief. However, the synthesis of pyrenochaetic acid A led to the synthesis of the regio-isomer and this, in turn, may lead to the production of other biologically active molecules. Indeed, the functional groups are present to make some pertinent derivatives.

Nothing is more frustrating to the natural products chemist than to read of the synthesis of a biologically active compound that is identical to the original natural metabolite and to then find that neither the precursors, or final product, or derivatives of the product have been tested in biological systems. Such is the case with the synthesis of pyriculol (Figure 5) isolated from the culture broth of *Pyricularia oryzae*, the organism responsible for rice blast. Pyriculol inhibits the growth of rice seedlings and induces necrotic lesions on the leaves. Four stereoisomers of pyriculol were synthesized, one of them was identical to the natural product, that is, 3'R, 4'S (16). Another synthesis that yielded derivatives that would predictably possess biological activity was that of (S)-(-)-vertinolide (Figure 6), a tetronic acid derivative obtained from the culture broth of *Verticillium intertextum*, a fungus isolated from wilted Japanese maple trees (17). While the original chemical structure was solved by X-ray crystallography (18), the absolute configuration had to await confirmation by synthesis. But again, biological data are lacking. Similar events surround the synthesis of gregatin B (Figure 7), a phytotoxic metabolite of *Cephalosporium gregatum* and *Aspergillus panamensis* (19).

In contrast, pyrenolide B (Figure 8), a ten-membered lactone ring, was originally isolated from the culture broth of the phytopathogen, *Pyrenophora teres* in conjunction with pyrenolides A and C (20,21). The preparation of (±)-pyrenolide B involved several synthetic intermediates and, of these, seven were tested against the fungi *Aspergillus niger*, *Cochliobolus miyabeanus* and the yeast, *Saccharomyces cerevesiae* (22). Both *A. niger* and *C. miyabeanus* were significantly inhibited by

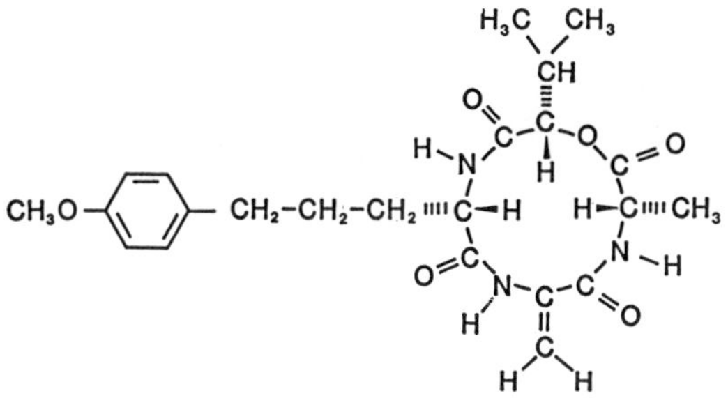

Figure 1. AM Toxin 1.

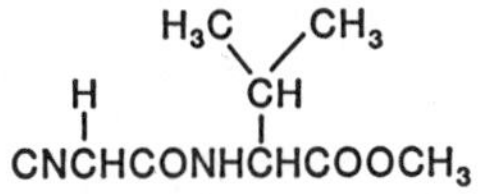

Figure 2. N-isocyanoacetyl-L-valine methyl ester.

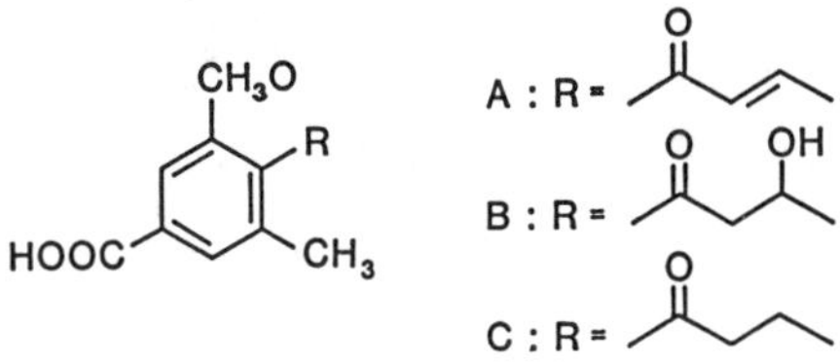

Figure 3. Pyrenochaetic acid.

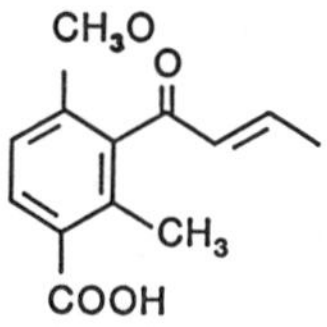

Figure 4. Pyrenochaetic acid, regio-isomer.

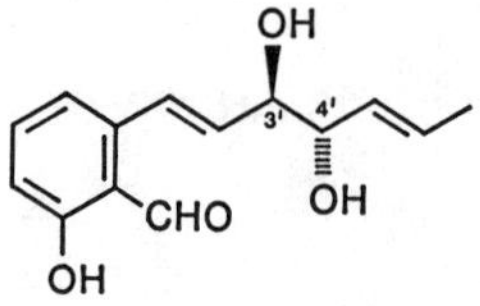

Figure 5. Pyriculol.

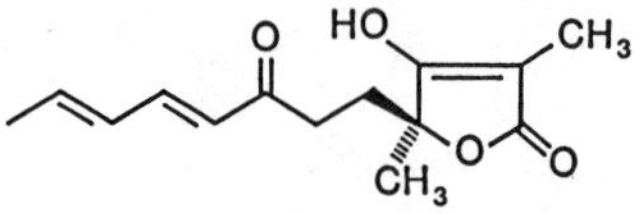

Figure 6. Vertinolide.

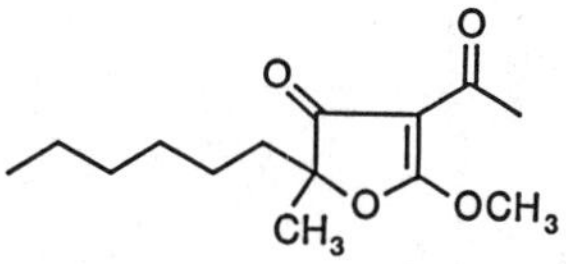

Figure 7. Gregatin B.

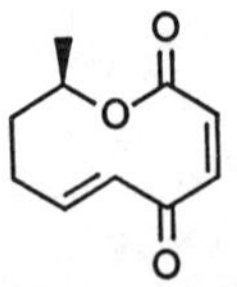

Figure 8. Pyrenolide B.

(±)-pyrenolide B at 100μg/disk and all microorganisms were slightly inhibited at 50μg/disk. But only one of the intermediates slightly inhibited the fungi at 50μg/disk and that intermediate was identical to (±)-pyrenolide B except for partial saturation of the ring and the double bond remaining intact in the dienone system. Also, (±)-pyrenolide B inhibited the growth of rice roots 40%, and shoot 30%, when applied at 100 ppm.

Other toxins have also been synthesized and include race T toxin, a corn specific toxin originally isolated from the phytopathogen Helminthosporium maydis, which is composed of an acyclic ß-polyketol with 35-45 carbons. To date, four constituents of C_{41} or C_{39} chain lengths which make up 70 to 90% of the natural toxin have been elucidated and each constituent has virtually the same biological activity as the native material. While shorter chain lengths of C_{15} to C_{26} have been synthesized (23,24) and indicate that the presence of a ß,ß'-dioxooxy group and linking of two sets of ketol groups by a $(CH_2)_5$ bridge are essential for toxicity, as in race T toxin, the increase in chain length of carbon chain or increase in the number of ketol groups had not been examined. The C_{25} chain length was the most effective of the set synthesized yet it was ten times less active than the natural toxin. Consequently, a C_{41} stereoisomeric mixture of (±)-8,16,26,34-tetrahydroxy-6,10,14,18,24,28,32,36-octaoxohentetracontane, containing four ß-ketol groups spaced by $(CH_2)_3$ and $(CH_2)_5$ bridges (which occur in race T toxin) was prepared. Also, a C_{23} analog with two of the ß-ketol groups spaced by a trimethylene bridge was made, that is, (±)-8,16-dihydroxy-6,10,14,18-tetraoxotricosane. When each synthetic product was bioassayed in leaves of susceptible corn cultivars it was shown that the C_{41} material, in fact a mixture of 10 stereoisomers, stimulated NADH oxidation by mitochondria and inhibited dark CO_2 fixation as effectively as the native toxin. The C_{23} product was approximately ten times less active than the native toxin in both these systems, and was as active as the C_{25} product discussed earlier (25). The same research group has also synthesized a stereoisomeric mixture of (±)-PM toxin B originally found in the plant pathogen, Phyllosticta maydis, which destroys corn that has Texas-male sterile cytoplasm. The toxin which is primarily composed of PM-toxin B (6,14,22,30,32-pentahydroxy-8,16,24-trioxotritriacontane) is toxic to the Texas-male sterile line at 10^{-8} to 10^{-9} M while corn with normal fertile cytoplasm is unaffected with treatments of 10^{-5} M. The mixture of stereoisomers comprising PM-toxins B and having the syn-1,3-hydroxy configuration at C_{30} and C_{32} was as specific in toxic action to corn as the native toxin (26). Two possibilities exist for each set of compounds. First, they may have application in other crops as herbicides but, as a minimum, the elucidation of the molecules that act as toxins in corn suggests that blocking agents may be found to protect susceptible cultivars.

One argument posed by critics for using natural products in agriculture concerns abscisic acid, the compound first isolated from cotton bolls, Gossypium hirsutum L., (27) and dormant Sycamore buds, Acer pseudoplatanus (28), which is a growth inhibitor in certain systems. To date, the compound has not been successfully used in agriculture even though there is ample evidence to support the fact that it mediates several important events in plant growth and development. It has been synthesized and is readily available from chemical supply houses. The 2-cis(+) form, which is the biologically active isomer, has been isolated from the fungi Cercospora cruenta (29), and Botrytis cinerea that has been irradiated with UV (30). Both the (2Z) and (2E)-deoxy-abscisic acid isomers were prepared (Figure 9) and tested against rice seedlings and lettuce seed. The (2Z) acid inhibited the growth of the seed leaf sheath of rice approximately 80% at 10^{-3} M, but lettuce seed germination was inhibited 100% by 10^{-4} M solutions. The (2E) acid inhibited rice second leaf sheaths only 20% at 10^{-3} M and lettuce seed germination was inhibited 100% at 5 X 10^{-4} M (31). It would appear that, eventually, a utilitarian homolog of abscisic acid will be found for use in agriculture.

Synthesis of compounds based on templates from higher plants. I first saw movies of J.W. Mitchell's work with the brassinosteroids in the late winter of 1963. At that time, it had been observed that an elongation response could be obtained in the bean second internode test upon treatment with extracts of rape pollen (Brassica napus L.) and some years later a formal report followed (32). The Japanese had independently observed novel growth regulatory responses with extracts of Distylium racemosum Seib. et Zucc. and these were reported as Distylium factors A and B, which produced a dramatic response in the rice lamina inclination test (33). However, HPLC developments had not been perfected and the work was temporarily set aside. The final characterization of the brassinosteroid structure appeared in 1979, specifically that of brassinolide from B. napus, as (22R,23R)-2α-3α,22,23 -tetrahydroxy-24S-methyl-β-homo-7-oxa-5α-cholestan-6-one (34) (Figure 10). But, as is usually the case with biologically active natural products, much work was accomplished between the initial discovery of brassinolide and the final structure determination concerning its biological properties. "Brassins" from rape pollen enhanced the production of vegetation and fruit when applied to bean plants (35). Later, brassins were shown to accelerate the growth of barley plants in greenhouse and growth chamber experiments when seed had been treated prior to sowing (36); and bean plants treated with brassinosteroids were shown to be affected by light quality which influenced the growth regulatory response (37). Further, it was demonstrated that brassinolide affects very specific tissues that are sensitive to indole-3-acetic acid-induced growth and tissues that are gravi-perceptive (38).

While many of the later studies appear to be a search for the mechanism of action of the brassinosteroids, the Japanese have intensively approached the problem from three angles. The isolation of new brassinosteroids, the synthesis of brassinosteroids and their analogs, and the practical application of these materials for agrochemical use. Insofar as the isolation of new brassinosteroids is concerned, two recent reviews already cover this topic in detail (39, 40) the former complete with references, unfortunately, in untranslated Japanese. Indeed, there presently exist twenty-two new brassinosteroids and one glucosidic conjugate (39). What is important is the practical application of these materials to crops. Both brassinolide and 24-epibrassinolide (Figure 11) have been used in field trials to promote the growth of young plants, according to seven Japanese references, and crop yields have been improved (41) in corn, cucumber, rice and sweet potato. Other applications have shown an increase in cold resistance in corn, cucumber, egg plant, and rice (42). And decreased injury by the herbicide simetryn, butachlor, and pretilachlor has been noted in treated rice, while wheat treated with simazine was more tolerant to that herbicide after brassinolide pre-treatment (42). Rice has been made less susceptible to salt following treatment (42). Another, as yet puzzling effect is the ability of brassinosteroids to enhance disease resistance in Chinese cabbage to soft rot and in rice to sheath blight (42). Some pictures of these effects were shown at the Joint Plant Growth Regulator Society of America - Japan Society for the Chemical Regulation of Plants by S. Marumo (39) in 1987 and the full impact of what these compounds may mean to agriculture became clear. The most remarkable slide was that of a brassinolide treated ear of corn (Zea mays L.) in which all the kernels had completely filled out to the extreme tip. The control, on the other hand, exhibited the usual 1-2 inches of totally immature kernels at the tip. This potential tip productivity represents as much as 5% of the total. Thus the potential economic impact is readily apparent and the accelerated effort by the Japanese in the area of synthesis is easily understood!

All brassinolide syntheses require as starting materials, natural sterols and their degraded products, stigmasterol, brassicasterol, pregnenolone, and dinorcholenic acid, (39). Improvements have been made on the synthesis of brassinolide (43) and other derivatives have been produced. In 1984, the synthesis of hexanor-brassinolide 22-esters was accomplished by German workers (44) and some were active in the bean second internode bioassay including the methyl, ethyl, and n-propyl ethers. The t-butyl and ethyl methyl ethers were inactive (Figure 12) and none of these derivatives contained the C_{22}, C_{23}, or C_{24} asymmetric centers. Homodolicholide and homodolichosterone (Figures 13 and 14), originally isolated from the immature seed of Dolichos lablab, are brassinosteroids that possess plant growth promoting properties. These, too, have been made (45) in a series of short-step

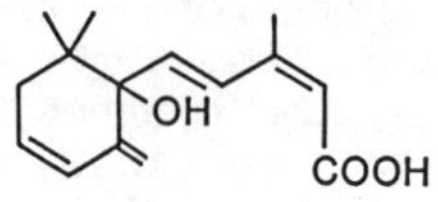

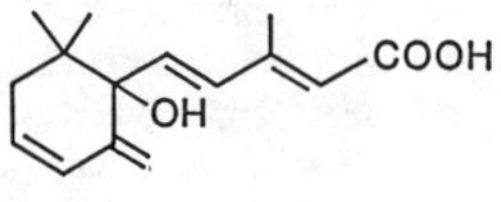

(2Z) (2E)

Figure 9. 2Z and 2E-deoxy-abscisic acid.

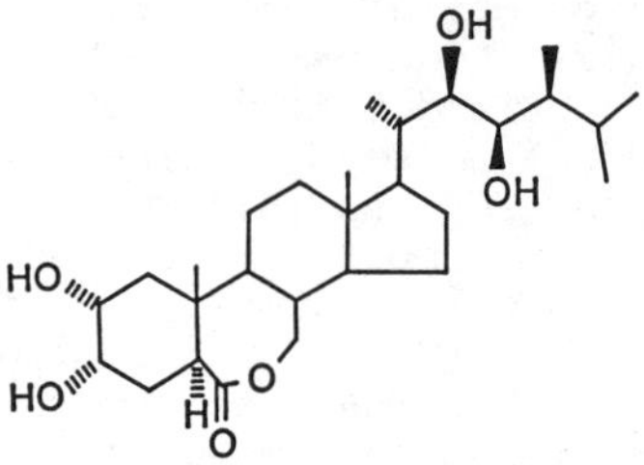

Figure 10. Brassinolide.

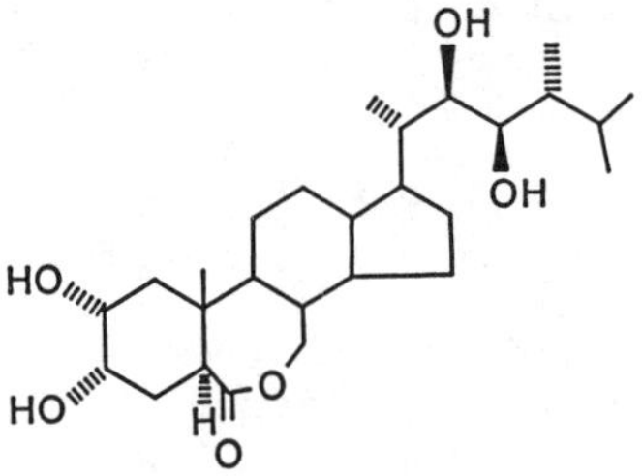

Figure 11. 24-epibrassinolide.

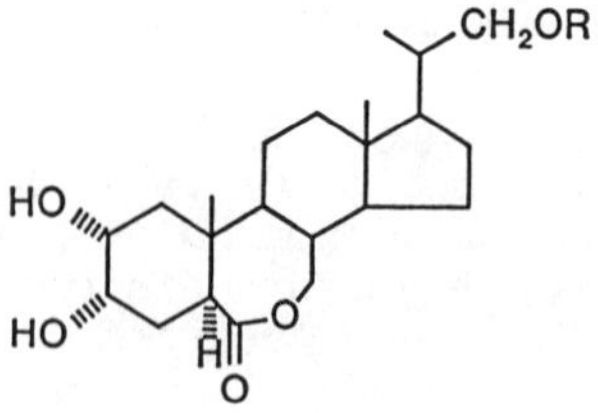

R

1. CH_3
2. CH_2CH_3
3. $CH_2CH_2CH_3$
4. $C(CH_3)_3$
5. $CH_2CH_2OCH_3$

Figure 12. Hexanor-brassinolide 22-esters.

syntheses. Recently, the facile synthesis of brassinolide, castasterone (Figure 15), teasterone (Figure 16), and typhasterol (Figure 17) has been accomplished using the key intermediate (22R,23R,24S)-3α,5-cyclo-22-23-diacetoxy-5α-ergostan-6-one (46). Fairly complex syntheses have been used to modify the 7 membered B ring of (22S,23S) and (22R,23R)- homobrassinolide to effect the production of the 7-aza, 7-thia, and 6-deoxo compounds (47) each of which exhibited weak biological activity in the rice-lamina inclination assay. From the frenetic activity surrounding the brassinosteroids it seems that a patented agrochemical may well be on the market within a few years and that product will most probably be a Japanese one that has broad spectrum application.

Another recent advance that has significance for the development of natural product agrochemicals are the photodynamic herbicides. While a recent comprehensive review has covered the topic (48), reference is given in this overview because of the implications that this new approach offers. Simply, the concept of photodynamic herbicides revolves around two principles. First, one of the vulnerable processes is that of greening during plant growth and development. Second, δ-aminolevulinic acid (Figure 18) is a natural amino acid that can supply all the atoms necessary to build protoporphyrin. Specifically, two molecules of δ-aminolevulinic acid may condense to form porphobilinogen and four molecules of this contribute to porphyrin synthesis. Chlorophyll is a magnesium-porphyrin derivative, a tetrapyrrole. It was postulated that if green plants could be chemically treated to induce large amounts of chlorophyll precursors, that is, tetrapyrroles, then a mechanism for the production of biodegradable herbicides existed. The theory being that magnesium-tetrapyrroles are type II photosensitizers (49,50,51) and absorb light energy to photosensitize the formation of singlet oxygen. This, in turn leads to a free radical chain reaction that destroys membranes, proteins, and nucleic acids (51).

Preliminary experiments were conducted with δ-aminolevulinic acid sprayed onto cucumber seedlings (Cucumis sativus L.) at rates of 525 gram/acre (48). The plants were covered with aluminum foil so that dark conversion of the δ-aminolevulinic acid could take place. Indeed, the experiments progressed as expected in that δ-aminolevulinic acid was converted to magnesium-protoporphyrins and protochlorophyllides. Upon exposure to light, treated plants rapidly degraded and died within a few hours, whereas those kept in the dark survived. It transpired that the choice of cucumber as a test plant was most serendipitous. When δ-aminolevulinic acid was tested against monocotyledonous plants such as barley, corn, oat, and wheat, the effects were negligible. Depending upon the species of plant treated, a Type I, II, or III response was obtained. Type I plants were those that died rapidly after treatment, as did cucumber. Type II plants, such as soybean, accumulated tetrapyrroles in leaves,

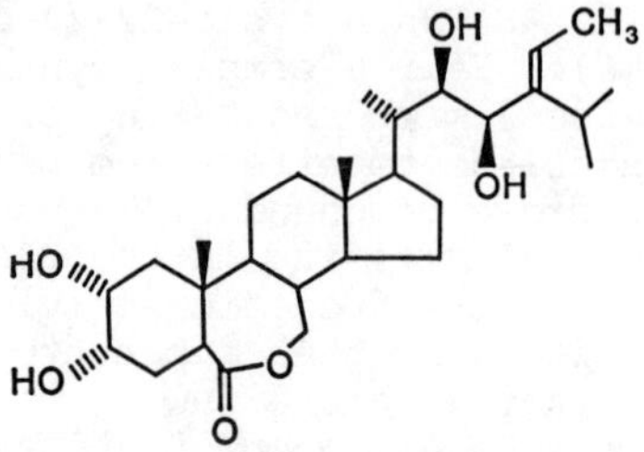

Figure 13. Homodolicholide.

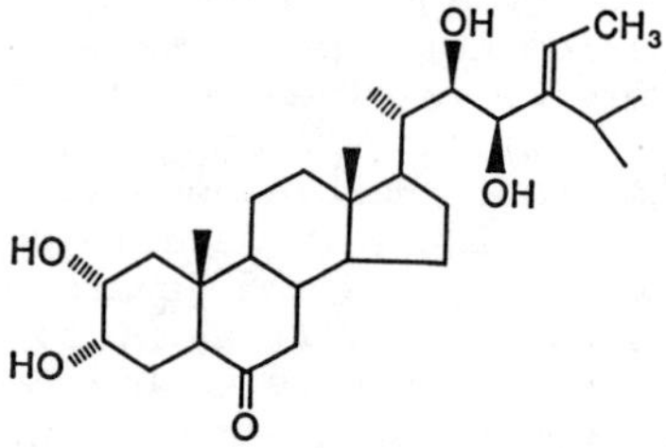

Figure 14. Homodolichosterone.

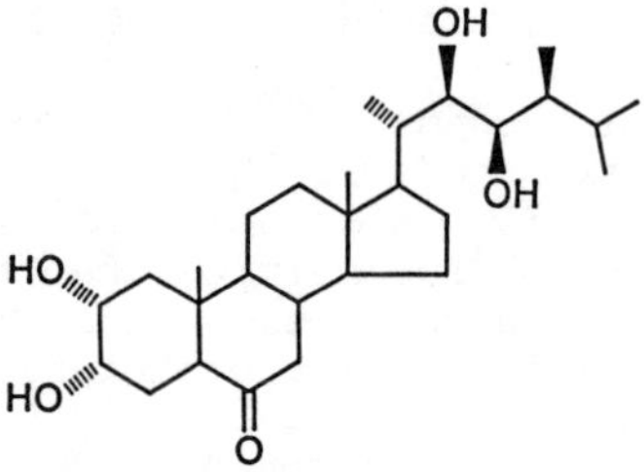

Figure 15. Castasterone.

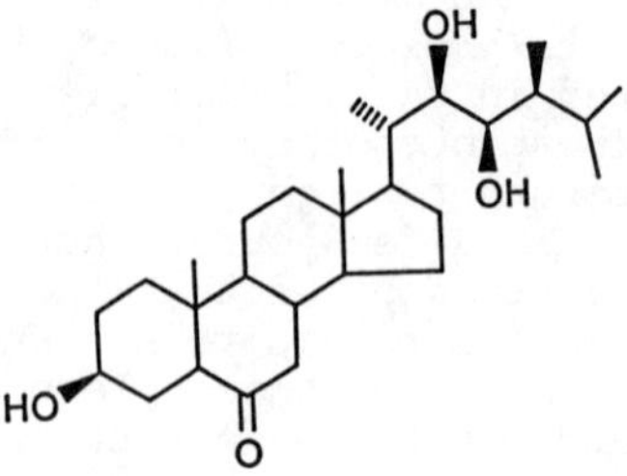

Figure 16. Teasterone.

but not cotyledons and stems and, upon exposure to light, there was photo-dynamic damage in the leaves only and plants soon recovered. Type III plants, including barley, oat, corn, and wheat, accumulated high levels of tetrapyrroles but photodynamic damage was minimal and these plants survived. Further examination revealed that there was a multibranched chlorophyll _a_ biosynthetic pathway consisting of six routes and the species-dependent photodynamic herbicide activity was directly dependent upon monocarboxylic routes 2 and 3, and 4 and 5 (48). It became apparent that certain compounds may act as photodynamic herbicide modulators and, depending upon the modulator used, the action of δ-aminolevulic acid could be enhanced. The search for modulators uncovered 13 compounds that acted in concert with δ-aminolevulinic acid and served to control the biosynthesis of chlorophyll _a_. These fell into three groups. The first group enhanced δ-aminolevulinic acid transformation to tetrapyrroles and were 2-pyridine aldoxime; 2-pyridine aldehyde; picolinic acid; 2,2'-dipyridyl disulfide, 2,2'-dipyridyl amine; 4,4'-dipyridyl, and phenanthridine. The second group were inducers of δ-aminolevulinic acid biosynthesis and subsequent tetrapyrrole accumulation and were comprised of 2,2'-dipyridyl and 1,10-phenanthroline. The final group inhibited monovinyl protochlorophyllide accumulation and these compounds were 2,3-dipyridyl; 2,4-dipyridyl; 1,7-phenanthroline; and 4,7-phenanthroline (48).

In a fine piece of scientific sleuthing, the authors put their findings to a practical test in a Kentucky bluegrass lawn that they had difficulty establishing because of the precocious growth of creeping charlie (_Glechoma hederacea_), common yellow wood sorrel (_Oxalis stricta_ L.), blackseed plantain (_Plantago rugelii_ Dene), dandelion (_Taraxacum officinale_ Weber), violet (_Viola adunea_) and musk thistle (_Carduus nutans_). After placement of each species into its type, it was decided that the best strategy would be to spray the plots with δ-aminolevulinic acid plus 2,2'-dipyridyl (524 gram of δ-aminolevulinic acid plus 403 gram 2,2'-dipyridyl/acre). With the exception of violet the control of all broadleaved weeds was excellent and, of course, the dynamics of the violet plant are presently under scrutiny (48). The lawn is becoming well established and the environment appears to be clean of any residual herbicides.

Synthesis of compounds based on templates from insects, or natural products that affect insects. Those compounds that have been isolated from insects, which possess biological activity and have subsequently been copied by chemical synthesis are, for the most part, the insect pheromones. Because this is the topic of another recent symposium honoring J. Tumlinson on the occasion of the Burdick and Jackson Award, it is not covered here. Neither is pyrethrin discussed even though it has been used as a model upon which the highly successful pyrethroids have been developed. Instead, some recent development of compounds that affect insects are illustrated.

Much has been written about the effects of black pepper (Piper nigrum L.) extracts on house flies (52,53,54) especially the compounds piperine, pellitorine, and pipericide. In a particularly imaginative piece of work, Nair et al., (55) isolated Z and E-fadyenolide (Figure 19) from Piper fadyenii with a view to accomplishing two goals. To open the fadyenolide ring at the oxygen bridge, and to replace the exocyclic carbonyl with a nitrogen thereby making a piperine analog. To produce the pyrethrin analog: that is, the structure would be almost identical to the piperine analog except that adjacent to the methyoxyl group originally present on the ring of fadyenolide there would now be a cyclopropane function. In fact, during the syntheses three hybrid compounds were produced which were amides of 3,5-dimethyloxy-4-oxo-5-phenylpent-2-enoic acid (Figure 20). Tests were conducted on adult cockroaches (Blatella germanica L.) and within 3 hours all the amides paralyzed the insects and within a day they were dead. Flour beetles (Tribolium confusum) were also challenged with amides A, B, and C and the LC_{50} was established as 30, 28, and 27 ppm, respectively. None of the products were active against cattle ticks (Boophilis microplus), but egg production appeared to be inhibited 50-70% with applications of 0.5 and 1.0μg/tick. One major drawback to the use of these compounds seems to be their photodegradation, though depending upon the circumstances this characteristic may be considered an attribute.

Other compounds of Piper species have served as templates for biologically active derivatives. For example (56), using the parent compound (2E,4E)-N-isobutyl-6-phenyl-hexa-2,4-dienamide, seven derivatives were synthesized that showed biological activity (Figure 21). Each of these analogs was made because of the relationship to tetrahydroanacyclin, a compound having weak biological activity and in which the acyclic diene system was seen as equivalent to a phenyl ring; this strategy paralleling that used in the synthesis of pyrethriods (57). All compounds (Figure 21) were up to four times more effective against super-kdr strains of adult houseflies (Musca domestica L.) than the susceptible strain. Thus, the results obtained with the N-alkylamides suggest that they may combat resistance conferred by the super-kdr genetic background which delays the onset of knockdown and kill by DDT, its analogs, and by the pyrethroids (58).

Even esoteric compounds, in the sense that they appear to be highly specific in activity against a relatively small insect population, have been synthesized. These include osmundalactone, the aglycone of osmundalin from the fern Osmunda japonia Thunberg. Osmundalactone is a congener of phomalactone, acetylphomalactone, asperlin, olguin, and phomopsolide: all oligoketides and fungal products. Also, osmundalactone is a feeding inhibitor for the larvae of the butterfly Eurema hecabe mandarina (59). Synthesis of osmundalactone, (5R,6S)- 5,6-dihydro-5-hydroxy-2H-pyran-2-one and its diasterosisomers (5S,6S)-5,6-dihydro-2H-pyan-2-one (Figure 22) was from 3-triethylsiloxy-1-propyne and (S)-1-

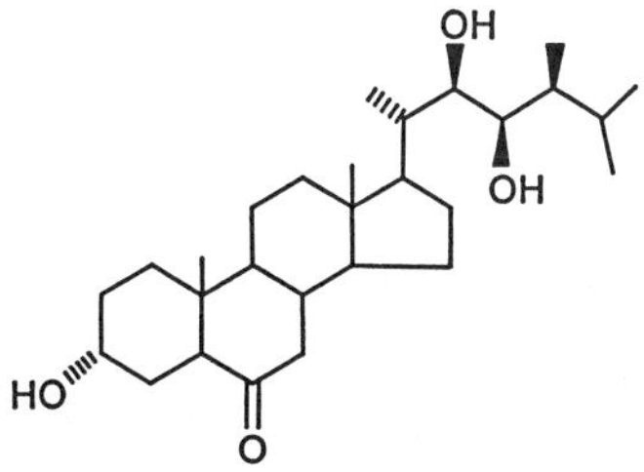

Figure 17. Typhasterol.

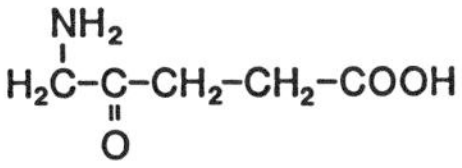

Figure 18. δ -aminolevulinic acid.

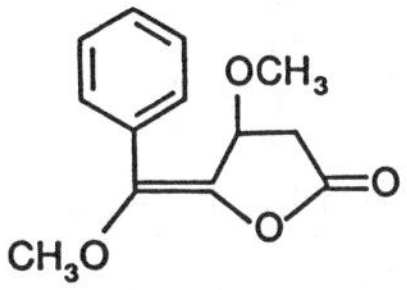

Figure 19. Z + E Fadyenolide .

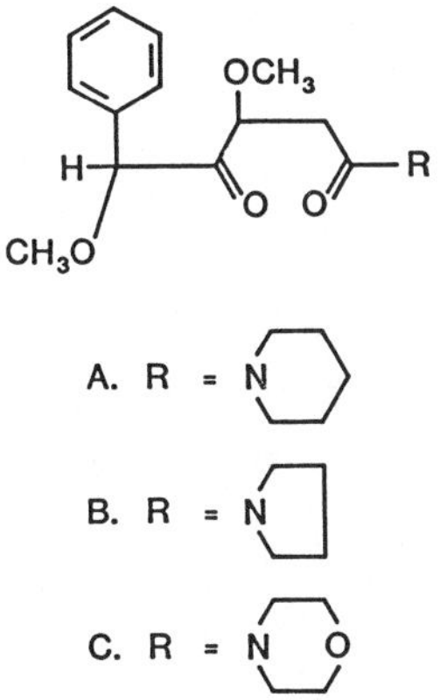

Figure 20. Amides of 3,5-dimethoxy-4-oxo-5-phenylpent-2-enoic acid.

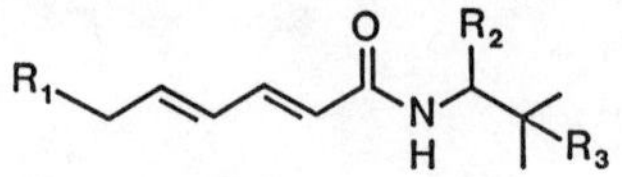

	R_1	R_2	R_3
1.	PHENYL	H	H
2.	3,5-DIFLUOROPHENYL	H	CH_3
3.	DIBENZOFURAN-3-YL	H	CH_3
4.	DIBENZOFURAN-3-YL	CH_3	H
5.	5-BROMONAPHTH-2-YL	H	CH_3
6.	5-BROMONAPHTH-2-YL	CH_3	H
7.	7-FLUORONAPHTH-2-YL	CH_3	H

Figure 21. (2E,4E)-N-isobutyl-6-phenyl-hexa-2,4-dienamide, and derivates.

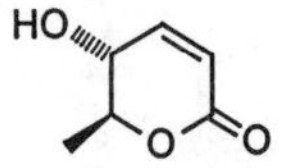

(5R,6S)-5,6-DIHYDRO-5-HYDROXY-2H-PYRAN-2-ONE

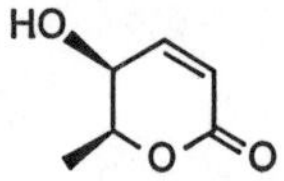

(5S,6S)-5,6-DIHYDRO-5-HYDROXY-2H-PYRAN-2-ONE

Figure 22. (5R,6S)-5,6-Dihydro-5-hydroxy-2H-pyran-2-one, (5S,6S)-5,6-Dihydro-5-hydroxy-2H-pyran-2-one.

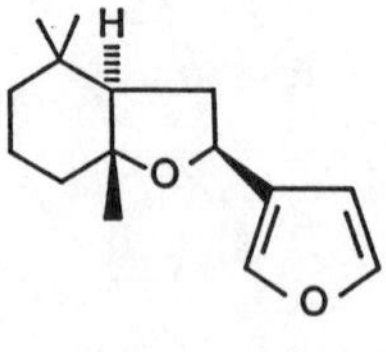

Figure 23. Ancistrofuran.

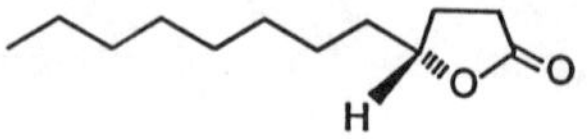

Figure 24. 4-Dodecanolide.

methyl-2-oxoethylbenzoate (60) to yield 60 mg of the former and 78 mg of the latter.

Defense secretions have also been the subject of isolation and synthesis. Two are given as recent examples. Ancistrofuran (Figure 23) is the major chemical secreted by the West African termite, Ancistrotermes cavithorax, soldier, (61). While previously somewhat complex syntheses have been reported, a simple procedure has been discovered in which 9-hydroxydendrolasin is cyclized to produce ancistrofuran and its C2 epimer (62).

Another defensive secretion has been isolated from the pygidial glands of the rove beetles, Bredius mandebularis and B. spectabilis (63), and identified as 4-dodecanolide (Figure 24). Oddly enough, this compound has also been isolated from assorted fruits (64,65,66) and butterfat (67,68). Synthesis of enantiomers of 4-dodecanolide was achieved from (S)- and (R)-2-aminodecanoic acid in gram quantities (69).

Conclusion. The wide array of secondary metabolites that possess biological activity is striking. While there is a certain degree of crossover, for example, Piper sp. natural products that affect insects, one is particularly struck by the relative lack of inter-disciplinary cooperation. But, as the complete synthesis of natural products becomes more common or genetic engineering gives rise to greater biosynthetic production of useful secondary metabolites, it is hoped that enough of the materials will be available for testing in several, apparently unrelated, systems. It is further hoped that promising lead chemicals will be synthetically adapted to fulfill specific roles.

Literature Cited

1. Chemical and Engineering News, 1988; p. 26.
2. Cutler, H.G. In "CRC Critical Reviews in Plant Science"; Conger, V., Ed.; CRC Press: Boca Raton, 1988; 6, 323.
3. Cutler, H.G. Proc. 12th Annual Meeting Plant Growth Regulator Soc. America, 1985; p. 160.
4. Cutler, H.G. In "Alleochemicals: Role in Agriculture and Forestry"; Waller, G.R., Ed.; ACS Symposium Series No. 330, American Chemical Society; Washington, DC 1987; p.23-38.
5. Gueldner, R.C.; Reilly, C.C.; Pusey, P.L.; Costello, C.E.; Arrendale, R.F.; Cox, R.H.; Himmelsbach, D.S.; Crumley, F.G.; Cutler, H.G. J. Agric. Food Chem. 1988; 36, 366.
6. Ueno, T.; Nakashima, T.; Hayashi, Y.; Fukami, H. Agric. Biol. Chem. 1975; 39, 2081.
7. Ueno, T.; Nakashima, T.; Hayashi, Y.; Fukami, H. Agric. Biol. Chem. 1975; 39, 1115.
8. Aoyagi, H.; Mihara, H.; Lee, S.; Kato, T.; Ueno, T.; Izumiya, N. Int. J. Pept. Protein Res. 1984; 25, 144.
9. Shimohigashi, Y.; Lee, S.; Kato, T.; Izumiya, N. Bull. Chem. Soc. Japan 1978; 51, 584.

10. Mihara, H. Aoyagi, H.; Lee, S.; Waki, M.; Kato, T.; Izumiya, N. Int. J. Pept. Protein Res. 1984; 23, 447.
11. Mihara, H.; Ikesue, K.; Lee, S.;Aoyagi, H.; Kato, T.; Ueno, T.; Izumiya, N. Int. J. Pept. Protein Res. 1986; 28, 141.
12. Hagashijima, T.; Shimohigashi, Y.; Kato, T.; Izumiya, N.; Ueno, T.; Miyazawa, T. Biopolymers. 1983; 22, 1167.
13. Matsumoto, K.; Suzuki, M.; Nunami, K.; Yoneda, N.; Takguchi, K. Bull. Inst. Chem. Res. Kyoto University 1983,; 61, 79.
14 Nunami, K.; Suzuki, M.; Matsumoto, K.; Yoneda, N.; Takiguchi, K. Agric. Biol. Chem. 1984; 48, 1073.
15. Ichihara, A.; Murakami, K.; Sakamura, S.; Agric. Biol. Chem. 1984; 48, 833.
16. Suzuki, M.; Sugiyama, T.; Watanabe, M.; Murayama, T.; Yamashita, K. Agric. Biol. Chem. 1987; 51, 2161.
17. Takaiwa, A.; Yamashita, K. Agric. Biol. Chem. 1984; 48, 961.
18. Trifonov, L.; Bieri, J.H.; Prewo, R.; Dreiding, A.S. Tetrahedron. 1982; 38, 397.
19. Takaiwa, A.; Yamashita, K. Agric. Biol. Chem. 1984; 48, 2061.
20. Nukima, M.; Sassa, T.; Ikeda, M. Tetrahedron Lett. 1980; 21, 301.
21. Nukima, M.; Ikeda, M.; Sassa, T. Agric. Biol. Chem. 1980; 44, 2761.
22. Suzuki, S.; Tanaka, A.; Yamashita, K. Agric. Biol. Chem. 1987; 51, 3095.
23. Suzuki, Y.; Knoche, H.W.; Daly, J.M. Bioorg. Chem. 1982; 11, 300.
24. Suzuki, Y.; Tegtmeier, K.J.; Daly, J.M.; Knoche, H.W. Bioorg. Chem. 1982; 11, 313.
25. Suzuki, Y.; Danko, S.J.; Kono, Y.; Takeuchi, S.; Daly, J.M.; Knoche, H.W. Agric. Biol. Chem. 1984; 48, 2321.
26. Suzuki, Y.; Danko, S.J.; Kono, Y.; Daly, J.M.; Takeuchi, S. Agric. Biol. Chem. 1985. 49, 149.
27. Addicott, F.T. In "Abscisic Acid", Praeger Pubs. (Holt, Reinhart and Winston): New York, 1983.
28. Cornforth, J.W.; Milborrow, B.V.; Ryback, G.; Wareing, P.F. Nature 1965; 205, 1269.
29. Oritani, T.; Ichimura, M.; Yamashita, K.; Agric. Biol. Chem. 1982, 46, 1959.
30. Marumo, S.; Kohno, E.; Natsume, M.; Kanoh, K. Proc. 14th Annual Meeting Plant Growth Regulator Soc. America, 1987; p. 146.
31. Takahashi, S.; Oritani, T.; Yamashita, K. Agric. Biol. Chem. 1986; 50, 3205.
32. Mitchell, J.W.; Mandava, N.; Worley, J.F.; Plimmer, J.R.; Smith, M.V. Nature 1970; 225, 1065.
33. Marumo, S.; Hattori, H.; Abe, H.; Nonoyama, Y.; Munakata, K. Agric. Biol. Chem. 1968; 32, 528.
34. Grove, M.D.; Spencer, G.F.; Rohwedder, W.K.; Mandava, N.; Worley, J.F.; Warthen, J.D., Jr.; Steffens, G.L.; Flippen-Anderson, J.L.; Cook, J.C., Jr. Nature 1979; 281, 216.

35. Mitchell, J.W.; Gregory, L.E. Nature New Biology 1972; 239, 253.
36. Gregory, L.E. Amer. J. Bot. 1981; 68, 586.
37. Krizek, D.T.; Mandava, N.B. Physiol. Plant. 1983; 57, 317.
38. Meudt, W.J. In "Ecology and Metabolism of Plant Lipids"; Fuller, G.; Nes, W.D.; Eds.; ACS Symposium Series No. 325, American Chemical Society: Washington, DC. 1987; p. 53-75.
39. Marumo, S. Proc. 14th Annual Meeting Plant Growth Regulator Soc. America, 1987; p. 174.
40. Hamada, K. In "Food & Fert. Tech. Ctr. Book Series No. 34, Plant Growth Regulators in Agriculture"; 1986; p. 190-196.
41. Luo, B.; Kumura, A.; Ishii, R.; Wada, Y. Proc. of the Crop Science of Japan; 1984; 53, 164.
42. Hamada, K.; Nishi, S.; Uezono, T.; Fujita, S.; Nakazawa, Y. Abstracts of the 12th IPGSA (Heidelberg). 1985; p. 43.
43. Sakakibara, M.; Mori, K. Agric. Biol. Chem. 1983; 47, 663.
44. Kerb; U.; Eder, U; Krähmer, H. Agric. Biol. Chem. 1986; 50, 1359.
45. Sakakibara, M.; Mori, K. Agric. Biol. Chem. 1984; 48, 745.
46. Aburatani, M.; Takauchi, T.; Mori, K. Agric. Biol. Chem. 1987; 51, 1909.
47. Kishi, T.; Wada, K.; Marumo, S.; Mori, K. Agric. Biol. Chem. 1986; 50, 1821.
48. Rebeiz, C.A.; Montazer-Zouhoor, A.; Mayasichi, J.M.; Tripathy, B.C.; Wu, S-M.; Rebeiz, C.C. In "CRC Critical Reviews in Plant Science"; Conger, V.; Ed.; CRC Press: Boca Raton, 1988; 6, 385.
49. Rebeiz, C.A; Montazer-Zouhoor, A.; Hopen, H.J.; Wu, S-M. Enzyme Microbiol. Technol. 1984; 6, 390.
50. Hopf, F.R.; Whitten, D.G. In "The Porphyrins" vol. 2. Dolphin D.; Ed.; Academic Press: New York, 1978; 161.
51. Foote, C.S. In "Porphyrin Localization and Treatment in Tumors". Alan R. Liss; New York, 1984; 3.
52. Harvill, E.K.; Hartzell, A.; Arthur, J.M. Contrib. Boyce Thompson Inst. 1943; 13, 87.
53. Jacobson, M. J. Am. Chem. Soc. 1953; 75, 2584.
54. Miyakado, M.; Nakayama, I.; Ohno, N.; Yoshioka, H. In "Natural Products for Innovative Pest Management" vol 2., Whitehead, L.; Bowers, W.S.; Eds.; Pergamon Press: Oxford, 1983; 369.
55. Nair, M.G.; Mansingh, A.P.; Burke, B.A. Agric. Biol. Chem. 1986; 50, 3053.
56. Elliott, M.; Farnham, A.W.; Janes, N.F.; Johnson, D.M.; Pulman, D.A.; Sawiki, R.M. Agric. Biol. Chem. 1986; 50, 1347.
57. Elliott, M.; Janes, N.F. Chem. Soc. Rev. 1977; 7, 473.
58. Sawiki, R.M. Nature 1978; 875, 4443.

59. Numata, A.; Hokimoto, K.; Takemura, T.; Fukui, S. Appl. Ent. Zool. 1983; 18, 129.
60. Murayama, T.; Sugiyama, T.; Yamashita, K. Agric. Biol. Chem. 1986; 50, 2347.
61. Baker, R.; Briner, P.H.; Evans, D.A. J. Chem. Soc. Chem. Commun. 1978; 410.
62. Saito, A.; Matsushita, H.; Kaneko, H. Agric. Biol. Chem. 1986; 50, 1309.
63. Wheeler, J.W.; Happ, G.M.; Araujo, J.; Pasteels, J.M. Tetrahedron Lett. 1972; 4635.
64. Broderick, J.J. Am. Perfumer Cosmet. 1966; 81, 43.
65. Willhelm, B.; Palluy, E.; Winter, M. Helv. Chim. Acta. 1966; 49, 65.
66. Tang, C.S.; Jennings, W.G. J. Agric. Food Chem. 1968; 16, 252.
67. Jurriens, G.; Oele, J.M. J. Am. Oil Chemist's Soc. 1965; 42, 857.
68. Forss, P.A.; Urbach, G.; Stark, W. Int. Dairy Cong. Proc. (Munich) 1966; 3, 211.
69. Sugai, T.; Mori, K. Agric. Biol. Chem. 1984. 48, 2497.

RECEIVED June 10, 1988

NATURAL PRODUCTS FROM MICROORGANISMS

Chapter 2

Tentoxin: A Cyclic Tetrapeptide Having Potential Herbicidal Usage

Alan R. Lax and Hurley S. Shepherd

Microbial/Plant Technology Research Unit, Southern Regional Research Center, U.S. Department of Agriculture, New Orleans, LA 70179

Tentoxin, a chlorosis-inducing cyclic tetrapeptide from *Alternaria alternata*, shows promise for use as a biorational herbicide and to identify novel target enzyme systems for the design of other selective herbicides. Specific effects on thylakoid and chloroplast envelope ATPases and on protein transport into the chloroplast have been identified as potential herbicidal target sites. A major impediment to development of tentoxin as a commerical herbicide is the limited production of the toxin by *A. alternata*. Identification of fungal viruses associated with tentoxin producing strains of the fungus provides some promise of increasing biosynthetic capabilities through traditional fermentation technology, semisynthetic production, or through bioengineering of organisms to provide economical levels of the toxin.

The widespread use of herbicides in the agriculture of industrialized nations is well established. Continued application of such compounds has led to the selection of weeds having resistance to established herbicides and to weeds physiologically more closely resembling the crop species they affect. Reduced tillage agricultural practices have created an increased reliance on herbicides while environmental concerns dictate a more judicious use of potentially detrimental compounds. The cost of herbicide development via traditional approaches has risen because of the extreme numbers of compounds which must be screened to yield a successful product. Increasingly industry is turning to natural products to provide new compounds or chemistries satisfying both efficacy and safety criteria for commercial development.

One new strategy is the use of pathogenic microorganisms in a biological control program (1). Two such herbicides, Collego and

Devine, have been commercially exploited but have experienced limited use because of the limited geographical distribution and economic importance of their weed hosts (1,2). Other constraints to direct microbial use involve the longevity of the propagules prior to application, possible environmental limitations, and storage space limitations (3). The potential for exploitation of these biological herbicides has been thoroughly reviewed and will not be considered further here (1-3).

The potential for the use of microbially derived toxins as pesticides has also been thoroughly reviewed recently (2,3). In the present review, we identify one such toxin for which commercial development has been suggested for examination in greater detail. This review is meant not to be encyclopedic or exhaustive, but rather to suggest and define areas of research which may permit further utilization of natural products in an integrated pest control program. Tentoxin has been selected for consideration because of its broad activity on a number of major weed species in corn and soybean without any discernable effect on those crops (2).

Toxins produced by plant pathogens

Many pathogenic organisms produce toxins having greater or lesser degrees of host specificity which cause or contribute to the pathogenicity of the producing organism (3). It is not a broad logical jump to consider the use of such toxins rather than the producing organism in a biorational pest control program. A number of phytotoxins produced by microorganisms are known which induce symptoms on only those species or even varieties within a species which are susceptible to the pathogen, thus host-specific toxins. Examples of these host specific toxins are AM and AK toxins produced by *Alternaria* causing disease symptoms on apple and pear, respectively, and HC and HMT toxins produced by *Helminthosporium* spp. causing symptoms on susceptible maize varieties (4,5). These toxins among many others are known primarily because of their association with disease-causing organisms of commercially valuable crop species, with few exceptions (6). Little is known of host specific toxins produced by pathogens of weeds, in part because of the paucity of physiological research on such host/pathogen relationships. Targeting weeds having major economic impact to discover such host-specific toxins should prove to be a fruitful area of research (6). Potential limitations to the herbicidal use of host-specific toxins lie in their extreme specificity with the probability of controlling only a single weed species, and the possibility of resistance within weed biotypes because of extensive co-evolution of such host-pathogen combinations. The discovery of such toxins could however, provide the basis for chemical manipulation to increased toxicity or decreased specificity resulting in an increased range of weeds that could be controlled (3). Structure/activity investigation and characterization of the molecular mechanisms underlying the specificities of such compounds could result in the design of simpler, more cheaply produced synthetic analogs as herbicides (3). Such studies also could provide novel target sites for the design of pesticides having the requisite specificities for closely allied crop/weed associations since most natural product toxins for which a molecular target site are known differ from sites affected by present commercial herbicides (3).

Non-host-specific toxins

Potentially more promising research lies in non-host selective toxins which affect species other than those infected by the producing organism. Several factors contribute to this probablity: 1) considerably more is understood chemically and physiologically about the non-host specific toxins and 2) the potentially broader range of weeds which may be controlled through a single application. The remainder of this discussion will focus primarily upon the discovery and characterization of one such toxin, tentoxin, and assessment of its promise as a herbicide. Tentoxin is a cyclic tetrapeptide produced by the fungus *Alternaria alternata* (4). Industry has expressed interest in commercial development of tentoxin, among other phytotoxins, and a thorough discussion of successes in understanding this toxin demonstrates potential for discovery and development of hitherto unknown compounds. As well, potential shortcomings demonstrate the problems which must be surmounted in the development of this under-represented class of pesticides.

While several important herbicides have been synthesized as analogues of natural products (3), herein we will consider only the activity and biosynthesis of the natural toxin. A full consideration of the synthesis of the toxin and its analogues as well as functional determinants of its toxicity is provided in another chapter of this volume.

Discovery

As with the discovery of most presently known phytotoxins the discovery of tentoxin resulted from its toxic effects on an important crop species. Cotton's susceptibility to chlorosis induced by tentoxin resulted in chlorotic seedlings and diminished stands of cotton (4,7). The causal organism associated with the seedling disease was *Alternaria tenuis* (=*A. alternata*) (4,7).

It was found that cell-free culture filtrates of the fungus when applied as a soil drench resulted in identical symptoms, thus establishing an extracellular metabolite as the causal agent of the chlorosis (4,7). The discovery of many other toxins which are primary determinants of plant disease have resulted from their effect on commercially important crop species. Examples of such discoveries include the host-selective toxins AM, AK, HMT, and HC toxins (3). To date, however, there have been only few of the many known fungal pathogen toxins described as a result of their effects on non-crop species. This lack of discovery probably reflects the limited research in this area rather than a paucity of toxins affecting weed species.

It is in this area that major studies may be made, especially through targeting of one or several particularly troublesome weeds and thorough examination of pathogens which infect these weeds. Industry is in fact dedicating some effort in the area of weed/pathogen research with the potential for discovery of new classical biological control organisms and/or toxins derived from them (3).

While many of the currently known toxins show extreme specificity within the genera or even within species which their producing microorganisns affect, we know of no reports of the systematic examination of the possibility that these toxins or their analogues may affect distantly related weed species.

Mode of action and identification of potential target sites.

After isolation and the discovery of the toxin structure, follows characterization of the molecular sites of toxicity. The nature of the site of action is known for several of the earlier mentioned phytotoxins while for others the molecular target remains unelucidated (2,3). Early results concerning tentoxin mode of action focused on its inhibition of the ATPase activity of chloroplast coupling factor 1 (CF_1) (8). Tentoxin at some concentrations was shown to inhibit ATPase activity of the CF_1 isolated from toxin-sensitive but not insensitive species (8). Moreover, in isolated chloroplasts tentoxin was shown to inhibit coupled photosynthetic electron transport thus acting as an energy transfer inhibitor (9). The resulting loss of metabolic energy was thought therefore to result in the lack of chloroplast development and chlorosis characteristic of tentoxin effects. Electron microscopic examination, however, revealed that tentoxin disrupted development in apical plastids and etioplasts, both tissues in which effects on CF_1-dependent phosphorylation should have no effects (10-12). Moreover, the tentoxin treated plastids morphologically resembled plastids of a mutant of Hosta which lacked polyphenol oxidase (PPO) (13). In a series of experiments, Vaughn and Duke demonstrated that PPO activity was absent from tentoxin treated plastids, etioplasts and apical plastids and that the nuclearly encoded, inactive enzyme accumulated at the chloroplast envelope (10-12). These results suggested that a process involved in the transport of PPO into the chloroplast may be involved in the toxin syndrome. That this transport inhibition is specific and not a result of overall energy (ATP) disruption is demonstrated by the presence in toxin-treated plastids of other nuclearly encoded proteins, notably ferredoxin-$NADP^+$ reductase (FNR) (14), light harvesting chlorophyll complex (LHC) (Lax unpublished) and small subunit of ribulose bisphosphate carboxylase/oxygenase (Lax unpublished), all plastidic enzymes which have been demonstrated to be transported into the chloroplast in an energy-dependent manner. Figure 1 shows that only minor differences exist in the polypeptide profile of etioplasts from control and tentoxin-treated plants, again indicative of a selective disruption of import of only one or a few nuclearly-encoded chloroplast proteins. That only minor differences are noted early in ontogeny prior to pleiotropic effects due to lack of chloroplast development strengthen the argument that only one or a few specific proteins are affected.

While some of the above effects of tentoxin could be mediated through inhibition of energy transduction by tentoxin's known effects on CF_1, several further lines of evidence indicate that the primary lesion caused by tentoxin treatment lies in other cellular processes. First, only one of three CF_1 isozymes isolated through electrophoresis show tentoxin sensitivity, and the sensitivity of

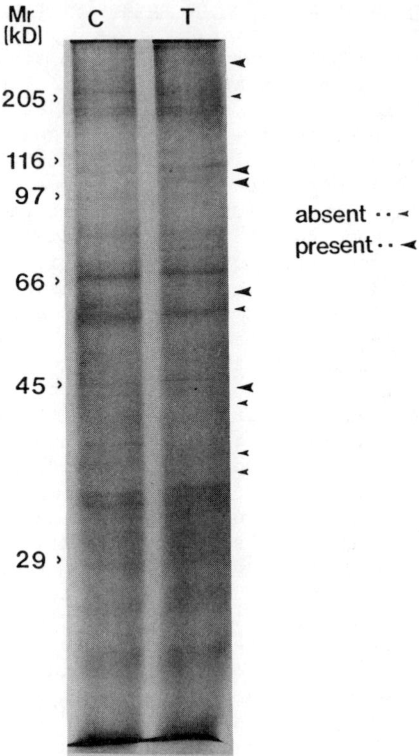

Figure 1. Polypeptide profile of control (C) and tentoxin-treated (T) etioplasts isolated from dark-grown lettuce cotyledons. Proteins were solubilized in lithium dodecyl sulfate and electrophoresed on polyacryamide slab gels. Proteins were silver stained. Note the minor differences in protein profiles. Small arrows indicate absence and large arrows denote presence of proteins in the tentoxin profile compared to the control profiles.

this isozyme is seen only at low toxin concentrations, while chlorosis mediated by tentoxin treatment shows no alleviation at higher toxin concentrations (15). Secondly, Zea mays, which is insensitive in vivo to chlorosis caused by tentoxin, has a CF_1 isozyme which is toxin sensitive. Finally, the most compelling argument against CF_1-mediated chlorosis is provided by the sensitivity to tentoxin of a CF_1 defective mutant of Oenothera hookeri (15). While tentoxin-mediated chlorosis does not appear to be solely or primarily dependent upon tentoxin's effects on CF_1 there is no argument that tentoxin and some of its analogues do possess a species specific effect on CF_1 which may prove valuable in the design of CF_1 effectors which could be effective as selective herbicides.

More recent evidence from our lab indicates that another molecular target site for tentoxin mediated chlorosis may exist at the chloroplast envelope, the known site of accumulation of inactive PPO in tentoxin-treated tissues. Figure 2 shows the sensitivity of one of two chloroplast envelope associated ATPases which can be separated electrophoretically. While it remains to be demonstrated that these envelope-associated ATPases are associated with processes involved in protein import into the chloroplast, it has been demonstrated for a number of nuclear-coded proteins that such import is an ATP-dependent process (16,17). That there are at least two envelope associated ATPases having different sensitivities to tentoxin could be invoked to explain the different sensitivities of the known imported proteins (e.g., PPO, FNR, SS) to inhibition by tentoxin. It is also possible that the above mentioned envelope-associated ATPases are unrelated to the import processes. Differential sensitivities of these envelope ATPases to tentoxin suggest yet another molecular site for biorational natural product (or analogue) herbicides (2).

Regardless of whether these envelope-bound ATPases are involved in the specific transport processes known to affect PPO, the tight correlation between PPO transport inhibition and chlorosis identifies this process itself as a potential herbicidal target site.

An important point here is that whether or not the natural product itself may be commercially exploited, research into its target site may provide clues to one or more target enzyme systems which may be exploited as herbicidal targets.

After the identification of potential herbicidal microbial products, field efficacy and production economics must be assessed. In the case of tentoxin these parameters are tightly coupled because of the generally low yields of the toxin in standard fermentation and the resultant expense of large scale field tests. In fact, to our knowledge tentoxin has never been tested in the field, although incorporation of crude culture filtrates into soil in growth chamber studies has demonstrated the potential efficacy of tentoxin's use as a herbicide in corn and soybean (2).

While the biosynthetic pathway to tentoxin is unknown, rather more attention has been paid to its biosynthesis than to field efficacy. Several methods have been utilized to quantitate toxin production including biological assay, HPLC, and thin layer chromatography (18), Using a semiquantitative seedling chlorosis bioassay, Alternaria isolate selection to maximize toxin production can be performed. Media composition and culture conditions to be discussed below have been demonstrated to have tremedous effect on toxin production, but to date the maximum yield of tentoxin has been 90 mg/l (19,20). Moreover these yields are realized only after prolonged still culture (18). By the addition of previously metabolized culture filtrates to Alternaria cultures, the toxin yield may be increased temporally, however, the final yield is little altered (18), suggesting the presence of a precursor(s) in the metabolized culture broths.

It is currently unknown whether tentoxin is synthesized on ribosomes as a higher molecular weight precursor which is subsequently cleaved, modified and cyclized or whether its synthesis procedes via enzymatic (such as ancovenin in Streptomyces sp. and subtilisin A in Bacillus subtilis) (21), or non-ribosomal synthesis as is known for gramicidin and tyrocidine (in bacteria) or Cyl-1 and HC toxins in fungi (21).

The addition of protein synthesis inhibitors such as cycloheximide and emetine had no effect on toxin biosynthesis, leading to the conclusion that tentoxin biosynthesis proceedes via a non-ribosomal pathway (18). However, in these experiments the inhibitors were added in late growth phase and the possibility of higher molecular weight, ribosomally synthesized precursor accumulation prior to inhibitor additions cannot be ruled out. Dot blots of culture filtrates of toxin producing strains of Alternaria using antisera to tentoxin (raised in rabbit against purified tentoxin) indicated greater amounts of immunologically reactive material than could be ascribed to the toxin itself (Lax unpublished). Western blotting, following electrophoresis of proteins isolated from various Alternaria strains, indicated the presence of high molecular weight proteins recognized by antisera to tentoxin in toxin producing, but not non-producing isolates. These data suggested the possible accumulation of a high molecular weight precursor to tentoxin which could be subsequently processed to yield the final product.

In our attempts to isolate mRNA from Alternaria to perform in vitro translations, we consistently found unique RNAs in strains which produced tentoxin. These RNA species were sensitive to RNAse digestion in low salt but not high salt, indicative of double stranded RNA characteristic of previously described fungal viruses (22).

Cesium chloride and sucrose density gradient ultra-centrifugation yielded purified isodiametric virus particles having a diameter of ~30 nm (23) also characteristic of previously described fungal viruses (22). Electrophoretic separation of the viral protein of two such viruses has demonstrated at least two proteins (Figure 3A) in each, at least one of which is recognized by antisera to tentoxin upon western blotting (Figure 3B). As indicated previously, tentoxin biosynthesis is dependent upon culture conditions such as temperature and aeration. Tentoxin biosynthesis is maximal at ~28°C (19) and is greatly reduced at temperatures up to 35°C (19) while the growth rate is unaffected. Toxin production is prevented in shake culture under conditions othewise identical with still cultures (Lax unpublished). Elevated temperatures and shaking also reduce levels of virus associated with toxin producing strains (Figures 4A, 4B). The addition of

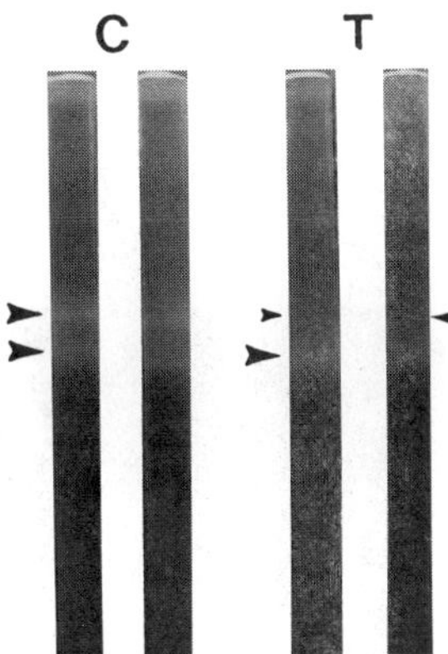

Figure 2. Differential sensitivity of chloroplast envelope ATPases of pea. Envelope membranes were isolated from intact pea chloroplasts and solubilized in CHAPS. Following electrophoretic separation ATPase activity was detected in the absence (C) or presence (T) of 10 μM tentoxin. Deposition of a white $Ca_3(PO_4)_2$ precipitate indicates activity. Note absence of activity of the more slowly migrating species in the presence of tentoxin.

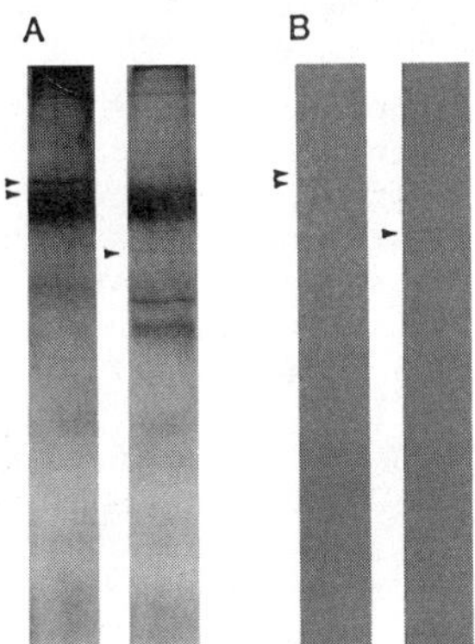

Figure 3. Analysis of proteins from virus particles of strains 1 and 2 of A. alternata purified by CsCl density gradient centrifugation. A) Proteins of lithium dodecyl sulfate-disrupted virus particles separated by polyacrylamide gel electrophoresis. Two major proteins are visible for each strain following silver staining. B) Western blot of A), probed with antisera to tentoxin. Note immunoreactivity of at least one band in each case.

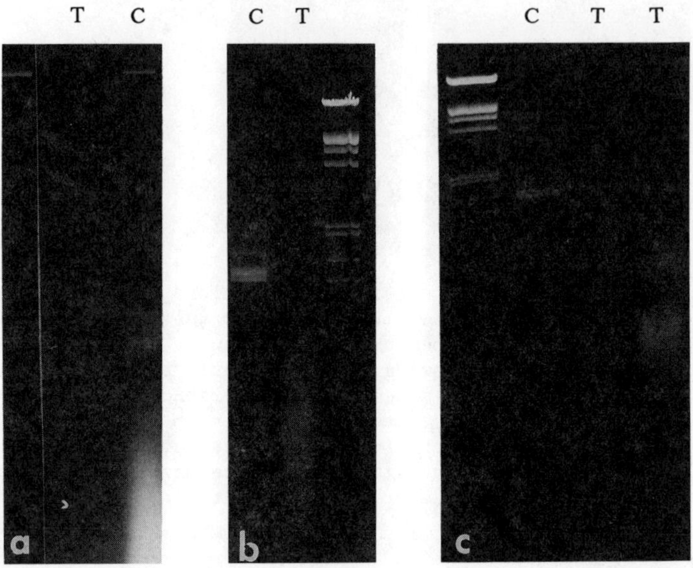

Figure 4. Double-stranded (viral) RNAs isolated from control (C) and treated (T) A. alternata cultures. All treatments were carried out two or more times on each of the strains. Representative results from strain 2 are shown. Tentoxin was detectable only in controls, except as noted. a) Double-stranded RNA from 28°C (C) and 34°C (T) cultures. b) Double-stranded RNA from Still (C) and Shaking (T) cultures. c) Double-stranded RNA from cultures grown in the absence (C) and presence (T) of cycloheximide, (0.2 mM on left, 0.4 mM on right. A low level of tentoxin was detected in the 0.2 mM cycloheximide treatment.

cycloheximide in early growth phase has been reported to reduce fungal virus (24); such addition to toxin-producing strains concomitantly reduces toxin biosynthesis and virus concentration (Figure 4C). All of these experiments have been carried out several times with at least two tentoxin-producing strains.

A firmer understanding of the biosynthetic processes will allow better assessment of the potential for increasing production to a commercially acceptable level. Whether this increase can be realized through traditional microbiological methods, the design of a semisynthetic pathway, or through more advanced molecular techniques remains to be seen. This appears to be the major hurdle to be overcome in the exploitation of tentoxin as a broad spectrum herbicide in the future.

From the foregoing it should be realized that deployment of tentoxin as a herbicide will at best be several years away. As with other chemically complex natural products a major limitation is production (2,3). Another major concern which is beyond the scope of this chapter is toxicological testing and licensing which are dependent upon regulatory agencies. Information regarding tentoxin's mode of action would indicate a low mammalian toxicity with all evidence pointing to an exclusively chloroplast localized activity. Tentoxin like other fungal natural products should be relatively non-persistent in the environment and therefore environmentally safer than other more persistent compounds.

Although only a few microbially derived herbicides are on the market today, there is great hope that natural products can fill the need for highly selective and safe herbicides. Tentoxin is but one of thousands of compounds that offers such potential. Certainly further research is needed and warranted to fulfill this vast untapped potential.

Literature Cited

1. Templeton, G.E.; Smith, R.J.; TeBeest, D.O. Rev. Weed Sci. 1986. 2, 1-14.

2. Duke, S.O.; Lydon, J. Weed Technology 1987. 1, 122-128.

3. Duke, S.O. Rev. Weed. Sci. 1986. 2, 15-44.

4. Templeton, G.E. In Microbial Toxins; Kadis, S; Ciegler, A; Ajl, S.J., Eds.; Academic: New York, 1972; Chapter 7, pp 169-192.

5. Luke, H.H.; Gracen, V.E. Jr.; In Microbial Toxins; Kadis, S; Ciegler, A., Ajl, S.J., Eds.; Academic: New York, 1972, Chapter 6, pp 139-168.

6. Robeson, P.; Strobel, G.; Matusumoto, G.K.; Fisher, E.L.; Chen, M.H., Clardy, J. Experentia 1984 40, 1248-50.

7. Fulton, N.O.; Bollenbacher, K.; Templeton, G.E. Phytopathology. 1965. 55, 49.

8. Steele, J.A.; Uchytil, T.F.; Durbin, R.D.; Bhatnagar, P.; Rich, D.H. Proc. Nat. Acad. Sci. U.S.A. 1976. 73, 2245-8.

9. Arntzen, C.J. Biochim. Biophys. Acta. 1972. 283, 539-542.

10. Vaughn, K.C.; Duke, S.O., Physiol. Plant. 1981, 53, 421-8.

11. Vaughn, K.C.; Duke, S.O., Photoplasma. 1982, 110, 48-53.

12. Vaughn, K. C.; Duke, S.O., Physiol. Plant. 1984, 60, 257-61.

13. Vaughn, K.C., Lax, A.R.; Duke, S.O. Physiol. Plant. 1988, 72, 659-665.

14. Lax, A.R.; Vaughn, K.C.; Sisson, V.A.; Templeton, G.E., Photosynth. Res., 1985, 6, 113-20.

15. Lax, A.R.; Vaughn, K.C., Physiol. Plant. 1986, 66, 384-391.

16. Schmidt, G.W.; Bartlett, S.G.; Grossman, A.R., Cashmore, A.R.; Chua, N.-H., J. Cell. Biol. 1981, 91, 468-78.

17. Pain, D.; Blobel, G., Proc. Natl. Acad. Sci. USA. 1987, 84, 3288-92.

18. Sheu, J.-T.; Talburt, D.E., Appl. Env. Microbiol. 1986, 51, 368-72.

19. Liebermann, B.; Oertel, B., Z. Allg. Mikrobiol., 1983, 23, 503-11.

20. Brücker, B.; Hänel, I.; Hänel, F.; Tröger, R., Z. Allg. Mikrobiol., 1983. 23, 549-56.

21. Kleinkauf, H.; vonDöhren, H., Ann. Rev. Microbiol. 1987, 41, 259-89.

22. Buck, K.W.; Ackermann, H.-W.; Bozarth, R.F.; Bruenn, J.A.; Koltin, Y.; Rawlinson, C.J., Ushiyama, R.; Wood, H.A. Intervirology, 1984, 22, 17-23.

23. Shepherd, H.S. Genetics. 1987, 116; S29.

24. Schmidt, F.R.; Lemke, P.A.; Esser, K. Appl. Microbiol Biotechnol. 1986, 24, 248-52.

RECEIVED May 26, 1988

Chapter 3

Approaches to Structure–Function Relationships for Naturally Occurring Cyclic Peptides

A Study of Tentoxin

Judson V. Edwards[1], Oliver D. Dailey, Jr.[1], John M. Bland[1], and Horace G. Cutler[2]

[1]Southern Regional Research Center, Agricultural Research Service, U.S. Department of Agriculture, New Orleans, LA 70179
[2]Richard B. Russell Research Center, Agricultural Research Service, U.S. Department of Agriculture, Athens, GA 30613

Approaches to studying structure/activity relationships in naturally occurring peptides with agronomic relevance are discussed. The cyclic tetrapeptide tentoxin has been examined for the relation of its conformational properties to biological activity and its use as a template for derivation of synthetic analogs possessing divergent phyto-active properties. The cyclic tetrapeptide HC toxin is discussed for similar properties. Approaches to amide bond and amino acid modifications are discussed; these include the incorporation of thiomethylene ether, retro-inverso, alkyl spacer, N-alkyl, and "cyclopropylog" modifications in tentoxin and biologically active synthetic fragments.

Microbial and Plant Peptides

The role of peptides in regulating fundamental biological functions in both animal (1) and microbial cells (2) is well documented. Structure-function studies employing synthetically modified analogs have proven an indispensible tool second only to the biological assay in evaluation of binding and transduction of peptides. In contrast, less attention has been given to structure and function of indigenous and microbially fixed peptides in higher plants (for a review see reference 3). Evidence for plant peptidic hormones (or fragments), analogous to mammalian hormones, which regulate vegetative growth and development is absent, but morphogenetic signal functions have been hypothesized for oligopeptides (4,5). Reports of biologically active peptides in plants and microbes have recently appeared (3,6-9). Of the peptides characterized, both cyclic and linear structures have been found, and in many cases unusual amino acids have been noted (10,11).

Microbial peptides that possess selective biological activity in roots or aerial tissues of plants have been isolated. The plant regulatory effects of the bicyclic pentapeptide malformin, a fungal metabolite, have been known for some time (7,9,12,13) and the structure-function relationships explored (14). Rhizobacteria have been found to enhance plant growth by production of peptide siderophores (15,16), and recent attention has been given to a group of peptides (Rhizobins) exuded by soybean nodules that may be involved in nitrogen fixation (9). A variety of cyclic tetrapeptides have demonstrated very potent and highly selective phytotoxic and plant growth regulating properties (11).

Agricultural Potential of Some Phyto-active Naturally Occurring Peptides

The potential use of biologically-active natural products as templates for the development of agricultural products has received increased attention in recent years. Our interest has been to study the structural features of fungally-secreted, cyclic tetrapeptides responsible for biological activity in plants. The studies described here deal principally with two types of approaches: analysis of predictable conformational features relevant to biological activity, and the use of cyclic peptides as a template for deriving bioactive fragments. The fungal metabolite tentoxin has been employed as a model compound.

An understanding of the effect of predictable alterations in secondary structure resulting from primary structural changes provides a better understanding of conformational requirements for biological activity. A thorough knowledge of the conformational requirements for activity in the plant increases understanding of the mode of action and furthers the potential for manipulating fungally secreted peptides of agricultural interest. For example the developments of agonists and antagonists from conformational studies on neuropeptide hormones have led to an increased understanding of sites of action and development of therapeutic drugs (17).

With the disclosure of a biologically active natural product the question often arises as to how the compound may be employed as a template for derivation of simpler bioactive fragments. Several examples of biological activity of subtly modified natural products or synthetic fragments have proven fruitful in this regard. The synthetic dipeptide carbobenzoxyprolylvalinol is a good example (18). The peptide is a synthetic fragment of HC toxin (a cyclic tetrapeptide which is a potent inhibitor of a genotype of corn roots) yet the fragment possesses an altogether different biological response. Treatment of corn plants with the synthetic fragment provides for increased yields in unfertilized corn fields.

We report here approaches to examining the potential use of the cyclic tetrapeptide tentoxin as a template for development of compounds of potential herbicidal and plant growth regulating activity. Both amide bond substitutions and novel amino acids have been used to assess the native conformation of tentoxin and for derivation of novel biologically-active peptides which are synthetic fragments of tentoxin.

The cyclic tetrapeptide tentoxin cyclo[N(Me)Ala-Leu-N(Me)Δ^{Z} Phe-Gly] (Figure **5**) is a secondary metabolite of Alternaria alternata. Tentoxin is of interest agriculturally for its selective herbicidal action. It induces chlorosis (a yellowing in plants due to depleted chlorophyll levels) in susceptible weed species while not effecting major crop plants such as corn and soybean (19).

As a cyclic tetrapeptide, tentoxin is among a class of fungally secreted peptides containing four amino acids. Within the past 10-15 years many of these analogs have been discovered and their phytotoxic and growth regulating effects elucidated (16). The structural and conformational characterization of a variety of these analogs has been achieved by extensive NMR, crystallographic, and theoretical calculations (21). Structural similarities among the pertinent peptides include aromatic groups, alkyl side chains, and a twelve atom peptide ring. Various novel amino acids are found, such as as dehydro amino acids (in tentoxin and the AM toxins) and the 8-oxo-9, 10-epoxydecanoic acid [in HC toxin and Cyl-2 (11)].

Five combinations of backbone conformation are possible along the ring of cyclic tetrapeptides including: all cis; all trans, 3 cis and 1 trans; 3 trans and 1 cis; and 2 cis and 2 trans. As a result of these backbone variations, distinctive conformational features are present which may provide useful leads in designing mimetic compounds. For example, the presence of a γ-turn in HC toxin, and cyclo-(L-Pro-D-Val-)$_2$ (found to promote stem growth of rice seedlings), resulting from a sequence specific reverse turn, suggests the possible use of reverse turn mimics to develop smaller non-peptidic molecules with analogous activity. A seven-membered ring compound of the type recently proposed by Huffman and Callahan (20) may be useful in the development of simplified non-peptide analogs. Previous work suggests that the cyclic peptide backbone serves as a carrier which induces binding, while a reactive site at a specific amino acid participates as the receptor ligand. Thus, the attachment of reactive ligands to carriers mimicking the peptide conformation may provide useful lead compounds.

Tentoxin - an Exception to Other Naturally Occurring Cyclic Tetrapeptides.

Since the activities of many of the cyclic tetrapeptides depend on their conformations, the relationship between primary structure and conformation is important to understand. Recently a useful set of empirical rules has been reported to predict the conformation of cyclic tetrapeptides on the basis of primary structure (21). The rules are based on the many conformations reported for cyclic tetrapeptides; these rules have proven useful in finding a starting conformation in energy calculations, and they take into consideration γ-turns, and the effect of D-amino acids on conformation. Tentoxin represents an interesting exception to the rule governing D-amino acids in that the effect of the N-methyldehydrophenylalanine is to adopt an all L configurational sequence. The shift to an L form would not be anticipated with tentoxin since glycine is treated as a D-amino acid according to the rules of Kato et al. It has been proposed that N-methyl-α, β-dehydroalanine has a deep energy minimum (-7 kcal with ϕ = -80 and Ψ = - 10) (22). Similar torsional angles have been found with all peptides having a C_2 symmetry conformation (21).

In assessing the relative conformational features of tentoxin we have employed rule 2 of Kato *et al*. (21): "in a right-handed peptide ring, the carbonyl group acylating a D residue is oriented to the upperside". The effect of the N-methylated amide bonds and substituted D-amino acids on biological activity was tested. In addition the relative role of the 12-membered ring with respect to the functional groups was considered. Tentoxin analogs were designed and synthesized to evaluate stereochemical and structural properties. Biological tests of the analogs demonstrated an interesting effect of D-amino acid-containing analogs on biological activity (23). Since tentoxin exhibited deviations from the empirical rules, only those analogs where dehydrophenylalanine had not been N-methylated were subject to conformational prediction when substituting D-amino acids. Substitution of D-amino acids for L-amino acids produced an increase in activity in one analog and a decrease in another. The increased activity was observed in the analog where a substitution of D-Alanine for L-Alanine was made with no N-methylation. On the other hand, a substitution of D-Leu for L-Leu in an N-methylated analog produced a decrease in activity. The predicted preferred conformational states for some chlorosis-inducing tentoxin analogs in order of their biological activity were calculated (Figure 1).

Approaches to Studying Amide Bond and Amino Acid Modifications in Tentoxin

The use of amide bond modifications in studying bioactivity profiles has been very useful in assessing biologically active peptides and the status of peptide backbone structure-function relationships. The following questions were asked: 1) To what extent are the alignment and stereochemistry of the peptide backbone critical? 2) How does the modification of the backbone affect the resistance toward enzymatic degradation? 3) To what extent are rigidity and flexibility of peptides manifested in the backbone? 4) Can agonists and antagonists be prepared by introducing peptide backbone modifications? These questions have been asked about animal peptide hormones, and they are equally applicable to plant-active cyclic tetrapeptides. We have studied the incorporation of a number of amide bond and amino acid modifications in both tentoxin and fragments of tentoxin. A number of amide bond modifications were considered (Figure 2).

A Thiomethylene Ether-Containing Pseudo-dipeptide

The thiomethylene ether substitution for an amide bond is characterized by its enzymatic resistance, compatibility with common peptide conformational features and synthetic accessibility from chiral amino acid precursors (24). This substitution has been explored in enkephalin analogs, and it has been found to both confer increased resistance to proteolysis as well as nonselectivity in cyclic enkephalin analogs. The preparation of peptides containing a thiomethylene ether substitution for an amide bond has previously been accomplished with a dipeptide surrogate (24). For the purposes of our studies we targeted the dipeptide surrogate Boc-ValΨ[CH_2S]Phe-OH (**1**), synthesized from the chiral precursors S-valinol and R-phenylalanine (Scheme 1). Bromination of phenylalanine results in retention of configuration. Preparation of the α-mercapto acid of phenylalanine from the bromo acid results in inversion of the configuration. Using the Boc-valinyl tosylate and the disodium salt

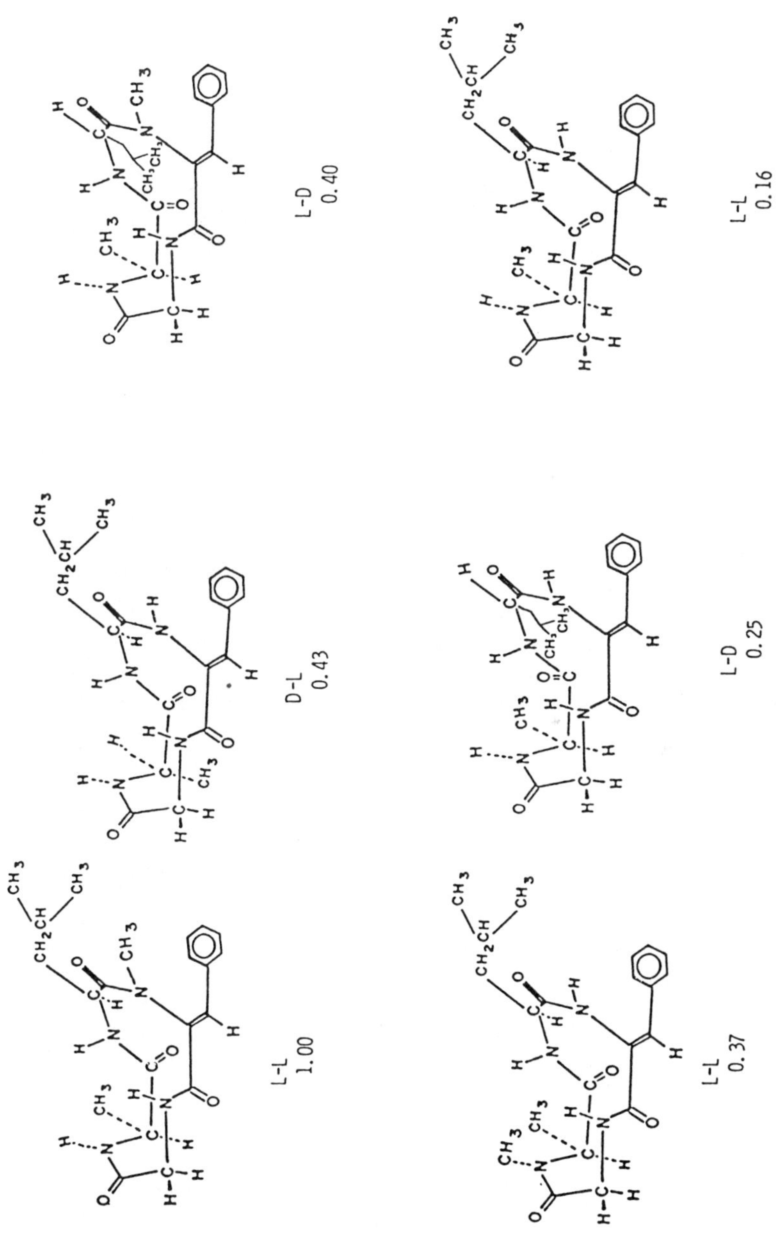

L-L 1.00

D-L 0.43

L-D 0.40

L-L 0.37

L-D 0.25

L-L 0.16

Figure 1 Analogs of tentoxin in their predicted conformation. Relative potency and the stereogenic configuration at alanine and leucine residues are noted. Data from ref. 23.

Peptide Mimetics

Replacing Substituent	Replaced Substituent	Proteolytic Resistance	Effect
CH_2S	CONH	XXX	Increased flexibility
NHCO	CONH	XX	Maintenance of side chain topology.
H_2C–$[CH_2]$–$_nCH_2$	$[Xxx]_nY_{yy}$	XXX	Simplification, maintain receptor affinity
CSNH	CO of amide	X	Potential inhibitor for metalloproteases. Conservative replacement.
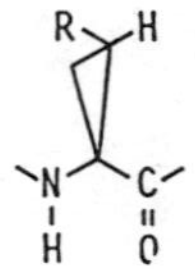 R H N C H O	αC-R	X	Conformational Rigidity χ, angle 0° (in _Z_), 120° (in _E_)
CONR	NH of amide	X	Many enhanced and equipotent analogs known.

[Degree of protease resistance ≅ Xs]

Figure 2 Amide bond and amino acid substitutions which confer proteolytic resistance and varying conformational and physical properties on peptides.

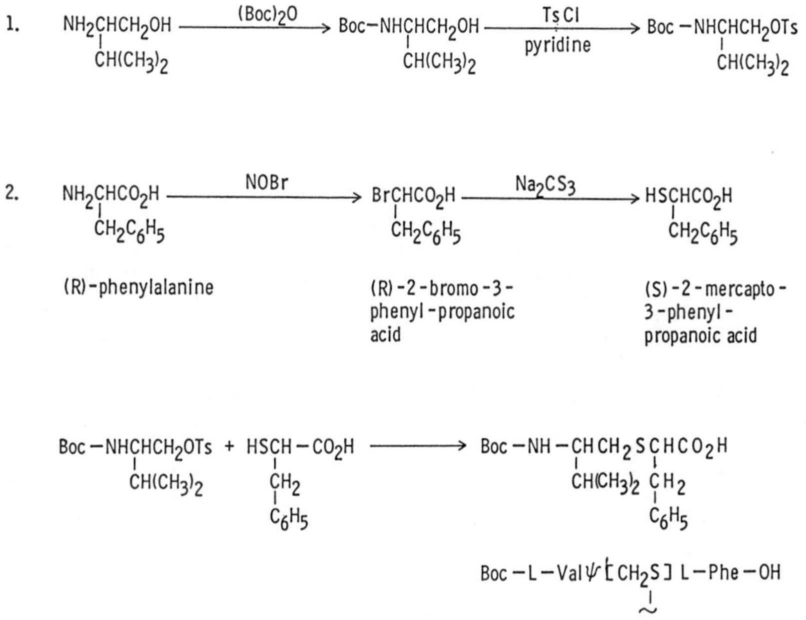

1. $NH_2CH(CH(CH_3)_2)CH_2OH \xrightarrow{(Boc)_2O} Boc\text{-}NHCH(CH(CH_3)_2)CH_2OH \xrightarrow[\text{pyridine}]{TsCl} Boc\text{-}NHCH(CH(CH_3)_2)CH_2OTs$

2. $NH_2CH(CH_2C_6H_5)CO_2H \xrightarrow{NOBr} BrCH(CH_2C_6H_5)CO_2H \xrightarrow{Na_2CS_3} HSCH(CH_2C_6H_5)CO_2H$

(R)-phenylalanine; (R)-2-bromo-3-phenyl-propanoic acid; (S)-2-mercapto-3-phenyl-propanoic acid

$Boc\text{-}NHCH(CH(CH_3)_2)CH_2OTs + HSCH(CH_2C_6H_5)\text{-}CO_2H \longrightarrow Boc\text{-}NH\text{-}CH(CH(CH_3)_2)CH_2SCH(CH_2C_6H_5)CO_2H$

Boc-L-Valψ[CH_2S] L-Phe-OH

Scheme 1 Synthesis of the pseudodipeptide Boc-Val[CH_2S]Phe-OH.

of α-mercaptophenylalanine, a coupling reaction was performed in dimethylsulfoxide. The reaction gave Boc-ValΨ[CH_2S]Phe-OH in 46% yield; this analog is a potent inhibitor of lettuce roots (J. V. Edwards, unpublished results). This analog, however failed to exert a herbicidal effection on morningglory and barnyardgrass, although an auxin-like response was observed with morningglory (J.V. Edwards and D. Carlson, unpublished results).

Retro-Inverso Modifications

In retro-inverso modified analogs of peptides, the direction of the amide bonds in the backbone has been reversed but the three-dimensional orientation of the side chains is maintained (25,26). The Russian group headed by Shemyakin utilized retro-inverso analogs in the development of a topochemical approach to the study of structure-activity relationships in peptide systems (27). Their studies dealt with cyclic peptides and depsipeptides, notably Gly^5, Gly^{10}-gramicidin S and enniatin antibiotics. Inversion of each chiral center (and concomitant reversal of the peptide backbone) resulted in enantiomers exhibiting exactly the same biological activity. Similar studies have dealt with linear bioactive peptides. For example, all D-retro-bradykinin (28) and an all D-retro analog of a melanocyte-stimulating (MSH) peptide (29) have been synthesized. However, the original spatial orientation of all side chains in the analogs was not retained since the C- and N-termini were reversed. The end-group problem can be circumvented by replacement of the C-terminal amino acid residue by a 2-alkylmalonyl residue and transformation of the N-terminal residue into a gem-diaminoalkyl residue (26,29).

Goodman and coworkers have conducted a number of studies on partially-modified, retro-inverso analogs of biologically active peptides in which a selected peptide bond is reversed through the incorporation of gem-diaminoalkyl and 2-alkylmalonyl derivatives. (26). Much of the research has dealt with partially modified retro-inverso analogs of the luteinizing hormone-releasing hormone (LH-RH) (30) and enkephalins (31-33). In many instances, the analogs have exhibited enhanced biological activity and increased stability toward enzymatic degradation (31,32).

The herbicidally active peptide tentoxin is an interesting model for incorporation of the retro-inverso modification. Conformational properties are associated with chlorosis induction and a proteolytically resistant peptide bond may enhance the biological activity of the naturally occurring toxin.

Synthesis of a Retro-Inverso Modified Analog of Tentoxin

Studies were carried out on a synthesis of diastereomers of cyclo [R,S-mLeu-N(CH_3)Δ^ZPhe-Gly-gAla] (compound **2** in Scheme 2) wherein the amide bond of the dipeptide unit Ala-Leu is reversed. The prefix m denotes the malonic acid derivative of the analogous amino acid and the prefix g denotes the gem-diamino-alkyl derivative of the corresponding amino acid (30-33).

The t-butyl monoester of isobutylmalonic acid, (**3**), malonic acid analog of leucine was prepared (Scheme 2). Diethyl isobutylmalonate was prepared in 74% yield in accordance with the literature method (35). Treatment with 1N ethanolic KOH in acetone afforded the monoester (32). Reaction with t-butanol and DCC in the presence of 4-dimethylaminopyridine (DMAP) afforded the t-butyl ethyl diester in 76% yield which was selectively hydrolyzed to **3** with 1N NaOH (aqueous) in methanol. There was no reaction upon treatment of the diester with 1N ethanolic KOH in acetone. Reaction of **3** with N-hydroxysuccinimide (HOSu) in the presence of DCC (32) provided the N-hydroxysuccinimide ester **4** which was obtained in 56% yield upon recrystallization from isopropanol. Reaction of **4** with d,l-3-phenylserine ethyl ester (**5**, obtained from the p-toluenesulfonate salt) (35) in the presence of 1-hydroxybenzotriazole (HOBt) provided **6** in 69% yield. Attempts to prepare **6** from **3** and **5** by the mixed anhydride method (35) gave poor (25% or less).

Selective saponification of **6** afforded the acid **7** in 94% yield. Treatment of **7** with excess acetic anhydride in the presence of sodium acetate (35) effected stereoselective dehydration forming the azlactone of t-BuO-R,S-mLeu-Δ^ZPhe **(8)** isolated in 96% yield as a yellow oil.

Reaction of **8** with a 20% excess of the dipeptide H-Gly-Ala-OMe (prepared as shown in Scheme 3) in refluxing ethyl acetate furnished the tetrapeptide t-BuO-R,S-mLeu-Δ^ZPhe-Gly-Ala-OMe **9.** Flash chromatography of the crude product afforded unreacted **8** (23% recovery) and **9** in 54% yield (71% based upon recovered starting material).

Compound **9** was selectively N-methylated at dehydrophenylalanine by treatment with excess K_2CO_3 and iodomethane in the presence of a catalytic amount of 18-crown-6 (35) to provide **10** in 98% yield. Tentoxin itself also contains an N-methylalanine moiety. However, the second N-methyl group has little or no effect on biological activity.

The conversion of the alanine residue of **10** to its gem-diamino derivative was effected by treatment of **10** with liquid ammonia in methanol with selective formation of the amide **11** in 93% yield. Treatment of **11** with [bis(trifluoroacetoxy)iodo] benzene (TIB) in 75% acetonitrile/water (36,37,32,33) afforded the amine salt **12** in 43% yield following work-up. The use of TIB in the conversion of amides to amines offers a significant improvement over earlier methods involving the Hofmann rearrangement, Curtius rearrangement, or Schmidt reaction.

In preparation of the final cyclization step, the t-butyl ester **12** was deprotected with 40% trifluoroacetic acid/dichloromethane, providing the acid **13** in quantitative yield. Compound **13** was treated with triethylamine in DMF at -20°C until the pH reached 8. Thereupon, 1.2 equivalents of diphenylphosphorylazide (DPPA), 1 equiv. HOBt, and 0.1 equiv. DMAP were added and the mixture was stored at -5° for four days (35). Isolation of the neutral product afforded the retro-inverso modified analog **2** in 41% yield. Treatment of lettuce seedlings with **2** resulted in nearly equivalent biological activity to the parent analog (**23**) at concentrations of 100 µM and 10 µM. Full chlorosis is maintained at 100 µM (O.D. Dailey and J. V. Edwards, unpublished results).

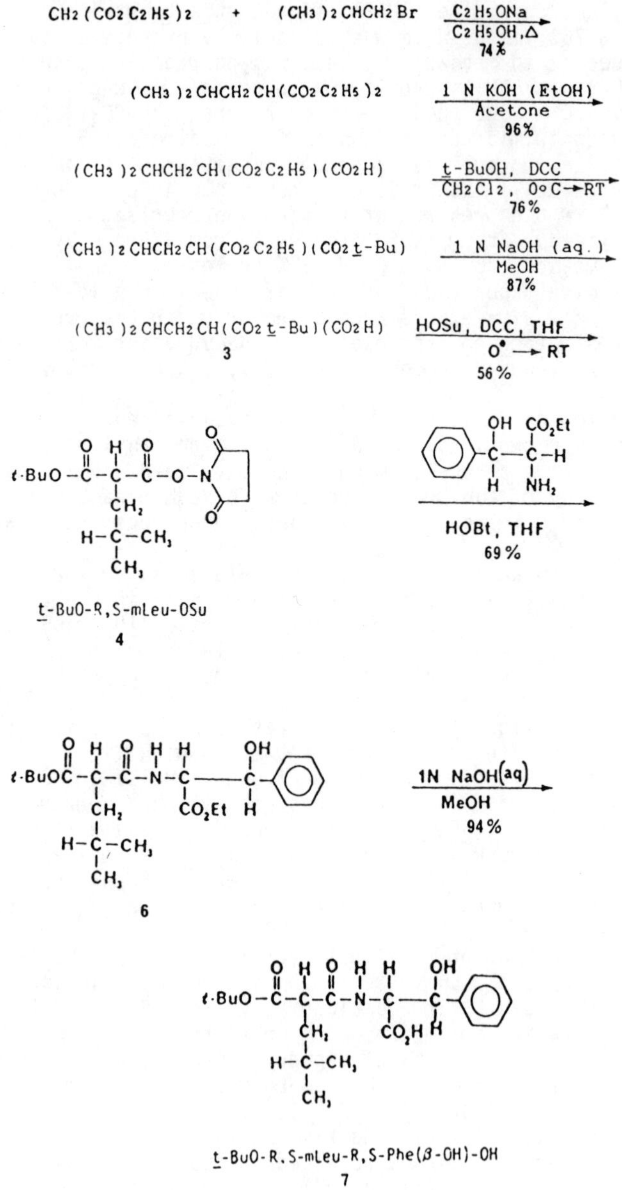

$CH_2(CO_2C_2H_5)_2$ + $(CH_3)_2CHCH_2Br$ $\xrightarrow[C_2H_5OH,\Delta]{C_2H_5ONa}$ 74%

$(CH_3)_2CHCH_2CH(CO_2C_2H_5)_2$ $\xrightarrow[Acetone]{1\ N\ KOH\ (EtOH)}$ 96%

$(CH_3)_2CHCH_2CH(CO_2C_2H_5)(CO_2H)$ $\xrightarrow[CH_2Cl_2,\ 0°C \rightarrow RT]{t\text{-}BuOH,\ DCC}$ 76%

$(CH_3)_2CHCH_2CH(CO_2C_2H_5)(CO_2t\text{-}Bu)$ $\xrightarrow[MeOH]{1\ N\ NaOH\ (aq.)}$ 87%

$(CH_3)_2CHCH_2CH(CO_2t\text{-}Bu)(CO_2H)$ **3** $\xrightarrow[0° \rightarrow RT]{HOSu,\ DCC,\ THF}$ 56%

t-BuO-R,S-mLeu-OSu
4

HOBt, THF
69%

1N NaOH(aq)
MeOH
94%

6

t-BuO-R,S-mLeu-R,S-Phe(β-OH)-OH
7

Scheme 2 Synthesis of a retro-inverso analog of tentoxin.

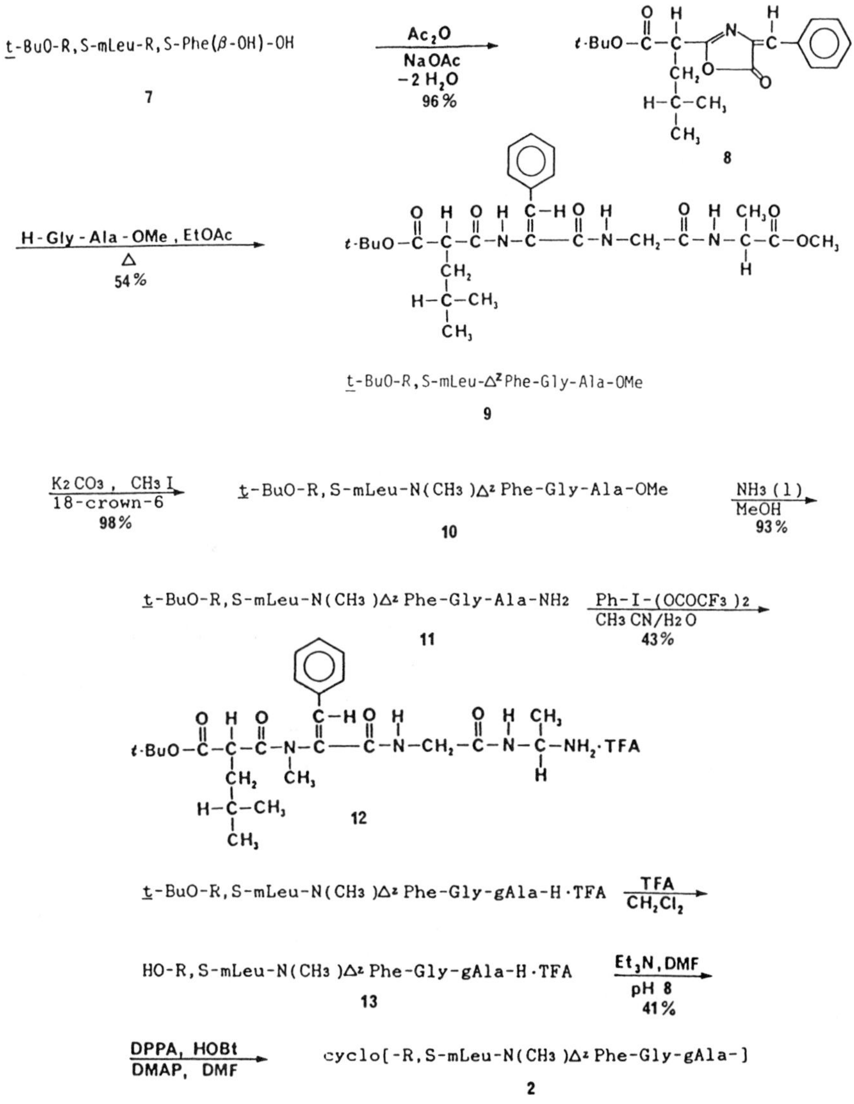

t-BuO-R,S-mLeu-R,S-Phe(β-OH)-OH

7

$\xrightarrow[\text{NaOAc, } -2\,H_2O,\ 96\%]{Ac_2O}$ 8

$\xrightarrow[\Delta,\ 54\%]{\text{H-Gly-Ala-OMe, EtOAc}}$

t-BuO-R,S-mLeu-Δ^ZPhe-Gly-Ala-OMe

9

$\xrightarrow[\text{18-crown-6},\ 98\%]{K_2CO_3,\ CH_3I}$ t-BuO-R,S-mLeu-N(CH_3)Δ^ZPhe-Gly-Ala-OMe $\xrightarrow[\text{MeOH},\ 93\%]{NH_3\ (l)}$

10

t-BuO-R,S-mLeu-N(CH_3)Δ^ZPhe-Gly-Ala-NH_2 $\xrightarrow[CH_3CN/H_2O,\ 43\%]{Ph\text{-}I\text{-}(OCOCF_3)_2}$

11

12

t-BuO-R,S-mLeu-N(CH_3)Δ^ZPhe-Gly-gAla-H·TFA $\xrightarrow[CH_2Cl_2]{TFA}$

HO-R,S-mLeu-N(CH_3)Δ^ZPhe-Gly-gAla-H·TFA $\xrightarrow[\text{pH 8},\ 41\%]{Et_3N,\ DMF}$

13

$\xrightarrow[\text{DMAP, DMF}]{\text{DPPA, HOBt}}$ cyclo[-R,S-mLeu-N(CH_3)Δ^ZPhe-Gly-gAla-]

2

Scheme 2 Continued.

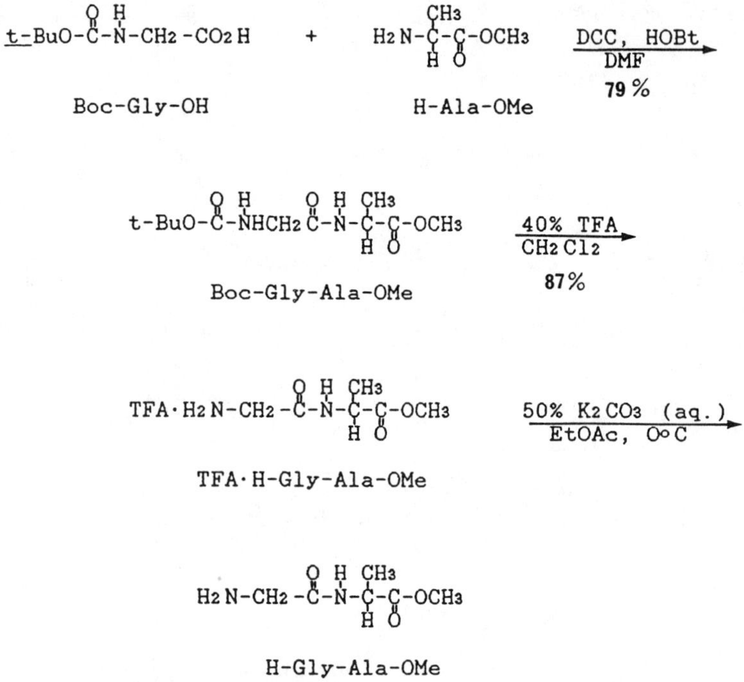

$$t\text{-}BuO\text{-}\overset{O}{\overset{\|}{C}}\text{-}\overset{H}{\overset{|}{N}}\text{-}CH_2\text{-}CO_2H \quad + \quad H_2N\text{-}\underset{H}{\overset{CH_3}{C}}\text{-}\underset{O}{\overset{}{C}}\text{-}OCH_3 \xrightarrow[DMF]{DCC,\ HOBt} $$

Boc-Gly-OH H-Ala-OMe 79 %

$$t\text{-}BuO\text{-}\overset{O}{\overset{\|}{C}}\text{-}\overset{H}{N}HCH_2\overset{O}{\overset{\|}{C}}\text{-}\overset{H}{N}\text{-}\underset{H}{\overset{CH_3}{C}}\text{-}\underset{O}{C}\text{-}OCH_3 \xrightarrow[CH_2Cl_2]{40\%\ TFA}$$

Boc-Gly-Ala-OMe 87%

$$TFA\cdot H_2N\text{-}CH_2\text{-}\overset{O}{\overset{\|}{C}}\text{-}\overset{H}{N}\text{-}\underset{H}{\overset{CH_3}{C}}\text{-}\underset{O}{C}\text{-}OCH_3 \xrightarrow[EtOAc,\ 0^\circ C]{50\%\ K_2CO_3\ (aq.)}$$

TFA·H-Gly-Ala-OMe

$$H_2N\text{-}CH_2\text{-}\overset{O}{\overset{\|}{C}}\text{-}\overset{H}{N}\text{-}\underset{H}{\overset{CH_3}{C}}\text{-}\underset{O}{C}\text{-}OCH_3$$

H-Gly-Ala-OMe

Scheme 3 Synthesis of the dipeptide H-Gly-Ala-OMe.

Use of Alkyl Spacers

The use of alkyl spacers for substitution of the peptide backbone in cyclic peptides gives structural simplification and it has been found in some cases to yield interesting selectivity and binding properties in studies on enkephalin analogs (38). Our goal in introducing alkyl spacers as part of the cyclic peptide tentoxin was to assess the role of lipophilicity, and amino acid side chains, relative to the peptide backbone and the N-methyl-dehydrophenylalanine in biological activity. Previous studies identified the key role of N-methyl dehydrophenylalanine in the biological activity of tentoxin (23,39). Therefore, we retained the functionality in the cyclic analog. A construction of molecular models containing 6 and 8 methylenes in cyclic peptides of 10 and 12 membered-rings (cyclo[8-amino-octanoyl-Δ^{Z}Phe-] **(14)** and cyclo-[6-aminohexanoyl-Δ^{Z}Phe-] **(15)**) demonstrated differences in symmetry and conformation.

The synthesis of the alkyl spacer analogs (Scheme 4) employed preparation of the C- and N-terminal protected dipeptides Boc-6-aminohexanoyl-β-phenylserine ethyl ester and Boc-8-amino-octanoyl-β-phenylserine ethyl ester. Following saponification, the dipeptides were converted to the azlactones in 56% yield with acetic anhydride and sodium acetate. Conversion of the azlactones to the dipeptide methyl esters (77% yield) was accomplished by refluxing in methanol with 0.1 equivalent of DMAP. Following selective alkylation and C- and N-terminal deprotection, cyclization of the peptides was accomplished as previously discussed in the section for synthesis of retro-inverso analogs. The products of the cyclization reactions were purified on RP-HPLC and an approximate 2:1 ratio of cyclic dimer to monomer was found following mass spectral determination of the molecular weight. Dimer/monomer mixtures of both alkyl spacer backbone substituted cyclic analogs yielded significant increases in chlorophyll levels of 170% at 10^{-6} molar concentrations when tested in lettuce seedlings (J.V. Edwards, unpublished results).

N-Alkyl Dehydrophenylalanine-containing Di- and Tripeptides

Previously we have demonstrated a structure/function relationship between N-alkylated dehydrophenylalanine-containing peptides and root growth promoting and inhibiting effects (19). By utilizing a selective N-alkylation reaction for the derivatization of dehydrophenylalanine-containing peptides with methyl and ethyl groups, synthetic linear fragments of tentoxin were prepared which demonstrated both root growth inhibiting and promoting effects. The wheat coleoptile assay provides an opportunity for detection of a broad range of biologically active substances such as bacteriostatic, fungistatic, antirickettsial, immunosuppresant, phytotoxic, plant growth inhibitors, plant growth promoters, and mycotoxins (9). The assay was employed to assess the biological activity of modified synthetic peptide fragments. Since the modified tentoxin fragments demonstrated an altogether different biological activity in plants than tentoxin, the assay also provided a means of further assessing the potential plant growth regulatory effects of peptide analogs.

Peptides were prepared through a synthetic route similar to one reported previously (Scheme 5). Structural and stereochemical variation was introduced in the amino acid side chain at the 1-position and in the N-alkyl group at the 2-position. Substitute synthesis at the 1-position is accomplished through use of D and L Boc-protected alkyl amino acids and at the 2-position through use of various alkyl iodides in the selective alkylation step.

Biological activity was determined for tripeptide derivatives and a cyclic tetrapeptide (Figure 3). The results of the wheat coleoptile assay are shown in Figure 4. The tripeptide Boc-Leu-Phe-Gly-OMe (**19**), containing no backbone modification was used as a standard for comparison; this analog was the least active (40% inhibition at 10^{-3} M) of the peptides containing the original fragment sequence of tentoxin. With conformational restriction imposed (40) by incorporation of the dehydroamino acid at the 2-position (**20**, Boc-Leu-Δ^ZPhe-Gly-OMe) a slight increase in inhibition was observed (74% at 10^{-3}). Leucine substituted with D-alanine (**21**) provided no biological activity. With further constraint imposed by N-methylation at the 2-position to give a modified fragment **(22)**, inhibition at both 10^{-3} and 10^{-4} M (44 and 17% respectively) was observed. It is interesting that analog **22** gave virtually the exact same inhibition profile as [Pro^1] tentoxin (**23**).

Substitution of the 2-position amide bond of **22** (Boc-Leu-N(CH_3)Δ^ZPhe-Gly-OMe) with an ethyl group at the nitrogen of dehydrophenylalanine (**24**), produced a shift from inhibition to promotion at 10^{-4} M (15%) with measurable effects at 10^{-5}M. It is interesting to note that prehelminthosporol, a naturally occurring phytotoxin and plant growth regulatory compound, has elicited a similar inhibition and promotion profile to **24** (41). Prehelminthosporol and compound **24** are the only compounds of over a thousand analogs tested which display an auxin-like activity in the wheat coleoptile assay. Olefinic functionalities in compound **24** and prehelminthosprol may play a critical role in the auxin activity. Plant growth promotion and inhibition in lettuce seedlings has been observed to result from geometrical and stereochemical factors in olefin-containing macrolide isomers (42). Plant growth promotion is not observed when D-Leu is substituted for L-Leu in **24** nor is it retained when the N-alkyl group is increased in size by an additional methylene (**25**, Boc-Leu-N(Pr)Δ^ZPhe-Gly-OMe).

To assess the effect of varying the type of alkyl side chain at the 1-position, valine analogs were tested. Substitution of valine at the 1-position while retaining N-methyl dehydrophenylalanine gave the analog Boc-Val-N(Me)Δ^ZPhe-Gly-OMe (**26**); this analog elicited a weaker inhibitory response in the wheat coleoptile assay than the leucine analog. However, the corresponding N-ethylated analog (Boc-Val-N(Et)Δ^ZPhe-Gly-OMe, **27**) gave a promotion effect at 10^{-4} M concentration. As observed with the leucine series, the promotion was not significant with an N-propyl group at the 2-position, (analog **28**).

These studies demonstrated the sensitivity of the wheat coleoptile assay to both side chain and backbone modifications in peptides. Though the peptides demonstrated relatively low specific activity compared to some other compounds tested in the wheat

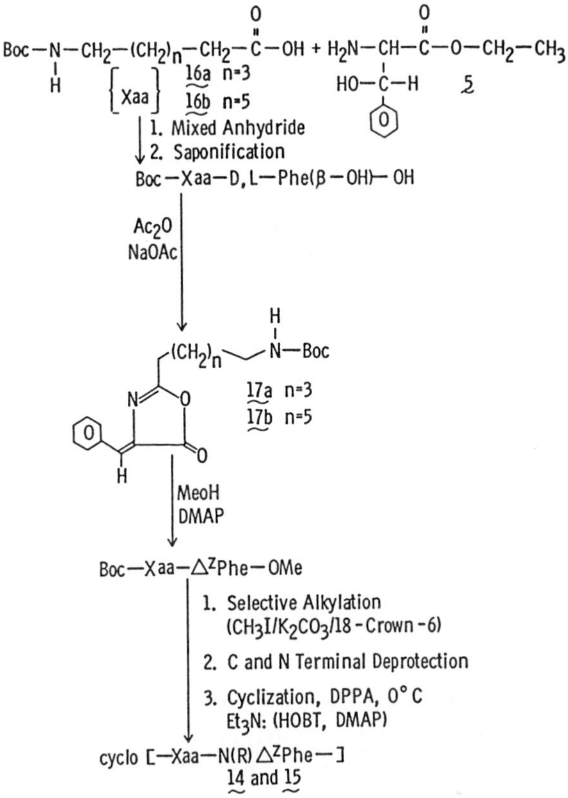

Boc—NH—CH₂—(CH₂)ₙ—CH₂—C(=O)—OH [Xaa] 16a n=3, 16b n=5 + H₂N—CH(C(OH)H—Ph)—C(=O)—O—CH₂—CH₃ 5

1. Mixed Anhydride
2. Saponification

Boc—Xaa—D,L—Phe(β—OH)—OH

Ac_2O, NaOAc

17a n=3, 17b n=5 (azlactone: (CH₂)ₙ chain with N—H—Boc)

MeoH, DMAP

Boc—Xaa—Δ^ZPhe—OMe

1. Selective Alkylation (CH_3I/K_2CO_3/18-Crown-6)
2. C and N Terminal Deprotection
3. Cyclization, DPPA, 0° C, Et_3N: (HOBT, DMAP)

cyclo [—Xaa—N(R)Δ^ZPhe—] 14 and 15

Scheme 4 Synthesis of alkyl spacer analogs of tentoxin.

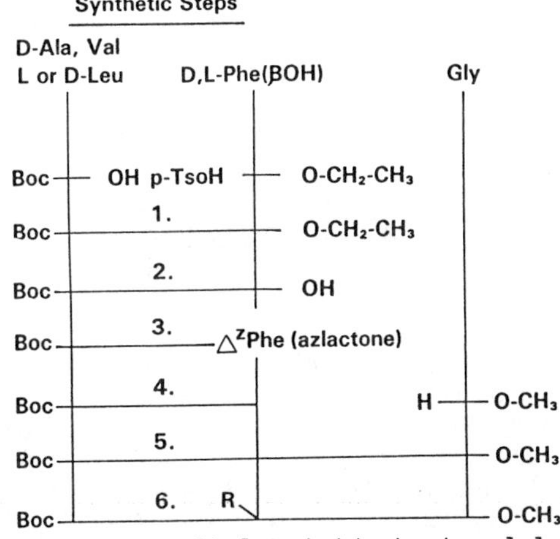

Scheme 5 Synthesis of N-alkylated dehydrophenylalanine-containing tripeptides. Details of the reaction steps have been previously reported. Data from ref. 23.

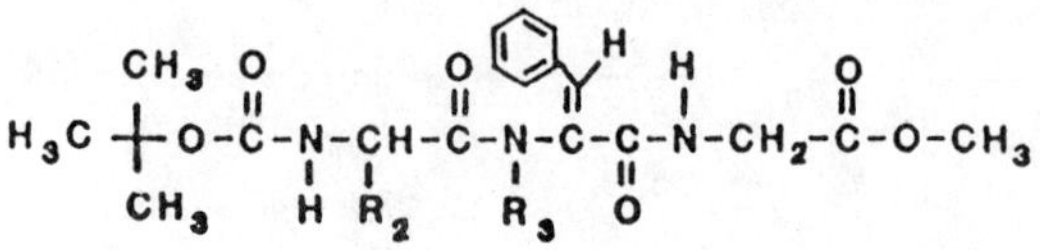

$$H_3C-C(CH_3)_2-O-C(=O)-NH-CH(R_2)-C(=O)-N(R_3)-C(=CH-C_6H_5)-C(=O)-NH-CH_2-C(=O)-O-CH_3$$

20. R_2=CH_2-CH(CH_3)-CH_3, R_3=H

21. R_2=CH_3, R_3=CH_3

22. R_2=CH_2-CH(CH_3)-CH_3, R_3=CH_3

24. R_2=CH_2-CH(CH_3)-CH_3, R_3=CH_2-CH_3

25. R_2=CH_2-CH(CH_3)-CH_3 , R_3=CH_2-CH_2-CH_3

26. R_2=HC(CH_3)$_2$, R_3=CH_3

27. R_2=HC(CH_3)$_2$, R_3=CH_2-CH_3

28. R_2=HC(CH_3)$_2$, R_3=CH_2-CH_2-CH_3

19. $H_3C-C(CH_3)_2-O-C(=O)-NH-CH(CH_2-CH(CH_3)_2)-C(=O)-NH-CH(CH_2-C_6H_5)-C(=O)-NH-CH_2-C(=O)-O-CH_3$

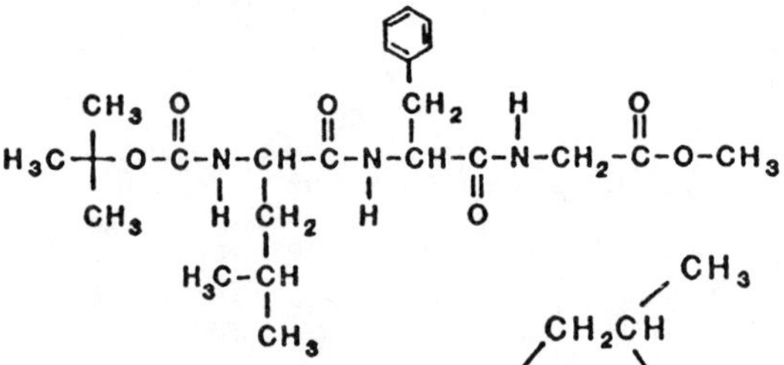

23.

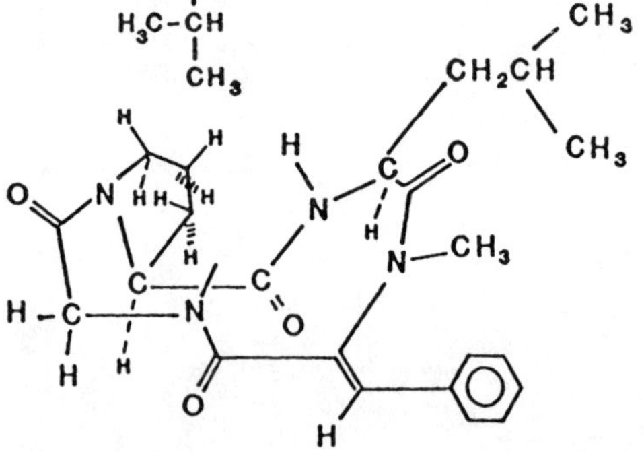

Figure 3 Structural formulas for analogs tested in the wheat coleoptile assay. Analogs of this tripeptide were substituted at the 1 and 2 position while retaining the dehydrophenylalanine ring and glycine methyl ester residues. The tripeptide tert-butyloxycarbonyl-leucyl-phenylalanyl-glycine methyl ester. Cyclo[Pro-Leu-N(Me)Δ^ZPhe-Gly-] or [Pro[1]] tentoxin.

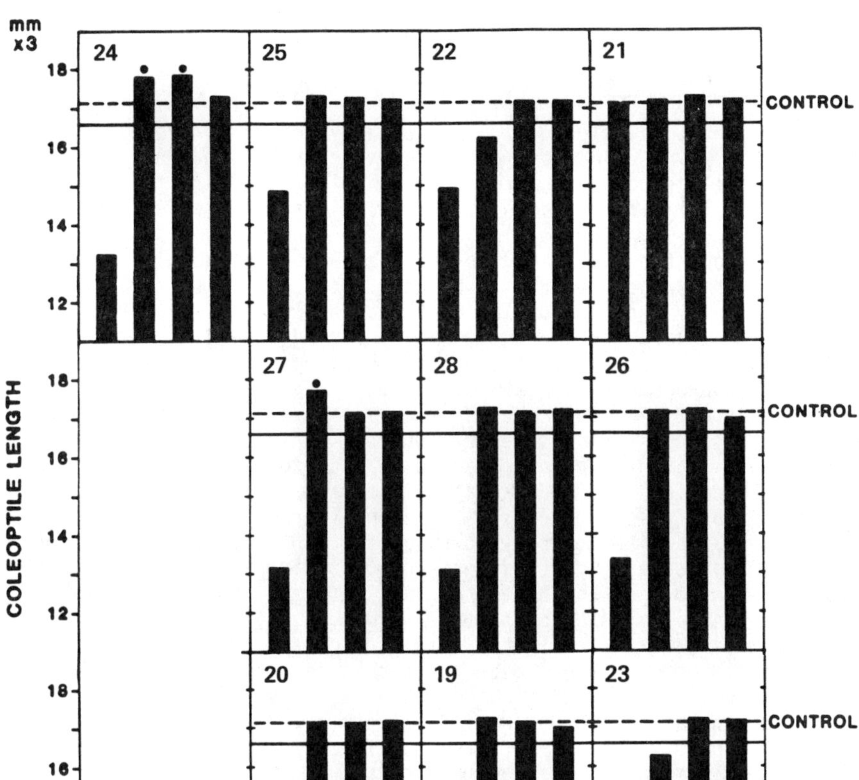

Figure 4 Wheat Coleoptile study.

coleoptile bioassay (8) the trend of promotion observed with the 2-position N-ethylated tripeptide, and the similarity in biological profiles between the synthetic fragment and [Pro^1] tentoxin demonstrate structure-function trends. Furthermore, our previous work (19) shows a similar plant growth regulating trend in dicotyledenous plants with peptides at lower concentrations (10^{-6} M) and corroborates the effectiveness of the wheat coleptile assay in assessing the plant growth regulating properties of peptides.

Cyclopropyl Amino Acids

A number of naturally-occurring microbially and plant-produced cyclopropyl amino acids have been discovered. Our interest in these types of analogs stems from the stereochemical similarity between the dehydrophenylalanine constituent in tentoxin and cyclopropylphenylalanine (Figure 5). The double bond at the alpha carbon restricts rotation about the C_α-C_β bond so that the beta functionality is fixed in space with respect to the amino acid moieties. With the substitution of a cyclopropylog of the amino acid in which the C_α-C_β bond forms one side of the three-membered ring a similar effect is achieved. The best known of the cyclopropylogs found in nature is 1-aminocyclopropane-1-carboxylic acid (ACC) which is a key intermediate in the biosynthesis of ethylene, a fruit ripening hormone in plants (43). Three cyclopropyl analogs [cis and trans-alpha-(carboxycyclopropyl)glycine and exo(cis)-3,4-methanoproline] were isolated and characterized from seeds of *Aesculus parviflora* and akee seed (44). Cis-alpha-(carboxycyclopropyl)glycine is a potent inhibitor of the growth of certain plant seedlings. Coronatine, produced by *Pseudomonas coronofacience*, is a toxin which induces chlorosis on the leaves of Italian rye grass and also expands potato cells at concentrations of 1×10^{-7} M (45). These types of analogs may be worthy of further exploration as potential crop protection agents. It has been noted that the pseudo-conjugation of the amino acid carboxyl, which occurs as a result of the sp^2 character of the cyclopropane ring, may function with the beta group of the cyclopropyl moiety serving as a reactive functionality of value in the design of enzyme inhibitors. ACC reacts readily with a pyridoxal dependent enzyme in *Pseudomonas* giving oxobutyrate and ammonia.

The use of cyclopropyl amino acids of the *E* and *Z* configurations give rise to a number of isomers depending on the nature of the R group; this process can be used to explore the role of R group orientation in bioactive amino acids and peptides. Our goal was the incorporation of cyclopropylphenylalanine into tentoxin analogs (Scheme 6). The analog cyclo-[Pro-Leu-∇^ZPhe-Gly] was synthesized by formation of the amide bond between glycine and proline. The precursor Pro-Leu-∇^ZPhe-Gly was synthesized by classical methods using active ester couplings. Beginning with HCl.∇^ZPhe-OMe (provided by Dr. C. H. Stammer), Boc-leucine N-hydroxy succinimide ester was coupled in 51% yield followed by hydrolysis of the methyl ester in 86% yield. Glycine methyl ester was then coupled to the dipeptide succinimide ester in quantitative yield. After removal of the Boc-protecting group, Boc-proline N-hydroxysuccinimide ester was coupled to the tripeptide in 80% yield. The methyl ester and

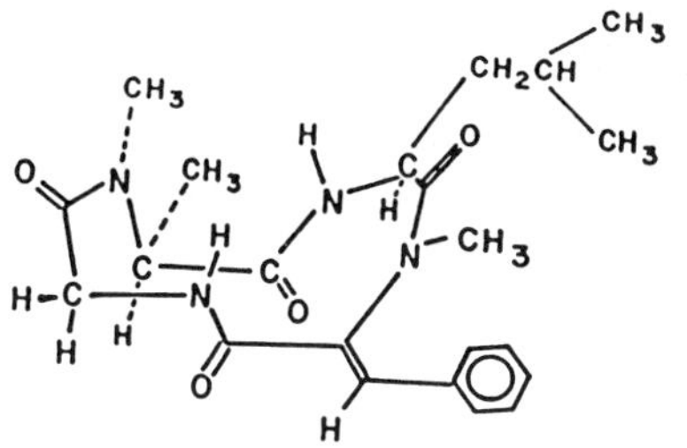

Tentoxin

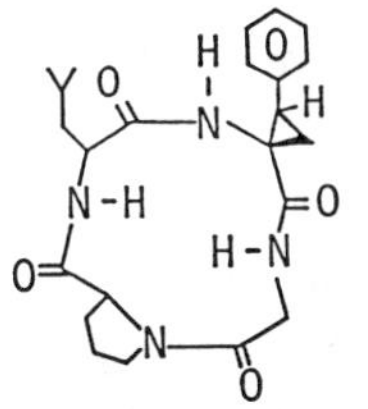

Cyclo-[Pro-Leu-∇^Z Phe-Gly]

Figure 5 Structures of tentoxin and [Pro^1, ∇^ZPhe^3] tentoxin.

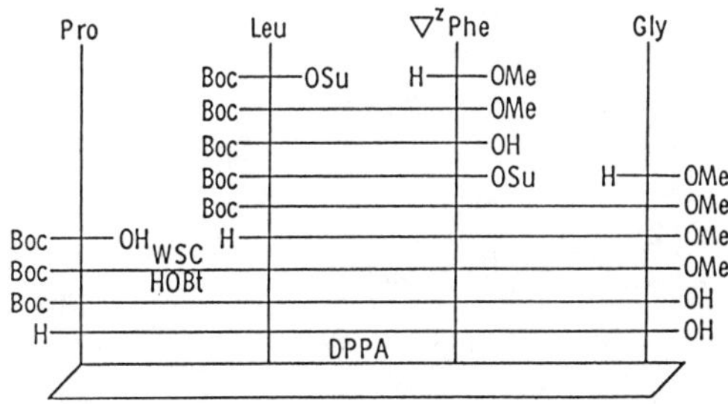

Scheme 6 Synthesis of [Pro^1,∇^ZPhe^3]tentoxin.

Boc-group were sequentially removed to give the unprotected tetrapeptide which was then cyclized using the DPPA procedure. HPLC purification isolated three major peaks. The results of a lettuce seedling assay involving treatment with the cyclopropylphenylalanine containing-tentoxin analogs showed no significant depletion of chlorophyll levels.

Summary

Naturally occurring cyclic tetrapeptides with phytoactive properties may be used as templates in designing compounds of potential use in agriculture. The bioactive cyclic tetrapeptides are amenable to computational chemistry analyses and their secondary structure is well characterized and may be mimicked by simpler nonpeptidic molecules. Tentoxin, HC toxin, and synthetic fragments have shown particular potential since they exhibit both herbicidal and growth promoting activity. The synthetic approaches to the introduction of amide bond surrogates may prove useful in developing pesticides or crop improvement agents from naturally occurring plant-active peptides. Many peptides with modified amide bonds offer increased proteolytic resistance, and assessing conformation properties versus biological activity may prove useful in the development of potential leads for new product lines.

Acknowledgment

The authors thank J. V. Kelly, J. Novak and R. Cougle for technical assistance, and special thanks go to Dr. Charles Stammer (Dept. of Chemistry, University of Georgia) for providing cyclopropylphenylalanine methyl ester.

Literature Cited

1. Schwyzer, R. Proc. R. Soc. London, B 1980, 210, 5-20.
2. Manson, M.D.; Blank, V.; Brode, G.; Higgins C.F. Nature, 1986, 321, 253-256
3. Higgins, C.F.,Payne, J.W. In Encyclopedia of Plant Physiology, 14A; Springer Verlag KG: Berlin, 1982, p. 438.
4. Klambt, D. J. Theor. Biol. 1983, 100, 435-441.
5. Klambt, D. Biol. Plant. 1985, 27, 204-208.
6. Bialek, K.; Cohen, J. D. Plant Physiol. 1986, 80, 99-104.
7. Burcklout, T. J.; Curtis R. W. Nature 1976, 260. 435-436.
8. Curtis, R.W., Plant Physiol. 1976, 57, 365-368.
9. Garay, A. S.; Ahlgren, J.A.; Gonzalez, M.A.; Stasney, M.A.; Mattes, P.C., Applied and Environmental Microbiology, 1986, 51, 457-461.
10. Payne, J.W.; Higgins, C.F. In Microorganisms and Nitrogen Sources; Payne, J.W., Ed.; John Wiley Sons Ltd: 1980, pp. 609-639.
11. Rich, D.H. In Toxins in Plant Disease. Mechanism of Action; Durbin, R.D., Ed.; Academic: New York, 1981. Physiological Ecology Series, Chapter 8.
12. Anderegg, R.J.; Biemann, K.; Buchi, G.; Cushman, M. J. Am. Chem. Soc. 1976, 98, 3366-3370.

13. Bodansky, M. and Stahl G.L., Proc. Natl. Acad. Sci. U.S.A. 1974 71, 2791-2794.
14. Bodansky, M.; Stahl, G.L. Bioorg. Chem. 1973, 4, 93.
15. Kloepper, J.W.; Leong J.; Teintze, M.; Schroth, M.N. Nature, 1980, 286, 885-886.
16. Schroth, M.N.; Hancock, J.G. Science, 1982, 216, 1376-2381.
17. Hruby, V.J. In Conformational Directed Drug Design; Vida, J.A.; Gordon, M., Eds.: ACS Symposium Series No. 251; American Chemical Society, Washington, D.C., 1984, pp 9-27.
18. Lin, W.; Kauer, J.C. Plant Physiol. 1985, 77, 403-406.
19. Edwards, J.V.; Cutler, H.G.; Zorner, P.S.; Coffman C.B. In Synthesis and Chemistry of Agrochemicals; Baker, D.R., Fenyes, J.G., Moberg, W.K.; Cross, B., Eds.; ACS Symposium Series No. 355; American Chemical Society, Washington, D.C., 1987; pp 151-160.
20. Huffman, W.F.; Callahan, J.F.; Eggleston, D.S.; Newlander, K.A.; Takata, D.T.; Codd, E.E.; Walker, R.F.; Schiller, P.W.; Lemieux, C.; Wire, W.S.; Burks, T.F. In Peptides Chemistry and Biology; Proceedings of the Tenth American Peptide Symposium; Marshall, G.R., Ed.; ESCOM, Leiden, 1988, pp 105-108.
21. Kato, T.; Lee, S.; Shimohigashi, Y; Tone, A.; Kodero, Y.; Izumiya, N. Int. J. Peptide Protein Res. 1987, 29 (1), 53-61..
22. Ajo, D.; Granozzi, G.; Tondello, E.; Del Pro, A. Biopolymers 1980, 19, 469-475.
23. Edwards, J.V.; Lax, A.R.; Lillehoj, E.B.; Boudreaux, G.J. J. Agric. Food Chem. 1987, 35(4), 452-456.
24. Spatola, A. F.; Edwards, J.V. Biopolymers 1986, 25, 229-244.
25. Goodman, M.; Chorev, M. Acct. Chem. Res. 1979 12, 1-7.
26. Goodman, M.; Chorev, M. In Perspectives in Peptide Chemistry; Eberle, A.; Geiger, R.; Wieland, T. Eds.; S. Karger: Basel, 1981; pp 283-294.
27. Shemyakin, M.M.; Ovchinnikov, Y.A.; Ivanov, V.T. Angew. Chem. Internat. Edit. 1969 8, 492-499.
28. Vogler, K.; Lanz, P.; Lergier, W.; Haefely, W. Helv. Chim. Acta 1966, 49, 390.
29. Chorev, M.; Willson, C.G.; Goodman, M. J. Am. Chem. Soc. 1977, 99, 8075-8076.
30. Chaturvedi, N.; Goodman, M.; Bowers, C. Int. J. Peptide Protein Res. 1981, 17, 72-88.
31. Chorev, M.; Shavitz, R.; Goodman, M.; Minick, S.; Guillemin, R. Science 1979, 204, 1210-1212.
32. Berman, J.M.; Goodman, M. Int. J. Peptide Protein Res. 1984, 23, 610-620.
33. Richman, S.J.; Goodman, M.; Nguyen, T.M.-D.; Schiller, P.W. Int. J. Peptide Protein Res. 1985, 25, 648-662
34. Fischer, E.; Schmitz, W. Chem. Ber. 1906, 39, 351-356
35. Edwards, J.V.; Lax, A.R.; Lillehoj, E.B.; Boudreaux, G.J. Int. J. Peptide Protein Res. 1986, 28, 603-612.
36. Radhakrishna, A.S.; Parham, M.E.; Riggs, R.M.; Loudon, G.M. J. Org. Chem. 1979, 44, 1746-1747.
37. Pallai, P.; Goodman, M. J. Chem. Soc., Chem. Commun. 1982, 280-281.
38. Schiller, P.W.; Nguyen, T. M-D.; Maziok, L.; Lemieux, C. Biochem. Biophys. Res. Commun. 1985, 127, 558-564.

39. Kono, Y.; Gardner, J.M.; Takeuchi, S. Agric. Biol. Chem. 1986, 50, 2401-2403.
40. Hruby, V.J. Life Sci. 1982, 31, 189.
41. Cutler, H.G., In Bioregulators: Chemistry and Uses; Ory, R.L.; Rittig, F.R., Eds.; ACS Symposium Series No. 257, American Chemical Society, Washington D.C. 1984, pp. 153-169.
42. Hirota, A.; Sakai, H.; Isogai, A. Agric. Biol. Chem. 1985, 49, 731-735.
43. Adaxus, D.O.; Yang S.F., Proc. Natl. Acad. Sci. U.S.A. 1979, 76, 170-174.
44. Millington, D.S.; Sheppard R.C., Phytochemistry, 1969, 8, 473-443.
45. Ichihara, A.; Shiraishi, H.S.; Sato, H.; Sakamiva, S.; Nishiyama, K., Sakai, R., Furusaki, A., Matsumoto, T. J. Am. Chem. Soc., 1977, 99, 636-637.

RECEIVED May 6, 1988

Chapter 4

Toxins of Phytopathogenic Microorganisms

Structural Diversity and Physiological Activity

S. Sakamura, A. Ichihara, and T. Yoshihara

Department of Agricultural Chemistry, Faculty of Agriculture, Hokkaido University, Sapporo 060, Japan

Plant pathogenic fungi and bacteria produce a number of phytotoxins and there is current interest in the use of their compounds and derivatives for agrochemicals. After commenting on the significance of toxin reseach, recent progress in the area of bacterial toxins and host-specific toxins is briefly mentioned. Recently, we have found several physiologically active compounds including novel ones, that is, betaenones and aphidicolanes from Phoma betae, reduced perylenequinones and anthraquinones from Stemphylium botryosum, and cyclopentanoid sesquiterpenes from timothy stalks infected with Epichloe typhina.

The chemistry of toxins produced by plant pathogens has made rapid progress in the last twenty years. At the present time, the number of toxins elucidated based on their chemical structures is more than 170 compounds and the number increases every year (1).

The chemical studies of toxins have been undertaken with the purpose of (a) understanding the causal factor in plant diseases and (b) discovering of physiologically active principles, including plant regulating substances.

In general, isolation of these compounds is carried out starting from culture broths or infected plant material. Toxins are structurally unique and belong to the class of secondary metabolites which almost all come from fungi except a few from certain bacteria. There has been no evidence that a toxin was confirmed from biotrophic fungi or plant material infected with these fungi.

Bacterial phytotoxins

Several phytotoxins have been isolated from the Pseudomonas group of plant pathogenic bacteria and their structures elucidated. Though some reviews on bacterial phytotoxins have appeared (2-4), particular points require summarization.

0097-6156/88/0380-0057$06.00/0

Pseudomonas syringae pv. tabaci, which causes wildfire disease of tobacco, produces tabtoxin (5). Recently the structure of tabtoxin was confirmed via total synthesis (6). The structure of phaseolotoxin (7), produced by Pseudomonas syringae pv. phaseolicola which causes halo blight of bean, has been revised because new mass spectrum data were provided (8). More recently interesting biosynthetic studies of coronatine (9) isolated from Pseudomonas syringae pv. atropurpurea, the causal agent of chocolate spot disease on Italian ryegrass, have shown that the acidic component, coronafacic acid, was derived from a branched polyketide with five acetate units and one pyruvate (10).

Host-Specific Toxins

Thirteen species of fungi which produce host-selective or host-specific toxins are known and they belong to the genus Alternaria (6 species) and Helminthosporium (4 species) and the others (11). These toxins have a specifically high toxicity toward the restricted hosts, such as cultivars and species of plant. Thus these compounds are regarded as a primary determinant in pathogenesis and are sometimes called pathotoxins. The principal toxin is structurally diverse, usually occurs together with closely related compounds, and sometimes in the company of non-host specific toxins as well. Based on the chemical structure, the toxins are classified as cyclopeptides, amino acid esters, aliphatic aminopolyol esters, polyalcohols, aliphatic polyol lactones, or sesquiterpene galactosides. The pathogens of different species also produce toxins analogus to each other in their structures. An example of this can be seen with HMT- and PM-toxins; they both possess an aliphatic polyalcohol structure (11).

Fungal Nonhost-Specific Toxins

A number of phytotoxic metabolites which are non-host specific have been isolated and identified by us from the following fungi, Phoma betae (synonym: Phyllosticta betae) and Stemphylium botryosum.

Betaenones and aphidicolanes (12-16). Phoma betae Fries, a causal agent of root rot and leaf spot diseases of sugar beets is grown on a potato-sugar medium, and the cultured filtrates are subjected to extraction. The individual toxins are isolated by extraction with EtOAc, silica gel chromatography and preparative TLC. A bioassay is performed which consists of measuring the growth inhibition of lettuce seedlings. As a result, the physiologically active compounds isolated were divided into two groups. One group belongs to the novel decalin derivatives and are named betaenones A(1), B(2), C(3), D(4), E(5) and F(6). The other was aphidicolin (7) and its newly identified analogs 3-deoxyaphidicolin (8), aphidicolin-17-monoacetate (9) and aphidicolin-3, 18-orthoacetate (10).

In our in vivo experiments (data not shown), aphidicolin and its analogs cause a marked inhibition of DNA synthesis and show selectivity for DNA polymerase α(17). On the other hand, betaenone C(3) strongly inhibited both the protein and RNA synthesis compared

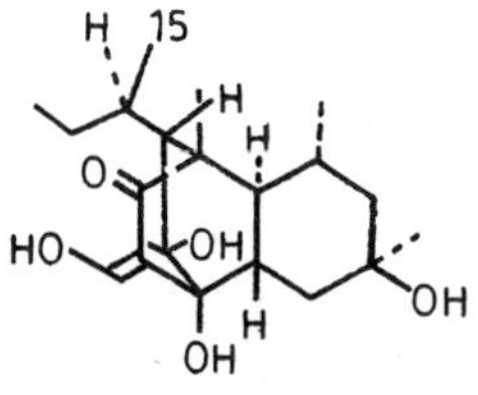

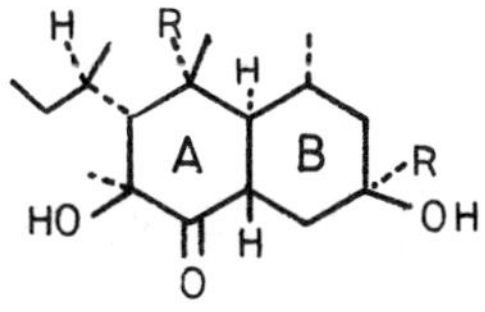

	R_1	R_2
2	$-COCH_2CH_2OH$	$-CH_3$
3	$-COCH{=}CHOH$	$-CH_3$
4	$-COCH_2CH_2OH$	$-CH_2OH$
5	$-COCH_3$	$-CH_3$
6	$-COCH_2CH_2OAc$	$-CH_3$

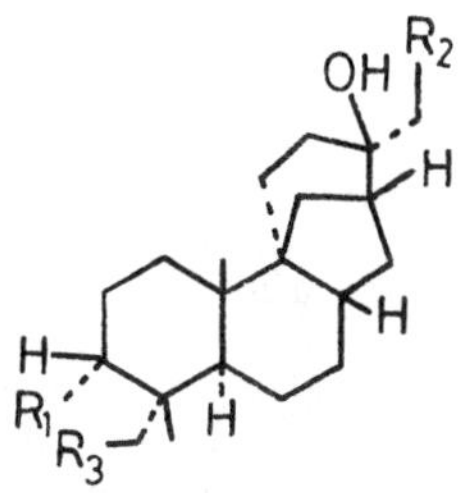

1

7 $R_2, R_1, R_3 = OH$

8 $R_1 = H$, $R_2, R_3 = OH$

9 $R_1, R_3 = OH$, $R_2 = OAc$

10 $R_1, R_3 = O\diagup\!\!\diagdown O$ (OH), $R_2 = OH$

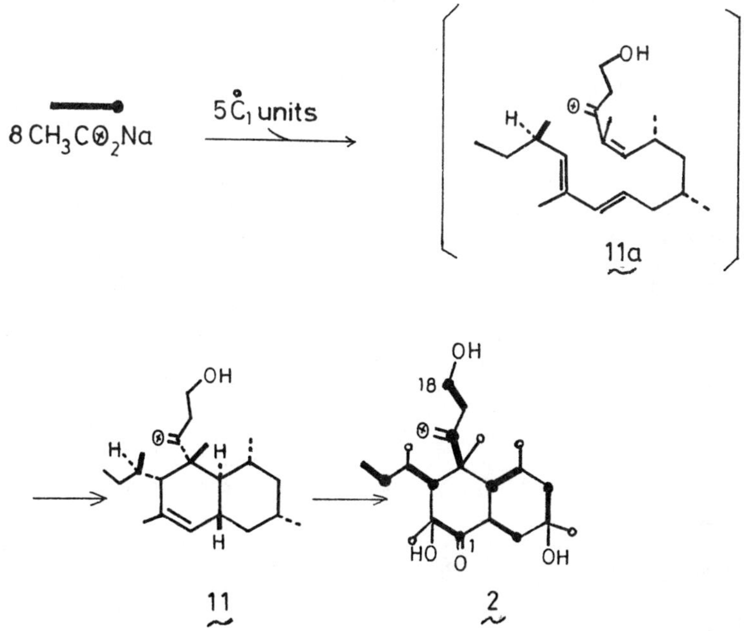

$8\ CH_3C^{\circ}O_2Na$ → $5\ \mathring{C}_1$ units → [11a] → 11 → 2

with A and B. Betaenone C, however, did not significantly inhibit DNA syntheses. Among the betaenones, C(3) was determined to have the most inhibitory effect as tested on growth of rice seedlings (18) (Table 1).

Table 1. Effect of betaenones and aphidicolanes on root elongation of rice seedlings

Toxin (10^{-4} M)	Root elongation (cm)	% of inhibition
Control	5.23	0
Betaenone A(1)	1.40	73
Betaenone B(2)	4.80	8
Betaenone C(3)	0.56	89
Aphidicolin(7)	1.64	69
3-deoxyaphidicolin(8)	2.17	59
Aphidicolin-17-monoacetate(9)	1.90	64
Aphidicolin-3, 18-orthoacetate(10)	1.75	67

a) Mean length, average of 20 seedlings

It should be noted that we established betaenone B(2) to be biosynthesized from eight acetate units via the polyketide pathway with the origin of the five branched methyl groups from methionine (16). Very recently feeding experiments with [1-^{13}C, ^{18}O] acetate revealed that in the ^{13}C-NMR spectrum of enriched 2, only the isotopic shifted signal was observed at C-16 [Δ + 0.05ppm], but not at C-1 and C-18. This means that the oxygen atom at C-1 does not originate from the acetate and absence of the expected isotopic signal ascribable to C-18 would be due to the rapid exchange of the labeled oxygen of the end carboxyl group of the polyketide chain. On the other hand, addition of ancymidol, a potent cytochrome P-450 inhibitor, into the culture has caused an inhibition of betaenone formation and a concomitant accumulation of a plausible intermediate, probetaenone I (11). These feeding and inhibitor experiments clearly show that 11 is an intermediate in the biosynthesis of 2 and presents a unique biosynthetic pathway involving an intramolecular Diels-Alder reaction of a triene 11a as a key step (Oikawa, H.; Ichihara, A.; Sakamura, S., submitted for publication).

More recently closely analogous metabolites to betaenones, stemphyloxins I and II have been identified from Stemphylium botryosum f. sp. lycopersici, and they correspond to betaenones C (3) and A(1) respectively. The compounds of both groups differ from one another in that the C_{15} methyl of betaenones is replaced with the hydroxymethyl of stemphyloxins.

Manulis et al. demonstrated, using growing tomato cells, that stemphyloxins and betaenones comparably inhibit protein synthesis (19). The data indicate that lower concentrations of stemphyloxins were required for 50% inhibition than for betaenones. The results parallel those from of the rice seedling test.

Reduced perylenequinones and anthraquinones. A strain, Stemphylium botryosum isolated from a beet plant with the leaf spot disease, was examined to obtain phytotoxins from the cultured broths. Procedures of the culture of the fungus, isolation of the phytotoxic compounds, bioassay and the structural determination were similar to those for Phoma betae. Several compounds were identified as the known compounds namely, scytalone (12), stemphyperylenol (13), dactylariol (14), macrosporin (15), and alterporiol A or B(16) (20,21). Two new compounds were named stemphylenols A(17) and B(18), both being atropisomers of each other. By using the pregerminated lettuce and beet seedlings bioassay, dactylariol (14) was more inhibitory in the letter seedling elongation test than the others, and an effective dose for inhibition was observed at a concentration of 12.5 ppm for beet. Stemphyperylenol (13) and scytalone (12) showed moderate inhibitory activity, whereas the dimeric anthraquinones alterporriol (16), stemphylenols (17, 18) and monomeric macrosporin (15) showed no significant effect (Teshima, Y.; Ichihara, A.; Sakamura, S. in preparation).

Fungitoxic compounds (22, 23), chokols and chokolic acid A. Choke disease fungi (Epichloe typhina) infect timothy plants and form stromata on them. The stroma is termed "choke" and impedes the development of the panicle. On the other hand, infected timothy plants acquire an induced resistance against another invader, the leaf spot disease pathogen (Cladosporium phlei). One of the resistance mechanisms is likely to be that some metabolite(s) of choke disease fungi have inhibitory effects against the leaf spot disease fungi. We have been exploring the fungitoxic compounds from the chokes to explain the resistance phenomenon. For monitoring the activity, TLC bioautography using Cladosporium herbarum was employed.

The 70% EtOH extracts of the fresh chokes were partitioned into n-hexane and EtOAc soluble fractions. Each fraction was chromatographed on silica gel and Sephadex LH-20, and further purified by HPLC. The fungitoxic compounds isolated were designated chokol A(19), B(20), C(21), D(22), E(23) and F(24) and chokolic acid A(25). The spectral data, together with spin decoupling experiments, suggested that the chokols possess the same five membered ring system with varied side chains. The relative stereochemistry of the common five membered ring system was determined by nuclear Overhauser effect difference spectroscopy of chokol C(21). The absolute configuration of chokol E(23) was confirmed by its CD spectrum with the chelating reagent $Eu(fod)_3$ and its chemical conversion to epicyclonerodiol oxide (24), the absolute configuration of which was evident. Thus the stereochemistry of chokol E(23) was determined to be 2S, 3R, 6R and 10R in the depicted structure.

Nerolidyl pyrophosphate is an intermediate in the biosynthesis (25) of cyclonerodiol which possesses a novel type skeleton with five carbon members in the sesquiterpenoids and the same absolute configuration as chokols. Biosyntheses of chokols and chokolic acid A(25) are likely to be via the nerolidyl pyrophosphate pathway.

Using TLC bioautography, a minimum quantity of the fungitoxic activity per spot was obtained for each compound; chokol A(25μg), B(5μg), C(5μg), D(5μg) and E(50μg).

Two reports on the synthesis of chokol A(19) have been published; one concerns the synthesis of the racemic compound (26) and the other concerns enantioselective synthesis (27).

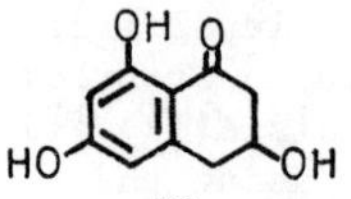

OH O
HO OH
12

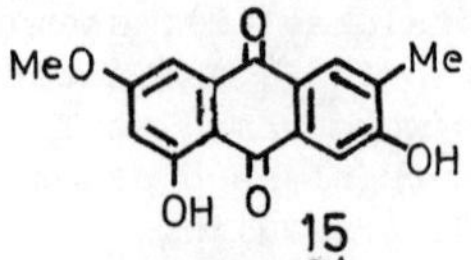

OH O
OH
H
H
HO
O OH
13

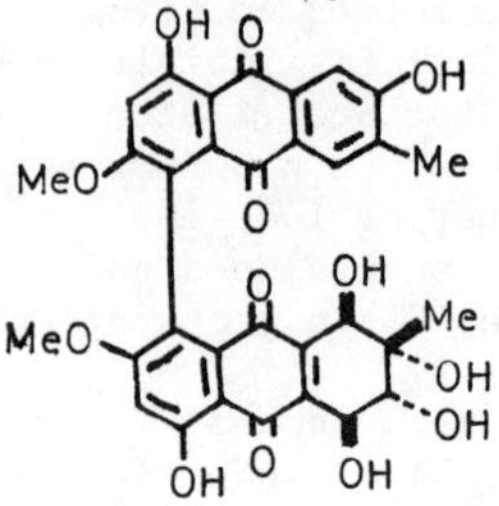

MeO OH Me OH OH OH O
14

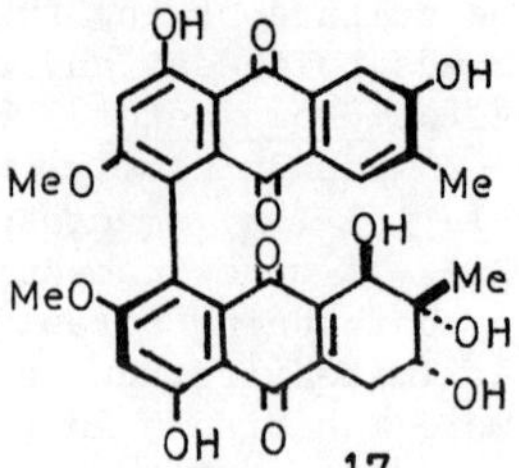

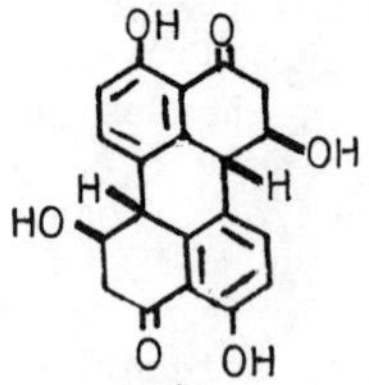

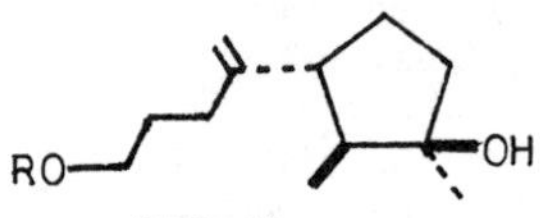

MeO Me OH O OH
15

OH O OH MeO Me O
MeO O OH Me OH OH OH O OH
16

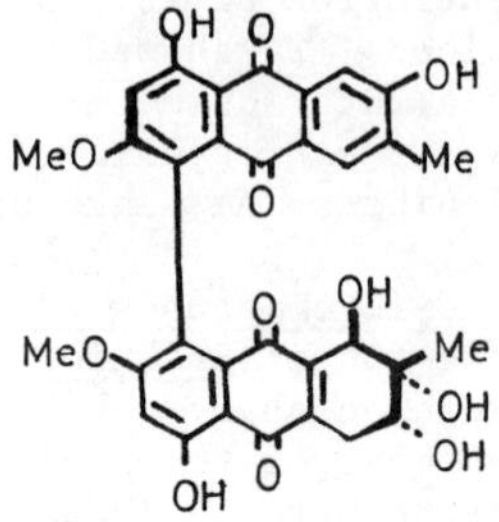

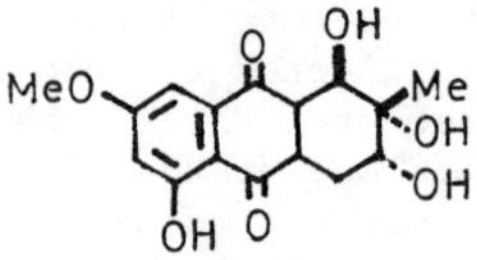

OH O OH MeO Me O
MeO O OH Me OH OH OH O
17

OH O OH MeO Me O
MeO O OH Me OH OH OH O
18

RO OH
19 R= H
24 R= Ac

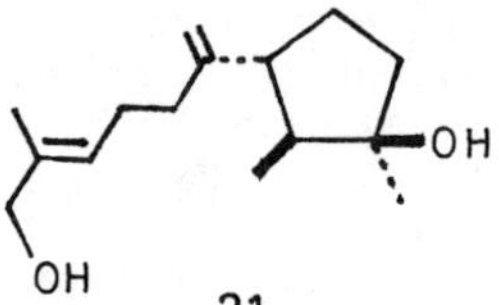

OH OH
21

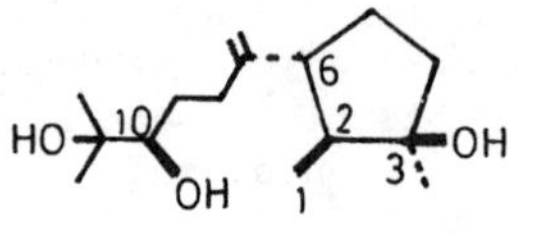

HO 10 OH 6 2 1 3 OH
23

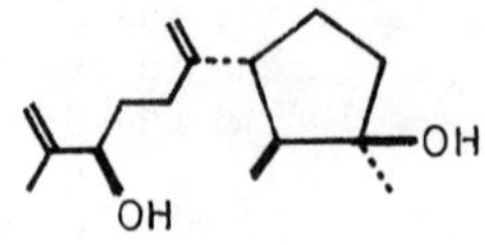

OH OH
20

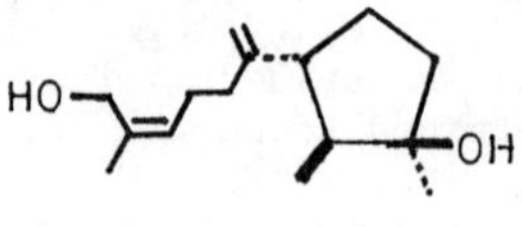

HO OH
22

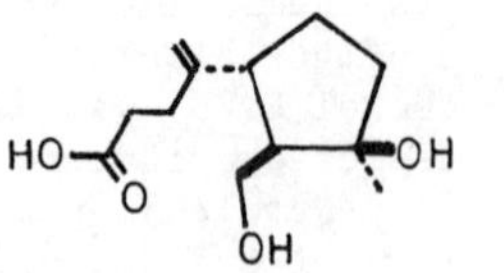

HO O OH OH
25

Significance and application of toxins

a) Selection of disease resistant plants (28). When resistant plants are treated with a specific toxin, the visible and physiological changes similar to those found in infected plants occur. Certain pathogen-produced metabolites, pathotoxins, could serve as valid substitutes for living pathogens to develop the disease symptom. Thus, it has been possible to identify disease-resistant individuals by selecting from populations those plants which are resistant to the toxins. In addition to these toxins, non-specific toxins that cause disease symptoms such as necrosis, browning leaf spots, chlorosis and cytotoxicity, may be used for the selection of disease resistant plants. Instead of using whole plants, the tests have been directed towards using protoplast, cell and tissue cultures. Extensive studies have been hindered by the limited supply of purified toxins but, with synthetic toxins materials and their mimics probably available, the difficulties should be reduced.

b) Utilization of toxins for agrochemicals (29). Since some of phytotoxins are known to have herbicidal activity as well as various other physiological activity, they may be potent for use as herbicides or plant growth regulating substances, though complete biological testing will be required. For example, although aphidicolin is noted as an antiviral and antimitotic compound, aphidicolin has recently been found to be very effective for inducing high synchronization of plant cells in liquid culture (30). Also those compounds are useful as lead or model compounds for the design of new agrochemicals.

LITERATURE CITED

1. Sakamura, S. KAGAKU TO SEIBUTSU 1985, 25, 289-298.
2. Mitchell, R. E. In Toxins in Plant Disease; Durbin, R. D., Ed,; Academic: New York, 1981; P259.
3. Macko, V. In Toxins and Plant Pathogenesis; Daly, J. M.; Deverall, B. J. Eds.; Academic: Australia, 1983; P41.
4. Ichihara, A. J. Synth. Org. Chem. 1987, 45, 357-368.
5. Stewart, W. W. Nature 1971, 229, 174-178.
6. Baldwin, J. E.; Bailey, P. D.; Gallacher, G.; Otsuka, M.; Singleton, K. A.; Wallace, P. M.; Prout, K.; Wolf, W. M. Tetrahedron 1984, 40, 3695-3708.
7. Mitchell, R. E. Phytochemistry 1976, 15, 1941-1947.
8. Moore, R. E.; Niemczura, W. P., Kwok, O. C. H., Patil, S. S. Tetrahedron Lett. 1984, 36, 3931-3934.
9. Ichihara, A.; Shiraishi, K.; Sato, H.; Sakamura, S.; Nishiyama, K.; Sakai, R.; Fukuoka, A.; Matsumoto, T. J. Am. Chem. Soc. 1977, 99, 636.
10. Parry, R. J.; Mafoti, R. J. Am. Chem. Soc. 1986, 108, 468-469.
11. Kono, Y.; Suzuki, Y.; Takeuchi, S. J. Synth. Org. Chem. 1985, 43, 980-989.
12. Ichihara, A.; Oikawa, H.; Hayashi, K.; Sakamura, S. J. Am. Chem. Soc. 1983, 105, 2907-2908.

13. Ichihara, A.; Oikawa, H.; Hayashi, K.; Hashimoto, M.; Sakamura, S.; Sakai, R. Agric. Biol. Chem. 1984, 48, 1687-1689.
14. Ichihara, A.; Oikawa, H.; Hashimoto, N.; Sakamura, S.; Haraguchi, T.; Nagano, H. Agric. Biol. Chem. 1983, 47, 2965-2967.
15. Oikawa, H.; Ichihara, A.; Sakamura, S. Agric. Biol. Chem. 1984, 48, 2603-2605.
16. Oikawa, H.; Ichihara, A.; Sakamura, S. J. Chem. Soc. Chem. Commun. 1984, 815.
17. Haraguchi, T.; Oguro, M.; Nagano, H.; Ichihara, A.; Sakamura, S. Nucleic acids Res. 1983, 11, 1197-1209.
18. Sakai, R.; Mino, Y.; Ichihara, A.; Sakamura, S. Ann. Phytopath. Soc. Japan 1985, 51, 219-222.
19. Manulis, S.; Netzer, D.; Barash, I. J. Phytopathology 1986, 115, 283-287.
20. Arnone, A.; Nasini, G.; Merlini, L.; Assante, G. J. Chem. Soc. Perkins I 1986, 525-530.
21. Suemitsu, R.; Sano, T.; Yamamoto, M.; Arimito, Y.; Morimatsu, F. ; Nabeshima, T. Agric. Biol. Chem. 1984, 48, 2611-2613.
22. Yoshihara, T.; Togiya, S.; Koshino, H.; Sakamura, S.; Shimanuki, T.; Sato, T.; Tajimi, A. Tetrahedron Lett. 1985, 26, 5551-5554.
23. Koshino, H.; Togiya, S.; Yoshihara, T.; Sakamura, S.; Shimanuki, T.; Sato, T.; Tajimi, A. Tetrahedron Lett. 1987, 28, 73-76.
24. Fujita, T.; Takaishi, Y.; Takeda, Y.; Fujiyama, T.; Nishi, T. Chem. Pharm. Bull. 1984, 32, 4419-4425.
25. Cane, D. E.; Iyengar, R.; Chiao, M-S. J. Amer. Chem. Soc. 1981, 103, 914-931.
26. Oppolzer, W.; Cunningham, A. F. Tetrahedron Lett. 1986, 27, 5467-5470.
27. Mash, E. A. J. Org. Chem. 1987, 52, 4142-4143.
28. Wenzel, G. Ann. Rev. Phytopathol. 1985, 23, 149-172.
29. Strobel, G.; Sugawara, F.; Clardy, J. In Allelochemicals: Role in Agriculture and Foresty; Waller, G. R., Ed. ACS Symposium Series 330; American Chemical Society: Washington, DC, 1987; P516-523.
30. Nagata, T.; Okada, K.; Takebe, I. Plant Cell Reports 1982, 1, 250-252.

RECEIVED April 1, 1988

Chapter 5

Production of Herbicidal and Insecticidal Metabolites by Soil Microorganisms

R. M. Heisey[1,3], S. K. Mishra[1], A. R. Putnam[1], J. R. Miller[1], C. J. Whitenack[1], J. E. Keller[1], and J. Huang[2]

[1]Pesticide Research Center, Michigan State University, East Lansing, MI 48824

[2]Rhone-Poulenc Ag Company, Research Triangle Park, NC 27709

A screening program directed toward discovery of soil microorganisms that produce herbicidal and insecticidal compounds is described. Microorganisms, especially actinomycetes, were isolated from soil samples, grown in broth, and bioassayed on seeds, seedlings, and mosquito larvae. Approximately 4-12% of the microbial isolates showed strong herbicidal activity in initial screens. A similar proportion was insecticidal. Several herbicidal (cycloheximide, geldanamycin, nigericin, (+)-epiepoformin, 3-hydroxybenzyl alcohol, 2-methylhydroquinone) or insecticidal (valinomycin) compounds produced by these microorganisms have been isolated, identified, and evaluated for potential as pesticides. The results demonstrate that soil microorganisms produce numerous pesticidal metabolites and support the validity of a microbial approach to discovery of new pesticides.

Many microorganisms indigenous to the soil, especially actinomycetes, produce biologically active secondary metabolites. The thousands of antibiotics discovered since the 1930's, when intensive screening for such compounds began, is impressive evidence. Recently, interest has focussed on microorganisms as a source of pesticidal compounds. A number of microbial metabolites exhibiting herbicidal, insecticidal, and nematocidal activities have been found (1-5).

The potential of the microbial approach to producing pesticides is evidenced by the avermectins, a group of metabolites from *Streptomyces avermitilis*. The avermectins, and their semisynthetic derivative ivermectin, are active against certain nematodes and arthropods at extremely low doses, but have relatively low mammalian toxicity (6-9). They have been highly effective in veterinary use and now show much promise for treating human infestations (10, 11).

Bialaphos is another microbial metabolite with interesting pesticidal potential (4, 5, 12). It is produced by *S. hygroscopicus*

[3]Current address: Department of Biological Sciences, Fordham University, Bronx, NY 10458

0097-6156/88/0380-0065$06.00/0

and S. viridochromogenes and has strong postemergence herbicidal activity against many plants. The chemical composition and phytotoxic effects of bialaphos are somewhat similar to those of the widely-used synthetic herbicide glyphosate, suggesting the possibility of using microorganisms to produce an agriculturally important herbicide.

We have been investigating soil microorganisms for the production of herbicidal and insecticidal metabolites for the past six years. Our efforts have concentrated on the actinomycetes because this group is known to produce many bioactive compounds. We have also screened numerous fungi and non-actinomycete bacteria. This chapter describes our screening program, summarizes results, and discusses some of the pesticidal compounds we have identified.

Materials and Methods

Isolation and Culture of Microorganisms. Soil samples collected from a variety of habitats and geographic locations were serially diluted and plated onto selective culture media (13, 14). The following were typically used: for actinomycetes, arginine-glycerol-salts agar (15), starch-casein agar containing antifungal and antibacterial antibiotics (16), NZ amine A agar, threonine agar, and asparagine-biphenyl agar (14); for non-actinomycete bacteria, nutrient agar containing the antifungal antibiotics nystatin and cycloheximide (17); for fungi, Czapek's agar containing the antibacterial antibiotic chloramphenicol (17). Inoculated petri dishes were incubated at 25-28 C. Colonies that developed on these plates within 3 or 4 weeks were transferred to other plates containing arginine-glycerol-salts agar, A-9 agar (18), yeast extract-malt extract-glucose agar (14) or NZ amine A agar. Identification and taxonomy were according to previously published criteria and methods (19-21).

Broth cultures for pesticide screening were grown in 0.5-L or 2-L, baffle-bottomed Erlenmeyer flasks containing 100 or 400-500 ml of A-9 (full or half-strength) or yeast extract-malt extract-glucose broth. Cultures were incubated in darkness for 5-14 days, depending upon the microorganism, at 25-28 C on orbital shakers at 200 (0.5-L flasks) or 110 (2-L flasks) rpm. The cultures were divided after incubation, with part of the broth going to the herbicide screens and part to the insecticide screens.

Herbicide Screening. Our tests for herbicidal compounds improved as we gained experience. Initially, microbial isolates were grown for two weeks on solid medium in square petri dishes and assayed with surface-sterilized seeds of garden cress (*Lepidium sativum*), barnyardgrass (*Echinochloa crusgalli*), and cucumber (*Cucumus sativus*)(13, 22).

The primary herbicide screen was later simplified to use liquid culture broth (13, 23). Seeds of garden cress and barnyardgrass were placed on filter paper in petri dishes and moistened with dilutions of microbial culture broth (typically 1:4, v:v, broth:distilled water). The bioassays were incubated in darkness at 25-28 C for 72 hr. Germination and radicle growth of the seeds were then evaluated relative to control seeds moistened with distilled water. In many bioassays, the dishes were subsequently incubated 72 hr more in light to observe chlorophyll development in the seedlings.

Microbial isolates that showed promising activity in the primary screens were subsequently evaluated in secondary screens. Dilutions of microbial broth were sprayed directly onto seedlings grown in soil in styrofoam (18x13x5-cm) flats in the greenhouse. Diluted, homogenized culture broth (typically 1:1 or 1:9, v:v, broth:distilled water, 7.5 ml per flat) containing 0.1% surfactant (Surfel, Union Carbide Corporation) was sprayed onto 10-day-old seedlings. Plant species used in secondary screens included barnyardgrass, large crabgrass (Digitaria sanguinalis), green foxtail (Setaria viridis), proso millet (Panicum miliaceum), redroot pigweed (Amaranthus retroflexus), purslane (Portulaca oleracea), and velvetleaf (Abutilon theophrasti). The effects were evaluated for 14 days after treatment.

Insecticide Screening. Tests for insecticidal metabolites used fourth instar mosquito larvae (Aedes aegypti, Rockefeller strain) in 180-ml polystyrene urine specimen cups (17). The larvae were placed into 90 ml of distilled water in the cups, 10 ml of microbial culture broth were added, and the cups were gently agitated. Controls received an identical amount of non-inoculated culture medium instead of microbial broth. The bioassays were maintained at room temperature and evaluated after 2, 24, and 48 hr. If the initial 0.1-strength dilution showed interesting activity, 0.01- and 0.001-strength dilutions were also tested. Evaluation was facilitated by holding the bioassay cups over a high-intensity lamp. Healthy mosquito larvae rapidly wriggled away from the light. Intoxicated larvae moved more slowly or not at all.

Identification of Pesticidal Metabolites. The larger volumes of microbial culture broth needed for isolation of the pesticidal compounds were typically produced with A-9 medium in shaken 2-L flasks or a 100-L stainless steel fermentor. Pesticidal compounds were extracted from the culture broth or cell cake with dichloromethane, dichloromethane/methanol (3:1, v:v), or diethyl ether. Extracts and fractions were concentrated at 55 C or lower with rotary evaporation in vacuo and/or under a stream of nitrogen gas. Fractionation of the crude extracts and purification of pesticidal components were done with solvent partitioning, column and thin-layer chromatography, and high pressure liquid chromatography. Bioassays on garden cress seeds or mosquito larvae were used to follow pesticidal activity throughout extraction and isolation procedures. Identification of the purified compounds was with mass spectrometry, ^{1}H and ^{13}C NMR spectroscopy, IR spectrophotometry, and co-chromatography with authentic samples.

Results and Discussion

Approximately 1500 microbial isolates were tested for production of pesticidal metabolites. About 4-12% caused strong inhibition of seed germination or seedling growth in the primary screens for herbicidal activity. A similar proportion was highly toxic to mosquito larvae. The percentage of microbial isolates showing strong herbicidal activity in the secondary screens was much lower, typically only 1-2% of the initial total.

To determine which microbial groups were most likely to yield pesticide-producing isolates, we conducted a study where actinomycetes and other microorganisms isolated from 18 soil samples were identified and tested for toxin production (17, 23). The actinomycete genus Streptomyces yielded the greatest number of pesticide-producing isolates (Table I). The percentage of pesticide-producing Streptomyces species, out of the total tested, was also high compared to most other groups. The actinomycete genera Actinoplanes, Nocardiopsis, and Streptoverticillium also yielded a comparatively high percentage of toxin-producing isolates, however, the absolute number obtained from these genera was lower.

Several of the pesticidal compounds and producer microorganisms are discussed below.

Cycloheximide. Figure 1 shows symptoms of phytotoxicity that occurred in secondary screens of certain actinomycetes. Damage was characterized by severe burning of leaves and stunting of plants and sometimes began to appear within 24-48 hr of spraying. Death ensued if injury was severe; less injured plants eventually recovered. Culture broth from these isolates also strongly inhibited seed germination and seedling growth in the primary screens. Cycloheximide (Figure 2) was primarily responsible for this injury and was found to be present in the culture broth of an appreciable proportion of our most phytotoxic Streptomyces isolates (13, 24). Cycloheximide is a potent inhibitor of seed germination and seedling growth, a concentration of 1 ug/ml causing 50% inhibition of radicle growth of garden cress (Figure 3). Cycloheximide can readily be partitioned from culture broth with dichloromethane and detected with thin-layer chromatography (13). Routine testing of phytotoxic actinomycetes is therefore useful for identifying cycloheximide producers early in the screening program.

Cycloheximide, an inhibitor of protein synthesis, is produced by Streptomyces griseus, noursei, naraensis, pulveraeus, albus, and ornatus (25-27). It has strong antifungal activity (28, 29) and has been used as an agricultural fungicide (30). It is also highly repellent to rats (31, 32). Toxicity tests indicate an LD50 (intravenous) to mice of 150 mg/kg (29). Cycloheximide has previously been found to inhibit germination of radish, pea, and wheat; to injure young leaves; and to cause toxicity to certain algae (33-35). It can apparently be absorbed and translocated (36, 37). Cycloheximide has been considered for use as a herbicide in Japan (5), but currently appears to have little potential for such use in the United States.

Geldanamycin and Nigericin. A strain of S. hygroscopicus (designated V-9 in 13) was isolated that strongly inhibited seed germination and seedling growth, but did not produce detectable amounts of cycloheximide. Geldanamycin and nigericin (Figure 2) were subsequently isolated and identified (38). Both were highly phytotoxic and caused 50% reduction of garden cress radicle elongation at concentrations of 3 ug/ml or less (Figure 3).

Geldanamycin is an ansamycin antibiotic. It was first reported in 1970 from S. hygroscopicus var. geldanus (39, 40). Structurally, it is identical to herbimycin B except it has a methoxyl group on carbon 17 (41, 42). The herbimycins are also ansamycin antibiotics produced by S. hygroscopicus that have strong herbicidal effects (41,

Table I. Pesticidal activity of actinomycetes and other micro-organisms isolated from 18 soil samples and tested in primary herbicide and insecticide screens

Genus	No. of isolates tested	No. of isolates showing strong pesticidal activity on: Cress	Barnyardgrass	Mosquito
Streptomyces	266/302*	47	24	26
Micromonospora	134	1	0	2
Actinomadura#	59	3	2	3
Actinoplanes	46	6	3	4
Rhodococcus	26	0	0	0
Micropolyspora	24	1	1	2
Nocardia	22	0	0	0
Nocardiopsis#	22	4	2	1
Streptosporangium	22	1	1	1
Oerskovia	21	0	0	1
Thermomonospora	21	0	0	1
Thermoactinomyces@	19	0	0	0
Streptoverticillium	18	1	0	3
Saccharomonospora	16	0	0	0
Promicromonospora	15	0	0	0
Dactylosporangium	11	0	0	0
Microbispora	11	0	0	0
Chainia	8	0	0	1
Pseudonocardia	7	0	0	0
Unknown actinomycetes	28	2	2	0
Non-actinomycete bacteria	40	1	0	3
Fungi	70	5	3	7

*Number of isolates tested = 266 in herbicide screens and 302 in insecticide screens.

#Use of these names does not imply authors' endorsement regarding validity.

@Organisms appeared similar to Thermoactinomyces peptonophilus.

Figure 1. Phytotoxicity of culture broth from a cycloheximide-producing actinomycete strain (right, T161). Microbial broth was sprayed onto foliage of cucumber (foreground), corn, and soybean (rear). Control (left) received an equivalent volume of previously sterile culture broth.

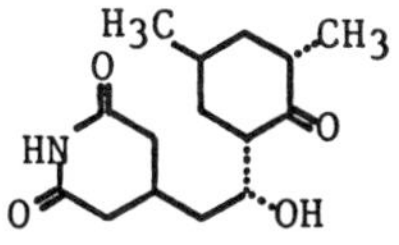

CYCLOHEXIMIDE

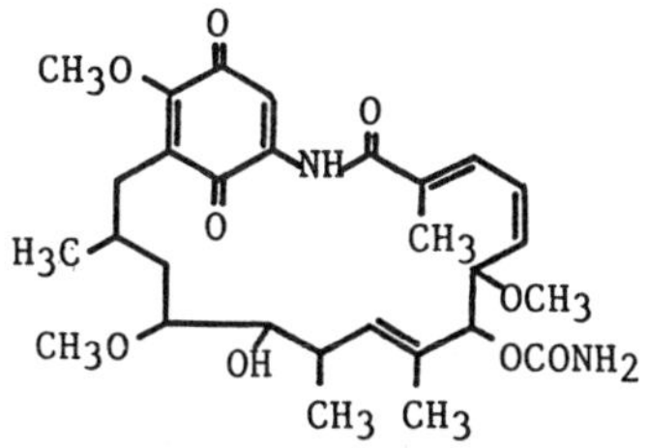

GELDANAMYCIN

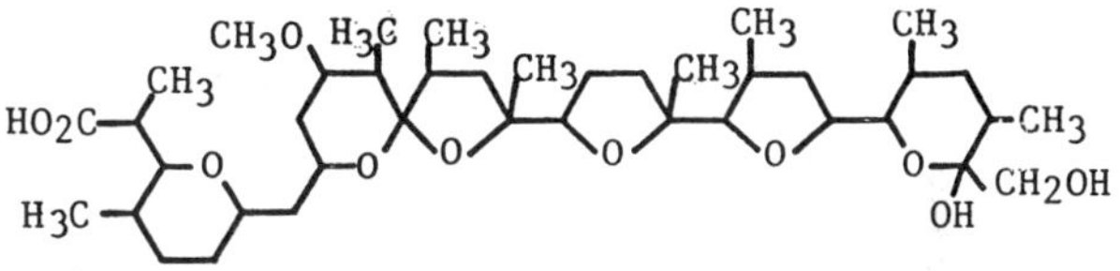

NIGERICIN

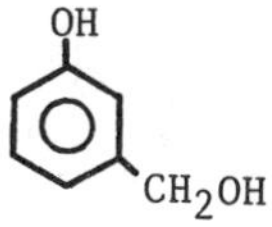

3-HYDROXYBENZYL ALCOHOL

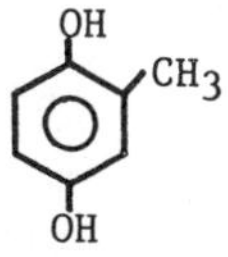

2-METHYLHYDROQUINONE

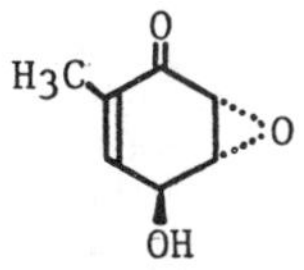

EPIEPOFORMIN

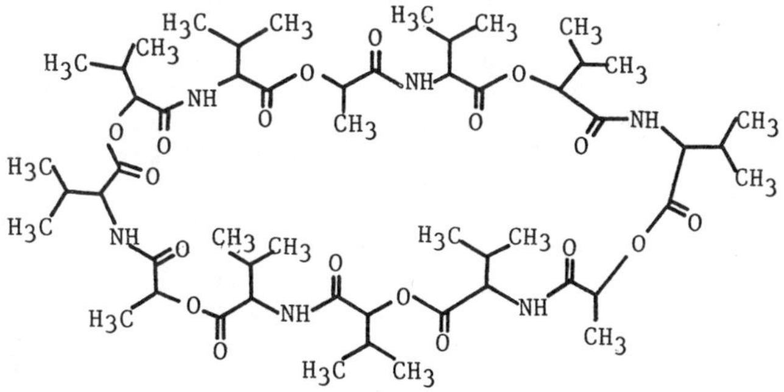

VALINOMYCIN

Figure 2. Structures of herbicidal and insecticidal metabolites recently isolated from microorganisms.

43). Geldanamycin inhibits DNA synthesis, is strongly active against tumor cells, and is moderately toxic to certain fungi, bacteria, and protozoa (39, 44). Its acute oral toxicity to rats is 2500-5000 mg/kg, and the intraperitoneal LD50 for mice is about 1 mg/kg (39).

Nigericin, a polyether antibiotic, was first reported in 1949 and was later found to be identical to polyetherin A, produced by S. hygroscopicus (45-47). Nigericin influences cation transport across membranes (48-52). It also inhibits photophosphorylation (48), Gram-positive bacteria, mycobacteria, and certain fungi (46, 53), and is patented as a feed additive to improve weight gain in cattle (54, 55). Its intraperitoneal LD50 ranges from 2.5 to 10-15 mg/kg for mice (46, 53).

We have evaluated geldanamycin and nigericin as pre- and postemergence herbicides. Tests were done in loamy sand soil in styrofoam flats (18x13x5-cm) in the greenhouse at applications of 0 (control), 0.3, 0.6, 1.1, 2.2, and 4.5 kg/ha. Applied preemergence, geldanamycin caused statistically significant inhibition of barnyardgrass, garden cress, proso millet, and redroot pigweed at 0.6 kg/ha or less (Table II). Large crabgrass and giant foxtail were less sensitive and were significantly inhibited only by 4.5 kg/ha. Velvetleaf was not appreciably inhibited by geldanamycin applied preemergence at rates up to 4.5 kg/ha. Nigericin was less effective than geldanamycin as a preemergence herbicide. Although it significantly inhibited cress at 1.1 kg/ha or higher, and large crabgrass at 4.5 kg/ha, it did not consistently inhibit any of the other species (Table II).

Postemergence, geldanamycin had little effect and did not statistically reduce growth of any of the assay species at doses up to 4.5 kg/ha (Table II). Nigericin was striking in its postemergence effect on velvetleaf and cress, which were significantly inhibited at rates as low as 0.3 kg/ha. Damage to velvetleaf was characterized by stunting and necrosis of leaves (Figure 4). In the species sensitive to geldanamycin and nigericin, most plants not initially killed resumed normal growth within 2-3 weeks following injury.

Epiepoformin, Hydroxybenzyl Alcohol, and Methylhydroquinone. An isolate of the fungus Scopulariopsis brumptii produced culture broth that strongly inhibited garden cress and barnyardgrass at 1:1 (v:v) dilution in the primary screens. A solvent extract of the culture broth, applied postemergence at 4.5 kg/ha, reduced growth of certain broadleaved species by 25-50% (56). Three phytotoxins were isolated: (+)-epiepoformin, 3-hydroxybenzyl alcohol, and 2-methylhydroquinone (Figure 2). Epiepoformin was most phytotoxic and showed strong activity on redroot pigweed and mustard; 3-hydroxybenzyl alcohol moderately inhibited redroot pigweed; 2-methylhydroquinone showed only weak activity on any of the assay species.

The three phytotoxins we isolated from Scopulariopsis brumptii have previously been reported from other fungi: 3-hydroxybenzyl alcohol and 2-methylhydroquinone from Penicillium urticae (57, 58), 2-methylhydroquinone from Nectrina erubescens (59) and Phoma species (60), and (+)-epiepoformin from an unidentified fungus (61). Two-methylhydroquinone is also produced by certain beetles as a defense compound (62-65). Phytotoxic or growth-regulatory effects have previously been reported for (+)-epiepoformin and 3-hydroxybenzyl

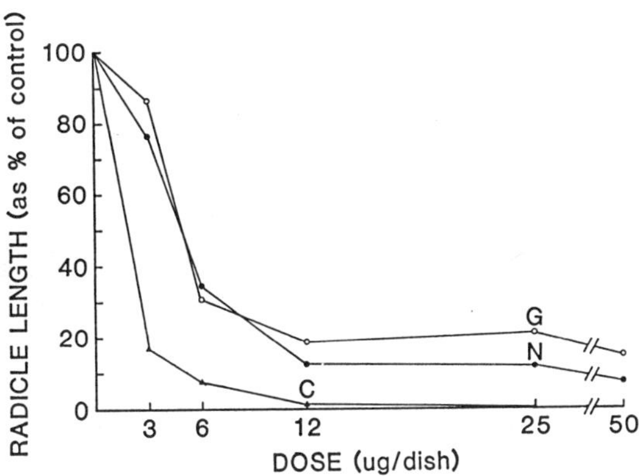

Figure 3. Effect of cycloheximide (C), geldanamycin (G), and nigericin (N) on radicle growth of garden cress. Dose is the amount of compound per 1.5 ml of distilled water applied to each bioassay dish.

Table II. Herbicidal effects of geldanamycin (G) and nigericin (N) to several plant species in soil. Rates of application were 0 (control), 0.3, 0.6, 1.1, 2.2, and 4.5 kg/ha

			Bioassay species@						
Application	Compound	Response	BYGR	GACR	GIFT	LACG	PRMI	RRPW	VELE
Preemergence	G	Yield*	49	32	75	140	37	13	89
		MIR#	0.6	0.6	4.5	4.5	0.3	0.3	>4.5
	N	Yield	84	2	65	70	79	38	91
		MIR	>4.5	1.1	>4.5	4.5	>4.5	>4.5	>4.5
Postemergence	G	Yield	95	122	107	96	97	--	104
		MIR	>4.5	>4.5	>4.5	>4.5	>4.5	--	>4.5
	N	Yield	108	52	77	81	119	--	53
		MIR	>4.5	0.3	>4.5	>4.5	>4.5	--	0.3

*Yield is oven-dry weight of shoots in treatments receiving 4.5 kg/ha expressed as percentage of control.

#MIR (minimum inhibitory rate) is the lowest application (in kg/ha) required to cause consistent statistically significant (5% LSD) inhibition relative to control.

@Bioassay species: BYGR = barnyardgrass, GACR = garden cress, GIFT = giant foxtail, LACG = large crabgrass, PRMI = proso millet, RRPW = redroot pigweed.

Figure 4. Effect of nigericin on velvetleaf seedlings (right) 8 days after postemergence application of 0.5 lb/acre (0.6 kg/ha). The control (left) received no nigericin.

alcohol (61, 66), whereas insecticidal (67-70) and antibiotic (71-73) activity have been reported for 2-methylhydroquinone.

Valinomycin. A strain of Streptomyces griseus was discovered that produced strongly mosquitocidal culture broth. Bioassays of the crude broth indicated an LC50 at 0.001- to 0.0001-strength dilution. The producing microorganism was atypical in spore chain morphology from previously described S. griseus (74). Two insecticidal compounds have been isolated from the microbial cells. One was identified as valinomycin (Figure 2) and found to have an LC50 of 2-3 ug/ml on mosquito larvae. The other has not yet been obtained in sufficient quantity for identification or accurate insecticidal evaluation.

Valinomycin has previously been reported as a metabolite of S. fulvissimus (75) and S. roseochromogenes (76). Our producing strain differs somewhat from these in both morphology and physiology (74). Valinomycin was patented for insecticidal, nematocidal, and acaricidal use in 1970 (76) and is active against Mycobacterium tuberculosis (75). Valinomycin influences cation movement across membranes (77, 78) and uncouples oxidative phosphorylation (79).

Conclusions

The results of our work indicate that soil microorganisms produce a variety of herbicidal and insecticidal metabolites. Continued screening of microbial isolates will undoubtedly reveal additional pesticidal compounds and producer organisms. We believe future work will eventually lead to discovery of new microbial metabolites that have potential for commercial development as pesticides.

Acknowledgments

We thank Dow Chemical Company and Union Carbide Agricultural Products Company for financial support.

Journal article No. 12572 of the Michigan Agricultural Experiment Station.

Literature Cited

1. Misato, T. J. Pestic. Sci. 1982, 7, 301-5.
2. Fischer, H.-P.; Bellus, D. Pestic. Sci. 1983, 14, 334-46.
3. Ando, K. In Pesticide Chemistry: Human Welfare and the Environment; Miyamoto, J.; Kearney, P. C., Eds.; Pergamon: New York, 1983; Vol. 2, pp 253-9.
4. Misato, T. In Pesticide Chemistry: Human Welfare and the Environment; Miyamoto, J.; Kearney, P. C., Eds.; Pergamon: New York, 1983; Vol. 2, pp 241-6.
5. Sekizawa, Y.; Takematsu, T. In Pesticide Chemistry: Human Welfare and the Environment; Miyamoto, J.; Kearney, P. C., Eds.; Pergamon: New York, 1983; Vol. 2, pp 261-8.
6. Burg, R. W.; Miller, B. M.; Baker, E. E.; Birnbaum, J.; Currie, S. A.; Hartman, R.; Kong, J. L.; Monaghan, R. L.; Olson, G.; Putter, I.; Tunac, J. B.; Wallick, H.; Stapley, E. O.; Oiwa, R.; Omura, S. Antimicrob. Agents Chemother. 1979, 15, 361-7.

7. Miller, T. W.; Chaiet, L.; Cole, D. J.; Cole, L. J.; Flor, J. E.; Goegelman, R. T.; Gullo, V. P.; Joshua, H.; Kempf, A. J.; Krellwitz, W. R.; Monaghan, R. L.; Ormond, R. E.; Wilson, K. E.; Albers-Schonberg, G.; Putter, I. Antimicrob. Agents Chemother. 1979, 15, 638-71.
8. Egerton, J. R.; Ostlind, D. A.; Blair, L. S.; Eary, C. H.; Suhayda, D.; Cifelli, S.; Reik, R. F.; Campbell, W. C. Antimicrob. Agents Chemother. 1979, 15, 372-8.
9. Campbell, W. C.; Fisher, M. H.; Stapley, E. O.; Albers-Schonberg, G.; Jacob, T. A. Science 1983, 221, 823-8.
10. Davies, H. G.; Green, R. H. Natural Product Reports 1986, 3, 87-121.
11. Time, 2 November 1987, p 78.
12. Seto, H.; Sasaki, T.; Imai, S.; Tsuruoka, T.; Ogawa, H.; Satoh, A.; Inouye, S.; Niida, T.; Otake, N. J. Antibiotics 1983, 36, 96-8.
13. Heisey, R. M.; DeFrank, J.; Putnam, A. R. In The Chemistry of Allelopathy; Thompson, A. C., Ed.; ACS Symposium Series No. 268; American Chemical Society: Washington, DC, 1985; pp 337-49.
14. Mishra, S. K.; Taft, W. H.; Putnam, A. R.; Ries, S. K. J. Plant Growth Regul. 1987, 6, 75-84.
15. El-Nakeeb, M. A.; Lechevalier, H. A. Appl. Microbiol. 1963, 11, 75-7.
16. Williams, S. T.; Davies, F. L. J. Gen. Microbiol. 1965, 38, 251-61.
17. Mishra, S. K.; Keller, J. E.; Miller, J. R.; Heisey, R. M.; Nair, M. G.; Putnam, A. R. J. Indust. Microbiol. 1987, 2, 267-76.
18. Warren, H. B., Jr.; Prokop, J. F.; Grundy W. E. Antibiot. Chemother. 1955, 5, 6-12.
19. Lechevalier, H. A.; Lechevalier, M. P. In The Prokaryotes; Starr, M. P.; Stolp, H.; Triiper, H. G.; Balows, A.; Schlegel, H. G., Eds.; Springer: New York, 1982; Vol. 2, pp 1915-22.
20. Mishra, S. K.; Gordon, R. E. In Infectious Diseases and Medical Microbiology; Braude, A. E.; Davis, C. E.; Fierer, J., Eds.; Saunders: Philadelphia, 1986; pp 371-81.
21. Mishra, S. K; Gordon, R. E.; Barnett, D. J. Clin. Microbiol. 1980, 11, 728-736.
22. DeFrank, J.; Putnam, A. R. Weed Sci. 1985, 33, 271-4.
23. Mishra, S. K.; Whitenack, C. J.; Putnam, A. R. Weed Sci. 1988, 36, 122-6.
24. DeFrank, J. Ph.D. Thesis, Michigan State University, 1983.
25. Whiffen, A. J. J. Bact. 1948, 56, 283-91.
26. Brown, R.; Hazen, E. L. In Antibiotics Annual 1955-1956; Welch, H.; Marti-Ibanez, F., Eds.; Medical Encyclopedia: New York, 1956; pp 245-8.
27. Jost, J. L.; Kominek, L. A.; Hyatt, G. S.; Wang, H. Y. In Biotechnology of Industrial Antibiotics; Vandamme, E. J., Ed.; Marcel Dekker: New York, 1984; pp 531-50.
28. Whiffen, A. J.; Bohonos, N.; Emerson, R. L. J. Bact. 1946, 52, 610-1.
29. Leach, B. E.; Ford, J. H.; Whiffen, A. J. J. Amer. Chem. Soc. 1947, 69, 474.
30. Ford, J. H.; Klomparens, W.; Hamner, C. L. Plant Disease Reporter 1958, 42, 680-95.

31. Traub, R.; DeWitt, J. B.; Welch, J. F.; Newman, D. J. Amer. Pharm. Assoc. (Scientific Edition) 1950, 39, 552-5.
32. Welch, J. F. J. Agric. Food. Chem. 1954, 2, 142-9.
33. Vaughn, J. R.; Lockwood, J. L.; Randwa, G. S.; Hamner, C. Mich. State College Agric. Exp. Sta. Quart. Bull. 1949, 31, 456-64.
34. Wallen, V. R.; Sutton, M. D.; Skolko, A. J. Phytopathology 1950, 40, 156-60.
35. Palmer, C. M.; Maloney, T. E. Ohio J. Sci. 1955, 55, 1-8.
36. Wallen, V. R.; Millar, R. L. Phytopathology 1957, 47, 291-4.
37. Lemin, A. J.; Magee, W. E. Plant Disease Reporter 1957, 41, 447-8.
38. Heisey, R. M.; Putnam, A. R. J. Nat. Prod. 1986, 49, 859-65.
39. DeBoer, C.; Meulman, P. A.; Wnuk, R. J.; Peterson, D. H. J. Antibiotics 1970, 23, 442-7.
40. DeBoer, C.; Dietz, A. J. Antibiotics 1976, 29, 1182-8.
41. Iwai, Y.; Nakagawa, A.; Sadakane, N.; Omura, S.; Oiwa, H.; Matsumoto, S.; Takahashi, M.; Ikai, T.; Ochiai, Y. J. Antibiotics 1980, 33, 1114-9.
42. Rinehart, K. L., Jr.; Shield, L. S. Fortschr. Chem. Org. Naturst. 1976, 33, 231-307.
43. Omura, S.; Iwai, Y.; Takahashi, Y.; Sadakane, N.; Nakagawa, A.; Oiwa, H.; Hasegawa, Y.; Ikai, T. J. Antibiotics 1979, 32, 255-61.
44. Yamaki, H.; Suzuki, H.; Choi, E. C.; Tanaka, N. J. Antibiotics 1982, 35, 886-92.
45. Harned, R. L.; Hidy, P. H.; Corum, C. J.; Jones, K. Proc. Indiana Acad. Sci. 1949, 59, 38.
46. Shoji, J.; Kozuki, S.; Matusutani, S.; Kubota, T.; Nishimura, H.; Mayama, M.; Motokawa, K.; Tanaka, Y.; Shimaoka, N.; Otsuka, H. J. Antibiotics 1968, 21, 402-9.
47. Stempel, A.; Westley, J. W.; Benz, W. J. Antibiotics 1969, 22, 384-5.
48. Shavit, N.; San Pietro, A. Biochem. Biophys. Res. Commun. 967, 28, 277-83.
49. Pressman, B. C.; Harris, E. J.; Jagger, W. S.; Johnson, J. H. Proc. Natl. Acad. Sci. USA 1967, 58, 1949-56.
50. Henderson, P. J. F.; McGivan, J. D.; Chappell, J. B. Biochem. J. 1969, 111, 521-35.
51. Rottenberg, H.; Scarpa, A. Biochemistry 1974, 13, 4811-7.
52. Sze, H. Proc. Natl. Acad. Sci. USA 1980, 77, 5904-8.
53. Harned, R. L.; Hidy, P. H.; Corum, C. J.; Jones, K. L. Antibiot. Chemother. 1951, 1, 594-6.
54. Raun, A. P. U.S. Patent 3 937 836, 1976.
55. Potter, E. L. U.S. Patent 4 405 609, 1983.
56. Huang, J.; Putnam, A. R.; Werner, G. M.; Mishra, S. K.; Whitenack, C. J. Weed Sci. (in press).
57. Rebstock, M. C. Arch. Biochem. Biophys. 1964, 104, 156-9.
58. Scott, A. I.; Yalpani, M. Chem. Comm. 1967, 945.
59. Carey, S. T.; Nair, M. S. R. J. Nat. Prod. 1979, 42, 231.
60. Sequin-Frey, M.; Tamm, C. Helv. Chim. Acta 1971, 54, 851-61.
61. Nagasawa, H.; Suzuki, A.; Tamura, S. Agric. Biol. Chem. 1978, 42, 1303-4.
62. Carlson, S. D. J. Econ. Entomol. 1967, 60, 878-9.
63. Englehardt, M.; Rapaport, H.; Sokoloff, A. Science 1965, 150, 632-3.

64. Ichihara, A.; Kimura, R.; Oda, K.; Moriyasu, K.; Sakamura, S. Agric. Biol. Chem. 1982, 46, 1879-83.
65. Dettner, K.; Schwinger, G. J. Chem. Ecol. 1982, 8, 1411-2.
66. Yoshino, Y.; Sato, C.; Maeda, S.; Japanese Patent 73 41 992, 1973; Chem. Abstr. 81, 59342p.
67. Bottger, G. T.; Yerington, A. P.; Gertler, S. I. U. S. Dept. Agric. Bur. Entomol. Plant Quarantine, E-826, 1951, 23 pp; Chem. Abstr. 46, 5778e.
68. Cutler, H. G. In The Science of Allelopathy; Putnam, A. R.; Tang, C. S., Eds.; Wiley: New York, 1986.
69. Questel, D. D.; Gertler, S. I. U. S. Dept. Agric. Bur. Entomol. Plant Quarantine E-840, 1952, 12 pp; Chem Abstr. 46, 8311d.
70. Takeda Chemical Industry Ltd. Japanese Patent 58 157 703, 1983; Chem. Abstr. 100, 30944m.
71. Basalkevich, E. D.; Cherepenko, T. I.; Vysotskii, N. N.; Shapovalenko, V. F.; Svishchuk, A. A. Fiziol. Aktiv. Veshchestva Respub. Mezhvedom. 1971, 3, 174-80; Chem. Abstr. 77, 61449q.
72. Owens, R. G. Contrib. Boyce Thompson Inst. 1953, 17, 273-82; Chem. Abstr. 48, 7243g.
73. Sun, H.; Chen, Z.; Xu, G.; Xu, L. Xaoxue Xuebao 1982, 17, 107-11; Chem. Abstr. 96, 192936m.
74. Heisey, R. M.; Huang, J.; Mishra, S. K.; Keller, J. E.; Miller, J. R.; Putnam, A. R.; D'Silva, T. D. J. J. Agric. Food. Chem. 1988, 36 (in press).
75. Brockmann, H; Schmidt-Kastner, G.; Chem. Ber. 1955, 88, 57-61.
76. Patterson, E. L.; Wright, P. U. S. Patent 3 520 973, 1970.
77. Kovac, L.; Bohmerova, E.; Butko, P. Biochim. Biophys. Acta 1982, 721, 341-8.
78. Ahmed, S.; Booth, I. R. Biochem. J. 1983, 212, 105-12.
79. McMurray, W. C.; Begg, R. W. Arch. Biochem. Biophys. 1959, 84, 546-8.

RECEIVED April 5, 1988

Chapter 6

Diverse but Specific Biological Activities of Four Natural Products from Three Fungi

Horace G. Cutler[1], Eberhard Ammermann[2], and James P. Springer[3]

[1]Richard B. Russell Research Center, Agricultural Research Service, U.S. Department of Agriculture, Athens, GA 30613
[2]Badische Anilin- und Sodafabrik Aktiengesellschaft, Landwirtschaftliche Versuchsstation, D–6703 Limburgerhof, Federal Republic of Germany
[3]Merck Therapeutic Institute, Rahway, NJ 07065

Biologically active natural products isolated from fungi fall into three general categories. Those isolated for medicinal antibiotic activity years ago that have not been tested for agricultural use; those recently isolated as mycotoxins that have not been tested in other biological systems; and novel products. Four examples are given. Cyclopenin and cyclopenol, isolated in 1954; 3,7-dimethyl-8-hydroxy-6-methoxyisochroman, isolated in 1979; and the novel metabolite cinereain, isolated in 1987. The diversity of structure, activity and possible applications of these products to agriculture are discussed.

One of the great problems in dealing with biologically active natural products is that individuals tend to compartmentalize their thinking when it comes to working with a specific compound and its congeners. There is a lack of perception that these natural substances may have very diverse activities in specific, but unrelated, biological systems. Those who deal with pharmaceuticals, for example antimicrobials, often fail to translate their discoveries to other fields, such as agrochemicals, with the consequence that much potentially valuable material is lost. For some reason, disciplines fail to communicate with each other and work performed within a discipline may be done for no apparent cause other than the pursuit of pure knowledge. Another difficulty arises in that a sense of defeat overcomes those who would venture into the isolation of new natural products from microorganisms; their first reaction is to say that the odds for discovering novel structures are very slim and their application for practical use is even slimmer. In this chapter three classes of compounds, all natural products from microorganisms, will be discussed relative to these general observations and should be viewed as cautionary verses (to avoid myopia) for those

0097–6156/88/0380–0079$06.00/0

who are interested in agrochemical development and those involved in a singular discipline who would like to diversify their scientific portfolio.

Cyclopenin, cyclopenol, diazepam, and related structures. While searching for biologically active structures, especially those that inhibit plant growth, we accessed a fungus, isolated from pecan kernels (*Carya illinoensis*), that had undergone insect damage. The organism, identified as an aberrant strain of *Penicillium cyclopium* (NRRL 6233) produced two metabolites, cyclopenin and cyclopenol (1), in addition to another metabolite, compactin (Figure 1), which showed no growth regulatory properties in the etiolated wheat (*Triticum aestivum* L., cv. Wakeland) coleoptile assay (2). Compactin later underwent independent development by the pharmaceutical industry for drug use to reduce blood serum cholesterol (3). However, it transpired that cyclopenin (Figure 2) significantly inhibited ($P<0.01$) the growth of etiolated wheat coleoptiles 100 and 33% at 10^{-3} and 10^{-4} M, respectively, while cyclopenol (Figure 2) inhibited coleoptiles only 20% at 10^{-3} M. The coleoptile bioassay has been described in detail (4,5). Thus, the two molecules which differ by a single hydroxyl group versus a proton at the C15 position have markedly different inhibiting properties by a greater than tenfold factor.

Because of the biological activities observed in the primary wheat coleoptile bioassay, the two metabolites were evaluated in greenhouse-grown intact plants. These included bean, corn, and tobacco. Solutions of cyclopenin and cyclopenol were formulated in 10% acetone plus Tween 20 (0.1%) in water, and plants were treated at 10^{-2}, 10^{-3} and 10^{-4} M. Nine-day old bean plants (*Phaseolus vulgaris* L. cv. Black Valentine) were sprayed with 1 ml per plant, in aerosol, of each concentration when they were in the first true leaf stage; four plants were included in each pot. Tobacco plants (*Nicotiana tabacum* L. cv. Hicks) were six-weeks old when individually sprayed with test solutions, using 1 ml per plant. Corn plants (*Zea mays* L. cv. Norfolk Market White), on the other hand, had 100 µl of each test solution introduced individually into leaf sheaths when one-week old.

The results of these tests demonstrated that tobacco plants were unaltered by applications of cyclopenin, or cyclopenol. However, cyclopenin induced malformations of the first trifoliates in bean which were pronounced at 10^{-2} M, but slight at 10^{-3} M, one week following treatment. In addition, growth of the upper trifoliates were inhibited relative to those of control plants. Cyclopenol treated plants resembled control plants. The effects in corn were noted within 24 hours with cyclopenin and there was necrosis and stem collapse at 10^{-2} M. After 48 hours, these effects were even more pronounced and at two weeks following treatment plants that had survived were stunted relative to control plants. The 10^{-3} M treatments were normal. In contrast, cyclopenol treated plants were unaffected. Two conclusions were drawn from these experiments. First, cyclopenin was more active than cyclopenol in plants. Second, cyclopenin selectively inhibited the growth of corn, a monocotyledon, induced morphological changes in bean plants, a dicotyledonous nitrogen-fixer, and produced no apparent changes in tobacco. Most importantly, cyclopenol was relatively inactive in plants.

It is important to highlight the fact that neither cyclopenin nor cyclopenol were novel metabolites. "Cyclopenin" (actually a mixture of cyclopenin and cyclopenol) had been described as early as 1954 by Bracken (6) when the mixture had been isolated from P. cyclopium. Clarification as to the exact nature of "cyclopenin" came with the work of Mohammed and Luckner in 1963 when the individual structures of cyclopenin and cyclopenol were elucidated as metabolites emanating from P. cyclopium and P. viridicatum (7). Nevertheless, cyclopenin had been shown to have slight antibiotic activity against Micrococcus pyogenes var. aureus and Escherichia coli (8). It has to be assumed that the original examination of these two organisms was prompted for two reasons. Primarily, the exercise was probably intended to discover new antibiotics for medicinal use but, as is so often the case, some rather intriguing chemistry which was essential to determine the absolute structure became the major focus. It is significant, as we shall see later, that the biological objectives remained relatively narrow. It is also significant that the plant growth regulatory, and other properties, were not noted earlier because exactly 20 years elapsed between the first report (6) of the isolation of cyclopenin and cyclopenol, and the disclosure of their relative activities in plant, and other systems (1).

During the course of our investigations we had two extraordinary pieces of good luck each unrelated to the other. Both metabolites had been tested in a few screening programs as potential agrochemicals but these were limited because of the small amounts of chemicals available. However, the first piece of luck was that we tested cyclopenin and cyclopenol in antifungal assays using greenhouse grown tomato plants that supported the plant pathogen Phythophthora infestans, the late blight of potato, under ideal conditions to propagate the disease. This organism, first recorded in the United States and Europe about 1840, was responsible for destroying the potato crop in Ireland in 1846-1847, the period known as the Great Famine, and prompted the subsequent emigration of one-quarter of the Irish population to all parts of the world, especially the United States. The disease is still a problem in Europe today, and in the United States the number of applications of fungicides to control this pest ranges from 8 to 18 times per season. The crop is grown on a little more than 1.0 million acres, mostly in northern states, to yield a basic commodity worth in excess of $1 billion per year. Unfortunately, there is presently a great deal of dispute concerning the use of synthetic fungicides relative to their potential harm to humans. A biodegradable fungicide would be of benefit to the food industry though it would have to retain its integrity long enough to control a pathogen to be of commercial value and would have to be non-carcinogenic. Cyclopenol controlled P. infestans in greenhouse experiments and closely paralleled responses obtained with the standard fungicide Metiram (Table I). Cyclopenin did not possess fungicidal properties. Neither metabolite was active against Plasmopora viticola, a disease of grapes. Note that the substitution at C15 is of paramount importance relative to specific biological activity.

Table I. Effectiveness of cyclopenin and cyclopenol in controlling Phytophthora infestans in greenhouse-grown tomato plants (Lycopersicon esculentum L.) relative to Metiram, 24 hours following treatment

Treatment	Concentration (%)	Disease Index (%)
Cyclopenin	0.025	15
Cyclopenol	0.025	5
Metiram	0.025	0
Control	0	70

The other serendipitous event was the testing of both metabolites for acute toxicity. Day-old-chicks were dosed, via crop intubation, with 1 ml of corn oil containing 25, 125, 250, and 500 mg/kg body weight of cyclopenin or cyclopenol (1). We reported that cyclopenin caused drowsiness in chicks within 2 hours following administration with the 250 mg/kg dose, but that they had all recovered within 18 hours. Planning to obtain an LD_{50}, the dose was increased to 500 mg/kg body weight and within 1 hour the chicks were intoxicated, and prostrate. They recovered with no apparent ill effects, other than slight drowsiness, in 18 hours. Cyclopenol induced none of these effects in chicks at these levels.

It was after these results had been published that the obvious structural relationship between cyclopenin and the benzodiazepine structure, 7-chloro-1,3-dihydro-1-methyl-5-phenyl-2H-1,4-benzodiazepin-2-one (diazepam, or valium), a tranquilizer that enjoyed great popularity in the 1960's and early 1970's (9), came to our attention. Comparison of that structure (Figure 3) with cyclopenin indicates the following differences. The methyl group attached to N1 is substituted by a single proton in cyclopenin. There is a common carbonyl at C2 in both substances. There is an epoxide at C3 in cyclopenin which is attached to a phenyl group. The N4 has an associated methyl group in cyclopenin. At N4, there is an adjacent carbonyl attachment in cyclopenin: that same nitrogen, in diazepam, has a double bond in conjunction with a benzene ring. An important addition in diazepam is the heavy atom, Cl, at the C7 position. The major difference between these metabolites is, then, the presence of a heavy atom in the one

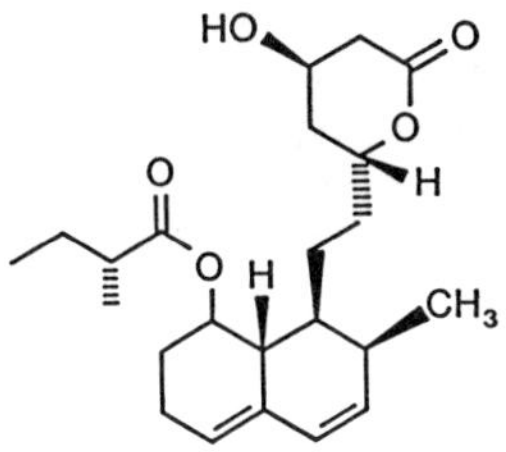

Figure 1. Compactin.

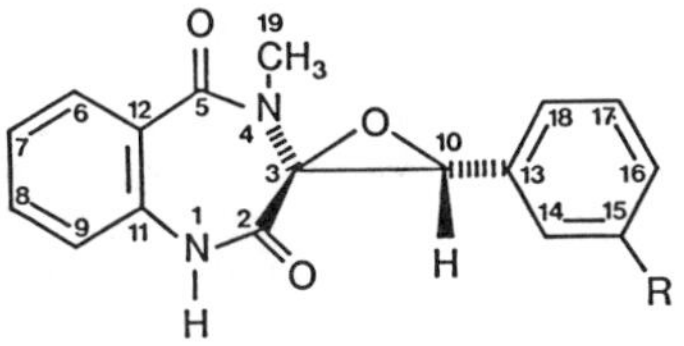

R = H CYCLOPENIN
R = OH CYCLOPENOL

Figure 2. Cyclopenin and cyclopenol.

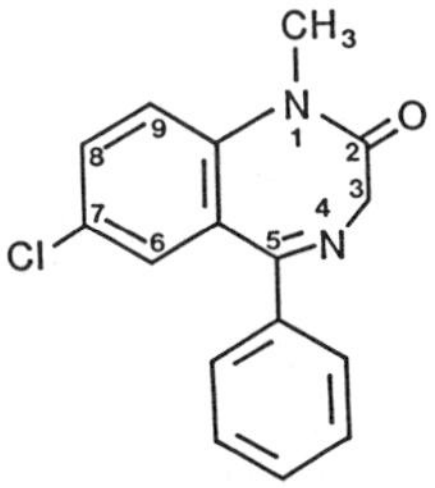

Figure 3. Diazepam (Valium).

and an epoxide in the other. Epoxides are notoriously active in biological systems, the case of the 12,13 epoxytrichothecenes being typical (10) and, in that case, the opening of the epoxide ring eliminates biological activity. The selective activity of each of these fungal metabolites raises a series of questions. Why is cyclopenin active against some plants (wheat, bean, corn) yet inactive against others (tobacco) and why does it possess tranquilizing properties against vertebrates? Why does it not control P. infestans? Conversely, why does cyclopenol inhibit the growth of wheat and P. infestans, yet it does not control the growth of the other plants tested nor does it appear to have any tranquilizing properties? There are very few answers at present, but one important observation is that the etiolated wheat coleoptile assay has been shown to detect plant growth promoters, inhibitors, antimicrobials, immunosuppressants, and mycotoxins (4,5). It does not, however, detect compounds that have neurological action (5).

Recently, when diazepam was tested against wheat coleoptiles, it inhibited their growth 100% at 10^{-3} M (11). This places the relative activity, in the same biological system, between cyclopenin and cyclopenol, so the situation now exists where a common pharmaceutical, an analog of cyclopenin and cyclopenol, has activity against plants. Obviously, further examination of the molecule will be carried out by us in both microbial and plant systems. But we now come to a further mystery. It would seem doubtful that the pharmaceutical industry synthesized only diazepam for clinical studies as a tranquilizer. Surely other derivatives were made, including heavy atom substituents; but were they tested in biological systems other than vertebrates? The circumstances surrounding cyclopenin and cyclopenol become even more puzzling because in 1968 there was a description of the synthesis of cyclopenin (12). A year later, a scheme appeared for the synthesis of cyclopenin and isocyclopenin (13) and confirmation of the spiro structure, postulated for cyclopenin, was made. By 1970, the sterospecific synthesis of dl-cyclopenin and dl-cyclopenol had transpired (14) so that all prior work was corroborated. Then, in 1974, derivatives of (±) cyclopenin were produced (15) and three of the four contained one or two chlorine atoms (Figure 4). Including the starting materials, there were 36 precursors. While the chemistry was elegant, not a single statement referred to testing in biological systems either at the time or possibly in the future. It seems probable that the relationship between cyclopenin and diazepam was missed by these authors because there were no chlorine substitutions at the C6, 7, 8, or 9 positions in cyclopenin. However, the chlorinated derivatives are excellent candidates for biological testing.

Almost 24 years have elapsed since the first disclosure of the "cyclopenins" from P. cyclopium and a fair volume of synthetic work has been accomplished. But apparently the various disciplines have not been communicating too well with each other. The evidence suggests that a few of the synthetic derivatives of cyclopenin, and its analogs, may be potential agrochemicals. At a minimum the basic template is a reasonable one for making other derivatives for use as plant growth regulators or antimicrobials.

3,7-Dimethyl-8-hydroxy-6-methoxyisochroman (3,7-DHMI). Even in supposedly well ordered management systems, communications between disciplines break down. Such was the case with 3,7-DHMI (Figure 5), a metabolite isolated from *Penicillium steckii* found growing on moldy millet hay (16) which was implicated in the death of a herd of cattle. Earlier, strains of *P. steckii* had been shown to be toxic to cultured cells and animals (17), while another isolate from moldy chocolate syrup killed brine shrimp and chick embryos (18). The toxin was not isolated from *P. steckii* until the moldy hay incident, but certain strains of *P. steckii* had been shown to produce citrinin (19) (Figure 6), a nephrotoxin which also inhibits the growth of etiolated wheat coleoptiles (11). When 3,7-DHMI was administered to day-old chicks, in corn oil via crop incubation, the LD_{50} of pure material was 800 mg/Kg which is a relatively high figure but, as the discoverers state, the situation was unique because the observed toxicity in cattle was due to the large quantities of the toxin produced by the organism. As a result of purification, the specific activity of the molecule was reduced because the relative solubility of the metabolite was decreased when introduced *in vivo* (16), a common phenomenon. No citrinin was found among the fermentation products.

Disclosure of the 3,7-DHMI toxin was published in 1979. Subsequently, in 1987, we were involved with an organism in which there was a contaminating fungus, *Penicillium corylophilum*, that was eventually isolated. Extracts of that organism yielded a metabolite that inhibited the growth of etiolated wheat coleoptiles. However, there was not, as we had earlier surmised, a biologically active natural product in the original fungal accession but the metabolite from the contaminating fungus inhibited wheat coleoptile growth by 100 and 43% at 10^{-3} and 10^{-4} M, respectively, relative to controls and had identical chemical and physical properties to 3,7-DHMI. It was further tested on greenhouse-grown bean, corn, and tobacco plants (*vide supra*) at 10^{-2}, 10^{-3}, and 10^{-4} M. Only corn exhibited leaf necrosis at 10^{-2} M within 20 hours following application (11). Two salient points emerged from our discovery. While we had been working very closely with the laboratory that had originally isolated 3,7-DHMI we had, for some reason, failed to include the metabolite in our discussions, resulting in a dormant period of 8 years during which some valuable agrochemical work might have been done. And while the molecule was not highly active against plants, it did exhibit selective herbicidal activity, and because of the relatively large quantity of material afforded by fermentation, there exists the possibility of synthesizing some active derivatives for testing in diverse systems.

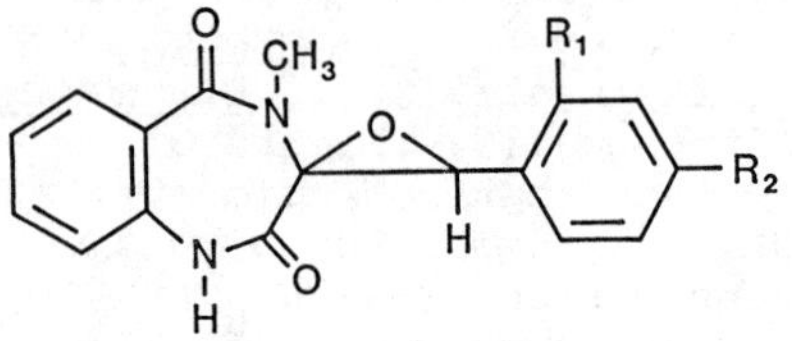

	R_1	R_2
1.	Cl	H
2.	H	Cl
3.	Cl	Cl

Figure 4. Synthetic cyclopenin derivatives.

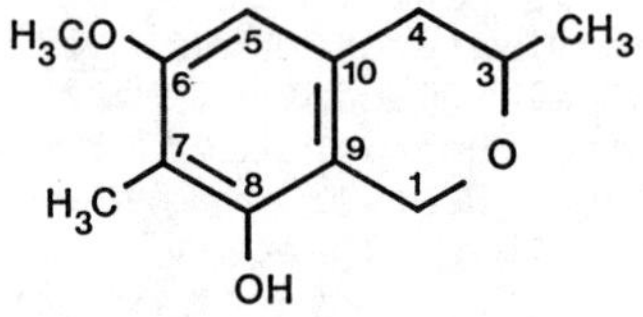

Figure 5. 3,7-dimethyl-8-hydroxy-6-methoxyisochroman.

Cinereain. A recurring criticism levied against the search for natural products from microorganisms is that the odds for discovering new structures are slim. Yet, in fact, dozens of new substances are discovered annually. Between 1971 and 1983, the number of newly elucidated fungal metabolites alone increased by 2000 according to one source (20). This does not include those that may have been acquired by fermentation industries and which are proprietary information, nor does it include bacterial products. Of these 2000, the number tested in biological systems seems to have been very limited. The problem in searching for new compounds is one of permutations coupled with the intuition to make the proper accessions from the right place, and luck. In addition, the fermentation process employed is also extremely important.

A new metabolite that has recently come to light is cinereain, isolated from *Botrytis cinerea* (Figure 7). The organism is generally associated, by oenophiles, with the disease commonly called the "noble rot" of grapes. Its presence causes the ripened grape to shrivel on the vine, increasing the relative amount of sugars and giving the vintage a distinctive flavor.The production of noble rot wine is generally limited to the sauternes and related wines; supplies are meagre and the cost is high. *B. cinerea* also attacks other fruits and vegetables both in the field and during storage (21). Our particular strain came from a bin of stored sunflower seed (*Helianthus annuus* L.), and the metabolite was subsequently obtained from solid fermentation on shredded wheat (22) supplemented with yeast extract, broth, and sucrose. Cinereain, which was obtained in limited quantity, was a bright red crystalline product whose structure was unequivocally determined by single crystal x-ray diffraction, supported by ^{1}H and ^{13}C NMR spectroscopy. The pure material inhibited the growth of etiolated wheat coleoptiles 100 and 34% at 10^{-3} and 10^{-4} M relative to controls. This level of activity is similar to that obtained with 3,7-DHMI. No effects were noted in experiments with greenhouse-grown bean or tobacco plants, but there were some slight necrotic and chlorotic effects induced in corn plants five days following treatment with 10^{-2} M solutions of the metabolite. Two weeks following treatment, the plant appeared to be growing normally. In antimicrobial assays with *Bacillus subtilis*, *B. cereus*, *Mycobacterium thermosphactum* (all Gram-positive), *Escherichia coli*, and *Citrobacter freundii* (both Gram-negative) there were no effects with concentrations of cinereain up to 500 ug/disk. Fungal assays were not included. It should be reiterated that the primary assay detects many types of biological activity and it is tempting to predict possible systems in which cinereain may eventually find a use. The curious structure is certainly novel: the seven-membered oxepine ring is rare among natural products and selective reduction of either the C2 or C14, or both, carbonyls and further derivatization should lead to some imaginative chemistry and compounds with selective biological activity.

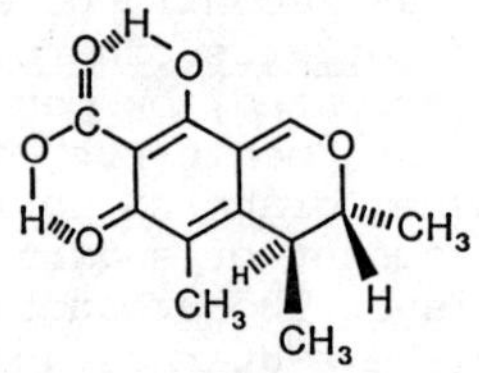

Figure 6. Citrinin.

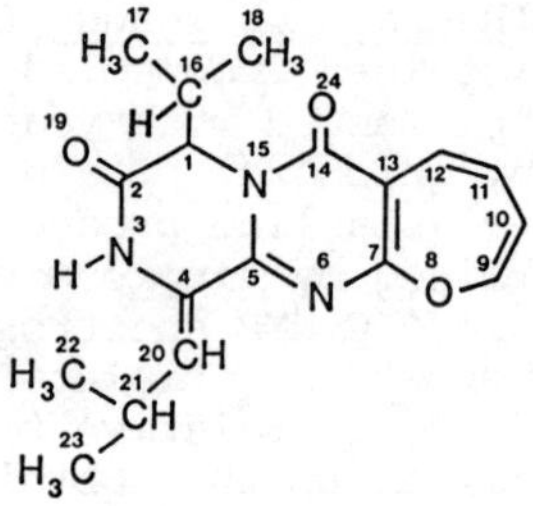

Figure 7. Cinereain.

In the final analysis, there are many microbial natural products that possess biological activity which may have potential as plant growth regulating substances and pesticides. This includes antimicrobials for use in crop protection. Many compounds, and their synthetic derivatives, appear not to have been tested in either singular or diverse biological systems (pharmaceutical or agrochemical). Perhaps what is now needed is a handbook of natural products listing their chemical and biological properties in multiple systems so that ready cross-reference can be made. In spite of the necessity of highly specific, high specific activity, biodegradable agrochemicals (especially fungicides) there is a lack of inter-disciplinary communication. Part of this is surely due to the secretive nature of commercial development, however, much remains to be accomplished and much of the accomplishment will be brought about by unifying the disciplines.

Literature Cited

1. Cutler, H.G.; Crumley, F.G.; Cox, R.H.; Wells, J.M.; Cole, R.J. Plant & Cell Physiol. 1984, 25, 257.
2. Doss, S.L.; Chu, C.K.; Mesbah, M.K.; Cutler, H.G.; Cole, P.D.; Arrendale, R.F.; Springer, J.P. J. Nat. Products 1986, 49, 363.
3. Endo, A; Kuroda, M.; Tsujita, Y. J. Antobiot. 1976, 29, 1346.
4. Cutler, H.G. Proc. 11th Ann. Meeting of the Plant Growth Regulator Soc. America 1984, 1.
5. Cole, R.J.; Cutler, H.G.; Dorner, J.W. In "Modern Methods in the Analysis and Structural Elucidation of Mycotoxins". R.J. Cole; Ed. Academic Press, Inc. New York 1986; pp. 1-28.
6. Bracken, A.; Pocker, A.; Raistrick, H. Biochem. J. 1954. 57, 587.
7. Mohammed, Y.S.; Luckner, M. Tetrahedron Lett. 1963. 28, 1953.
8. Birkinshaw, J.H.; Luckner, M.; Mohammed, Y.S.; Mothes, K.; Sticking, C.E. Biochem. J. 1963, 89, 196.
9. In "Merck Index". M. Windholz; Ed. Merck and Co.; Inc. Rahway, New Jersey, 1976 Monograph number: 2961, p. 394.
10. In "Protection Against Trichothecene Mycotoxins". National Academy Press, Washington, DC, 1983, pp. 73-75.
11. Cutler, H.G. Unpublished.
12. Smith, H.; Wegfahrt, P.; Rapoport, H. J. Am. Chem. Soc. 1968, 90, 1668.
13. Martin, P.K.; Rapoport, H.; Smith, H.W.; Wong, J.L. J. Organic Chem. 1969, 34, 1359.
14. White, J.D.; Haeflinger, W.E.; Dimsdale, M.J. Tetrahedron 1970, 26, 233.
15. Richter, H.; Winter, K.; Elkousy, S.; Luckner, M. Pharmazie 1974, 29, 506.
16. Cox, R.H.; Hernandez, O.; Dorner, J.W.; Cole, R.J.; Fennell, D.I. J. Agric. Food Chem. 1979, 27, 999.

17. Saito, M.; Enomoto, M.; Umeda, M.; Ohtsubo, K.; Ishiko, T.; Yamamoto, S.; Toyokawa, H. In "Mycotoxins in Human Health". I.H.F. Purchase; Ed. MacMillan Press Ltd., London, 1971; p. 181.
18. Davis, N.D.; Wagener, R.E.; Dalby, D.K.; Morgan-Jones, G.; Biener, U.L. Appl. Microbiol. 1975, 30, 159.
19. Krog, P. In "Mycotoxins". I.H.F. Purchase; Ed. Elsevier, New York, 1974, p. 425.
20. Turner, W.B.; Aldridge, D.C. In "Fungal Metabolites II". Academic Press, New York, 1983; Preface, p. v.
21. Stakman, E.C.; Harrar, J.G. In "Principles of Plant Pathology". The Ronald Press Co., New York, 1957. p. 392.
22. Kirksey, J.W.; Cole, R.J. Mycopathol. Mycol. Appl. 1974, 54, 291.

RECEIVED May 1, 1988

Chapter 7

Avermectins: Biological and Pesticidal Activities

J. R. Babu

Agricultural Research and Development, Merck Sharp & Dohme Research Laboratories, Three Bridges, NJ 08887

The avermectins, which show highly potent, broad spectrum of activities against plant parasitic mites, insects and nematodes were discovered in a screening program for natural products of microbial origin. The successful characteristics of this program are discussed. The GABAergic mode of action of the avermectins is unique. Other novel biological properties include: rapid photodegradation of foliar surface deposits, translaminar activity which maintains a pesticidal reservoir within the leaf; sublethal effects on organisms, such as debilitated feeding and reduction in fecundity; and virtual immobility and microbial decomposition in soil.

The discovery of the avermectin family of macrocyclic lactones produced by the soil actinomycete *Streptomyces avermitilis* marks an instructive chapter in the search for natural products of microbial origin. They were not found in a generalized, broad spectrum screen, but in one which had demonstrable elements of rationale and specific objectives. In discussing the characteristics of the successful screening program which led to the discovery of the avermectins at the Merck Sharp and Dohme Research Laboratories, Campbell *et al.* (*1*) point out that the discovery "..was by no means serendipitous; those who were seeking found what they sought". Their account and those of Stapley and Woodruff (*2*) and Woodruff and Burg (*3*) give the details of the initiation and organization of the screening efforts and their denouement as the avermectins.

0097–6156/88/0380–0091$06.00/0

The avermectins were discovered in an anthelmintic screening program in which microbial fermentation broths were tested in mice against the nematode *Nematospiroides dubius*. Two characteristics of this screening program are worth mentioning. First, instead of *in vitro*, "rationalist" tests such as target enzyme inhibition or receptor binding, the fermentation broths were tested by being administered in diet to nematode-infested mice; even though such an *in vivo* approach was expensive, it simultaneously tested for efficacy against a parasite and toxicity to the host, which contributed to the speed of further work on the active entities. Second, a deliberate choice was made to emphasize the selection of microorganisms of unusual morphological traits and nutritional requirements (2). Indeed, the cultures of *S. avermitilis* have characteristics which -- including brownish-grey spore masses, smooth spore surface, sporophores forming compact to slightly open spirals and presence of melanoid pigments -- are unlike those of any other previously described species of *Streptomyces*. Burg *et al.* (4) have given the taxonomic description and the fermentation procedures for *S. avermitilis*.

Among the avermectins, avermectin B_1, and to a lesser extent avermectin B_{2a}, have been studied with reference to plant parasitic mites, insects and nematodes. Since the introductory summary by Putter *et al.* (5), a limited review of the agricultural miticidal and insecticidal activities of avermectin B_1 has been published by Dybas and Green (6). More recently, Strong and Brown (7) have compiled a comprehensive review of the literature on the agricultural and veterinary *insecticidal* activities of the avermectins. In the context of this symposium, I intend this article not as a comprehensive review but as a way of discussing some of the unique biological properties of the avermectins in relation to their activities against nematodes, mites and insects of agricultural importance.

Chemistry and Nomenclature

The avermectins comprise a complex of 8 discrete but closely related macrocyclic lactones. Within this complex there are four major components --- avermectins A_{1a}, A_{2a}, B_{1a} and B_{2a} and four minor homologous "b" components-A_{1b}, A_{2b}, B_{1b} and B_{2b}. Mixtures of the homologous substances containing approximately 80% or more of the "a" and 20% or less of the corresponding "b" components are usually referred to as avermectin A_1, avermectin B_1, avermectin A_2, and avermectin B_2.

A composite structure of the avermectins is given in Figure 1. It consists of a rigid 16-membered lactone ring system, a spiroketal forming two 6-membered rings, and a cyclohexene diol or methoxycyclohexenol cis-fused to a five-membered cyclic ether. In addition, the structures are characterized by a disaccharide substituent consisting of two identical monomers, alpha-L-oleandrose, coupled to carbon-13 through an oxygen bond.

The large "A" designation refers to the avermectin components in which a methoxy group is present at C-5 and the large "B" refers to the corresponding C-5 hydroxy analog. The subscript "1" is used to identify those components with a 22,23-double bond. The subscript "2" identifies those components with a 23-hydroxyl group. Both "A" and "B" series of components are further characterized by the presence of a secondary-butyl substituent in the 25-carbon position, while the minor homolog contains an isopropyl substituent.

Details of the steps leading to the elucidation of the structures of the avermectins have been published earlier (8,9). Fisher and Mrozik (10) give a comprehensive review of the chemistry of the avermectins and of a complex of 13 closely related compounds known as milbemycins which were isolated from *S. hygroscopicus* by Japanese researchers (11, 12, 13). A notable difference between the milbemycins and the avermectins is the absence of the 13-hydroxy disaccharide substituent and saturation at the 22,23-positions in all the reported milbemycin compounds. Another major difference is the presence of methyl and ethyl groups attached to C-25 of the milbemycins while the avermectins have secondary butyl and isopropyl groups.

While the avermectins as a class are active against nematodes, insects, mites and other arthropods, they show differences in terms of degrees of activity (6). Components of the "B" series are more biologically active than those of the "A" series. Among the "B" series, avermectin B_1 (containing 80% or more of avermectin B_{1a} and 20% or less of avermectin B_{1b}) has been predominantly studied as an agricultural acaricide and insecticide. Avermectin B_{2a}, and its soil metabolite known as avermectin B_{2a}-23-ketone have been studied for their soil nematicidal activities. A synthetic derivative of avermectin B_1, 22,23-dihydroavermectin B_1, known by the generic name ivermectin has been developed for veterinary and human health uses (14, 15, 16, 17).

The word "ABAMECTIN' has been accepted as the nonproprietary common name for avermectin B_1 (18, 19). It is currently marketed as Avid and Vertimec which are the trade

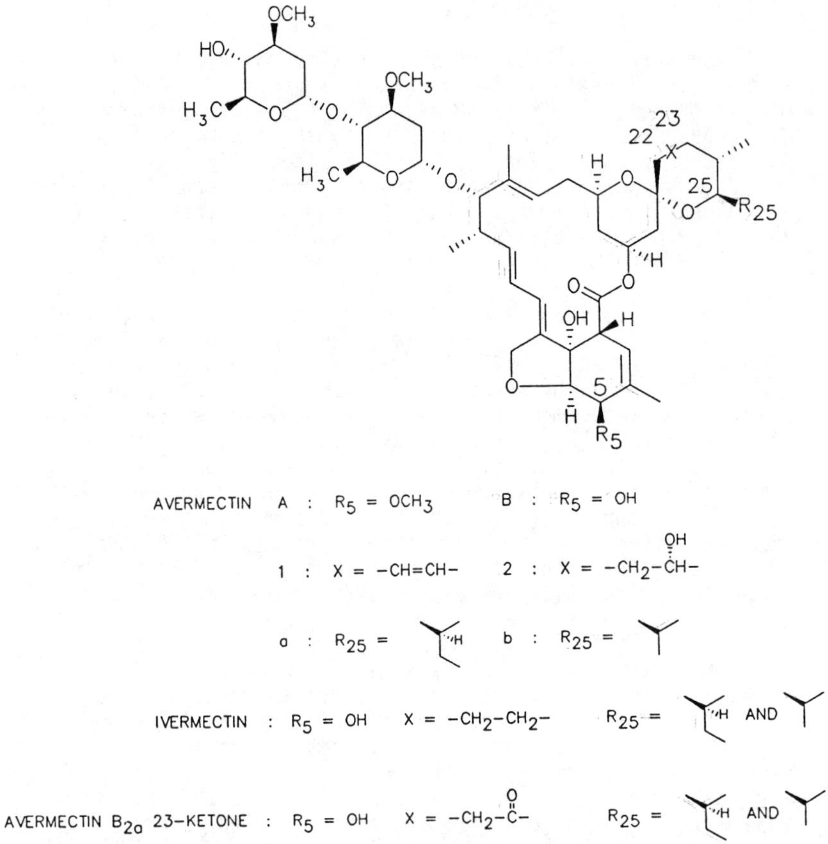

AVERMECTIN A : R_5 = OCH_3 B : R_5 = OH

1 : X = $-CH=CH-$ 2 : X = $-CH_2-CH(OH)-$

a : R_{25} = (sec-butyl) b : R_{25} = (isopropyl)

IVERMECTIN : R_5 = OH X = $-CH_2-CH_2-$ R_{25} = (sec-butyl) AND (isopropyl)

AVERMECTIN B_{2a} 23-KETONE : R_5 = OH X = $-CH_2-C(=O)-$ R_{25} = (sec-butyl) AND (isopropyl)

Figure 1: Structures of the avermectins (Courtesy of H. Mrozik, Merck and Co., Inc.).

names for an emulsifiable concentrate containing 1.8% w/v of abamectin for use against plant parasitic mites and insects in agriculture. Affirm is the trade name for a bait formulation containing 0.011% w/w of abamectin for use against the red imported fire ant *Solenopsis invicta*. It is also used as a cattle anthelmintic and ectoparasiticide as a 1% w/v injectable under the trade name Avomec in Australia.

Early pharmacological, biophysical and biochemical studies used several avermectins such as avermectin B_1, the B_{1a} component alone and ivermectin (20). Observations from these studies and the structure-activity relationships (10) show there is no evidence for qualitative differences in the mode of action of the avermectins. Therefore, in this discussion, the word 'avermectins' will be used as a generic reference.

Mode of Action

Early observations on the selective biological activities of the avermectins were instrumental in understanding their mode of action. While showing strong anthelmintic, insecticidal and acaricidal activities, they were ineffective against Platyhelminths such as flukes and tapeworms. Avermectins were also inactive against bacteria, yeasts and protozoa. An earlier report that a methanol extract of a culture of *S. avermitilis* inhibited some filamentous fungi by interfering with chitin biosynthesis (21) has been proved to be erroneous by Onishi and Miller (22) who showed that oligomycin (23) and an antifungal polyene produced by the organism accounted for the antifungal activity and that pure avermectin B_1 did not affect the fungi or their chitin metabolism. Similarly, Gordnier *et al.* (24) have been unable to detect inhibition of chitinase or chitin synthetase derived from a variety of insects.

Thus the conclusion was that there must be a specific target for the avermectins' activity in nematodes and arthropods which is either absent or inaccessible in fungi, bacteria, and Platyhelminths. Wang (25) gives an excellent, protagonist's account of the efforts which led to the conclusion that the avermectins act on invertebrates by potentiating the activity of gamma-aminobutyric acid (GABA) which is an inhibitory neurotransmitter in their nervous systems. Wright (20) gives a comprehensive review of the work done on understanding the mode of action of the avermectins and Stretton *et al.* (16) have reviewed the mode of action with specific reference to nematodes. Since a detailed analysis of the work done to understand the mode of action is out of the scope of this article, I will provide a summary of the current knowledge on the mode of action in relation to the observed biological activities.

Electrophysiological research done soon after the discovery of the avermectins with the nematode Ascaris (26) and the crustacean arthropod lobster (27) showed that avermectins functioned as post-synaptic agonists of GABA, potentiating the GABA-mediated chloride ion channel conductance. Corollary work (28) demonstrated that avermectins markedly stimulated the release of GABA from rat brain synaptosomes which had been preloaded with radiolabelled GABA. Further work on the GABA-receptor preparations derived from mammalian brain (28, 29) showed that avermectins bind to GABA-receptors and also increased the affinity of these receptors for benzodiazepams (30). It is also known that GABA itself also stimulates the binding of benzodiazepams to the GABA-receptors (31). Since both GABA and the avermectins stimulate benzodiazepam binding, one can infer that avermectins, like GABA, open the chloride channel of the GABA-receptor and thus avermectins behave as GABAergic agonists.

Therefore, with reference to nematodes and insects, the current explanation for the mode of action is that avermectins stimulate the release of GABA from nerve endings and then enhance the binding of GABA to receptor sites on the post synaptic membrane of an inhibitory motoneuron in the case of nematodes, and on the post-junction membrane of a muscle cell in the case of insects and other arthropods. The enhanced GABA-binding results in an increased flow of chloride ions into the cell with consequent hyperpolarization and elimination of signal transmission. An indirect yet strong evidence that GABA plays a role is that the effects of avermectins on invertebrates can be reversed by the chemicals picrotoxin or bicuculline (32) which act as GABA-antagonists by slightly different mechanisms (33, 34). Picrotoxin noncompetitively antagonizes GABA (35), while bicuculline competitively antagonizes by displacing GABA from the binding sites on the receptor (36).

While the GABAergic mechanism has been largely accepted, there are also indications that avermectins have more than one mode of action. At least two distinct sites of action, differing in their location, pharmacological behavior or both, have been recognized in arthropods (37, 38) and nematodes (26, 39). Stretton *et al.* (16) conclude: "Whether there are only two sites in each system, and whether the two sites are comparable in different phyla (*Nematoda and Arthropoda*) is not clear. A common thread in many cases is that there are correlations between AVM (*sic*) sensitive loci and the presence of (gamma)-aminobutyric acid (GABA) sensitive mechanisms involving a chloride ion permeability change".

Activity Against Soil Nematodes

The biological activities of the avermectins on plant parasitic nematodes have mostly been studied in terms of their gross effects on the movement and infective behavior of the juveniles of the rootknot nematodes *Meloidogyne* spp. Juveniles of *M. incognita* exposed to a 120-mM aqueous solution of abamectin or B_{2a}-23-ketone showed a three-phase response consisting of initial loss of movement within 10 minutes while being responsive to touch, partial recovery within 30 minutes of exposure and irreversible loss of movement after 2 hours (32). It is not known if the intervals between responses are dependent upon the concentration of the chemical in the solution. The initial loss of movement in *M. incognita* may be reflective of the avermectins' activity as GABA-agonists at the inhibitory synapses in nematodes (40); this possibility has been supported by the observations, as mentioned earlier, of Wright *et al.* (32) that GABA antagonists picrotoxin and bicuculline counteracted the inhibitory effects of avermectins on the locomotion of the juveniles. Under soil free conditions B_{2a}-23-ketone reduced the invasion of cucumber roots by *M. incognita* juveniles and their further development at concentrations much lower than were needed to immobilize them. It has been proposed in this context that avermectins affect the root-seeking behavior of juveniles, a mode of action also suggested for organophosphorus and carbamate nematicides (41). Avermectin B_{2a} did not affect the post-invasion development of *M. incognita* juveniles in tomato roots exposed to 1.0 ppm w/v solutions 48-72 hours post-invasion (16). Abamectin and avermectin B_{2a} showed only limited downward movement: spraying of aerial parts of tomato plants with 1000 ppm w/v solutions resulted in only minor inhibition of root galling (Merck and Co., Inc., unpublished). Eggs of *M. incognita* placed in a 0.1 ppm w/v aqueous solution of avermectin B_{2a}-23-ketone failed to hatch, but when they were rinsed in water 96 hours later, hatching occurred, which Wright *et al.* (41) suggest indicates that embryogenesis proceeded normally and that hatching was halted by the immobilization of the juveniles by the chemical. Avermectins begin to immobilize nematodes within 10 minutes of exposure (42, 32), while acetylcholinesterase inhibitors such as oxamyl cause hyperactivity. This is probably the reason behind the reduced oxygen consumption by juveniles of three *Meloidogyne* spp. exposed to 0.05 ppm solutions of avermectin B_{2a} (43).

When incorporated into soil, avermectin B_{2a} was slightly more potent than abamectin and was about 10-30 times more potent than several organophosphate and carbamate nematicides against *M. incognita* (44, 45). The longer soil residual activity of avermectin B_{2a}, with a half-life in soil of

2.5-3.0 days, has been ascribed to its conversion by soil microorganisms to avermectin B_{2a}-23-ketone (5), itself having a soil half-life of about 30 days (46). Interestingly, Preiser et al. (44) determined that the nematicidal potency of B_{2a}-23-ketone was slightly greater than that of B_{2a}; it is possible that the greater soil nematicidal potency B_{2a} is a combination of its own nematicidal activity combined with that of its soil metabolite.

Stretton et al. (16) have reviewed the microplot and large scale field trials done with avermectin B_1, B_{2a} and B_{2a}-23-ketone. The salient observation was that at soil application rates ranging from 0.168 to 1.52 kg ai/hectare all the avermectins were effective in controlling the rootknot nematodes. However, the differences in efficacy among the avermectins observed in the controlled greenhouse experiments do not obtain in the large scale field trials, at least among the limited number done so far. In this context, the influence of the physico-chemical properties on the behavior of the avermectins in soil should be considered. The water solubility of abamectin is 7.8 ppb w/v and its leaching potential through many types of soil is extremely low (47, 48), with the result that the chemical does not move into the rhizosphere readily unless mechanically incorporated to a sufficient depth. These factors have limited the successful use of the avermectins as soil nematicides. Paradoxically however, the physico-chemical properties also confer many potential advantages upon the use of the avermectins as nematicides. Their rapid degradation and poor mobility suggest that field applications would not result in persistent residues or contamination of ground water (48).

Photodegradation

Extensive and rapid photodegradation after application to plant surfaces appears to be a prominent characteristic of abamectin.

Studies on the fate of tritiated abamectin after application to cotton leaves showed that the compound was rapidly degraded on leaf surfaces; at 48 hours post-treatment, only 18.4% of the recovered radioactive material was abamectin (48). Jenkins et al. (49) applied abamectin to greenhouse-grown chrysanthemums at 22.4 and 44.8 g ai/hectare and determined that dislodgeable abamectin residues on the leaf surfaces were reduced by 90-98% by 24-72 hours postapplication (Figure 2). While the dynamics of the degradation would vary depending upon the morphology of the leaf surface and the intensity of the light, it is clear that the surface deposits of abamectin are degraded rapidly.

The rapid disappearance of the surface deposits of abamectin is an advantage in terms of nontarget, beneficial organisms. An example of this can be seen in the case of foraging honeybees (Figure 3). Field-grown alfalfa foliage treated with abamectin at different rates was collected at various post-treatment intervals. Bees were then kept in continuous contact with this foliage for 24 hours at which time their mortality was assessed. There was a steady decline in mortality, resulting in virtually no mortality at 36 hours post-treatment (50).

Translaminar Activity

In spite of the observed rapid degradation of the surface deposits, abamectin shows high post-application residual activity on leaves. This anomaly can be explained as being due to the translaminar activity of abamectin.

Translaminar activity of chemical refers to the movement of chemical from the treated surface into the leaf so that insect or mite pests feeding on the untreated surface would be affected (51, 52). In the context of abamectin, it can be proposed that while the surface deposits are quickly depleted (48, 53), the amount which has penetrated into the leaf forms a within-the-leaf reservoir, which give abamectin its residual miticial activity (6).

Wright *et al.* (54) have demonstrated such translaminar activity in bean, cotton and chrysanthemum leaves. In their experiments, abamectin was applied to the upper or lower surface of the leaf and mites were confined on the opposite surface. The differences in the activity of abamectin among the three plants are possibly due to structural differences in the cuticular waxes; Wright *et al.* (54) do point out that bean leaves have the least and chrysanthemum the most waxy cuticle. There were no significant differences in translaminar movement whether the chemical was applied to the upper and lower surface. This observation is of particular interest because penetration of chemicals into leaves is usually assumed to be greater through the lower than the upper leaf surface (55, 56).

Little is known about abamectin's patterns of movement within the leaf after it has penetrated the cuticle or whether its presence in the leaf mesophyll is apoplastic or symplastic or both. However, plant parasitic mites are destructive feeders and withdraw the contents of the palisade cells and those of the mesophyll (57, 58) and thus seem to ingest sufficient amounts of abamectin associated either with the cytoplasm or the cell walls. The depth of penetration need not extend from the treated upper epidermis to the untreated lower epidermis (or vice versa) where mites feed: it is known that the stylets of the tetranychid mite

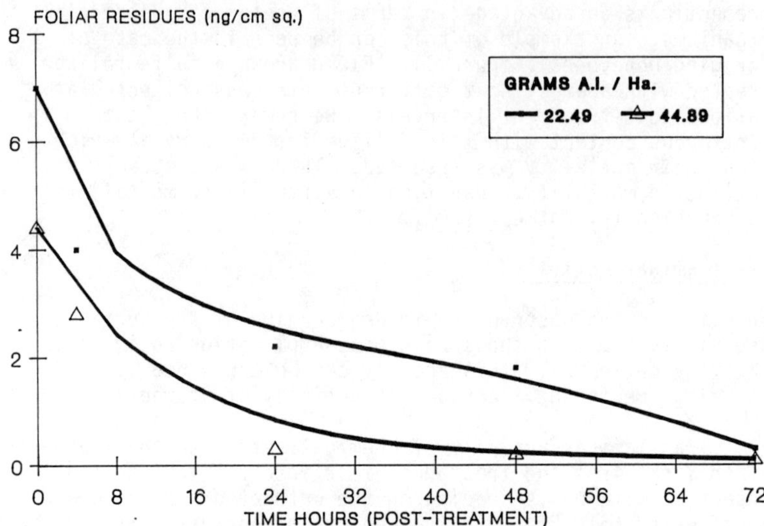

Figure 2: Foliar dislodgeable residues (expressed as nanograms/cm^2) of abamectin applied at two different rates to chrysanthemum leaves, determined at different times after application.(Adapted from ref. 49).

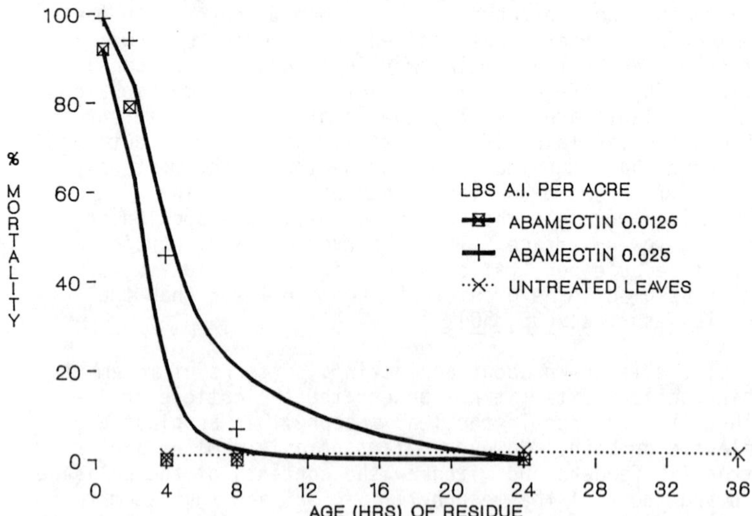

Figure 3: Effects on honeybees (Apis mellifera) of residues of abamectin applied at two rates onto alfalfa foliage. Bees were introduced onto treated foliage bearing residues aged for different time periods (Adapted from ref. 50).

Tetranychus urticae can penetrate the leaf to a depth of 70-100 um on a bean leaf of approximately 180 um thickness (58).

Behavior in Soil

Gullo *et al.* (46) discovered that avermectin B_{2a} incubated in a sandy loam soil under greenhouse conditions was rapidly degraded, with a half-life of 2.5 to 3.0 days, to a metabolite identified as a 23-keto derivative formed by soil. This metabolite was formed by soil microorganisms: in sterilized moist soil after 13 days, less than 1% of the added B_{2a} was converted to the metabolite compared with 44% under non-sterile conditions. At least three microorganisms capable of such transformation have been reported (46).

Similar studies on the fate of tritium-labeled avermectin B_{1a} in three kinds of soil have shown that under aerobic conditions it was degraded at a rapid rate, with a half-life in sandy loam of 14-28 days, in clay 25-56 days, and in coarse sand 56 days. The major soil degradation product was an equilibrium mixture with a ratio of 1:2.5 of the 8-alpha-hydroxy derivative and the corresponding open-ring aldehyde derivative (48). No further work regarding the nematicidal activity of these metabolites has been reported.

Lethal And Sublethal Activities

Avermectins differ fundamentally from other neuroactive pesticides, probably in keeping with their GABAergic mode of action, in that they do not cause hyperactivity in the affected organisms. Immobilization of nematodes soon after exposure to avermectins as reported by Wright *et al.* (32) was also reported earlier (5) in the case of mites and insects. The effects of such a mode of action, depending upon the degree of exposure to the chemical, can be lethal or sublethal.

As Strong and Brown (7) observe, there is no satisfactory definition of a "lethal effect"; however, death in the sense of an irreversible effect, described, for example, by Deecher *et al.* (59) as failure to respond to tactile stimulus, occurs in 72-120 hours after sufficient exposure to abamectin. The comparative symptoms of toxicity of abamectin and the pyrethroid cypermethrin when applied topically to the lepidopteran insect cotton bollworm (*Heliothis zea*) larvae are illustrative (B. I. Goll, personal communication). In the case of abamectin, there was flaccid paralysis, cessation of feeding, arrested ecdysis manifested as the presence of head capsule at the tip of the mandibles and a silvery grey color of an otherwise unaffected body with the presence of heartbeat observable in the third posterior abdominal segment. The presence of heartbeat has also been observed in *H. zea* adults which have ingested abamectin (60). Cypermethrin, in contrast, resulted in rapid convulsions and a shrunken larval body accompanied by an intense darkening of the cuticle.

Apart from the direct toxic effects resulting in mortality, an insecticide or a miticide can have other nonlethal, yet deleterious effects on the organism. Moriarty (61) described many such effects induced by a number of earlier insecticides. Kumar and Chapman (62) recently reported on such sublethal effects such as the inhibition of feeding, developmental disturbances and reduction in fecundity in the diamondback moth *Plutella xylostella*. Knowles (63) has described the effects, other than lethality, of some formamidines on plant parasitic mites.

Among such sublethal effects of the avermectins, reduced fecundity has attracted particular attention. Lofgren and Williams (64) observed that abamectin, when fed to the colonies, inhibited the reproductive capacity of the queens of the red imported fire ant *S. invicta*, with the resultant truncation of the colonies. Subsequently, Glancey *et al.* (65) histologically examined the ovaries of queens from colonies treated 22-weeks earlier and described the damage as: hypertrophy of the squamous epithelium which sheathes the ovarioles and pycnosis of the nurse cell nuclei which resulted in complete absence or reduction in the numbers and size of eggs produced.

There have also been reports of reduced fecundity among survivors of lepidopteran insects exposed to abamectin. Adults of the codling moth *Cydia pomonella* developing from larvae exposed to abamectin produced significantly fewer eggs (66). Beach and Todd (67) fed abamectin to male and female adults of soybean looper *Pseudoplusia includens* and subsequent matings of such insects resulted in reduced fecundity and fertility. A variation has been that of Robertson *et al.* (68) where fertility but not fecundity was affected in matings of males and females of the western spruce budworm *Choristoneura occidentalis* developed from larvae exposed to abamectin. Sublethal effects have been reported to include post-embryonic development of some insects; *Heliothis virescens* and *H. zea* larvae which survived abamectin treatment continued molting but did not survive pupation (69). The consequence was that even at doses below LD_{50}, few adults emerged from the pupae. Sublethal doses resulted in prevention of pupation in the eastern yellow jacket *Vespula maculifrons* (70) and extension of the pupal period in codling moth *C. pomonella* larvae (66).

An insecticide's effect on reproductive potential can be caused by a lowered incidence of mating, a shortened life span, the suppression of the reproductive organs and a direct toxic effect on the eggs (62). There is no evidence that abamectin affects insect or mite reproductive tissues directly. Even the ovarian regression in *S. invicta* observed by Glancey *et al.* (65) should be considered from the perspective that at the time of their observations 22 weeks had elapsed since treatment. The observed effects could have been due to starvation of the queen and the resultant dysfunctional changes in the general metabolism. The introduction of controls consisting of untreated but starved queens and histological observations at shorter post-treatment intervals would clarify this report. In the absence of direct observations on oogenesis or embryogenesis, the reduced effects of abamectin on fecundity will have to follow the obverse of the hormoligosis hypothesis proposed by Luckey (71) in that sublethal poisoning is likely to reduce the general fitness of the insects or mites and thereby their reproductive capacity and even the post-embryonic development.

A result of immobilization is the cessation of feeding, which has been termed as feeding inhibition. Strong and Brown (7) object to such a description on the reasoning that antifeedant properties should not be ascribed to a chemical unless such properties can be separated from general toxicity and debilitation. In my experience, in all instances where immobilization or cessation of feeding have been observed, death eventually resulted. It appears that an observed sublethal effect is really the beginning of the lethal effect. The relationships between the concentration of the chemical, speed of activity on different insects and the consequences of debilitated feeding, for example, the reduction in the leaf area consumed, are yet to be studied.

Differential Toxicities

Among the insects which were recorded as being affected by abamectin were three lepidopteran insects against which abamectin showed differential toxicity (5). The foliar ingestion toxicity LC_{90} values (for neonate larvae) were 0.02 ppm for the tomato hornworm *Manduca sexta*, 0.75-1.2 ppm for the cabbage looper *Trichoplusia ni* and 1.5 ppm for the southern armyworm *S. eridania*. Further research on the toxicity of abamectin to lepidopteran insects has shown some interesting characteristics of abamectin.

Anderson *et al.* (72) confirmed the observation by Putter *et al.* (5) that *S. eridania* was less sensitive to abamectin than *H. virescens*, both in treated foliage ingestion by neonate and topical applications to third instar larvae among the *Heliothis* spp. *H. virescens* is more susceptible to abamectin than *H. zea* (73, 74). Bull (69) confirms the differences in the sensitivities of H. virescens and H. zea and also demonstrates the marked insensitivity of the fall armyworm S. frugiperda.

The reasons for the differences in the toxicity of abamectin to the different lepidopteran insects are not clear. Bull (69), by using orally administered tritiated-B_{1a}, observed that physiological processes like absorption from the larval midgut, metabolism and excretion of the metabolites were slower in H. virescens than in H. zea or S. frugiperda. Similarly, significantly more tritiated avermectin B_{1a} was recovered from the heads of H. virescens. Neither slower metabolism nor faster accumulation in the head can categorically explain the greater sensitivity of H. virescens, considering that there were no significant differences between H. zea and S. frugiperda in the metabolism of avermectin B_{1a}. There were substantial differences between them in their susceptibility to avermectin B_{1a}. This aspect can be better understood by testing the hypothesis that differential sensitivity is related to differences in the affinity to, and therefore the accumulation of, abamectin at the GABA-receptors in the insects (69).

Conclusion

In summarizing the biological-pesticidal activities of the avermectins, two paradoxical properties can be noted. First, foliar surface deposits are rapidly degraded with the result that many beneficial organisms do not encounter the toxic entity significantly. However, avermectins seem to penetrate the leaf lamellae and be available as a pesticidal reservoir against mites and insects. Second, while they are potent against soil nematodes, they are also nearly insoluble in water and have an extremely low leaching potential. A clearer understanding of the mechanisms behind the leaf cuticular penetration would be helpful in finding ways to increase the toxic reservoir within the leaf while the surface deposits remain low, with the attendant advantages. Similarly, defining the mechanisms of soil binding would be helpful in finding ways to increase the avermectins' mobility and presence in soil water to act against nematodes.

The sublethal activities of the avermectins also need further scrutiny. For example, is the reduction in fecundity a direct effect or an indirect one due to the effect of the avermectins on the 'fitness' of the organism? The consequences of reduced fecundity and debilitated feeding need to be quantified in terms of the effects on the pest population dynamics. More research along the lines that Bull (69) suggests can explain the mechanisms behind the lesser susceptibility of some economically important lepidopteran insects.

The avermectins mark an important event in the search for natural products of microbial origin which are useful in agriculture. However, they may be only the beginning. Given their broad range of pesticidal activities, it is possible that future screening programs or semisynthetic modifications will yield entities which have one or more characteristics such as exclusive miticidal or insecticidal activity, enhanced cuticular penetration, resistance to photodegradation and greater soil mobility.

Acknowledgments

Many thanks to Connie M. Graf for her help in preparing this paper and to John G. MacConnell for his critical comments.

Literature Cited

1. Campbell, W. C.; Burg, R. W.; Fisher, M. H.; Dybas, R. A. In Pesticide Synthesis Through Rational Approaches, Magee, P. S.; Kohn, G. K.; Menn, J. J.; Eds.; ACS Symposium Series No. 255; American Chemical Society, Washington, D.C., 1984, pp. 5-20.
2. Stapley, E. O.; Woodruff, H. B. In Proceedings, An International Conference on Trends in Antibiotic Research, Umezawa, H.; Demain, A. L.; Hata, J.; Hutchinson, C. R., Eds.; Japan Antibiotics Research Association, Tokyo, 1982, pp. 154-70.
3. Woodruff, H. B.; Burg, R. W. In Discoveries in Pharmacology, Volume 3: Pharmacological Methods, Receptors & Chemotherapy; Parnham, M. J.; Bruinvels, J., Eds.; Elsevier Science Publishers, 1986, pp. 303-51.
4. Burg, R. W.; Miller, B. M.; Baker, E. E.; Birnbaum, J.; Currie, S. A.; Hartman, R.; Kong, Y. L.; Monaghan, R. L.; Olson, G.; Putter, I.; Tunac, J. B.; Wallick, H.; Stapley, E. O.; Oiwa, R.; Omura, S. Antimicrob. Agents Chemother. 1979, 15, 361-67.
5. Putter, I.; MacConnell, J. G.; Preiser, F. A.; Haidri, A. A.; Ristich, S. S.; Dybas, R.A. Experientia 1981, 37, 963-64.
6. Dybas, R. A.; Green, A. St. J. In: 1984 British Crop Protection Conference. Pests and Diseases. British Crop Protection Council, Croydon, U.K., Volume 3, pp. 947-54.
7. Strong, L.; Brown, T. A. Bull ent. Res. 1987, 77, 357-89.
8. Albers-Schonberg, G.; Arison, B. H.,; Chabala, J. C.; Douglas, A. W.; Eskola, R.; Fisher, M. H.; Lusi, A.; Mrozik, H.; Smith, J. L.; Tolman, R. L. J. Am. Chem. Soc. 1981, 103, 4216-21.
9. Springer, J. P.; Arison, B. H.; Hirschfield, J. M.; Hoogsteen, K. J. Am. Chem. Soc. 1981, 103, 4221.
10. Fisher, M. H.; Mrozik, H. In Macrolide Antibiotics, Academic Press, New York, 1984, pp. 553-606.

11. Aoki, A.; Fukuda, R.; Nakayabu, T.; Ishibashi, K.; Takeichi, C.; Ishida, M. J. Antibiotics. 1976, 29, 76-14, 76-35.
12. Mishima, H. In IUPAC Pesticide Chemistry 2, Miyamoto, J.; Kearney, P. C.; Eds.; Pergamon Press, Oxford, 1983; pp. 129-34.
13. Mishima, H.; Kurabayashi, M.; Tamura, C.; Sato, S.; Kuwano, H.; Saito, A. Tetrahedron Lett. 1975, p. 711.
14. Campbell, W. C.; Fisher, M. H.; Stapley, E. O.; Albers-Schonberg, G.; Jacob, T. A., Science 1983, 221, 823-28.
15. Campbell, W. C.; Benz, G. W. J. Vet. Pharmacol. Therap., 1984, 7, 1-16.
16. Stretton, A. O. W.; Campbell, W. C.; Babu, J. R. In Vistas on Nematology; Veech, J. A.; Dickson, D. W., Eds.; Society of Nematologists, Inc., Hyattsville, Maryland, 1987; pp. 136-46.
17. Aziz, M. A. Rev. Infect. Dis. 1986, 8, 500-04.
18. Anonymous WHO Chronicle 1985, 39, Supplement.
19. USAN Council. Clin. Pharm. Ther. 1988, 43, 91-92.
20. Wright, D. J., In Neuropharmacology and Pesticide Action, Ford, M. G.; Lunt, G. G.; Deay, R. D.; Usherwood, P. N. R., Eds.; Ellis Horwood, Chichester, 1986, pp. 174-202.
21. Calcott, P. H.; Fatig, III, R. O. J. Antibiot. 1984, 37, 253-59.
22. Onishi, J. C.; Miller, T. W. J. Antibiot. 1985, 38, 1568-72.
23. Umezawa, H., In Index of Antibiotics from Actinomycetes, University Park Press, Tokyo, 1967; p. 476.
24. Gordnier, P. M.; Brezner, J.; Tanenbaum, S. W. J. Antibiot. 1987, 40, 110-12.
25. Wang, C. C. Proceedings of Satell. Symp. 1984. Plenum, New York, 1986, pp. 125-38.
26. Kass, I.S.; Wang, C. C; Walrond, J. P.; Stretton, A. O. W. Proc. Natl. Acad. Sci. U.S.A. 1980, 77, 6211-15.
27. Fritz, L. C.; Wang, C. C.; Gorio, A. Proc. Natl. Acad. Sci. U.S.A. 1979, 76, 2062-66.
28. Pong, S. S.; Wang, C. C. Neuropharmac. 1980, 19, 311-17.
29. Pong, S. S.; Wang, C. C. In Membranes and Genetic Disease. Sheppard, J. R.; Anderson, V. E.; Eaton, J. W., Eds.; Liss, New York, 1982; pp. 373-95.
30. Pong, S. S.; DeHaven, R.; Wang, C. C. Biochimica et Biophysica Acta. 1981, 646, 143-50.
31. Tallman, J. F.; Thomas, J. W.; Gallager, D. W. Nature 1978, 274, 383-85.
32. Wright, D. J.; Birtle, A. J.; Roberts, I. T.J. Parasitology 1984, 88, 375-82.
33. Takeuchi, A.; Takeuchi, N. J. Physiology 1969, 205, 377-91.
34. Olsen, R. W.: Snowman, A. M. J. Neurochemistry 1985, 44, 1074-82.

35. Trifiletti, R. R.; Snowman, A. M.; Snyder, S. H. European J. of Pharmacol. 1985, 106, 441-47.
36. Olsen, R. W.; Wong, E. H. F.; Stanber, G. B.; King, R. G. Federation Proceedings 1984, 43, 2773-78.
37. Duce, I. R.; Scott, R. H. British J. Pharmacol. 1985, 85, 395-401.
38. Gregory, E.; Foreyt, W. J.; Breeze, R. Veterinary Medicine February, 1985, pp. 114-17.
39. Martin, R. J. Biochem. Soc. Trans. 1987, 15, 61-5.
40. Mellin, T. N.; Busch, R. D.; Wang, C. C. Neuropharmacology 1983, 22, 89-96.
41. Wright, D. J.; Birtle, A. J.; Corps, A. E.; Dybas, R. A. Ann. appl. Biol. 1984, 103, 465-70.
42. Sano, M.; Terada, M.; Ishi, A. I.; Kino, H. Experientia 1981, 37, 844-47.
43. Nordmeyer, D.; Dickson, D. W. J. Nematology 1981, 13, 452-53.
44. Preiser, F. A.; Babu, J. R.; Dybas, R. A.; Haidri, A. A.; Putter, I. J. Nematology 1981, 13, 457 (Abstr.).
45. Sasser, J. N.; Kirkpatrick, T. L.; Dybas. R. A. Plant Disease 1982, 66, 691-93.
46. Gullo, V. P.; Kemf, A. J.; MacConnell, J. G.; Mrozik, H.; Arison, B.; Putter, I. Pesticide Science 1983, 14, 153-57.
47. Bull, D. L. Southwestern Entomologist Supplement No. 7, 1985, pp. 2-10.
48. Bull, D. L.; Ivie, G. W.; MacConnell, J. G.; Gruber, V. F.; Ku, C. C.; Arison, B. H.; Stevenson, J. M.; Vandenheuvel, W. A. J. J. Agric. Fd. Chemistry 1984, 32, 94-102.
49. Jenkins, J. J.; Rosenthal, H. S.; Mollet, J.; Brown, R. D.; Norton, J.; Dybas, R. A. In American Chemical Society-Agrochemicals, Proceedings, 193rd ACS. National Meeting, April 1987, Denver, Colorado.
50. Atkins, E. L., In Annual Report, Dept. of Entomology, University of California, Riverside, 1980, p. 693.
51. Cooke, V. A. Annals of appl. Biol. 1963, 51, 485-88.
52. Harrison, I. R.; Kozlik, H.; McCarthy, J. F.; Palmer, B. F.; Weighton, D. M. Meded. Rijksfac. Landbouwwet. Gent. 1972, 37, 765-75.
53. Iwata, Y.; MacConnell, J.; Flor, J. E.; Putter, I.; Dinoff, T. M. J. Agric. Fd. Chem. 1984, 33, 467-71.
54. Wright, D. J.; Loy, A.; Green, A. St. J.; Dybas, R. A. Med. Fac. Landbouww, Rijksuniv. Gent. 1985, 50/2b, 595-601.
55. Hull, H. M., In Residue Reviews, Gunther, F. A.; Gunther, J. D. Eds.; Springer-Verlag, New York, 1970, pp. 1-155.
56. Sargent, J. A., Ann. Rev. Plant Physiol. 1965, 11, 1-12.
57. Baker, J. E.; Connell, W. A. Annals of the Entomological Society of America 1963, 56, 733-36.
58. Mothes, U.; Seitz K. Acarologia 1982, 23, 149-57.

59. Deecher, D. C.; Brezner, J.; Tanenbaum, S. W. J. econ. Entomol. 1987, 80, 1284-87.
60. Agee, H. R. Agric. Entomology 1985, 2, 325-36.
61. Moriarty, F. Biological Reviews 1969, 44, 321-57.
62. Kumar, K.; Chapman, R. B. Pesticide Science 1984, 15, 344-52.
63. Knowles, C. A. In Sites of Action for Neurotoxic Pesticides, Hollingworth, R. M.; Green, M. B., Eds.; ACS Symposium Series No. 356; American Chemical Society, Washington, D.C., 1987; pp. 174-90.
64. Lofgren, C. C.; Williams, D. F. J. econ. Entomol. 1982, 75, 798-803.
65. Glancey, B. M.; Lofgren, C. S.; Williams, D. F. J. Med. Entomol. 1982, 19, 743-47.
66. Reed, D. K.; Tromley, N. J.; Reed, G. L. J. econ. Entomol. 1985, 78, 1067-71.
67. Beach, R. M.; Todd, J. W. J. econ. Entomol. 1985, 78, 1125-28.
68. Robertson, J. L.; Richmond, C. E.; Preisler, H. K. J. econ. Entomol. 1985, 78, 1129-32.
69. Bull, D. L. J. Agric. Fd. Chem. 34, 74-78.
70. Parrish, M. D.; Roberts, R. B. J. econ. Entomol. 1984, 77, 769-72.
71. Luckey, T. D. J. econ. Entomol. 1968, 61, 7-12.
72. Anderson, T. E.; Babu, J. R.; Dybas, R. A.; Metha, T. J. econ. Entomol. 1986, 79, 197-201.
73. Wright, J. E.; Jenkins, J. N.; Villavaso, E. J. Southwestern Entomologist Supplement No. 7, 1985, pp. 11-16.
74. Wolfenbarger, D. A.; Johnson, A. W.; Herzog, G. A.; Tappan, W. B. Southwestern Entomologist Supplement No. 7, 1985, pp. 17-26.

RECEIVED May 23, 1988

Chapter 8

Fungal Elicitors of Phytoalexins and Their Potential Use in Agriculture

Jack D. Paxton

Department of Plant Pathology, University of Illinois, Urbana, IL 61801

Fungal elicitors of plant phytoalexins [natural plant antibiotics] (1) have the potential of becoming a new class of plant protectants and herbicides. Progress has been made recently to characterize elicitors chemically and further study their application to agriculture. The best characterized elicitor at present is from the cell walls of the Oomycete *Phytophthora megasperma* f. sp. *glycinea*. The smallest active, and potentially the most useful, fragment of this elicitor is a heptaglucan with specific structural requirements (2). Other fungal polysaccharides also elicit phytoalexin production in plants but their structures are not as well characterized. Fungal pectinases also have been implicated in phytoalexin elicitation. These enzymes appear to elicit phytoalexin production by releasing pectic fragments from the cell walls of plants (3).

The use of elicitors in agriculture holds exciting promise. Because phytoalexins have an important role in plant disease and pest resistance, their controlled elicitation could be used to stimulate natural disease and pest resistance without the use of environmentally damaging compounds. The toxicity of phytoalexins toward the plant in which they are produced might also be used to create a new class of herbicides, elicitors that cause the plant to 'self destruct'.

Plant diseases are caused by a range of organisms including bacteria, fungi, insects, nematodes and viruses. The diseases caused by these pathogens

0097–6156/88/0380–0109$06.00/0

collectively cause significant yield losses on crops around the world. Plant pathogens have evolved highly specific mechanisms to recognize and attack many of our crop plants and even circumvent our attempts at controlling them. Several methods of controlling plant diseases have been devised. These include breeding for plant disease resistance, chemical control either before or after the plant is attacked, agri-cultural practices, and biological control.

Each method of disease control has problems associated with it. New races of pathogens emerge to defeat genes for resistance. Some fungicides and pesticides have resulted in the build-up of pathogens and pests resistant to these chemicals. Adverse environments can stress a normally resistant plant and cause it to become susceptible to specific microorganisms and pests. Therefore, new methods of disease control are needed to reduce crop losses.

The mechanisms of plant disease resistance have been studied extensively to improve control of plant diseases. For this purpose a model system of an important disease, Phytophthora root rot, on an important U.S. crop, soybeans, was developed. Soybean is a major export crop worth $10 billion per year in the U.S., and it has been variously estimated that Phytophthora root rot can reduce the yield of this crop by 1 to 48% (4). This represents a $100 million to $1 billion loss per year! The pathogen which causes this severe disease is quite variable; 25 races of the pathogen have been reported (5). Other _Phytophthora_ species have developed resistance to the best chemical control, metalaxyl (6). Therefore, it becomes imperative to understand plant disease and pest resistance better, and to learn how fungal elicitors might be used in plant disease and pest control.

By studying Phytophthora root rot of soybean, a better understanding has been gained of how plants can recognize pathogens (by their elicitors) and what they often do after a pathogen is recognized (produce phytoalexins).

Phytoalexins and their elicitors

Phytoalexins are low molecular weight, antimicrobial compounds that are both synthesized by and accumulated in plants after exposure to microorganisms [1]. Several lines of evidence suggest that these compounds have an important role in plant disease and pest resistance [7-9].

The accumulation of phytoalexins in plants can be evoked by biologic or non-biologic treatments [10-13]. Examples of non-biologic treatments that lead to the accumulation of phytoalexins are treatment with ultraviolet light, salts of heavy metals, and freezing of tissues. Elicitors are compounds that are capable of evoking the accumulation of phytoalexins in plants

[14-19]. Biologic elicitors can be divided into endogenous, such as fragments of pectin molecules from plant cell walls [3], and exogenous, such as carbohydrate and glycoprotein molecules from fungal cell walls [20-24]. In this paper I will confine the discussion to recent work on fungal biologic elicitors of phytoalexins.

Phytoalexins from a large number of different plants have been chemically characterized. These phytoalexins include isoflavanoid-derived pterocarpan compounds characteristic of the Leguminosae, sesquiterpenoid compounds characteristic of the Solanaceae, phenanthrene compounds characteristic of the Orchidaceae and acetylenic compounds characteristic of the Compositae (Figure 1).

Glyceollin, a phytoalexin of soybeans, occurs as a series of isomers first identified by Lyne et al.[25]; the three most common isomers are shown in Figure 2. Plants frequently produce a series of active isomers and related phytoalexins [26].

Elicitors

Elicitors were first discovered by Cruickshank [27], who found a protein, Monilicolin A (Mr=6K), produced by *Monolinia fructicola*. This protein elicits the phytoalexin phaseollin in garden beans but not the related phytoalexins in pea or broad bean. Another protein, found only recently to be produced by *Phytophthora parasitica* var. *nicotianae*, is active at 20 ng/cotylendon [$4x10^{-13}$mole]. This protein (Mr=46K) may be a ß 1-4 endoxylanase [28]. Many of the elicitors studied are carbohydrates [29,30]. That carbohydrates carry 'recognizable information' is well known. An example of specific recognition of subtle changes in surface carbohydrates is the human blood group antigens (31). 'Recognition' of microorganisms is being studied in several plant-microorganism interactions [32]. Another area of interest is how the elicitor signal is transduced to the plant nucleus for directed processing of nucleic acids and/or de novo production of proteins [33].

An elicitor produced by *P. megasperma* f. sp. *glycinea* was identified by Frank and Paxton [34] as a glycoprotein. Ayers et al [35-38] isolated four different carbohydrate fractions from *Phytophthora megasperma* f. sp. *glycinea* that were active in eliciting phytoalexin production by soybean suspension cells [25] and cotyledons, using a bioassay developed by Frank and Paxton [34]. An elicitor from this fungus has been characterized [39] and synthesized [40,41], and is shown in Figure 3. The soybean plant contains enzymes capable of releasing active fragments from the walls of Oomycetes [42,43], which also elicit in other plant systems [44,45].

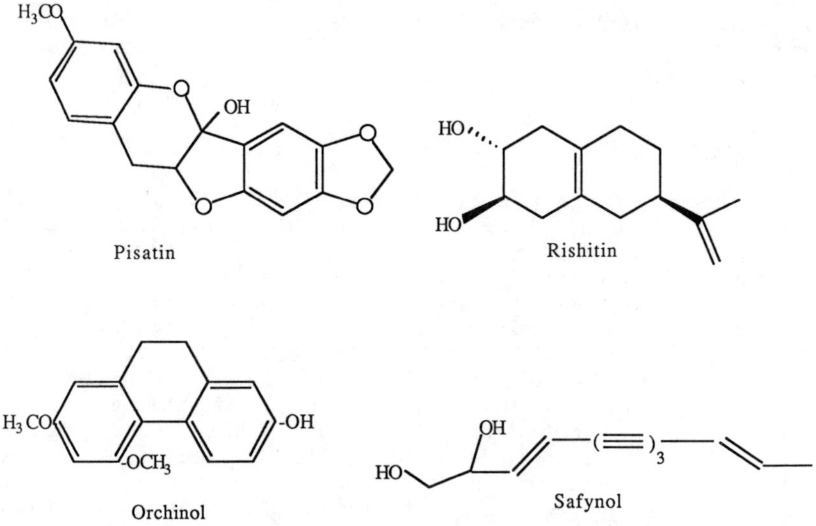

H_3CO OH Pisatin

HO HO Rishitin

H_3CO -OH $-OCH_3$ Orchinol

OH HO $(\equiv)_3$ Safynol

Figure 1. Phytoalexins from: Leguminosae, pisatin from peas; Solanaceae, rishitin from potato; Orchidaceae, orchinol from Orchis spp.; Compositae, safynol from safflower.

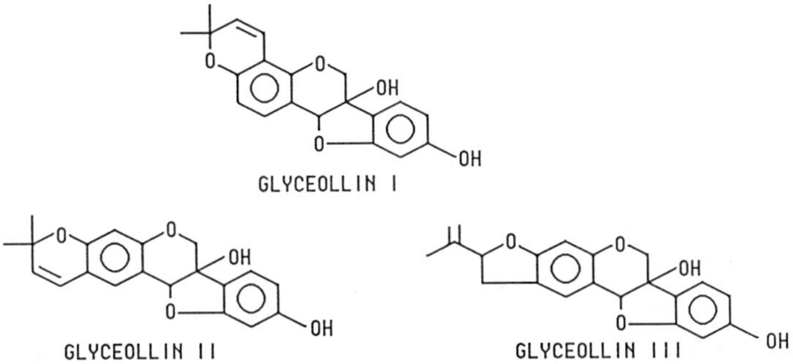

GLYCEOLLIN I

GLYCEOLLIN II

GLYCEOLLIN III

Figure 2. Phytoalexin isomers from soybean.

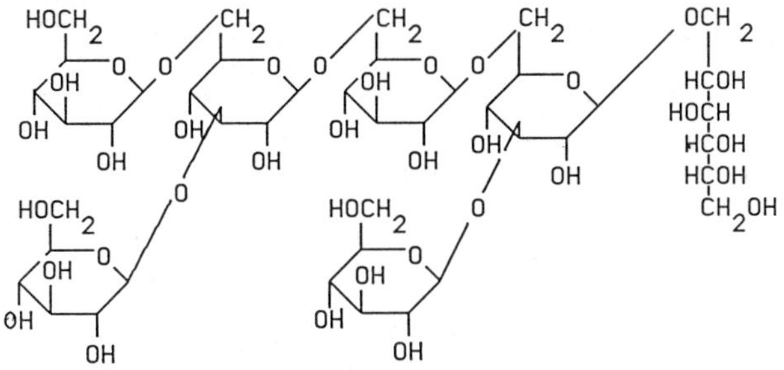

Figure 3. Elicitor derivative from *Phytophthora megasperma* f. sp. *glycinea*.

Anderson [46] found that Colletotrichum lindimuthaneum, the causal agent of bean anthracnose, produced a compound in culture which was active in a specific manner even at 10 ng/ml. Hamdan and Dixon [47] found that this fungus produces a mannose-rich polysaccharide that elicits three enzymes in phytoalexin synthesis and found a polysaccharide with lower mannose content, that preferentially induced one of these enzymes, chalcone synthase.

Chitin, an important component of nematode eggs, insect exoskeletons and the cell walls of many plant pathogenic fungi, also is an effective elicitor. Hadwiger and colleagues [48] found that chitosan, a deacylated fragment of chitin, at 10 ug/ml elicits phytoalexin accumulation in peas, and acts as a fungicide against Fusarium solani, a pathogen of peas. The heptamer was the smallest active polymer size.

Various researchers have studied the elicitation of potato phytoalexins by arachidonic and eicosapentaenoic acids [24,49-51]. These compounds are released by Phytophthora infestans, elicit phytoalexin accumulation in potato at 10 ug/tissue slice, and are potentiated by glucans from this fungus [52].

Verticillium albo-atrum releases a glucan [Mr <1.8 Kd] with ß 1-3 linkages, that elicits phytoalexin production in tomato [53].

Endogenous elicitors

Endogenous elicitors are compounds that either exist pre-formed in the plant or are fragments of pre-formed compounds, and can elicit phytoalexin accumulation by the plant. The best known of these elicitors is pectin or its fragments [54-57]. Pectin fragments also can potentiate the activity of other elicitors [58,59]. This suggests that damage to a plant cell wall may activate several defense responses including enhanced phytoalexin accumulation. Many plant pathogens produce various enzymes capable of degrading pectin, presumably as a source of nutrients, and these enzymes have been shown to be good elicitors of phytoalexin accumulation [56,60,61].

Vaillincourt [62] recently devised a bioassay for glyceollin elicitors in soybean roots, that presumably does not require wounding.

Specificity and Elicitors

One disconcerting aspect of most of the elicitors studied so far is the lack of confirmed specificity in phytoalexin elicitation. This is in contrast to the clear host specificity shown by the pathogen. The availability of such specific elicitors would be useful because it would allow rather specific targeting of plants to be protected or killed.

Some work suggests that this is possible [63-65]. De Wit and Spikman obtained a glycoprotein from the intercellular fluids of tomato tissue infected by incompatible races of Cladosporium fulvum, that elicited necrosis characteristic of an incompatible interaction [resistance] as early as 7 hours after treatment [66]. This glycoprotein did not cause necrosis in the susceptible cultivars of tomato. The same fungus grown on tomato extracts produced only non-specific elicitors. Colletotrichum lindemuthaneum produced race-specific elicitors, active at 0.01 ug/ml on garden bean [67]. Keen and others reported these elusive specific elicitors in Phytophthora megasperma f. sp. glycinea [68]. The elicitors appeared to be glycoproteins, but were neither chemically characterized nor their activity confirmed. Plant cell membranes undoubtly have receptors for these elicitors and attempts have been made to identify these receptors [69].

Crop Protection

Cross protection of plants against disease has been demonstrated several times. This protection is effected by first inoculating plants with an avirulent microorganism; then, after a period of time, ranging up to several days, inoculation with a normally virulent microorganism will not cause disease. This phenomenon may be akin to the vaccination of humans with the cowpox virus to protect them from smallpox.

The use of elicitors of phytoalexin production in plants as protectants was first suggested by Paxton [70]. Since that time several companies have become interested in applying this knowledge to create a new type of 'fungicide' that uses the plant's own defense mechanisms to protect it from disease. These 'fungicides', which might be relatively stable polysaccharides or glycoproteins, would be welcome additions for plant protection, because, unlike most of the other fungicides, these compounds generally would have low impact on the environment and non-target organisms.

Elicitors, as 'fungicides', would need to be formulated to remain on the surface of the plant until the plant is wounded, or pathogens or insects introduce the compound into the cell during attack. For use as herbicides, elicitors would need to be formulated to enter the plant immediately after application.

Suppressors of Phytoalexin Production

Suppressors are compounds produced by fungi, that prevent elicitation of phytoalexin accumulation. An extracellular mannan-glycoprotein (presumably an invertase) produced by Phytophthora megasperma f. sp. glycinea supresses the response of soybean plants to the glucan elicitor produced by the same fungus [71]. This suppression is of interest, especially because of its race specific nature. The races of *Phytophthora megasperma* f. sp. *glycinea*, that attack a given set of soybean cultivars, produce compounds that suppress phytoalexin accumulation (that could be elicited by the glucan) only on those cultivars. Doke and Tomiyama [72] reported that ß 1-3 glucan molecules of 17-23 glucose units from *Phytophthora infestans* suppressed phytoalexin accumulation in an anticipated race-specific manner in potato cultivars.

Suppressors may be important for virulence of fungal pathogens by allowing a pathogen to attack its host rapidly. They conceivably could be used as herbicides [for example suppression of nightshade (*Solanum nigrum*) in crops] since they should increase the aggressiveness of natural pathogens on target plants. Specific suppressors could prove useful as selective herbicides, but first they must be chemically characterized and their activity confirmed.

Commercial Potential for Elicitors in Plant Protection

The commercial application of elicitors and suppressors in plant protection is suggested by the interest that several companies have in developing carbohydrates for applications in medical science. Bicarb AB of Lund Sweden and Chembiomed of Edmonton, Canada are two such companies engaged in R&D and marketing of biologically active carbohydrates. Carbohydrates International AB of Arlov, Sweden [a subsidiary of Volvo AB] also is looking for novel ways to synthesize complicated carbohydrates found in human membranes.

IGENE Biotechnology of Columbia, MD recently developed a chitin nematicide, CladoSan, which has been successfully used in nematode control by R. Rodriquez-Kabana, at Auburn University. Other companies also are looking into the production of chitin for disease control. Large masses of crab shells, which are predominantly chitin, are wasted presently.

Future Research

Sugar polymers contained in cell walls of plants, animals and microorganisms are important regulators of plant cell activities, including accumulation of phytoalexins effective against invading insects, bacteria and fungi.

The application of our knowledge of elicitors and fungicide formulation is an area that needs our immediate attention. Understanding how these molecules function in plant disease and pest resistance will significantly improve our lives by increasing crop yields and lowering food costs.

Acknowledgments

Certain aspects of the research work presented in this report were supported in part by the Illinois Agricultural Experiment Station. Lisa Vaillancourt helped draw the chemical structures.

Literature Cited

1. Paxton, J.D. Phytopath. Zeit. 1981, 101, 106-9.
2. Sharp, J.K.; McNeil, M.; Albersheim, P. J. Biol. Chem. 1984, 259, 11321-6.
3. Robertsen, B. Physiol. Mol. Plant Pathol. 1986, 28, 137-48.
4. Schmitthenner, A.F. Plant Disease 1985, 69, 362-8.
5. Layton, A.C.; Athow, K.L.; Laviolette, F.A. Plant Disease 1986, 70, 500-1.
6. Davidse, L.C.; Looijen, D.; Turkensteen, L.J.; van der Wal, D. Neth. J. Plant Pathol. 1981, 87, 65-8.
7. Ebel, J.; Ayers, A.R.; Albersheim, P. Plant Physiol. 1976, 57, 775-9.
8. Hart, S.V.; Kogan, M.; Paxton, J.D. J. Chem. Ecol. 1983, 9, 657-72.
9. Smith, D.A. In Phytoalexins; Halstead Press 1982, pp.218-52.
10. Moesta, P.; Grisebach, H. Nature 1980, 286, 710-1.
11. Moesta, P.; Grisebach,H. Arch. Biochem. Biophys. 1981, 211, 39-43.
12. Yoshikawa, M. Nature 1978, 275, 546-7.
13. Hadwiger, L.A.; Beckman, J.M. Plant Physiol. 1980, 66, 205-11.
14. Yoshikawa, M. In. Biochemical Plant Pathology; John Wiley and Sons, Ltd. 1983, pp. 267-98.
15. Osswald W.F.; Zieboll, S.; Elstner, E.F. Z. Naturforsch. 1985, 40c., 477-81.
16. Tietjen, K.G.; Hunkler, D.; Matern, U. Eur. J. Biochem. 1983, 131, 401-7.
17. Tietjen, K.G.; Matern, U. Eur. J. Biochem. 1983, 131, 409-13.
18. Walker-Simmons, M.; Hadwiger, L.; Ryan, C.A. Biochem. Biophys. Res. Commun. 1983, 110, 194-9.
19. Young, D.H.; Kohle, H.; Kauss, H. Plant Physiol. 1982, 70,1449-54.
20. Ebel, J. ACS Symp. Ser. 1984, 257, 257-71.
21. Anderson, A.J. Phytopathology 1978, 68, 189-94.
22. Anderson, A.J. Can. J. Bot. 1978, 26, 1473-9.
23. Anderson-Prouty, A.J.; Albersheim, P. Plant Physiol. 1975, 56, 286-91.
24. Bostock, R.M.; Laine, R.A.; Kuc, J. Plant Physiol. 1982, 70, 1417-24.
25. Lyne, R.L.; Mulheirn, L.J.; Leworthy, D.P. JCS Chem. Comm. 1976, 1976, 497-8.
26. Ingham, J.L.; Keen, N.T.; Mulheirn, L.J.; Lyne, R.L. Phytochem. 1981, 20, 795-8.

27. Cruickshank, I.A.M.; Perrin, D.R. Life Sci. 1968, 7, 449-58.
28. Farmer, E.E.; Helgeson, J.P. Plant Physiol. 1988, in press
29. Cline, K.; Wade, M.; Albersheim, P. Plant Physiol. 1978, 62, 918-21.
30. Keen, N.T.; Bruegger, B. ACS Symp. Ser.62 1977, 62, 1-26.
32. Halverson, L.J.; Stacey, G. Microbiological Reviews 1986, 50, 193-225.
33. Strasser, H.; Hoffmann, C.; Grisebach, H.; Matern, U. Z. Naturforsch. 1986, 41c, 717-24.
34. Frank, J.A.; Paxton, J.D. Phytopathology 1971, 61, 954-8.
35. Ayers, A.R.; Ebel, J.; Finelli, F.; Berger, N.; Albersheim, P. Plant Physiol. 1976, 57, 751-9.
36. Albersheim, P.; McNeil, M.; Darvill, A.G.; Valent, B.S.; Hahn, M.G.; Robertson, B.K.; Aman, P. Recent Advances in Phytochemistry 1981, 15, 37-58.
37. Ayers, A.R.; Valent, B.; Ebel, J.; Albersheim, P. Plant Physiol. 1976, 57, 766-74.
38. Ayers, A.R.; Ebel, J.; Valent, B.; Albersheim, P. Plant Physiol. 1976, 57, 760-5.
39. Sharp, J.K.; Albersheim, P.; Ossowski, P.; Pilotti, A.; Geregg, P.; Lindberg, B. J. Biol. Chem. 1984, 259, 11341-5.
40. Ossowski, P.; Pilotti, A.; Garegg, P.; Lindberg, P. Angew. Chem. Int. Ed. Engl. 1983, 22, 793-5.
41. Ossowski, P.; Pilotti, A.; Garegg, P.; Lindberg, B. J. Biol. Chem. 1984, 259, 11337-40.
42. Yoshikawa, M.; Matama, M.; Masago, H. Plant Physiol. 1981, 67, 1032-5.
43. Keen, N.T.; Yoshikawa, M.; Plant Physiol. 1983, 71, 460-5.
44. Hauffe, K.D.; Hahlbrock, K.; Scheel, D. Z. Naturforsch. 1986, 41c, 228-39.
45. Kuhn, D.N.; Chappell, J.; Boudet, A.; Hahlbrock, K. Proc. Natl. Acad. Sci. 1984, 81, 1102-6.
46. Griffiths, H.M.; Anderson, A.J. Can. J. Bot. 1987, 65, 63-8.
47. Hamdan, M.A.M.S.; Dixon, R.A. Physiol.Mol.Plant Pathol. 1987, 31, 91-103.
48. Kendra, D.F.; Hadwiger, L.A. Phytopathology 1987, 77, 100-6.
49. Bostock, R.M.; Kuc, J.A.; Laine, R.A. Science 1981, 212, 67-9.
50. Bostock, R.M.; Schaeffer, D.A. Physiol. Mol. Plant Pathol. 1986, 29, 349-60.
51. Kurantz, M.J.; Osman, S.F. Physiol. Plant Pathol. 1983, 22, 363-70.
52. Bostock, R.M.; Nuckles, E.; Henfling, J.W.D.M.; Kuc, J.A. Phytopathology 1983, 73, 435-8.
53. Woodward, S.; Pegg, G.F. Physiol. Mol. Plant Pathol. 1986, 29, 337-47.

54. Hahn, M.; Darvill, A.; Albersheim, P. Plant Physiol. 1981, 68, 1161-9.
55. Nothnagel, E.; McNeil, M.; Albersheim, P.; Dell, A. Plant Physiol. 1983, 71, 916-26.
56. Lee, S.-C.; West, C.A. Plant Physiol. 1981, 67, 33-9.
57. Lyon, G.D.; Albersheim, P. Plant Physiol. 1982, 70, 406-9
58. Davis, K.R.; Darvill, A.G.; Albersheim, P. Plant Mol. Biol. 1985, 6, 23-32.
59. Davis, K.R.; Darvill, A.G.; Albersheim, P.; Dell, A. Z. Naturforsch. 1986, 41c, 39-48.
60. Davis, K.R.; Lyon, G.D.; Darvill, A.G.; Albersheim, P. Plant Physiol. 1984, 74, 52-60.
61. Lee, S.-C.; West, C.A. Plant Physiol. 1981, 67, 640-5.
62. Vaillincourt, L.J. A New Biological Assay for the Elicitation of Glyceollin in Soybeans. Ph. D. Thesis, University of Illinois, Urbana, 1987.
63. Wade, M.; Albersheim, P. Proc. Natl. Acad. Sci. USA 1979, 76, 4433-7.
64. Bruegger, B.B.; Keen, N.T. Physiol. Plant Pathol. 1979, 15, 43-51.
65. Keen, N.T. Science 1975, 187, 74-5.
66. De Wit, P.J.M.; Spikeman, G. Physiol. Plant Path. 1982, 21, 1-11.
67. Tepper, C.S.; Anderson, A.J. Physiol.Mol.Plant Pathol. 1986, 29, 411-20.
68. Keen, N.T.; Legrand, M. Physiol. Plant Pathol. 1980, 17, 175-92.
69. Yoshikawa, M.; Keen, N.T.; Wang, M. Plant Physiol. 1983, 73, 497-506.
70. Paxton, J.D. Ill. Res. 1973, 15, 13.
71. Ziegler, E.; Pontzen, R. Physiol.Plant Pathol. 1982, 20, 321-31.
72. Doke, N.; Tomiyama, K. Physiol. Plant Path. 1980, 16, 177-86.

RECEIVED May 5, 1988

Chapter 9

Endophytic Bacteria for the Delivery of Agrochemicals to Plants

Jed W. Fahey

Crop Genetics International, 7170 Standard Drive, Hanover, MD 21076

A novel delivery system (InCide) utilizing naturally occuring plant endophytic bacteria is being developed for the systemic delivery of agrichemicals *in-planta*. Host crops are systemically colonized by the endophyte, thus providing an environmentally safe vehicle with which to deliver potent, targeted, biologically-derived agrichemicals to the plants. InCide products, including those having insecticide, fungicide, bactericide, viricide, nematicide and plant growth enhancer activity are designed to be applied as a seed, seedling or young plant treatment only once during the life of a plant. They exploit the biological characteristics of a natural endophytic microbe by systemically colonizing the xylem and achieving sustained levels in their host plants. Because of their *in-planta* growth, effects on non-target organisms and the environment are minimized and significant advantages are afforded over current externally applied agrichemicals. The first InCide product is an endophyte of corn that has been genetically modified by insertion of a gene from *Bacillus thuringiensis* encoding for the production of a highly specific insecticidal protein (delta-endotoxin) active against the European Corn Borer. Refinement and scale-up of techniques for inoculation of this product into corn seeds will permit rapid commercialization of this new delivery system.

An examination of plant-associated microbes will illustrate the diversity of relationships existing between plants and "internally dwelling" plant-neutral or beneficial microbes. Just how some of these relationships evolved is not fully understood, but there are a variety of stable, mutualistic or symbiotic relationships between

0097–6156/88/0380–0120$06.00/0

higher plants and microbial endophytes. Endophytes are microbes which live within plants. Maintenance of all plant-endophyte relationships require that although the plant may recognize the invader, it does not trigger any overt pathogenic response to the microbe or that the microbe successfully avoids the plant's defense reactions and does not cause disease. Numerous plant-endophyte interactions have been identified over the past century. To date, only the *Rhizobium* species, not even true endophytes but mild plant pathogens of the legume family, have been commercialized for their agricultural potential as symbionts.

Naturally Occurring Endophytes

The nodulation of legumes like soybean, clover and alfalfa by the gram-positive *Rhizobium* & *Bradyrhizobium* is by far the most thoroughly characterized plant-microbe relationship and the only one which has been exploited commercially. *Rhizobium* species in symbiotic association with plants are responsible for the conversion of as much as 2 x 10^7 tons of atmospheric N_2 per year to ammonium (1). This "fixed" nitrogen is then utilized by plants for the synthesis of amino acids and protein. In *Rhizobium* symbioses, the bacteria infect plant roots and are then contained within specialized structures called root nodules. A highly integrated expression of genes from both the plant and the bacteria lead to the formation of a morphologically distinct nodule, with its own meristematic regions and vascular innervation. A molecule very similar to the hemoglobin found in blood is synthesized to protect the oxygen-sensitive enzyme, nitrogenase, and the bacteria actually alter their morphology once established in a developing nodule. Without the protective environment and carbon supply afforded the bacteria by the plant, the bacteria would not perform the energy-intensive process of nitrogen reduction. Without the bacteria's significant nitrogen input, the plants would not, in many cases, be able to prosper in an otherwise nitrogen poor environment. This natural plant-bacterial symbiosis can therefore save farmers growing many leguminous crops from the considerable expense associated with fertilizer nitrogen input and may permit certain species to thrive in areas which they might otherwise not occupy.

Certain non-legumes also form root nodules which are responsible for nitrogen fixation. The microbe found in all non-legume root nodules (actinorhizal plants) studied to date is the actinomycete, *Frankia*. Most actinorhizal symbyotic associations are found in woody plants in the temperate or cold zones of the Northern Hemisphere. Over 178 species in 20 different genera have been documented to contain actinorhizal symbionts (2). There are two major morphological types of actinorhizal nodules: The *Alnus*-type or coralloid nodules are short, stubby, dichotomously branched growths on lateral roots. *Myrica*-type or lobed nodules are thin, and produce negatively geotropic rootlets at their apex (3).

Stem nodules, too, are formed on certain members of the Leguminosae family by Rhizobium bacteria. These nitrogen-fixing symbioses have been documented in three genera of predominantly tropical or hydrophytic plants --- Aeschynomene, Sesbania and Neptunia, which are represented by about 200, 70 and 15 species respectively (4). Nitrogen fixation by the endophyte in stem nodules and the protection of the oxygen-sensitive microbial enzyme seems to be accomplished via a similar strategem of leghemoglobin and nodule morphology to that adopted by the legume root nodules.

Leaf nodules occur predominantly in the families Rubiaceae and Myrsinaceae which contain over 400 nodule-forming species (5). These families of predominantly sub-tropical shrubs are nodulated by a gram negative, pleiomorphic bacterium in an association which in at least one species (Psychotria bacteriophila) has been shown to be obligate (6). In addition, one family of monocots, Dioscoreaceae has been identified as having a species with leaf nodules colonized by endophytic bacteria (7). Although the nature of these symbioses are not fully understood, it is hypothesized that the bacteria have a growth promoting effect.

There is at least one well-characterized association of a heterocystous, nitrogen-fixing blue green alga (cyanobacterium) with a plant. The cyanobacterium Anabaena sp. develops in a mutualistic relationship with the water fern (Azolla sp.). Anabaena is a filamentous, photosynthetic cyanobacterium containing heterocysts at intervals along a chain of vegetative cells. These heterocysts contain the requisite enzymes for di-nitrogen fixation. Azolla provides "safe harbour" for the N_2-fixing symbiont within cavities or nodules on the undersides of its leaves.

Another important and extremely widespread group of endophytes is the mycorrhyzal fungi. These fungi ramify throughout the root cells of plants and are hypothesized to function by aiding in the mobilization of relatively insoluble soil nutrients such as phosphate and zinc to the plant. Ascribing a generalized function to these endophytes is the object of considerable controversy. Endo-mycorrhyzae penetrate and ramify throughout certain root cells. Ecto-mycorrhyzae are those whose hyphae do not penetrate the root cells, but either envelop the roots and/or enter the space between root cells, usually outside the endodermis. The majority of higher plant taxa form associations with mycorrhyzal endophytes.

The Acremonium-type fungi are endophytes of certain types of grasses. They have been responsible for significant outbreaks of livestock toxicity in forage grasses, however, they are also responsible for conferring beneficial qualities to certain of their hosts. Enhanced performance, enhanced insect resistance and improved persistance of Lolium and Festuca species have been documented in turfgrass stands colonized by species of Acremonium (8).

And finally, there are the xylem-inhabiting bacteria. These bacteria are not ubiquitous, but certain species can reliably be recovered from their host plant species in a pattern which suggests that these bacteria may be part of the normal microflora of those hosts. Bacteria from 13 different genera were isolated from the xylem of healthy Citrus trees (9) which were colonized at levels ofup to 2 x 10^4 CFU/g. The presence in Citrus of large numbers of diverse endophytic bacteria has been confirmed in subsequent studies (Zablotowicz, R.M., Allelix, personal communication),although this phenomenon appears to be the exception and not the rule. Xylem inhabiting endophytic bacteria may be responsible for the frequently observed inability of plant tissue culturists to "disinfest" cultures of rigorously surface-sterilized explant material. The failure of certain species to grow in-vitro could in fact be related to a stimulatory effect of an endophytic microbe which cannot be reproduced in the culture of isolated, bacteria-free explants. It is these xylem-inhabiting endophytes which we are exploiting.

Crop Protection

Today's crop protection market was created and is presently dominated by the chemical industry. Biologicals account for only $100 million of this $14 billion market. Although synthetic chemicals presently dominate the market, the safety of these chemicals has been called into question. Biological pesticides are known to be safe but have lacked the efficacy of synthetic chemicals. The safety issue of chemicals and the efficacy problem of biological pesticides are consequences of their external application. Currently utilized methods for application of either chemicals or biologicals (e.g. foliar sprays, soil applications or seed treatments), all involve placing the treatment outside the plant even if the treatments themselves have systemic activity.

Without conventional pest control, food production in the U.S. alone could drop by as much as one-third. Despite the presence of chemical pesticides, more than one-third of the world's potential crop yield is still lost to fungal diseases and insect pests. Some chemical products leave toxic residues on the crop, leach into groundwater and are toxic to farm workers. While chemicals are effective against a broad spectrum of plant pests, they can destroy natural pest predators and other beneficial organisms. For many pesticides and for externally delivered biologicals, multiple applications are required because the products are diluted by rain, dissipated by wind and degraded by sunlight and microorganisms. The effectiveness of chemicals may decline over time due to the development of resistance by the target pests. There is much less risk of target pests developing resistance to endophyte-delivered agrichemicals because of the manner in which they are exposed and the vastly smaller quantities of active ingredient which thus need to be present in the crop. Finally, government regulations protecting consumers, workers and the environment are increasingly restricting the use of presently-registered chemical products and are making the introduction of new chemical products more difficult. The regulatory ground-rules for the introduction of biologicals are currently being written as numerous groups are in the process of attempting to field-test and register such products.

Endophytic Bacteria, A New Delivery System

InCide biopesticide technology is a system of crop protection in which biology is substituted for chemistry: Incide products are microorganisms genetically engineered to be both environmentally safe and efficacious. InCide products are designed to function internally in a plant's vascular system. They involve the use of naturally-occuring endophytes for the production and delivery of crop protectants and growth enhancers. Methods have been developed to screen, identify, recover and characterize endophytic microorganisms. CGI has conducted an extensive search for and analysis of endophytes capable of colonizing the major crops and now has a large collection of endophytes. Microorganisms are selected for the Company's collection based on ability to live inside, and inability to live outside, the target crops. CGI has indentified endophytes capable of colonizing corn, cotton, soybeans, wheat and rice as well as other major crops. InCide products are being developed to solve many of the problems associated with externally applied chemicals and biologicals. There are numerous economic, environmental and technological advantages to be realized by using endophytic bacteria for the delivery of agrichemicals to plants. Some of the advantages of this technology over conventional delivery of pesticides includes:

Economic Advantages.

- Single Application. Seed inoculation or inoculation of the juvenile plants with an appropriate endophyte results in colonization of that plant. The endophytes live inside the plant and are thus protected from the external environment.
  Externally applied chemicals and biologicals are unprotected and often require multiple applications.

- Minute Dosage. Seed application of endophytes can be accomplished with only milligram quantities of bacteria per acre. After application, the endophytes multiply inside each plant so that the final manufacturing step occurs after the point of sale.
  Conventional externally applied chemicals are generally applied in pounds per acre.

- Sustained Potency. Endophytes can thrive and produce the desired agrichemical for the duration of the plant's life.
  Externally applied products are adversely affected or rendered ineffective by the environment and by subsequent plant growth.

Environmental Advantages.

- o Contained Activity. Endophytes survive and function only within the plant which they protect or enhance. Externally applied products are dispersed widely each time they are applied.

- o Plant Dependency. Endophytes, by nature, do not surviveoutside the plant and therefore do not multiply or spread in the environment or remain active after harvest. Externally applied microorganisms must survive in the environment to be effective.

- o No Toxic Residues. Endophyte products can be designed to be environmentally safe and degradable. Most externally applied pesticides must resist degradation to be effective.

Technological Advantages. Certain endophyte products may have the following technological advantages over genetically improved plants:

- o Rapid Development. Changing the genetics of microorganisms is a more rapid and simpler process than changing the genetics of plants.

- o Early Commercialization. Development of endophyte-based products does not require multi-year plant breeding programs.

- o Wide Applicability. Endophyte-based products can be designed to function in a wide range of commercially useful varieties of the targeted crop.

- o Yield. Endophyte delivery has minimal impact on host plant physiology (e.g. little or no effect on yield, vigour or quality). In many instances, changing the genetics of plants can reduce yield substantially.

- o Repeat Sales. Many endophytes are not seed transmitted and farmers will need to purchase products based on these organisms each growing season. New plant varieties normally have repeat sales opportunities only when the purchased seed is a hybrid.

Progress is underway to develop the InCide delivery system for the delivery of insecticides, fungicides, nematicides, viricides and bactericides to protect corn, cotton, soybeans, rice and wheat, as well as other vegetable, forestry, and horticultural crops. We are developing a family of genetically engineered biopesticides using endophytes that colonize the major crops and genes that protect against major plant pests.

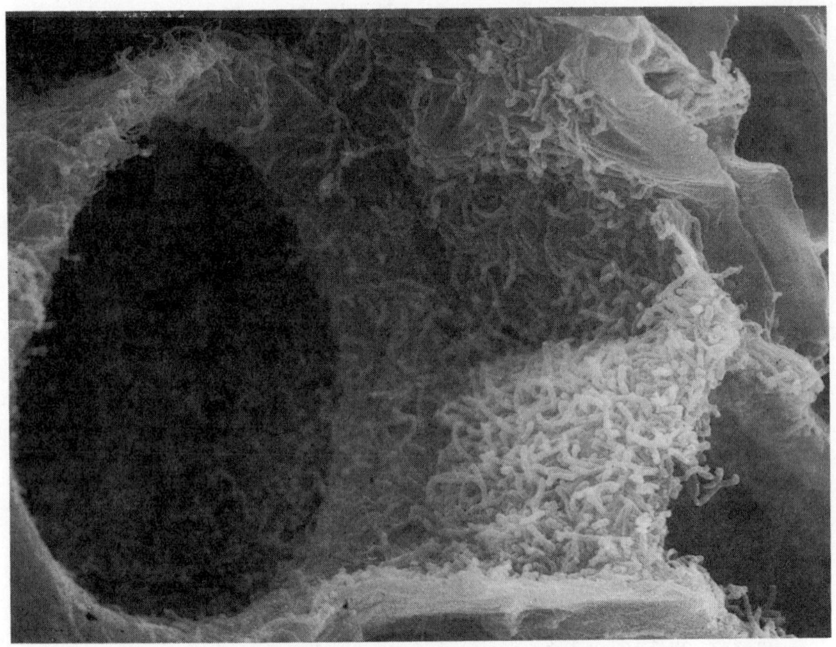

Figure 1. Lumen of xylem aerenchyma of _Zea mays_ var. FR632 colonized with endophytic bacterium _Clavibacter xylii_ subsp. _cynodontis_. Freeze-fracture surface reveals annular rings on left of plate. Magnification = 4250 X

The first InCide product under development is an insecticide for the United States and French corn market. We are conducting extensive field and greenhouse trials with our best corn endophyte. These tests are designed to accumulate data for product registration in the United States and France. We are seeking regulatory approval to field test our first recombinant corn insecticide product in the United States and France in the spring of 1988. This product uses the endophyte, Clavibacter xyli subsp. cynodontis, a Coryneform bacterium which was originally isolated from a non-crop plant and can be reproducibly introduced into corn (Fig. 1). This bacterium has been transformed by inserting a gene encoding for production of an insecticidal protein from the bacterium Bacillus thuringiensis (B.t.). B.t. has experienced decades of safe use as an insecticidal product. B.t. was first sold in France in 1939 and has been a leading biological insecticide in the U.S. since registration in 1961. There are more than one thousand B.t. isolates, each with one or more genes having a specific spectrum of insecticidal activity. Some B.t. genes are active against certain lepidoptera (caterpillars) and others are toxic to certain coleoptera (beetles). For its initial products, the Company purchased the rights to a B.t. gene effective against the caterpillar stage of European Corn Borer. Insect feeding trials have shown that the toxin produced by the B.t. gene is lethal to the European Corn Borer.

The wild-type endophyte rapidly colonizes the xylem of inoculated corn plants and achieves average levels of up to 1 x 10^8 CFU/g fresh weight of tissue. It systemically colonizes the roots, stem, leaves and husks of inoculated plants and can be detected within a week of inoculation but it does not transmit via the seed of colonized plants.

Endophytes can be found throughout the plant kingdom. Natural endophytes exist which provide manifold benefits to the plants with which they are associated. We are using the tools of biotechnology to add specific beneficial qualities to carefully chosen endophytes. By selective enhancement, endophytes can be engineered to help solve some of agriculture's most pressing problems using a biological system which has been around for millions of years.

Literature Cited

1. Burris, R.H. In Free-living Systems and Chemical Models. Nitrogen Fixation; Newton, W.E., Orme-Johnson, W.H. Eds.; University Park Press: Baltimore, 1980; Vol. 1, pp 7-16.
2. Moiroud, A. & Gianinazzi-Pearson, V. In Genes Involved in Microbe-Plant Interactions; Verma, D.P.S. & Hohn, T., Eds.; Springer-Verlag: New York, 1984; pp 205-223.
3. Becking, J. In Dinitrogen Fixation; Hardy, R., Silver, W. Eds.; Wiley & Sons Inc.: New York, 1977; Vol. 2, pp 185-275.
4. Legocki, R.P. & Szalay, A.A. In Genes Involved in Microbe-Plant Interactions; Verma, D.P.S. & Hohn, T., Eds.; Springer-Verlag: New York, 1984; pp 225-268
5. Lersten, N.R. & Horner, H.T. Bot. Rev. 1976, 42, 146-214.

6. Lersten, N.R. & Horner, H.T. J. Bacteriol. 1967, 94(6), 2027-2036.
7. Schaede, R. Jarhb. f. wiss. Botanik. 1939, 88, 1-21.
8. Funk, C.R., Halisky, P.M., Ahmad, S. & Hurley, R.H. In Proc. 33rd Short Course for Seedsman, Seed Technology Lab, Mississippi State University, 1985, pp 85-97.
9. Gardner, J.M., Feldman, A.W. & Zablotowicz, R.M. Appl. Environ. Microbiol. 1982, 43(6), 1335-1342.

RECEIVED April 12, 1988

NATURAL PRODUCTS FROM HIGHER PLANTS

Chapter 10

Phenol Glycosides in Plant Defense Against Herbivores

Paul B. Reichardt[1], Thomas P. Clausen[1], and John P. Bryant[2]

[1]Department of Chemistry, University of Alaska, Fairbanks, AK 99775–0520
[2]Institute of Arctic Biology, University of Alaska, Fairbanks, AK 99775–0180

Phenol glycosides are commonly found as plant metabolites, but little is known about their functional significance in the plants which produce them. We have found that several phenol glycosides are important components of the chemical defenses of two woody plants, *Populus balsamifera* and *P. tremuloides*. In these cases the phenol glycosides contribute to plant defense by converting to active defensive chemicals in damaged plant tissues. These findings suggest that phenol glycosides may play similar roles in many plants but that their contributions to defense have gone undetected because of their indirect involvement. Consideration of the phenol glycosides found in agricultural plants as potential precursors of defensive metabolites could lead to a new appreciation of their roles in crop resistance to pests.

Phenol glycosides are among the most common and wide-spread of all plant metabolites. From their first isolation from plants by Piria in 1845 (1), they have occupied a prominent place in phytochemical investigations. Several hundred of these substances are now known (e.g. 2), ranging in structure from very simple to very complex (Figure 1).

Even though the term "phenol glycoside" has been used by chemists for over a century, there remains some ambiguity over its exact meaning. Some authors have used the term to refer to any natural product having a structure which includes a phenolic residue bonded to a carbohydrate while others have excluded any compounds which can be classified as flavonoids. In this paper we will employ the latter, more restrictive, definition.

0097–6156/88/0380–0130$06.00/0

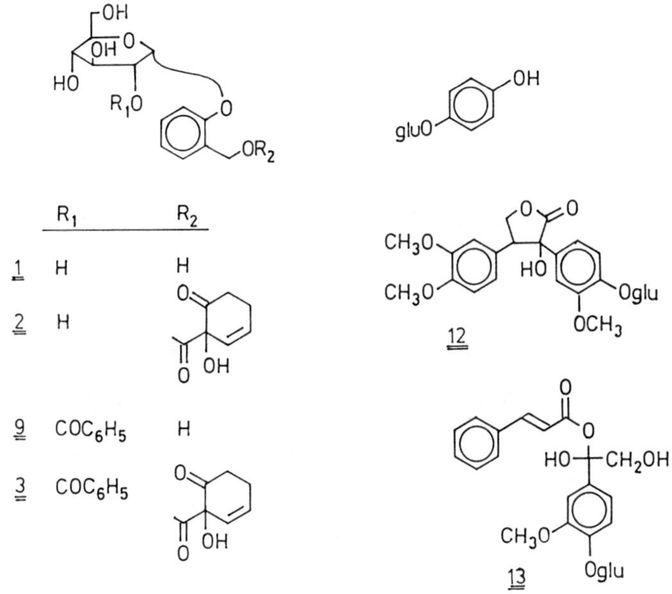

Figure 1. Examples of phenol glycosides produced by plants.

Chemistry

Perhaps phenol glycosides have attracted the attention of chemists primarily because of the challenges involved in their characterization. Many of them are labile substances which are difficult to isolate and purify. For many years isolation of phenol glycosides was accomplished by tedious gravity column chromatography on polyamide (3) or gel filtration (4) supports followed by difficult recrystallizations. Often time-consuming pretreatment of extracts (countercurrent extraction and lead subacetate treatments) were required prior to chromatography, and these pretreatments commonly resulted in the isolation of artifacts (5). The recent development of flash chromatography (6) has dramatically simplified isolation procedures (7), and modern chromatography coupled with enzymatic transformations offers promise for further simplification of the isolation protocols (Clausen, T.P., University of Alaska Fairbanks, unpublished data).

Once purified, phenol glycosides remain reluctant to reveal their chemical structures. Classically, structure elucidation has rested upon chemical or enzymatic hydrolysis followed by separate structural determinations of the aglycone ("genin") and carbohydrate. Repetition of the procedure employing a properly derivatized phenol glycoside is then usually required to locate the site at which the glycoside is linked to the genin (8). Recent advances in spectroscopy have, however, presented more direct methods for structure elucidation. Molecular formulas and fragmentation patterns can now be obtained from high resolution fast atom bombardment and field desorption mass spectral analyses (9) of these non-volatile compounds. Atomic connectivities and, in some cases, complete structures can be determined by analysis of ^{1}H- and ^{13}C-NMR spectra (7), especially with the aid of 2D NMR techniques (10).

Similarly, analysis of phenol glycoside mixtures has been dramatically facilitated by modern chromatographic techniques. The older methods of paper or thin layer chromatography coupled with spray reagents for detection (11) have been replaced by gas chromatographic analyses of derivatized phenol glycosides (12) or direct chromatographic analyses of phenol glycosides by high performance liquid chromatography (13). The latter method is particularly effective when a diode array detector is employed because the UV/Vis spectra obtained can be used to substantiate structural assignments for peaks otherwise based solely on retention times (14).

Biosynthesis

It appears that the most common biosynthetic route to phenol glycosides involves a final step which couples the preformed genin (usually a shikimate) and carbohydrate. The final step is catalyzed by a transferase which usually also has hydrolytic ("glycosidase") activity (15).

Biological properties of phenol glycosides

Two general biological properties seem to be associated with phenol glycosides: 1) they have a bitter taste, at least to humans (16) and 2) many phenol glycosides or their aglycones are toxic to a variety of organisms (15,17). Other specialized properties ascribed to individual phenol glycosides include enzymatic inhibition (18), phytoalexic properties (19), and regulation of plant growth (15,20,21).

Agricultural relevance of phenol glycosides

Phenolic glycosides have been found in a wide variety of edible plants - both wild and cultivated - as well as in many forage crops (22,23). Interest in the agriculturally-relevant phenol glycosides has centered on their generally bitter taste (16), a property which renders them undesirable constituents of foodstuffs. However, it is clear that bitterness is not a uniform trait of dietary phenol glycosides, as strikingly exemplified by the facile conversion of naringin (a bitter principle of grapefruit) to the sweet-tasting isomer, naringin chalcone (24).

Ecological relevance of phenol glycosides

Glycosylation as the most common final step in the biosynthesis of most phenol glycosides and the biological activities ascribed to many phenol glycosides obviously raise the question of their *raison d'etre* (especially *vis a vis* the phenolic genins). Certainly this question has not lacked for answers. Over one hundred years ago Errara (25) first answered this question by proposing that phenol glycosides protect plants from "the voracity of animals". However, the intervening years have seen the emergence of a number of competing proposals including: secondary food reserves (26), waste metabolites (27), detoxified derivatives of phytotoxic aglycones (15), protectors of phenols from oxidation (28), and--in some cases--plant growth regulators (20,21).

In this paper we wish to return to Errara's's initial suggestion and evaluate the current status of his proposal. Compared to the often accepted maxim of phenol glycosides constituting a common mode of chemical defense, hard data to support this contention are sparse.

There is some evidence to suggest that phenol glycosides constitute a plant defense against polyphagous (generalist) and some oligophagous herbivores, but it is largely correlative in nature. Markham (16) and Edwards (29) demonstrated the aversion of opossums to *Populus* and *Salix* species containing relatively high levels of salicin (**1**) and its derivatives and ascribed the protection to the bitter taste of phenol glycosides. Tahavanainen et al. (30) likewise discovered a negative relationship between palatabilities of *Salix* species to the mountain hare (*Lepus timidus*) and phenol glycoside content. They additionally demonstrated the unpalatabilities of phenol glycoside-containing extracts of Salix to L. timidus. Lindroth et al. have demonstrated the negative effects of phenol glycosides from Populus tremuloides on the performance of Papillo glaucus larvae (31) and have ascribed the effect to the levels of salicortin (2) and tremulacin (3) in the leaves (32). Zucker (33) has tenuously suggested that phenol glycosides of Populus angustifolia leaves inhibit the successful colonization of this plant by a gall-forming aphid (Pemphigus betae).

Some oligophagous and monophagous (specialist) insect herbivores, however, apparently utilize phenol glycosides as positive cues for feeding. Tahavanainen et al. (34) have reported that four leaf beetle species select their favored host plants (Salix sp.) based upon the plant's suite of phenol glycosides. When the favored host plant was removed in feeding trials, insects shifted to the Salix species having a phenol glycoside content most like the preferred host.

A very revealing study of exploitation of phenol glycosides by herbivores has been reported in a series of papers by Rowell-Rahier and Pasteels (35-41). They discovered several oligophagous chrysomelid beetles which prefer Salix hosts rich in salicin (37,40) and one beetle which utilizes a salicortin-containing Salix species (35). In these cases the phenol glycoside does not serve as a feeding cue; but it is metabolized by the beetle to produce salicaldehyde (4), a defensive chemical utilized by the insect, and glucose, which serves as a significant energy source (41). Smiley et al. (42) have extended this work to a North American ecosystem. They found that the larvae of the Californian Sierra Nevada beetle (Chrysomela aenicollis) prefer Salix leaves which have high salicin levels and that larvae placed on Salix leaves with high salicin content have a higher survival rate than larvae placed on leaves low in salicin.

In summary, we may say that the few reports on phenol glycosides as mediators of plant/herbivore interactions suggest that Errara's's proposal has some merit but that its generality remains to be determined. Furthermore, beyond the work of Rowell-Rahier and Pasteels there is a general lack of information on the chemistry behind the observed responses of herbivores to phenol glycosides.

Phenol glycosides and defense of Alaskan woody plants

Over the past several years we have investigated the roles of phenol glycosides in two woody plant/herbivore interactions which have significant implications in high-latitude ecosystems of North America. In each case we were drawn to study phenol glycosides by results obtained from bioassay experiments, and we have been able to define the ecological roles of these compounds only by paying attention to their detailed structures and the dynamic nature of their existence in plants.

Our initial encounter came about during attempts to chemically define the reasons for the selective use of balsam poplar (Populus balsamifera) parts and growth stages by the snowshoe hare (Lepus americanus). Field observations (Bryant, J.P., University of Alaska Fairbanks, unpublished data) indicated that internodes (stems between buds) and bark are the only parts of winter-dormant balsam poplar eaten by hares. Furthermore, only older internodes from mature saplings are utilized; all internodes from juvenile plants and current-year growth internodes from mature plants are ignored by the hares.

Guided by palatability bioassay results from extracts of poplar, we tentatively concluded that ether soluble metabolites in the internodes (stems between buds) were responsible for hares' selective use of this potential food source. Subsequently we found trichocarpigenin (benzyl gentisate, 5) to be a major component of this extract, as had been previously reported by Pearl and Darling (43). We were encouraged when we found this substance, which has a peppery flavor and causes numbness to the mouth and gums when tasted, to be a deterrent to hare feeding. However, a problem developed when we discovered that hares readily consume poplar internodes which our analyses showed contained concentrations of 5 well above those which caused complete aversion in the bioassays (Mattes, B.R., University of Alaska Fairbanks, unpublished results). More careful investigation revealed that little, if any, free 5 occurs in the plant but that it is produced from trichcocarpin (6) during extraction of fresh plant material with diethyl ether. Further experiments revealed that this hydrolysis during extraction is apparently catalyzed by a plant-contained enzyme which is presumably segregated from the apparently palatable substrate (6) but is released during the disruptive events associated with the extraction process (44). We concluded, however, that this potential latent defense of poplar to hares is ineffective due to the relatively slow rate of reaction experienced in plant material during herbivory.

Further investigation of the lipids and phenol glycosides of P. balsamifera provided an even more interesting and ecologically relevant aspect of phenol glycoside biochemistry of this plant. Among the ether-soluble metabolites of balsam poplar internodes, we discovered 6-hydroxycyclohexenone (7). A second major phenol glycoside proved to be salicortin (2) (44). Based upon structural considerations and the report by Pearl and Darling (45) that 2 produces catechol (8) upon basic hydrolysis, we investigated the biochemical relationship between 2 and 7. We were gratified to find that an enzyme preparation from poplar internodes readily converts 2 to 7 and 1. In this case, we found that the unpalatable product (7) is not an artifact. Thus the phenol glycoside (2) is a biosynthetic precursor of a defensive chemical as well as a latent additional source of the deterrent which is produced from 2 at some unknown rate during herbivory. Furthermore, the other product from this reaction (1) could well serve as the biosynthetic precursor of another defensive metabolite (4) of P. balsamifera (Clausen, T.P., University of Alaska Fairbanks, unpublished results).

At this point our work on the defensive chemistry of P. balsamifera paid dividends in our parallel study of the large

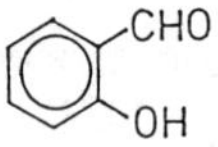

CHO
OH
OH
O
O
OR

4

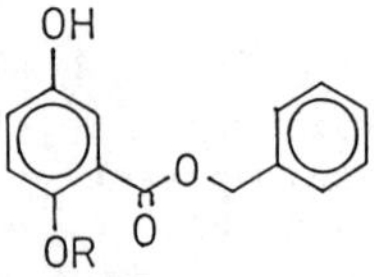

5 R = H
6 R = glu

aspen tortrix (Choristoneura conflictana (Walker))/quaking aspen (P. tremuloides) interaction. In this study we had set out to examine the reasons behind the population cycles of this moth specialist. Our hypothesis was that herbivory-induced changes in host plant chemistry causes a drop in dietary quality of the leaves, leading to a precipitous decline in moth populations in years following severe defoliation of the plants.

Our discoveries that the major phenol glycosides of quaking aspen foliage were salicin (1), salicortin (2), tremulacin (3), and tremuloiden (9) and that the same leaves contained 7 led us to evaluate the importance of their biochemical relationships to the suitability of quaking aspen foliage as food for the large aspen tortrix larvae. Initially we demonstrated that phenol glycoside-containing fractions from aspen leaves reduced the performance of C. conflictana larvae (46). Subsequently we discovered that all phenol glycosides except 9 adversely affect larval performance but that 2 and 3 are particularly detrimental in this regard (Bryant, J.P., University of Alaska Fairbanks, unpublished results). Furthermore, disruption of aspen leaf tissue leads to the production of 7 from both 2 and 3. The product of these transformations apparently reduces leaf quality to the larvae by disrupting the insects' metabolism due to its reactivity as an electrophile or after its facile conversion to phenol or catechol (Figure 2).

Experiments with intact aspen saplings revealed another facet of the role phenol glycosides play in the plant's response to leaf damage. Perforation or clipping the edges of leaves, as might occur during insect attack, can elicit the translocation of 2 and 3 from internodes to leaves. Thus the plant's response to leaf damage is at least two-fold: 1) conversion of two phenol glycosides to a defensive chemical and 2) replacement and enhancement of phenol glycoside reserves in damaged leaves by translocation (Clausen, T.P., University of Alaska Fairbanks, unpublished results).

It is interesting to consider these findings in light of the results of two other studies. Among the insect-mediated transformations described by Rowell-Rahier and Pasteels is the conversion of 2 to 4 and glucose by P. vitellinae (36). Based upon this information, and our findings, one can surmise that 7 is also produced during these transformations. What would be the effect of its production on the insect? The two options seem to be that 1) its adverse effects are the price the insect pays for metabolic energy and defensive chemical (4) or that 2) the insect incorporates 7 into its defensive secretion, thus making three-fold use of dietary 2.

The relevance of our findings to the P. glauca/P. tremuloides system studied by Lindroth et al. (31) seems more clearcut. They report that of the four phenol glycosides (1, 2, 3, and 9) found in P. tremuloides foliage, only two (2 and 3) cause decreases in performance of P. glauca larvae (32). It is highly likely that the biochemical transformations of 2 and 3 to 7 described here explain this observation.

A proposal for the defensive significance of phenol glycosides in plants

If there is any generality to a defensive role of phenol glycosides in plants, it appears to be in their ability to function as mobilizable defenses. Glycosylation of phenolic metabolites offers two advantages to biological systems: 1) generally, although not always, it increases the water solubility of the metabolite (15) and 2) it attenuates the toxic (15,17) and destructive properties (15) of the free phenol. Thus the phenol glycosides can readily be translocated by plants and enzymatically converted to defensive substances at the site of attack. While this proposal is based upon data from only three phenol glycosides (2, 3, and 6), one can envision at least one other mechanistic scenario, exemplified by salicin (1) - perhaps the most commonly found phenol glycoside in plants. Biochemical conversion of 1 to 10 (with R = PO_3H^{-1}, for example) could easily lead to 11 as depicted in Figure 3. This product would undoubtedly be an excellent electrophile (Figure 3; 47-49) and thus interfere with herbivore metabolism in a manner similar to that ascribed to sesquiterpene lactones (50) and other Michael acceptors (49). The potential importance of the process depicted in Figure 3 goes well beyond its application to salicin in that many of the known phenol glycosides contain part structures (e.g. 12 and 13) which could serve as precursors to analogues of 11 (quinonemethides; 48).

An appreciation of the practical consequences of the biochemical lability of phenol glycosides allows one to consider their roles in plant defense from a new perspective. We can move from the classical approach of correlating herbivore behavior with chemical constituents of plants to one which is based, at least in part, upon a fundamental understanding of the chemistry behind the interactions. The potential payoff of this approach is that it has *predictive* as well as retrospective properties. Thus we believe that Errara's's (25) old proposal that phenol glycosides provide plants with defense from herbivores can only be fully evaluated by studies which are based upon detailed considerations of phenol glycoside structure and elucidation of biochemical transformations.

The application of such an approach to agricultural plants could have a dramatic effect upon the present view that phenol glycosides are generally undesirable constituents of crops. Consider, for example, the potential ecological roles of two phenol glycosides, vicine and convicine, in fava beans (*Vicia faba*). These two substances rapidly undergo both enzymatic and chemical hydrolysis to β-D-glucose and aglycones - divicine and isovamil, respectively. The latter two substances are potent reducing agents (51,52) and cause declines in glutathione and ATP levels in human blood cells (53). Their roles in the acute human hemolytic anemia known as favism have been extensively investigated (52,54,55), and they are known to inhibit the growth of certain plant pathogens

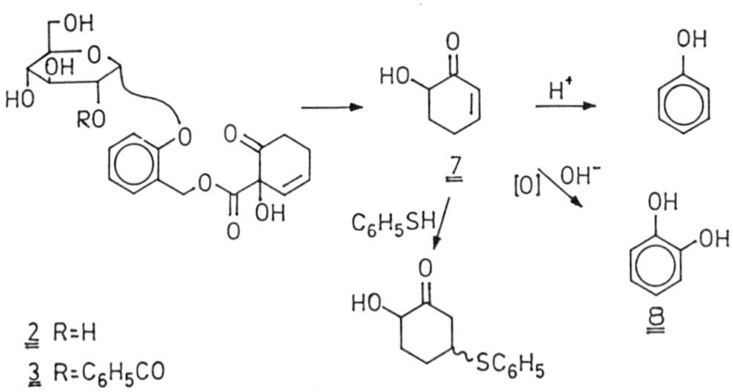

Figure 2. Ecologically relevant aspects of the chemistry and biochemistry of 6-hydroxycyclohexenone.

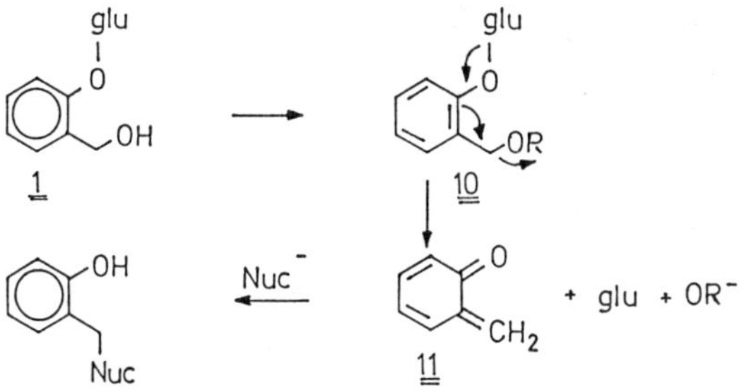

Figure 3. Hypothetical pathway for biochemical mobilization of salicin for defensive purposes.

(56). However, the generality of the roles of these four compounds in the defense of Vicia faba is unknown.

As agricultural practice moves away from use of "hard" pesticides, the efficacies of intrinsic defensive systems of crops will become more important. If phenol glycosides prove to be significant components of the chemical defenses of cultivated plants (as they appear to be in at least some wild plants), no longer will they be viewed as totally undesirable constituents. In fact, they could become important ingredients of integrated pest management systems.

Acknowledgments

The experimental work described here has been supported by the National Science Foundation (BSR8416461 and BSR8500160).

Literature Cited

1. Armstrong, E. F.; Armstrong, K. F. The Glycosides; Longmans: London, 1931; p 12.
2. Devon, T. K.; Scott, A. I. Handbook of Naturally Occurring Compounds; Academic Press: New York, 1975; Vol. 1.
3. Estes, T. K.; Pearl, I. A. Tappai 1967, 50, 318-323.
4. Repas, A.; Nikolin, B.; Dursun, K. J. Chromatog. 1969, 44, 184-187.
5. Steele, J. W.; Bolan, M.; Audette, R. C. S. J. Chromatog. 1969, 40, 370-376.
6. Still, W. C.; Kahn, W.; Mitra, A. J. Org. Chem. 1978, 43, 2923-2925.
7. Lindroth, R. L.; Hsia, M. T. S.; Scriber, J. M. Biochem. Syst. Ecol. 1987, 15, 677-680.
8. Pridham, J. B. In Adv. Carbohydrate Chem.;. Wolfram, M. L.; Tipson, R. S., Eds.; Academic Press: NY, 1965; Vol. 20, pp 371-408.
9. Howe, I.; Jordan, M.; In Fortshritte der Chemie organischer Naturstoffe; Herz, W.; Grisebach, H.; Kirby, G. W.; Tamm. Ch., Eds.; Springer-Verlag: NY, 1985; Vol. 47, pp 107-152.
10. Benn, R.; Gunther, H. Agnew. Chem. Int. Ed. 1983, 22, 350-380.
11. Audette, R. C. S.; Blunden, G.; Steele, J. W.; Wong, C. S. C. J. Chromatog. 1966, 25, 367-372.
12. Julkunen-Tiitto, R. J. Agric. Food Chem. 1985, 33, 213-217.
13. Meier, B.; Sticher, O.; Bettschart, A. Deutsche Apotheker Zeitung 1985, 125, 341-347.
14. Meier, B.; Lehmann, D.; Sticher, O.; Bettschart. A. Pharm. Acta Helv. 1985, 60, 269-275.
15. Hopkinson, S. M. Quart. Rev. (London) 1969, 23, 98-124.
16. Markham, K. R. N. Z. J. Sci. 1971, 14, 179-186.
17. Vickery, M. L.; Vickery, B. Secondary Plant Metabolism; Macmillan: Baltimore, MD, 1981.

18. Akinrimis, E. O.; Akinwanoe, A. I. West Afr. J. Pharmacol Drug Res. 1976, 3, 141-148.
19. Flores, G.; Hubbes, M. Eur. J. For. Pathol. 1980, 10, 95-103.
20. Kefeli, V. I.; Turetskaya, R. Kh.; Kof, E. M.; Kutucek, M.; Protasova, N. M.; Sidorov, K. K.; Lozhnikova, V. N.; Khlopenkova, L. P.; Yakovleva, L. V.; Chumakovskii, N. N. Fenol'yne. Soedin. Ikh Fiziol. Svoistva Mater. Vses. Simp. Fenol'nym Soedin. 2nd 1971 1973, 41-45. Chem. Abstr. 1974, 81, 148598u.
21. Goris, A. Rev. Gen. Sci. Pure Appl. 1921, 32, 337-342.
22. Meyer, L. H. Food Chemistry; Reinhold Publishing Co.: New York, 1960; pp 218-292.
23. Hodge, J. E. In Principles of Food Science. Part 1. Food Chemistry; Fennema, O. R., Ed.; Marcel Dekker Inc.: New York and Basel, 1976; p 66.
24. Horowitz, R. M. In Biochemistry of Phenolic Compounds; Harborne, J. R., Ed.; Academic Press: New York, 1964. p 545.
25. Errara, L. Bull Soc. Roy. Bot. Belg. 1886, 25, 86.
26. Pfeffer, W. In Planzenphysiologie 2nd Ed.; Engelmann, W., Ed.; Leipzig, 1897; Vol 1, p 492.
27. Baruah, P.; Swain, T. J. Sci. Food. Agr. 1959, 10, 125-129.
28. Roberts, E. A. H.; Wood, D. J. Nature 1951, 167, 608.
29. Edwards, W. R. N. N. Z. J. Sci. 1978, 21, 103-106.
30. Tahavanainen, J.; Helle, E.; Julkunen-Tiitto, R.' Lavola, A. Oecologia (Berl.) 1985, 65, 319-323.
31. Lindroth, R. L.; Scriber, J. M.; Hsia, M. T. S. Oecologia (Berl.) 1986, 70, 13-19.
32. Lindroth, R. L.; Scriber, J. M.; Hsia, M. T. S. Abstr. 72nd Ann. Mtg. Ecol. Soc. Amer. 1987, p 352.
33. Zucker, W. V. Ecology 1982, 63, 972-981.
34. Tahavanainen, J.; Julkunen-Tiitto, R.; Kettunen, J. Oecologia (Berl.) 1985, 67, 52-55.
35. Rowell-Rahier, M.; Soetens, P.; Pasteels, J. M. In Insects-Plants; Labeyriie, V.; Lachaise, D.; Eds.; Dr. W. Junk Publishers: Dordrecht, 1987; pp 91-95.
36. Rowell-Rahier, M.; Pasteels, J. M. Proc. 5th Int. Symp. Insect-Plant Relationships; Wageningen, Pudoc, 1982; pp 73-79.
37. Rowell-Rahier, M. Oecologia (Berl.) 1984, 64, 375-380.
38. Rowell-Rahier, M. Oecologia (Berl.) 1984, 62, 26-30.
39. Pasteels, J. M.; Gregoire, J.-C. J. Chem. Ecol. 1984, 10, 1693-1700.
40. Pasteels, J. M.; Rowell-Rahier M.; Braekman, J. C.; Dupont, A. Physiol. Entomol. 1983, 8, 307-314.
41. Rowell-Rahier, M.; Pasteels, J. M. J. Chem. Ecol. 1986, 12, 1189-1203.
42. Smiley, J. T.; Horn, J. M.; Rank, N. E. Science 1985, 229, 649-651.
43. Pearl, I. A.; Darling, S. F. Phytochemistry 1969, 7, 1855-1860.

44. Mattes, B. R.; Clausen, T. P.; Reichardt, P. B. Phytochemistry 1987, 26, 1361-1366.
45. Pearl, I. A.; Darling, S. F. Tetrahedron Lett. 1970, 3827-3830.
46. Bryant, J. P.; Clausen, T. P.; Reichardt, P. B.; McCarthy, M. C.; Werner, R. A. Oecologia (Berl.) 1987, 73, 513-517.
47. Lin, A. J.; Cosby, L. A.; Sartorelli, A. C. Cancer Chemother. Rep. 4 1974, 23, 23-25.
48. Wagner, H.-U.; Gompper, R. In The Chemistry of the Quinonoid Compounds; Patai, S., Ed.; John Wright and Sons Ltd.: Bristol, 1974; pp 1145-1178.
49. Moore, H. W. Science 1977, 197, 527-532.
50. Kupchan, S. M. Trans. N. Y. Acad. Sci. 1970, 32, 85-106.
51. Bendich, A.; Clements, G. Biochim. Biophys. Acta 1953, 12, 462-477.
52. Albano, E.; Tomasi, A.; Mannuzzu, L.; Arese, P. Biochem. Pharmacol. 1984, 33, 1701-1704.
53. Mager, J.; Glaser, G.; Razin, A.; Izak, G.; Bien, S.; Noam, M. Biochem. Biophys. Res. Commun. 1965, 20, 235-240.
54. Mager, J.; Razin, A.; Hershko, A. In Toxic Constituents of Plant Food stuffs; Liener, I.E., Ed.; Academic Press: NY, 1969; pp 293-318.
55. DeFlora, A.; Benatti, V.; Guida, L.' Zocchi, E. Glucose-6-Phosphate [Lect. Int. Symp.] 1985, 153-177. Chem. Abstr. 1987, 106, 173816f.
56. Bjerg, B.; Heide, M.; Knudsen, J.C.N.; Soerensen, H. Z. Planzenkrankh. Pflananschute 1984, 91, 483-487.

RECEIVED May 6, 1988

Chapter 11

Biosynthetic Relationship Among Cyanogenic Glycosides, Glucosinolates, and Nitro Compounds

Eric E. Conn

Department of Biochemistry and Biophysics, University of California–Davis, Davis, CA 95616

Three different groups of natural products — cyanogenic glycosides, glucosinolates and nitro compounds — are known to be derived from amino acids. Studies indicate that the initial biosynthetic reactions for all three types of compounds are N-hydroxylation followed by oxidative decarboxylation of the N-hydroxyamino acid so formed to yield the corresponding aldoxime. The biosynthetic pathways then diverge at this common precursor to produce the different natural products.

This paper describes the biosynthetic relationship between cyanogenic glycosides, glucosinolates and nitro compounds. These well-known nitrogenous natural (secondary) plant products share the following properties: they exist in the plants as glycosides; their respective aglycones derive their nitrogen and most of their carbon atoms from protein amino acids; and they exert deleterious effects on animals that ingest the plant that contains the compounds. These effects range from acute toxicity often resulting in death to indirect effects caused by metabolites produced in the animal after ingestion.

Structures and Distribution

The cyanogenic glycosides, which have the general formula shown here

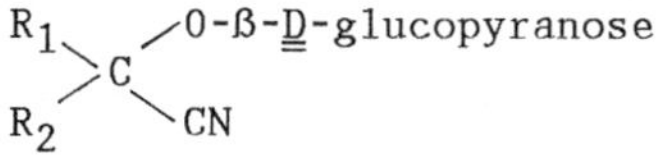

$$\begin{matrix} R_1 \searrow & & \nearrow O\text{-}\beta\text{-}\underline{\underline{D}}\text{-glucopyranose} \\ & C & \\ R_2 \nearrow & & \searrow CN \end{matrix}$$

have the widest distribution of the three types of compounds under discussion. On hydrolysis they yield a sugar, usually D-glucose, and an α-hydroxynitrile (cyanohydrin) which subsequently dissociates to form HCN and a carbonyl compound. The latter may be either aromatic or aliphatic (1). Approximately 2000 species of plants, including both gymnosperms and monocotyledenous and dicotyledenous

0097–6156/88/0380–0143$06.00/0

angiosperms representing more than 100 families, are known to be cyanogenic. The presumed physiological role of these compounds is that of defense against herbivores, and there is significant evidence to support this role (2,3).

The glucosinolates or mustard oil glycosides have the general formula shown here. Again the aglycone moiety may be either

$$R—C—S—glucose$$
$$\|$$
$$N—OSO_3^-$$

aliphatic or aromatic. These compounds, however, have a much more limited distribution being confined almost entirely to members of the *Cruciferae*.

Compounds bearing an aliphatic nitro group have been reported in only a few plants, mainly legumes (4). 3-Nitropropanol exists as a ß-glucoside in *Astragalus* where it has been given the name miserotoxin (5). 3-Nitropropanoic acid (NPA) is found esterfied to glucose and may exist as the mono-, di-, tri- and tetra esters (6). In addition to these aliphatic compounds, 1-nitro-2-phenylethane has been reported in two species of South American *Lauraceae* (7) and *Tropaeolum majus* (8). 1(4'-Hydroxyphenyl)-2-nitroethane (HPNE) has been isolated as its glucoside from *Thalictrum aquilegifolium* (9). These nitro derivatives therefore have the most restricted distribution of the three groups of compounds under discussion.

Biosynthesis of Cyanogenic Glycosides

The biosynthetic origin of the cyanogenic glycosides has been extensively investigated. Work primarily in our laboratory and in that of G. Butler in New Zealand initially demonstrated the precursor-product relationship shown in Figure 1 (1,10). In this conversion, it was early established that the carboxyl carbon atom of the precursor amino acid was lost, but the carbon-carbon bond between C_2 and C_3, as well as the carbon-nitrogen bond between C_2 and the amino nitrogen remained intact during biosynthesis. Thus all intermediates in the biosynthetic pathway had to be nitrogenous, and oxidations resulting in a nitrile carbon and an oxygenated carbon atom at C_3 must occur during biosynthesis.

A large body of evidence is available which support the biosynthetic pathway shown in Figure 2 (1,10,11). Initially, the amino nitrogen of the precursor amino acid is oxygenated to form the corresponding N-hydroxyamino acid in a reaction requiring NADPH. The next step involves conversion of the N-hydroxyamino acid to the corresponding aldoxime. The overall reaction is a 2-electron oxidative decarboxylation and neither the corresponding N-hydroxyamine nor keto acid are intermediates. Dehydration of the aldoxime, in a reaction requiring NADPH, yields the nitrile, and stereospecific oxygenation of the latter will produce the cyanohydrin.

These four reactions are catalyzed by a membrane-bound enzyme system which exhibits many of the properties (channeling, catalytic facilitation) of a multi-enzyme complex. The last step of the biosynthetic sequence, which is glucosylation of the cyanohydrin, is catalyzed by a soluble UDPG-glucosyl transferase. Membrane-bound

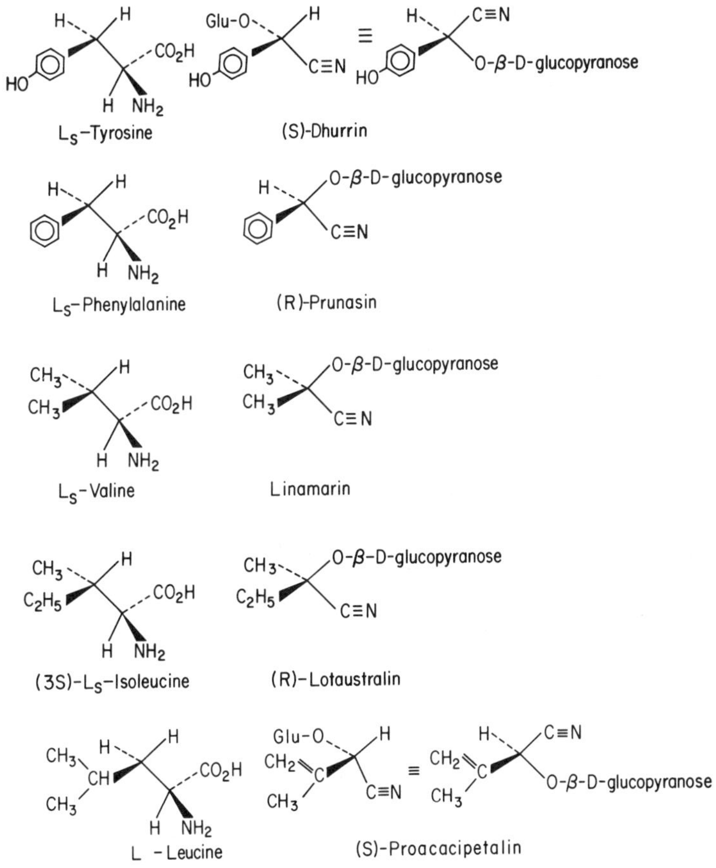

L_S–Tyrosine

(S)-Dhurrin

L_S–Phenylalanine

(R)-Prunasin

L_S-Valine

Linamarin

(3S)-L_S–Isoleucine

(R)–Lotaustralin

L -Leucine

(S)-Proacacipetalin

Figure 1. The precursor-product relationship between certain amino acids and cyanogenic glycosides. (Reproduced with permission from Fed. Proc. 1982 41, 2639. Copyright 1982 Fed. Amer. Soc. Exptl. Biology)

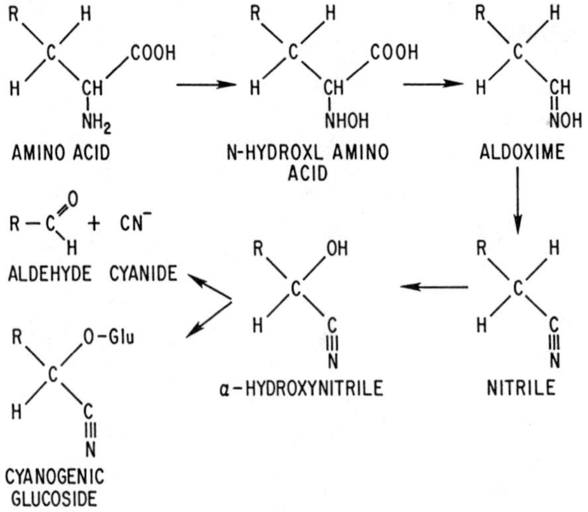

AMINO ACID → N-HYDROXL AMINO ACID → ALDOXIME → NITRILE → α-HYDROXYNITRILE → ALDEHYDE + CYANIDE ($R-CHO + CN^-$); α-HYDROXYNITRILE → CYANOGENIC GLUCOSIDE

Figure 2. The biosynthetic pathway for cyanogenic glycosides. (Reproduced with permission from Ref. 10. Copyright 1979 Springer.)

enzyme systems have been described for *Sorghum bicolor* (12), *Triglochin maritima* (13), *Linum usitatissimun* (14), and *Trifolium repens* (15). The pathway is unusual in that it involves nitrogenous compounds (N-hydroxyamino acids, aldoximes, and nitriles) not normally encountered in amino acid metabolism.

Biosynthesis of Glucosinolates

A close relationship between cyanogenic glucosides and glucosinolates was established when E.W. Underhill, L. Wetter and their associates showed that aldoximes are intermediates in the biosynthesis of glucosinolates. There is extensive literature which supports the role of the amino acid, N-hydroxyamino acid, aldoxime, thiohydroximic acid and desulfoglucosinolate in the biosynthetic pathway shown in Figure 3 (16,17). The other compounds postulated as intermediates have not been tested primarily because of their instability. The ketoxime, which is tautomeric with the α-nitroso acid, is not an effective precursor of the glucosinolate (16). In the case of cyanogenic glycosides, the ketoxime is also considered not to be an intermediate because the α-hydrogen atom is retained when the precursor amino acid is oxidatively decarboxylated to form the aldoxime (B.L. Møller, unpublished).

Biosynthesis of Nitro Compounds

Fungal biosynthesis. The biosynthesis of 2-nitropropanoic acid (NPA) has been the subject of numerous studies. The earliest papers were concerned with the origin of the nitro group with some investigators favoring an oxidized, inorganic nitrogen acid (HNO_3 or HNO_2) as the donor. Others postulated an organic nitrogen compound as the precursor (reviewed in 18) and in the specific case of NPA, evidence supporting aspartic acid was acquired (19). After nearly two decades during which little work was done, Baxter and associates have recently shown that the carbon (C_2)-nitrogen bond in aspartic acid is conserved during the biosynthesis of NPA from that amino acid in *Penicillium atrovenetum* (20). More recently, the same laboratory has demonstrated that both oxygen atoms in the nitro group of NPA arise from molecular O_2, but did not speculate on the nature of any possible intermediates in the fungal biosynthesis (21).

Biosynthesis in plant cell cultures. 1(4'-Hydroxyphenyl)-2-nitroethane (HPNE) has recently been observed in cell suspension cultures of California poppy (*Eschscholtzia californica*) which have been osmotically stressed (22). These experiments were conducted to see whether such stressed cells might produce cyanogenic glycosides because Berlin and associates some time earlier had demonstrated the production of isoquinoline alkaloids in stressed cells of *E. californica* (23). Since this plant is also highly cyanogenic, it was hoped that this species might also produce a cyanogenic substance under stress.

Cells were grown to stationary phase and then placed in 0.6 M sorbitol, the same osmoticum used by Berlin *et al.* to induce the formation of alkaloids. After periods ranging from one to 11 days, cells were harvested, lysed and almond emulsin was added to

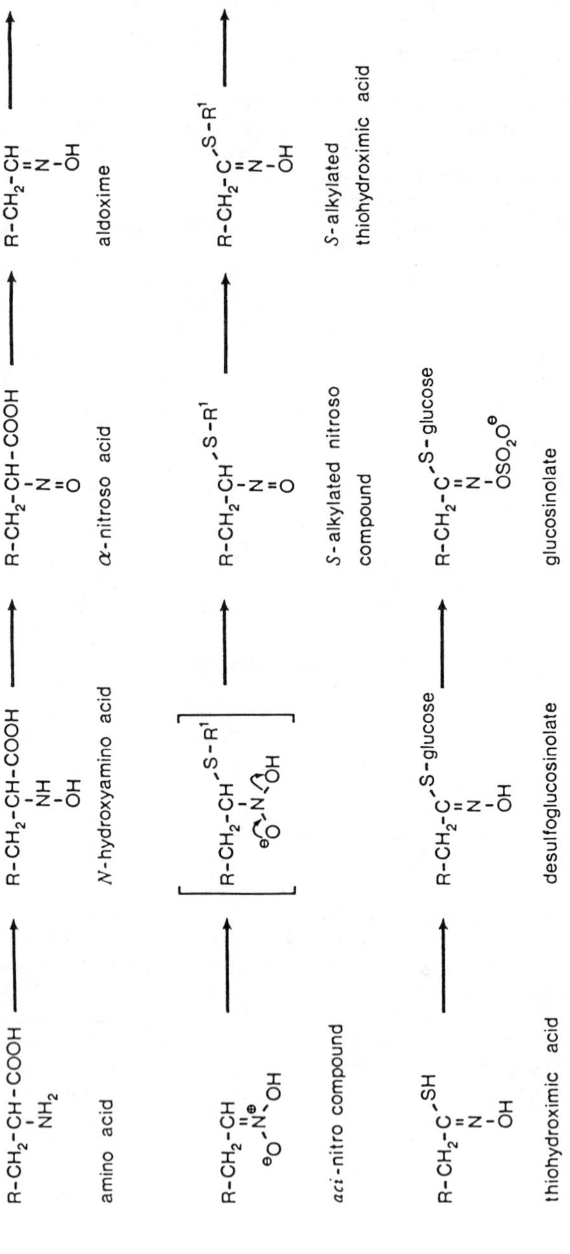

Figure 3. Proposed biosynthetic pathway for glucosinolates. (Reproduced with permission from Ref. 17. Copyright 1981 Academic.)

hydrolyze any cyanogenic glycoside that might be present. (Differentiated poppy plants contain dhurrin and triglochinin; these are cyanogenic glycosides derived from tyrosine.) As shown in Figure 4, HCN was released from a bound form and could be quantified although the amount formed was less than 1% of that occurring in differentiated plant tissues. The amount of HCN formed by stressed cells was dependent on the concentration of the osmoticum but independent of the nature of the osmotic agent.

In an effort to identify the source of the HCN released by emulsin treatment, microsomes were isolated from stressed cells and incubated with L-(U-^{14}C)-tyrosine and NADPH. This action was taken when L-(U-^{14}C)-tyrosine was administered to the stressed cells and failed to produce any labelled dhurrin or triglochinin. It was felt that microsomal preparations, enriched in the biosynthetic enzymes, might allow detection of cyanogens. When extracts of such incubations were processed and analyzed by HPLC, small amounts of radioactivity were detected in 4-hydroxyphenylacetonitrile, an intermediate in dhurrin biosynthesis, but much more was located in an unknown that was subsequently identified as HPNE (Figure 5). In the presence of UDP-glucose, the HPNE was glucosylated on the aromatic hydroxyl group to form the corresponding glucoside. The microsomal synthesis was strongly dependent on NADPH; NADH was less than half effective. Such preparations will not utilize other amino acids (e.g. aspartic) but will utilize 4-hydroxyphenylacetaldoxime and NADPH as substrates and to form HPNE. Unstressed cells failed to yield active microsomal preparations.

This finding raises the question of whether HPNE occurs naturally in differentiated poppy plants. Such has never been reported but some preliminary evidence suggests that such a compound is present at very low concentration in flowers of California poppy. On the other hand, HPNE has been isolated from *Thalictrum aquilegifolium*, a cyanogenic species that contains cyanogens derived from tyrosine (9).

Summary

Figure 6 summarizes the central role of aldoximes in the biosynthesis of cyanogenic glycosides, glucosinolates and nitro compounds. The evidence supporting aldoximes (together with N-hydroxyamino acids and nitriles) as intermediates in the formation of cyanogenic glycosides is extensive, both at the level of the intact plant and in cell free enzymatic studies (1). The role of the aldoxime as an intermediate in glucosinolate biosynthesis is also documented, although much remains to be learned about other intermediates and the enzymes involved (16,17). The role of 4-hydroxyphenylacetaldoxime in the formation of HPNE by microsomes isolated from stressed cell suspension cultures of California poppy is also clear, although the natural occurrence of HPNE in this plant has not been observed (22).

These three groups of natural products have in common an oxidized nitrogen atom that is derived by the oxidation of the nitrogen of an amino acid. The corresponding N-hydroxyamino acid is established as a precursor in the case of cyanogenic glycosides and glucosinolates. Oxygenation of the aldoxime nitrogen would yield

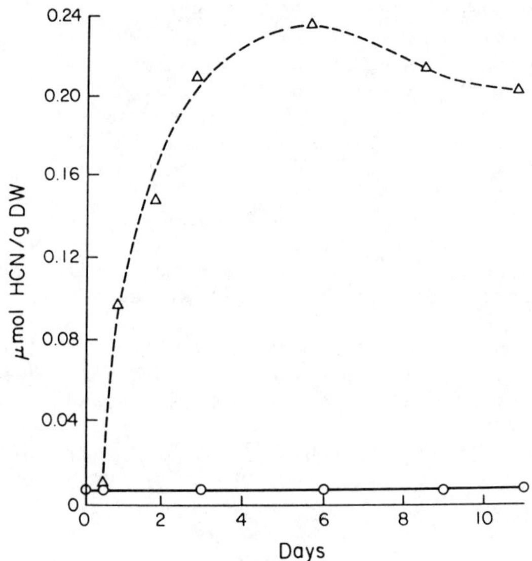

Figure 4. Release of HCN from cell-suspension cultures of *E. californica* cultivated in B5 medium in the presence and absence of 6% sorbitol for 11 d. Solid line, cells in B5 medium. Broken line, cells in B5 medium plus 6% sorbitol. The cells did not increase in fresh weight under these conditions. (Reproduced with permission from Ref. 22. Copyright 1985 Springer.)

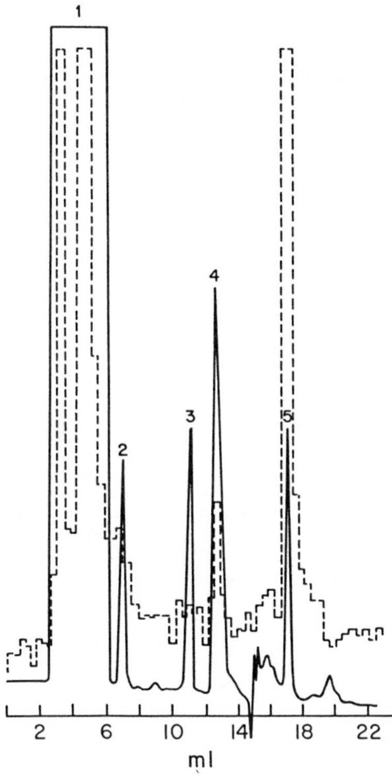

Figure 5. Separation by HPLC of the products formed when L-(U-^{14}C) tyrosine and NADPH were incubated with microsomes isolated from sorbitol-treated cell-suspension cultures of *E. californica*. Reference compounds: (1) L-tyrosine (part of the large peak of polar compounds); (2) 4-hydroxyphenylacetaldoxime; (3) 4-hydroxybenzaldehyde; (4) 4-hydroxyphenylacetonitrile; (5) 1-(4'hydroxyphenyl)-2-nitroethane. Solid line, absorption at 276 nm; broken line, radioactivity. (Reproduced with permission from Ref. 22. Copyright 1985 Springer.)

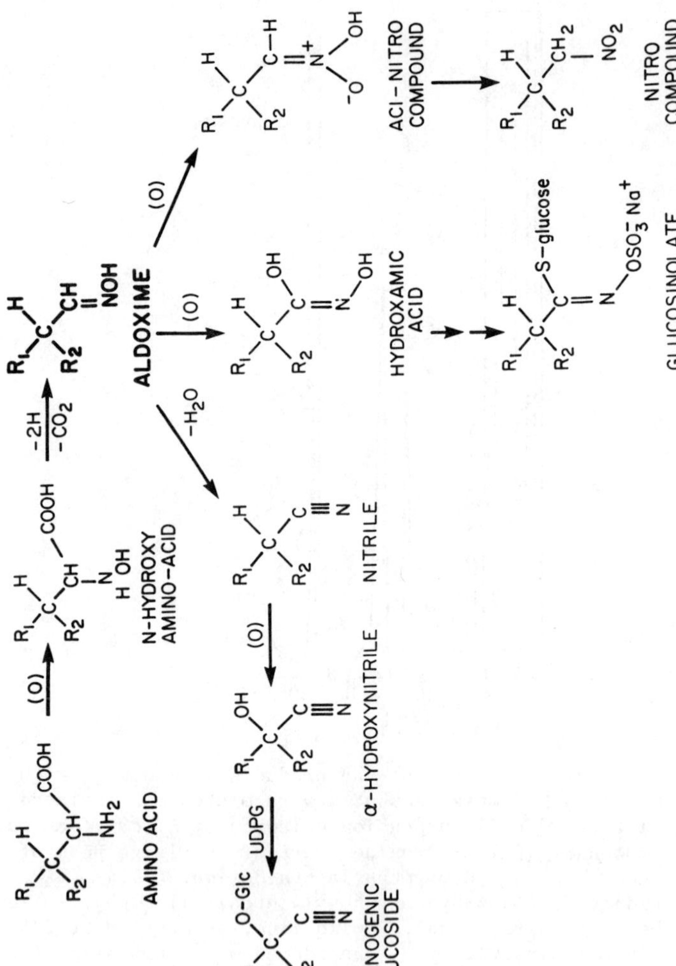

Figure 6. The aldoxime pathway for biosynthesis of cyanogenic glycosides, glucosinolates, and nitro compounds derived from a common precursor amino acid.

the α-nitroso-compound that is tautomeric with the nitro compound. This mechanism is consistent with the recent finding by Baxter *et al.* (21) showing that both oxygens in the nitro group of NPA are derived from molecular oxygen.

While oxygenation of the aldoxime carbon atom would yield a hydroxamic acid, such compounds are not effective precursors of glucosinolates (16). The oxidation of aldoximes to further intermediates in the formation of glucosinates may therefore involve thiol adducts as shown in Figure 3. Also, the enzymatic dehydration of aldoximes to form nitriles in the formation of cyanogenic glycosides is known to require NADPH (12). Finally, it has recently been observed that three oxygen atoms are required when L-tyrosine is oxidized to 4-hydroxymandelonitrile by sorghum microsomes (24). All of these findings indicate that the reaction sequences shown in Figure 6 may undergo modification, although it is unlikely that the role of the aldoxime as a common intermediate will change in view of its demonstrated effectiveness both *in vivo* and *in vitro*.

Acknowledgments

The author's research on cyanogenesis in plants has been generously supported by grants from the Metabolic Biology Program of the National Science Foundation since 1971.

Literature Cited

1. Conn, E. E. In Biochemistry of Plants; Conn, E. E., Ed.; Academic: New York 1981; Vol. 7, p 112.
2. Jones, D. A. In Biochemical Aspects of Plant and Animal Coevolution; Harborne, J. B., Ed.; Academic: New York, 1978; Chapter 2.
3. Jones, D. A. Am. Nat. 1978, 113, 445
4. Williams, M. C. Can. J. Bot. 1982, 60, 1956; Agron. J. 1983, 75, 520.
5. James, L. F. In Plant and Fungal Toxins; Keeler, R. F.; Tu, A. T., Eds.; Dekker: New York, 1983, p 445.
6. Gnanasunderam, C.; Sutherland, O. R. W. Phytochemistry 1986, 25, 409.
7. Gottlieb, O. R.; Magalhaes, M. T. J. Org. Chem. 1959, 24, 2070.
8. Matsuo, M.; Kirkland, D. F.; Underhill, E. W.; Phytochemistry 1972, 11, 697.
9. Iida, H.; Kikuchi, T.; Kobayashi, K.; Ina, H. Tetrahedron Lett. 1980, 21, 759.
10. Conn, E. E. Naturwissenschaften 1979, 66, 28.
11. Conn, E. E. Annu. Rev. Plant Physiol. 1980, 31, 433.
12. Møller, B. L.; Conn, E. E. J. Biol. Chem. 1980, 255, 3049.
13. Hösel, W.; Nahrstedt, A. Arch. Biochem. Biophys. 1980, 203, 753.
14. Cutler, A. J.; Sternberg, M.; Conn, E. E. Arch. Biochem. Biophys. 1985, 238, 272.
15. Collinge, D. M.; Hughes, M. A. Arch. Biochem. Biophys. 1982, 218, 38.

16. Underhill, E. W.; Wetter, L. R., Chisholm, M. D. Biochem. Soc. Symp. 1973, 38, 303.
17. Møller, B. L. In Cyanide in Biology, Vennesland, B.; Conn, E. E.; Knowles, C. J.; Westley, J.; Wissing, F., Eds.; Academic: New York 1981, p 197.
18. Shaw, P. D.; DeAngelo, A. B. J. Bact. 1969, 99, 463.
19. Shaw, P. D.; McCloskey, J. A. Biochemistry 1967, 6, 2247.
20. Baxter, R. L.; Abbot, E. M.; Greenwood, S. L.; McFarlane, I. J. J. Chem. Soc., Chem. Commun. 1985, p 564.
21. Baxter, R. L.; Greenwood, S. L. J. Chem. Soc., Chem. Commun. 1986, p 176.
22. Hösel, W.; Berlin, J.; Hanzlick, T. N.; Conn, E. E. Planta 1985, 166, 176.
23. Berlin, J.; Forche, E.; Wray, V.; Hammer, J.; Hösel, W. Zeit. Naturforsch. Teil C 1983, 38, 346.
24. Halkier, B. A.; Møller, B. L. In Cyanide Compounds in Biology, Ciba Found. Symp. No. 140, Harnett, S., Ed.

RECEIVED October 3, 1988

Chapter 12

Biological Effects of Glucosinolates

F. S. Chew

Department of Biology, Tufts University, Medford, MA 02155

The glucosinolate-thioglucosidase system potentially generates hundreds of compounds, but knowledge of autolysis conditions permits prediction of the products. How these products are generated, how their precursors are distributed and compartmentalized, and how these products affect herbivorous insects, pathogens, other plants, and potential symbionts are examined.

The glucosinolate-thioglucosidase system has long been the defining phytochemical character of the order Capparales (1,2). This system and its components have been preeminent subjects of chemical investigations of species in this order.

Comprehensive reviews of many aspects of the glucosinolates and their hydrolysis products have been published relatively recently (3,4,5). In addition to then-current discussions of glucosinolate hydrolysis, these reviews include especially useful compendia of the distribution of glucosinolates in food and forage sources, and discussions of the effects of glucosinolates and their hydrolysis products on domesticated birds and mammals.

Much evidence suggests that glucosinolates and their hydrolysis products are biologically active, but the conditions that determine these effects are not yet well understood. We know little of the conditions under which components of this system are delivered to the sites of biological activity in target organisms. In this paper I explore the causes for the diversity of effects observed for glucosinolates and their hydrolysis products on interactions of crucifers with other oranisms, especially herbivorous insects and fungi. I focus primarily on the following questions: i) How does the glucosinolate-thioglucosidase system produce diverse hydrolysis products? ii) How is this system distributed and compartmentalized? and iii) How does this system and its products affect herbivorous insects, other plants, and fungi?

0097–6156/88/0380–0155$07.75/0

The Glucosinolate-thioglucosidase System

The hydrolysis of glucosinolates (I, Figure 1) involves enzymatic attack on the thioglucoside bond to yield D-glucose, acid sulfate ion, and an unstable aglycone (II, Figure 1) that rearranges non-enzymatically to form any of several products. Skeletons of some products are shown in Figure 1 (III, IV, V, VI). Given an estimated activation energy of 22.6 Kcals/mole for thermal degradation of allyl glucosinolate (6) it is doubtful that glucosinolate hydrolysis occurs to any significant extent under natural thermal conditions without involving the enzyme. Non-enzymatic thermal degradation of the few glucosinolates investigated so far produces chemical species identical to products of enzymatic hydrolysis (7, 8).

Components of the endogenous plant system comprise both the thioglucoside glucohydrolase enzymes (EC 3.2.3.1) -- a family of glycoprotein enzymes, and its β-D-thioglucoside substrates. In addition, this enzyme-substrate system sometimes includes an unusual third component, a specifier protein (9) which does not attack the substrate itself, but whose presence directs the hydrolysis reaction. Other components that may be endogenous include catalytic amounts of ferrous ions and L-ascorbic acid, which activates many thioglucosidase isoenzymes (10).

The most frequent fate of the unstable aglycone generated by initial glucosinolate hydrolysis is to undergo rearrangement spontaneously via one of two pathways: a proton-independent Lossen rearrangement with a concerted loss of sulfate to yield an isothiocyanate, and a competing proton-dependent desulfuration to yield a nitrile. The structural diversity of products generated by glucosinolate hydrolysis is due to three major causes: a) the structural diversity of the precursors--the glucosinolates themselves; b) the specific endogenous or exogenous conditions under which components of the system interact (pH, ferrous ions); and c) the presence of epithiospecifier protein (ESP). A fourth set of factors, enzyme characteristics (substrate affinity, temperature and pH optima), may alter relative proportions of products by causing some glucosinolates with different R-groups to be hydrolyzed at different rates.

R-Groups. Glucosinolate R-groups generate corresponding aglycones so that series of one aglycone type--isothiocyanates, for example--are to be expected. In addition, certain structural groups are required for rearrangement to some aglycone products. Aglycones from only three naturally-occurring glucosinolates rearrange to form organic thiocyanates. Aglycones from indolyl and *p*-hydroxybenzyl glucosinolates are thought to form unstable isothiocyanates (IX, Figure 2) and thiocyanate ion. Indolyl alcohols may then condense to form diindolylmethane (XI, Figure 2). Aglycones from glucosinolates that contain β-hydroxyl-substituted side-chains may spontaneously cyclize to form the corresponding oxazolidine-2-thiones (e.g. XIV, Figure 3). Terminally unsaturated alkenyl groups are required for formation of epithionitriles (e.g. XVI, Figure 3).

pH, Ferrous Ions. In general, autolysis under neutral or near-neutral pH results in the formation of isothiocyanate. In contrast, the predominant product of enzymatic hydrolysis at acid pH is the

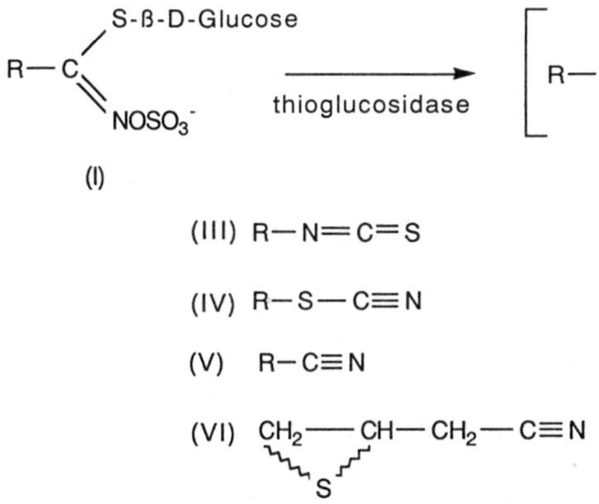

$$R{-}C(S{-}\beta{-}D{-}Glucose){=}NOSO_3^- \xrightarrow{\text{thioglucosidase}} \left[R{-}C(SH){=}NOSO_3^-\right]$$

(I) (II)

(III) $R{-}N{=}C{=}S$

(IV) $R{-}S{-}C{\equiv}N$

(V) $R{-}C{\equiv}N$

(VI) $CH_2{-}CH{-}CH_2{-}C{\equiv}N$ (with S bridging the first CH_2 and CH)

Figure 1. A glucosinolate (I) is enzymatically hydrolyzed to an unstable intermediate (II). Subsequent rearrangement depends on R-group characteristics and hydrolysis conditions and may yield the corresponding isothiocyanate (III), organic thiocyanate (IV), nitrile (V), or epithionitrile (VI).

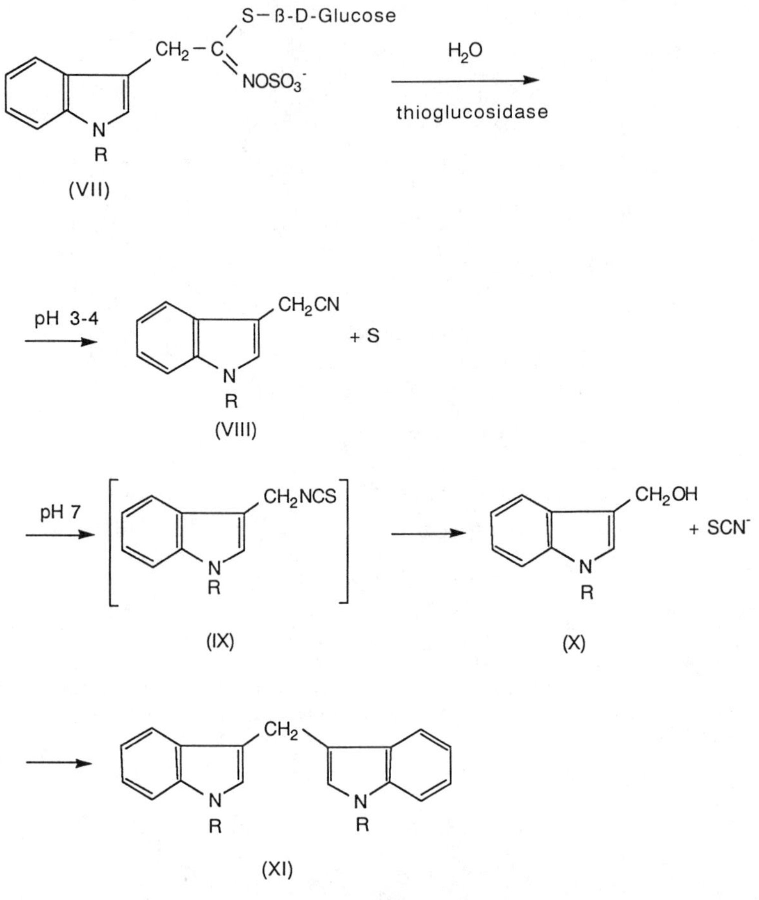

S−β-D-Glucose
CH$_2$−C
NOSO$_3^-$
N
R
(VII)
H_2O
thioglucosidase

pH 3-4
CH$_2$CN
+ S
N
R
(VIII)

pH 7
CH$_2$NCS
N
R
(IX)
CH$_2$OH
+ SCN$^-$
N
R
(X)

CH$_2$
N R
N R
(XI)

Figure 2. p-hydroxybenzyl and indolyl glucosinolates (VII) yield the corresponding indolylnitrile (VIII) at acid pH; at neutral pH, an unstable isothiocyanate (IX) forms and yields thiocyanate ion and indolyl alcohol (X) which may condense to form diindolylmethane (XI).

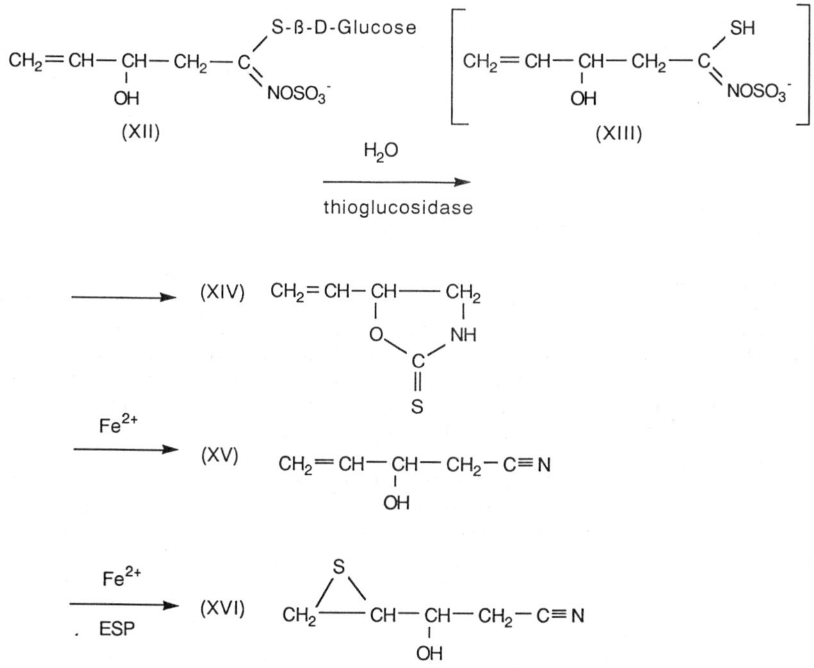

$CH_2{=}CH{-}CH(OH){-}CH_2{-}C(S\text{-}\beta\text{-D-Glucose}){=}NOSO_3^-$ (XII)

$\xrightarrow[\text{thioglucosidase}]{H_2O}$

$[CH_2{=}CH{-}CH(OH){-}CH_2{-}C(SH){=}NOSO_3^-]$ (XIII)

$\longrightarrow$ (XIV) $CH_2{=}CH{-}CH{-}CH_2$ / O, NH / C / S (oxazolidine-2-thione ring: O–C(=S)–NH)

$\xrightarrow{Fe^{2+}}$ (XV) $CH_2{=}CH{-}CH(OH){-}CH_2{-}C{\equiv}N$

$\xrightarrow[\text{ESP}]{Fe^{2+}}$ (XVI) $CH_2{-}CH$ (S bridged) ${-}CH(OH){-}CH_2{-}C{\equiv}N$

Figure 3. A glucosinolate with a β-hydroxylated side chain (XII) is enzymatically hydrolyzed to form an unstable intermediate (XIII). Subsequent rearrangement involves three alternative pathways mediated by the presence or absence of ferrous ions and epithiospecifier protein: spontaneous cyclization yields the oxazolidine-2-thione (XIV); ferrous ions mediate rearrangement to the corresponding nitrile (XV) or epithionitrile (XVI) if the glucosinolate contains a terminally unsaturated group and epithiospecifier protein is present.

nitrile. In a model system of purified enzyme from *Brassica napus* and synthesized 2-hydroxy-3-butenyl glucosinolate, the major glucosinolate found in *B. napus*, the pH of reaction mixtures and the nature and relative proportions of the products were monitored (11). In an unbuffered aqueous system, pH decreased as the hydrolysis progressed, from 6.94 at time zero to 6.03 when the amount of substrate became limiting. In this pH range, only the oxazolidine-2-thione was formed, presumably from spontaneous cyclization of the unstable isothiocyanate initially formed. In systems buffered to specific pH ranging from 2.6 to 9.7, nitrile was the sole product at pH below ca. 3.4, and oxazolidinethione was the sole product at pH ca. 5.4 and above. At intermediate pH, there was a gradual shift in relative proportions of the two products. This variation in relative product composition with varying reaction time and pH is similar to results using allyl and 2-phenylethyl glucosinolates with enzyme preparations from commercial mustard seed (12).

The addition of ferrous ions to reaction mixtures promotes formation of the nitrile from the aglycone (13). The effects of ferrous ion on the degradation of [illegible]yglucosinolate and extracted alkenyl glucosinolate mixtures were examined over a pH range from 3.5 to 7.5, using crude extracts of enzyme from *Brassica juncea* and *B. napus* (14, 15). In the absence of ferrous ion, the formation of alkenyl isothiocyanates and alkenyl nitriles was a function of pH of the reaction mixture. At pH 3.5, nitrile dominated the mix of hydrolysis products, but at pH 7.5, only alkenyl isothiocyanates were formed. With the addition of 2.5 mM ferrous ion, the relative amount of nitrile increased at all pH values from 3.5 to 7.5, but the isothiocyanate still predominated at pH 7.5. At low pH, a proton may block Lossen rearrangement of the aglycone, thus promoting formation of the nitrile. The ferrous ion may serve a similar function in the rearranging aglycone (12, 14, 15). Ferrous ions might also act by complexing ascorbic acid (a co-factor of some thioglucosidases), making it unavailable to isoenzymes that promote isothiocyanate production (16). This idea is consistent with the early observation (9) that ferrous ions catalyse nitrile formation directly without mediation by pH. On the basis of evidence from degradation of alkenyl glucosinolate mixtures and silver sinigrate, a model analog of the aglycone from allyl glucosinolate, it was shown that common thiol compounds such as L-cysteine and glutathione similarly accelerate nitrile formation (15). This catalysis of nitrile formation by thiols, in the presence of ferrous ions, was attributed to rapid desulfuration competing with the Lossen rearrangement (15). Variation in endogenous iron pools among species may account for the difference between species that form isothiocyanates upon autolysis versus those that yield nitriles (17).

Epithiospecifier protein. Formation of epithionitriles represents a third pathway for alkenyl glucosinolates with terminally unsaturated carbon. This outcome is mediated by epithiospecifier protein (ESP) and the specific conditions under which the thioglucoside cleavage occurs. ESP is a small protein (MW 30,000-40,000) first isolated from *Crambe abyssinica* seed meal (9). While this protein does not itself have thioglucosidase activity, it interacts with thioglucosidase to promote sulfur transfer from the S-glucose moiety to the terminal alkenyl moiety. Combinations of thioglucosidase and ESP from various plant sources produced the same products, demonstrating that epithionitriles are formed from glucosinolates by the same mechanism in all crucifers tested (18). These findings were extended to the thioglucosidase-producing fungus Aspergillus sydowi, suggesting that thioglucosidase enzymes from both plant and fungal sources function similarly (19). This similarity prompted speculation that amino acid sequences near the active sites may be highly conserved evolutionarily (19). The intramolecular transfer of sulfur may occur in an enzyme cleft of a thioglucosidase-ESP complex, and work demonstrating the failure of competing alkenyl substrate to inhibit ESP-mediated addition of sulfur to the alkenyl group supports the idea of intramolecular transfer (20). Direct evidence was provided by the demonstration that labelled glucosinolate ^{35}S was transferred to the corresponding epithionitrile rather than to competing unlabelled produce (21). Kinetic studies show that ESP inhibits thioglucosidase activity non-competitively, providing evidence that that ESP interacts at a site on the thioglucosidase other than the substrate binding site (17). Addition of exogenous ferrous ion to ESP-mediated hydrolysis in autolytic reaction mixtures containing synthesized 2-hydroxy-3-butenyl glucosinolate resulted in epithionitrile increase at the expense of the oxazolidine-2-thione (cyclized from the isothiocyanate) (17). This effect was proportionate to ferrous ion concentration, and demonstrated that ferrous ions are essential to epithionitrile formation, not merely a co-factor that stabilizes the ESP (17). In some B. napus cultivars, ESP was present but inactive until exogenous ferrous ions were added, after which epithionitriles became the major hydrolysis product (17). With allyl glucosinolate as a substrate, the epithionitrile was formed at the expense of both the isothiocyanate and the alkenylnitrile, suggesting that the three products represent competing pathways from a common aglycone precursor (18). With 2-hydroxy-3-butenyl glucosinolate as a substrate, although oxazolidine-2-thione decreased with increasing ESP activity, the ratio of alkenylnitrile (1-cyano-2-hydroxybut-3-ene in this case) to the epithionitrile (1-cyano-2-hydroxybut-3,4-epithiobutane) remained constant (about 1 : 2.5) (17). This observation supports the possibility that ferrous ions promote formation of a nitrile precursor at the expense of isothiocyanate (which in this case cyclizes to the 5-vinyl-oxazolidine-2-thione) and that subsequent partition to alkenyl and epithionitriles is then mediated by ESP (22,17).

The possibility of stereoisomeric interaction of ESP was investigated by using ESP from plant sources with stereoisomeric substrates: Crambe abyssinica where the natural substrate is epi-progoitrin ((S)-2-hydroxy-3-butenyl glucosinolate; and from Brassica campestris where the natural substrate is progoitrin, or (R)-2-hydroxy-3-butenyl glucosinolate. Thioglucosidases from several sources were used: Aspergllus sydowi, where there is no endogenous natural substrate, and from the three crucifers, C. abyssinica, B. campestris, and Sinapis alba (23). No clear difference between the ESPs was demonstrated. In nearly all combinations studied, the ratio of the diastereomeric products, erythro/threo cyanoepithiobutanes (XX,XXI, Figure 4), was the same, about 1.4 to 1. This ratio compares to the 1.25 to 1 ratio reported for autolysis of the C.

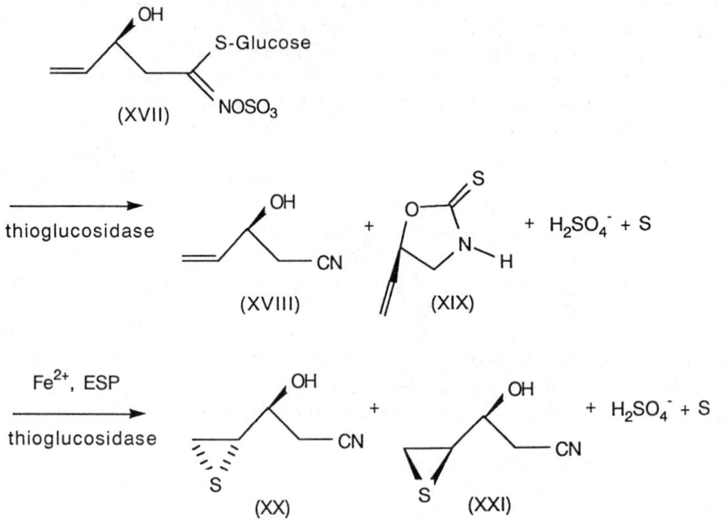

(XVII) S-Glucose, $NOSO_3$, OH

thioglucosidase → (XVIII) OH, CN + (XIX) S, O, N, H + $H_2SO_4^-$ + S

Fe^{2+}, ESP, thioglucosidase → (XX) OH, CN, S + (XXI) OH, CN, S + $H_2SO_4^-$ + S

Figure 4. (S)-2-hydroxy-3-butenyl glucosinolate (epi-progoitrin) (XVII) may be enzymatically hydrolyzed by two alternative pathways: to (S)-1-cyano-2-hydroxy-3-butene (XVIII) and the oxazolidine-2-thione, (R)-goitrin (XIX); alternatively, in the presence of ferrous ions and epithiospecifier protein, hydrolysis yields the diastereomer epithionitriles, erythro-1-cyano-2-hydroxy-3, 4-epithiobutane (XX) and threo-1-cyano-2-hydroxy-3, 4-epithiobutane (XXI).

abyssinica seed (24). A more extreme ratio in the case of S. alba thioglucosidase combined with ESP from B. campestris may reflect steric constraints: in epiprogoitrin, the Si-Si face of the vinyl group is on the same side as the 2-hydroxy group; this arrangement may affect the interaction with the ESP-thioglucosidase complex (23).

Thiocyanate formation. Organic thiocyanates (distinct from ionic thiocyanate liberated by rearrangement of aglycones from indolyl and p-hydroxybenzyl glucosinolates), are a fourth potential hydrolysis product, formed in a few natural systems by only three glucosinolates. The first is allyl (in Thlaspi arvense, [25]). The second is benzyl (in seeds, but not leaves of Lepidium sativum, leaves of Coronopus didymus, [25,26], leaves and seeds of Lepidium ruderale and L. virginicum [27]). The third is 4-methylthiobutyl (in leaves of Diplotaxis tenuifolia and Eruca sativa [28], though failure to confirm these observations suggests sporadic occurrence [27], and leaves of Alyssum species [27]).

The formation of organic thiocyanates and cyanoepithioalkanes share several similarities (9,16,20): both processes are restricted to substrates with specific structures, both are heat-labile, and both are susceptible to oxidation. Formation of organic thiocyanates is restricted to glucosinolates with R-groups that form stable cations, R^+ (a characteristic of the three that are found in natural systems). Evidence for this substrate restriction is provided by failure to produce organic thiocyanates from glucosinolates similar to the natural substrates (28,29). Degradation of synthesized propylglucosinolate by extracts from Thlaspi arvense seeds was studied, but no propylthiocyanate was found (29). Crude enzyme preparations from L. sativum seed convert pure benzylglucosinolate to the thiocyanate as well as the isothiocyanate and nitrile, but convert the structurally similar 2-phenylethyl glucosinolate only to the isothiocyanate and nitrile (30).

The distribution of thiocyanate-forming activity among crucifer species and plant parts suggests that some endogenous factor with restricted distribution is involved, because a structural substrate restriction alone could not account for the finding of organic thiocyanates in leaves of some Lepidium species but only the seeds of others (31). By monitoring autolysates from Lepidium sativum seeds heated for varying periods from a few minutes to several hours, it was shown that in these seeds, thiocyanate-producing activity is more heat-labile and is lost prior to thioglucosidase activity (31). This unknown thiocyanate-forming factor apparently acts catalytically to redirect the Lossen rearrangement that would otherwise occur spontaneously. Although not yet isolated, this detection of a thiocyanate-forming factor as distinct from thioglucosidase activity is a promising development in the study of a difficult problem for which numerous solutions have been considered (20,25,28,29,31,32).

Enzyme characteristics. Beyond the structural diversity of products potentially generated from a single glucosinolate, the co-occurrence of different glucosinolates in plants and plant parts permits considerable diversity of aglycone profiles. This diversity of profiles reflects both differential accumulation of specific glucosinolates in plants and plant parts and differential activity of thioglucosidases towards some substrates (33). Broad substrate specificities are observed for thioglucosidase activity from many sources, but the few thioglucsidases examined so far do hydrolyze some glucosinolates at different rates (seed thioglucosidases of Brassica napus [11,34], Sinapis alba [34]). In particular, glucosinolates with more hydrophilic side-chains (such as hydroxyl-substituted groups), produce aglycone products at slower rates, and rate differences may reflect non-specific hydrogen-bonding between glucosinolate and enzyme that hinders binding at the active site (11). None of these differences in substrate affinity are associated with individual isoenzymes isolated from Sinapis alba and Brassica napus (11,34). However, correspondence between relative proportions of glucosinolate precursors and the resulting aglycone products may thus be mediated by differential thioglucosidase binding specificity as well as differential accumulation in plant parts.

Knowledge of the relative proportions of specific glucosinolates may be a poor indicator of the relative proportions of aglycone species presented to target organisms. For example, in a Brassica napus cultivar, 2-hydroxy-3-butenyl glucosinolate was reported as the predominant glucosinolate, but thioglucosidase activity from that cultivar was greater towards allylglucosinolate than towards the hydroxy-substituted compound (11). The 2-hydroxy-3-butenyl glucosinolate is the major natural substrate (3,35), but glucosinolate profiles in cultivars and local populations of wild Brassica species vary widely (36,37,38,39). Given the thioglucosidase activities observed however, aglycones from the several alkenyl glucosinolates and the allyl precursor 3-methylthioglucosinolate (5) reported in B. napus (3) may be more important in the initial products of hydrolysis than their in situ concentrations would indicate. Such effects could be responsible for the observation that autolysis products change as autolysis proceeds (32). In another example, autolysis of allyl glucosinolate in Farsetia spp. forms primarily the isothiocyanate (37), but in Brassica oleracea, the product is primarily nitrile (11,38).

Botanical occurrence and compartmentation

Glucosinolates co-occur with thioglucosidases among species of the order Capparales, as far as is known. Glucosinolates have been recorded sporadically in representatives of a number of plant families outside the order Capparales (2), but the status of accompanying thioglucosidase activity in these plants is less certain. Reports of trace amounts of glucosinolate hydrolysis products (0.1-5 ug/g wet tissue) in volatiles of three unrecorded sources--Plantago major and an unidentified Plantago species (Plantaginaceae [27, 40]), an edible fungus (41) and cocoa beans (Sterculiaceae [42])--suggest more widespread occurrence of glucosinolates and thioglucosidase activity. However, these reports were not confirmed in P. major (43) or in cocoa beans (44). No explanation for the apparent discrepancies has been proposed, and resolution awaits further scrutiny.

System components. A large number of R-group substituents (approaching 100) is reported from plants of the Capparales (3). Most R-groups derive from carbon skeleons of protein amino acid precursors (5). Glucosinolates derived from chain-lengthened homologs of amino acids are common, but the distribution of these homologs is apparently confined to the Capparales (e.g., 45). Some unusual R-group modifications include recently reported hydrox- and methox- and ring oxygenated indole glucosinolates (46-50) and the β-glycosylated side chains of R-groups, prominent in representatives of the Resedaceae and Moringaceae (51,52). Glucosinolates within plant species and individual plants usually form series related by homologization, hydroxylation, or oxidation (e.g. 53-57), with characteristic quantitative profiles for different plant parts (3,58,59).

Thioglucosidase isoenzymes from plant sources show substantial diversity in physico-chemical characteristics (33), but broad substrate specificity is the rule (3,33). Distribution of isoenzymes is organ-specific. Electrophoretic examination of isoenzymes from many plant organs or tissues demonstrates that the distinctive patterns vary by species (11,33,34,60,61), organ (62,63,64), and age (62). Little has been inferred from the organ-specific distributions of isoenzymes, but such distinctive patterns beg explanation. One important question is to determine whether the biochemical behavior of the particular isoenzymes corresponds to endogenous conditions found in that plant tissue (e.g., pH optima), or to conditions in the target organism, or to particular glucosinolates that dominate the profile of that tissue (62). Thioglucosidase activity in tissues declines with age (65-67; Bones, A. Proc 7th Int'l. Rapeseed Cong., 1988, in press).

Enzymes with thioglucosidase activity are unusual outside the glucosinolate-containing angiosperms. Such activity has been observed in the fungi *Aspergillus syndowi* (68) and *Aspergillis niger* (69,70), the intestinal bacteria *Enterobacter cloacae* (71) and *Paracolobactrum aerogenoides* (72) and some muscles of the head and thoracic regions of the crucivorous aphids *Brevicoryne brassicae* and *Lipaphis erisimi* (73). These aphid thioglucosidases hydrolyze 2-hydroxy-2-phenylethyl glucosinolate *in vitro* and exhibit electrophoretic mobilities distinct from those of isoenzymes isolated from their hostplants. This distinctive biochemical behavior does not rule out the possibility that the thioglucosidases are modified from enzymes acquired from the hostplant, in contrast to the *de novo* synthesis involved in endogenous thioglucosidases in the bacteria and fungi. This possibility would require testing in model hostplant systems because these aphids feed only on crucifers, which, of course, also contain thioglucosidases. Glucosinolates have been found in aphid honeydew (74).

The occurence of the third component, the epithiospecifier protein, has been sought so far only within a small number of crucifer species since its isolation from *Crambe abyssinica* seed meal (9). Specifier protein has been found in growing parts of *Brassica oleracea* cultivars (white cabbage and Brussels sprouts), and seeds of *Brassica campestris* (75), *Crambe abyssinica* and turnip (18), but not in *Sinapis alba* seed or horseradish root (18). ESP activity has been identified by studying formation of epithionitrile by endogenous enzyme systems in seeds of *B. campestris*, *B. napus*, and *Lepidium sativum* using reaction mixtures from which endogenous glucosinolates were removed and replaced by known amounts of 2-hydroxy-3-butenyl glucosinolate (17). Similar activity was sought but not found in seeds of *Nasturtium officinale* (38). Various species in other crucifer genera produce epithionitriles upon autolysis, and if formation of this product always involves an ESP (4,75), then these species may be expected to contain proteins with ESP activity. These include (in addition to sources listed above) leaves of C. abyssinica and B. napus (76), Descurainia sophia (77), and leaves of representatives of genera Cardamine, Hirschfeldia, Lobularia, Sisymbrium (27). The finding of ESP activity in seed samples of Lepidium sativum (containing small amounts of allyl glucosinolate), but not in N. officinale seeds (containing no alkenyl glucosinolates), suggests that ESP co-occurs with terminally unsaturated alkenyl glucosinolates and may be limited to species containing these substrates (17).

Component compartmentation. Little definitive evidence is available concerning how the glucosinolate-thioglucosidase system is held latent until the tissue is disrupted. In mature seeds of Carica papaya (Caricaceae), benzyl glucosinolate is contained in the seed endosperm and thioglucosidase is found only in the sarcotestae. In immature fruit, the benzyl glucosinolate was found primarily, if not exclusively, in the latex, but no thioglucosidase activity was found in the latex (78). In Cruciferae and other glucosinolate-containing plants, the situation is unresolved. Intact crucifers release few volatiles (79-81;82 cited in 83). Enzyme and substrate are thus clearly separated, but localization of thioglucosidase activity in particular is uncertain.

Discussion has focused on storage of glucosinolates and thioglucosidases in specialized cells, and alternatively, subcellular compartmentation of enzyme and substrate. Morphologically and sometimes histochemically distinct cells, known as idioblasts, were observed to contain protein and for nearly a century were assumed to contain thioglucosidases (84-86). In some cases levels of thioglucosidase activity in the particular plant tissue paralleled idioblast density, providing correlative evidence for the localization of thioglucosidase in specialized cells (86,87). Contradictory evidence suggested that thioglucosidase activity was observed in the absence of cytologically distinct cell types. Idioblasts were absent from primary roots of plants that show thioglucosidase activity (Lepidium sativum [87], Brassica chinesis, B. napus, B.oleracea, Iberis amara, Raphanus sativus, and Sinapis alba [86]). The correlation of thioglucosidase activity and protein-rich idioblasts was similarly poor in rootlets and flowers of Sinapis alba and Raphanus sativus (62,63), and in pith cells of horseradish roots (88). Cytochemical work at the ultrastructural level showed that a majority of cells in rapeseed cotyledons show thioglucosidase activity (89). This suggested that activity was associated with particular organelle fractions, implicating subcellular compartmentation associated with the plasmalemma (89), smooth endoplasmic reticulum (64), dictyosomes (64), and mitochondria (90).

The enzyme and its substrates have proved extremely difficult to localize cytologically, however, owing to the lack of specific histochemical stains (84,91,92), and the lability of the components of the system. Isolation and culture of protoplasts (66) suggested that thioglucosidase activity, if contained in these cells, decreased rapidly with age. Cell fractionation and centrifugation in density gradients showed broad non-specific distribution of thiogluosidase activity with many fractions (85).

Grob and Matile (88,93) provided compelling evidence that glucosinolates were contained in vacuoles of horseradish root cells. Isolating vacuoles by using acid phosphatase (an enzyme associated with the vacuolar membrane, the tonoplast) as a marker, they demonstrated that vacuoles contained glucosinolates and L-ascorbic acid, the latter in amounts shown to catalytically activate horseradish thioglucosidase >800 fold, the highest activation factor yet reported (cf. 13). Positive correlations between ascorbic acid and glucosinolate levels during germination have been reported in *Brassica napus*, *B. juncea* and *Eruca sativa* (94). Yet a further unexpected finding was reported: a considerable proportion of the thioglucosidase activity was found in subcellular fractions containing these vacuoles but no hydrolysis products were produced until the vacuoles were disrupted (88).

An attractive hypothesis that accounts for these discrepancies is that thioglucosidases are cytosolic enzymes that have a marked tendency to adhere to membranes (85). Addition of exogenous glucosinolates yielded hydrolysis products in subcellular fractions containing intact vacuoles, if the vacuoles were not bathed in high concentrations of NaCl (85). NaCl is normally used to maintain osmotic balance, but the concentrations used are inhibitory to thioglucosidase activity (95). When the vacuoles were disrupted by dilution of the osmoticum, the thioglucosidases adhering to the vacuole surfaces were no longer inhibited (85). Under these osmotic conditions however, the vacuoles burst, preventing unambiguous association of thioglucosidase activity with the outside surface of the tonoplast. Substitution of sorbitol for the salt showed that hydrolysis products were produced when exogenous glucosinolates were added to solutions of intact vacuoles, demonstrating that the enzyme activity was associated with the outer surface of the vacuole (85). Further, washing membrane preparations resulted in removal of almost all thiogluosidase activity after three washings, providing evidence that binding occurs, but is loose. Association of thioglucosidases with cytosolic surfaces of membranes of diverse function may thus reflect the *in situ* location of this enzyme (85). Further development of specific tools for unambiguously identifying thioglucosidases (e.g. monoclonal antibodies [Bones, A.; Espevi, T. Proc. 7th Int'l. Rapeseed Cong. 1988, in press]) and substrates, (e.g. immunoassays for glucosinolates [J.P. Sang, State Chemistry Laboratory Dept. Agric. Rural Affairs, East Melbourne, Autralia, personal communication]) show promise for definitive localization of system components.

Effects of glucosinolates and their products on insects and fungi

Interest in the effects of glucosinolates and their hydrolysis products has been manifest at least since Verschaffelt (96) induced *Pieris brassicae* and *P. rapae* caterpillars to feed on non-crucifers and paper by painting them with juice of *Bunias orientalis* (a crucifer) and a solution of sinigrin itself. Glucosinolates or their hydrolysis products have been implicated in a variety of interactions between crucifers and their potential herbivores, pathogens, competitors, and symbionts.

Responses to volatiles. Volatiles from intact or wounded crucifers act as attractants for crucivorous insects seeking food or egg-laying sites. Among crucivorous insects, little unambiguous information is available, owing partly to the difficulty of bioassays using volatiles (83), and to the often unknown or unpredictable nature of volatile hydrolysis products produced by plants *in situ*. Rapidly growing plants may induce low levels of mechanical injury during cell transformations, and volatile hydrolysis products might be formed most frequently during those times. Finch (83), who has recently reviewed insect host-seeking behavior in relation to volatiles, points out that under laboratory conditions, the most rapidly growing cultivated crucifers tested were those preferred by adults of the fly *Delia brassicae*, the moth *Plutella xylostella*, and the beetle *Phaedon cochlaeriae*. Allyl isothiocyanate attracted adult flies of *Delia (Hylemya) brassicae* to its source in gelatin-cellulose blocks (97), and field trials of isothiocyanates demonstrate their attractiveness to adult flea beetles, *Phyllotreta cruciferae* and *P. striolata* (98), in contrast to allyl thiocyanate (99). Differential responses of adult flea beetles to allyl isothiocyanate were observed in experiments to design traps (100): traps baited with allyl isothiocyanate attracted about twice as many *P. cruciferae* as *P. striolata*, but did not attract *Psylliodes punctulata*. *Pieris rapae* larvae were attracted to several natural and synthetic isothiocyanates (allyl, benzyl, phenyl and phenylethyl) (101). Laboratory observations of alate female aphids (*Brevicoryne brassicae*) to allyl glucosinolate and the corresponding isothiocyanate showed the aphids were weakly attracted to the isothiocyanate, but thioglucosidase-generated hydrolysis products of several other glucosinolates were not more attractive than water (102). Allyl glucosinolate in water solution was attractive to the midge *Dasineura brassicae*, but since the attraction was based on olfactory reaction of the midge rather than contact chemoreception, the result is puzzling (103); glucosinolate degradation could have occurred during the experimental choice period, but no information on this point is available. Allyl nitrile has been suggested as a specific attractant for egg-laying adults of the butterfly *Pieris brassicae* (104), but this hypothesis has not been tested. That specific hydrolysis products such as isothiocyanates are perceived as parts of volatile profiles is demonstrated by the several-orders-of-magnitude reduction in attractiveness of allyl isothiocyanate alone compared to volatile hostplant extracts (105,106). Such field results on the fly *Delia brassicae* are corroborated by electrophysiological work demonstrating that while antennal receptors are sensitive to allyl isothiocyanate, they are similarly responsive to other compounds found in headspace volatiles of their foodplants (107). The suggestion that microorganisms may mediate plant presentation of volatiles (108), underscores the likely complexity of profiles to which herbivorous insects are attracted.

Lepidopteran larval behavior. Glucosinolates, rather than hydrolysis products, are the active contact stimuli in the feeding specificity of crucivorous insects. Larvae of the crucifer-feeding moth *Plutella maculipennis* were fed cellulose-agar gels containing a variety of glucosinolates isolated from Cruciferae and Tropaeolaceae

(109,110). All the glucosinolates stimulated feeding activity in these cabbage-reared larvae, especially when accompanied by nutrients such as glucose, but allyl isothiocyanate alone neither stimulated feeding nor increased feeding rates in combination with allyl glucosinolate. Minimum concentrations that elicit feeding responses were reported as 8 ppm (109), a concentration comparable to thresholds observed in electrophysiological studies of another crucifer specialist, the butterfly *Pieris brassicae* (111). Exogenous methyl and allyl glucosinolates in semi-synthetic diets of the crucifer specialist *Pieris brassicae* similarly elicits larval feeding in the absence of crucifer leaves (112). *P. rapae* larvae were undeterred by unnaturally high concentrations of allyl glucosinolate infused into leaves (113), and variation in concentrations of allyl glucosinolate in plants did not affect larval feeding behavior (114). Such behavior can probably be anticipated with all crucivorous pierid larvae.

In non-crucifer-adapted larvae, responses vary. Caterpillars of *Papilio polyxenes*, an Umbelliferae feeder, were deterred by allyl glucosinolate infused into foodplant leaves (>500 ppm of leaf fresh weight), while the generalist *Spodoptera eridania* was unaffected until doses increased above 870 ppm of leaf fresh weight (113). In contrast, larvae of laboratory-reared and field-collected *Manduca sexta*, a Solanaceae feeder that can grow on crucifers (115), were undeterred by up to 2000 ppm allyl glucosinolate in semi-synthetic agar diets (Chew F.; Mechaber, W., Tufts Univ., unpublished data).

Larval lepidopteran behavior differs towards glucosinolates with different R-groups. In *Plutella maculipennis*, although allyl glucosinolate was the most highly stimulatory at the lowest concentration presented (4 mg/10 ml water or somewhat less than 400 ppm w/v in the cellulose-agar gels), other glucosinolates, notably 2-hydroxy-3-butenyl (the major glucosinolate constituent of cabbage [3]), elicited greater response at higher concentrations (110). These results suggest that glucosinolates could provide a basis for caterpillar preference among crucifer parts containing several glucosinolates in different proportions. Curiously, *P. maculipennis* larvae were stimulated by low concentrations of 3-butenyl and 2-phenylethyl glucosinolates (both found in cabbage), but in contrast to findings for *P. rapae* (113), dramatic evidence of toxicity to larvae was observed: 50% or more of the larvae tested on these compounds at 30-50x the concentration in cabbage died within the 18 hour test period (110). Short-chain and unsaturated alkyl R-groups gave the strongest responses in electrophyiological studies of larvae of another lepidopteran crucifer-specialist, *Mamestra brassicae* (116). Change in the relative effects of different glucosinolates when offered at several concentrations was also obesrved in *Pieris brassicae* (111), as well as in flea beetles (below).

Adult beetle responses. The feeding behavior of crucivorous beetles also shows differential response to R-groups (117-119). A minimum threshold concentration of 15.3 - 153 ppm allyl glucosinolate elicited feeding in *Phyllotreta cruciferae* adults (117) a species that is more responsive to this compound than *P. striolata* (100). Benzyl glucosinolate was the most highly stimulatory compound for several flea beetle species examined: *Phyllotreta cruciferae* (117), *P.*

nemorum, P. undulata, P. testrestigma, and Phaedon cochleariae (118). More recent work shows the most stimulatory compound was indolylmethyl glucosinolate, which had not been previously tested (119). In contrast to results from flea beetles, laboratory tests with the seed weevils Ceutorhynchus assimilis showed that the longer chain alkenyl glucosinolates and benzyl and p-hydroxybenzyl glucosinolates stimulated greater feeding activity than did allyl glucosinolate. Other glucosinolates were more stimulatory than water but less than allyl. Surprisingly, indolylmethyl glucosinolate was not stimulatory compared to water. Glucosinolates with alkyl R-groups dominate seed profiles, while indolyl glucosinolates, usually absent from seeds, are often major components of green-part profiles (3); these findings are roughly consistent with the microhabitat preferences of these beetles, but the critical plant parts are yet to be examined. In general, beetle feeding and oviposition activities are poorly correlated with seed glucosinolate profiles; clearly it will be important to obtain data on the profiles of specific plants parts, e.g. the seed pod wall through which seed weevils oviposit (119). Differences in sensitivity and specificity to glucosinolates was observed among cells in the medial galeal chemosensillum of the red turnip beetle, Entomoscelis americana (121). Some glucosinolates are not stimulatory to crucifer specialists, e.g. indolylmethyl glucosinolate to seed weevils. Similar negative results are reported for E. americana, which does not respond to 2-hydroxy-3-butenyl glucosinolate (120), and one strain of the moth Mamestra brassicae, whose sensilla do not resond electrophysiologically to benzyl glucosinolate (116). Mitchell reviews the role of glucosinolates in host-plant recognition by adult chrysomelid beetles (Mitchell, B.K., J. Insect Physiol. 1988, in press).

Larval lepidopteran chemoreceptors. Electrophysiological work on lepidopterous larvae shows that contact chemoreceptors in the mouthparts respond to glucosinolates but are only weakly responsive to isothiocyanates (111). In Pieris brassicae and P. rapae, lateral and medial maxillary sensilla of each species show different response profiles towards a range of glucosinolates tested (methyl, allyl, benzyl, p-hydroxybenzyl, and sinalbin [where the p-hydroxybenzyl anion is paired with the sinapine cation]) (122). Two different strains of the moth Mamestra brassicae also showed differential sensitivity to sinalbin and benzyl glucosinolate--one not responding to benzyl glucosinolate (116). The lack of response can be attributed to either cations associated with the two glucosinolates or structural diferences in the R-groups, or both, since commercial preparations of benzylglucosinolate are the tetramethylammonium salt. In both P. brassicae and P. rapae, while both lateral and medial sensilla are sensitive to the benzyl glucosides, only the lateral sensilla exhibit much sensitivity to methyl and allyl glucosinolate (van Loon, J.J.A., Fdn. Agric. Plant Breeding, Wageningen, The Netherlands, personal communication). Behavioral consequences of some of these chemosensory differences remain to be studied. Considering that the dose-response curve for P. rapae lateral sensilla tested on allyl glucosinolate reaches saturation at 0.01mM sinigrin (well below concentrations found in cultivated cabbages [31]), and that rather weak dose-dependent effects are observed in P. brassicae larvae (111), the response to glucosinolates may be all

-or-none (van Loon, J.J.A., personal communication). For very common (e.g. allyl) or quite rare glucosinolates (e.g. methyl), presence or absence may convey sufficient information.

Oviposition responses. Adult females of *Pieris brassicae* similarly possess tarsal contact chemoreceptors that are sensitive to glucosinolates and are probably used during oviposition (122). At least one contact chemoreceptor type responds to glucosinolate solutions, with thresholds of 10^{-5} to 10^{-4} M. No consistent differences were observed in receptor responses to allyl glucosinolate, or to the sinapine and tetramethylammonium salts of *p*-hydroxybenzyl and benzyl glucosinolates respectively (122). Adult females of *Delia brassicae* flies also possess tarsal chemoreceptors that respond to sinigrin at similar thresholds, but not to the corresponding isothiocyanate (123,124). When ovipositing, both insects exhibit "drumming" behavior using the tarsi; this behavior may enable insects to contact compounds in plant saps beneath dry, waxy surfaces (125,126).

The role of different glucosinolates as egg-laying cues for pierid butterflies is not well understood. *Pieris brassicae* will lay eggs on non-host surfaces painted with allyl glucosinolate (126) or if their tarsi are brushed with the compound (127). Correlative evidence that egg-laying adults of *P.napi* distinguish among groups of species that show consistent differences in glucosinolates profiles suggests the glucosinolates may be used as oviposition cues (128). In *Pieris rapae* and *P. oleracea*, however, allyl glucosinolate generally stimulated oviposition activity, but did not promote discrimination among plants (129; Renwick, J.A.A.,; Radke, C.D. *J. Insect Physiol*. 1988, in press). Stimulant and deterrent fractions were isolated from non-host crucifers such as *Erysimum cheiranthoides*, and their relative effects correspond to pierid butterfly behavior towards the intact plant (Renwick, J.A.A., Boyce Thompson Institute, Cornell Univ., personal communication).

Examination of the relative effects of primary nutrients and glucosinolate content in plant choice by larvae (114,130) and adults of *P. rapae* (131), suggests that glucosinolates may often be of secondary importance to primary nutrients in food choice. Field experiments examining the relative effects of glucosinolates and primary nutrients on insect fauna of native crucifers similarly suggest complex patterns (132,133; Collinge, S.K.; Louda, S..M. *Oecologia* 1988, in press; Collinge, S.K.; Louda, S.M., *Ann. Ent. Soc. Amer.* 1988, in press) whose resolution awaits additional information, especially on mechanisms by which individual insects respond.

Physiological responses of larval Lepidoptera. Little is known of how crucifer specialists are adapted physiologically to glucosinolates and their hydrolysis products, partly because the most accessible species are those with the broadest natural diets (e.g. for pierid butterflies, *P. rapae* and *P. brassicae* [134]), and therefore least likely to exhibit differential responses that can be studied. Such differential responses to glucosides account for toxicity effects in other species (135). Few data are available concerning crucifer suitabilities for non-pest insects. However, the coincidence of poor suitability and a characteristic compound for a particular genus make a potential relationship worth testing (Chew, F.S. In *Chemical Mediation of Coevolution*; Spencer, K.C., Ed.

Academic Press: New York, 1988, in press). Such cases are exemplified by the genera Allyssum, Erysimum, and Cheiranthus. Representatives of these genera were refused by caterpillars of seven pierid species (134). Representatives of these genera accumulate unusual glucosinolates thought to be characteristic (136), but some also contain other classes of compounds, e.g. cardenolides. These other classes of compounds are effective deterrents to various flea beetle species (118), but the only test done on pierid caterpillars (137) provides negative evidence that is inconclusive because the most europhagous pierid species was tested in its most euryophagous instar (138).

Methanolic extracts of pupae of P. brassicae show they contain both allyl glucosinolate and allyl isothiocyanate, but attempts to detect these compounds in adults and eggs of this species, and life stages of P. rapae were unsuccessful (139). It is presumed that these compounds are acquired from the crucifer food plants, but evidence on this point is inconclusive: butterflies captured in fields of low-glucosinolate rapeseed plants contained no glucosinolates or isothiocyanates. Whether this situation reflects a poor plant source or sporadic sequestration among butterflies is not known (139). Chewing insects probably ingest primarily glucosinolate rather than hydrolysis products, since bites will not disrupt all cells and the reaction in situ in disrupted plant tissue is relatively slow (109,117). Among pierid butterfly species, the most europhyagous (P. rapae and P. brassicae) have lower midgut pH than P. napi (140), whose foodplant range is more restricted (138). If the midgut is a major site of glucosinolate hydrolysis in pierid caterpillars, these conditions might favor formation of nitriles at the expense of isothiocyanates and their derivatives in the more broadly-feeding species. Another lepidopteran crucifer-feeder, Plutella maculipennis, has midgut pH comparable to that of P. rapae (140). Allyl, benzyl and 2-phenylethyl isothiocynates are all acutely toxic (>50 ppm) to first-instar Spodoptera frugiperda, although they had no adverse developmental effects on survivors. Allyl isothiocyanate was also acutely toxic to final-instar larvae of moths S. frugiperda, Trichoplusia ni, and Anticarsia gemmatalis (>approx. 1000-2000 ppm) (Wadleigh, R.W.; Yu, S.J. J. Chem. Ecol., 1988, in press).

In non-crucifer-adapted Lepidoptera, Papilio polyxenes larvae feeding on leaves infused with low concentrations of allyl glucosinolate were sluggish and 50% died within the 24-hour test period, but Spodoptera eridania growth efficiencies were affected only at very high concentrations (113). The smaller effect on S. eridania may be due to induction of microsomal mixed-function oxidases (113). These enzymes are induced by and metabolize several isothiocyanates, including allyl isothiocyanate, as well as the potential hydrolysis products indole-3-acetonitrile and indole-3-carbinol in the moth caterpillars Spodoptera frugiperda and Anticarsia gemmatalis (141). Some glucosinolate-related compounds including indole-3-acetonitrile are effective inducers of this enzyme activity (141), but induction is not clearly established in other cases (142). Further, some isothiocyanates inhibit these enzymes (e.g. 2-phenylethyl, 143). In addition, gluthathione transferase activity (analogous to glutathione transferase dexotification of isothiocyanates in mammals (144,145) towards allyl and benzyl isothiocyanates was induced by these compounds in S. frugiperda and the cabbage looper Trichoplusia ni in assays at pH 6.5 (141). Whether these enzyme systems could metabolize isothiocyanates before being attacked by them in vivo, given the high pH of many lepidopteran larval midguts (140) and the binding affinities of isothiocyanates for proteins, especially at high pH (146,147) is uncertain. The acute isothiocyanate toxicities cited earlier suggest these compounds reach their sites of action despite high pH in the midget lumen. Curiously, the Solanaceae-specialist Manduca sexta not only ate diets containing allyl glucosinolate, but also excreted large quantities of intact allyl glucosinolate in their frass pellets (Chew, F.S.; Mechaber, W., Tufts Univ., unpublished data). The highly efficient excretory system of M. sexta, which rapidly excretes the nicotine normally encountered by these animals in their tobacco foodplant (148-151), apparently also handles allyl glucosinolate.

Egg susceptibility. Eggs of herbivorous insects are susceptible to isothiocyanates produced by wounded tissue near oviposition sites. Benzyl isothiocyanate is toxic to eggs and first instars of three fruitfly pests of Carica papya at levels comparable to those produced by ripening fruit (152). The effect of glucosinolates and hydrolysis products on eggs of the midge Dasineura brassicae was studied by placing eggs on Sephadex "beds" which contained representative hydrolysis products of allyl glucosinolate and 2-phenylethyl glucosinolate, including isothiocyanates and an epithionitrile (153,154). These were chosen because allyl glucosinolate is a major component in Brassica juncea and B. nigra, which are less suitable hosts of the midge, and phenylethyl glucosinolate was present in all Brassica species investigated (153). Allyl isothiocyanate was the most toxic, lethal at 10 ppm; this concentration is well within the range of the corresponding parent allyl glucosinolate in Brassica juncea cultivars (55). The effect is described as "fumigant", similar to early findings of the effects of isothiocyanates on vinegar flies (Drosophila melanogaster) (155,156). The nitrile 1-cyano-2-phenylethane was less toxic than allyl isothiocyanate, and the glucosinolate sinigrin was not toxic at any concentration comparable to its occurrence in green plant material. Oviposition sites for these midges may accumulate hydrolysis products because the midge used holes made by a seed weevil (155). Female midges may avoid laying eggs in recently-made holes, thereby permitting hydrolysis to approach completion and the products of hydrolysis to diffuse before placing eggs in them (154). The midge prefers B. napus to B. juncea under field conditions (153) and these preferences are correlated with consequences for larval development on the hosts (157). Among cultivars of a single species, e.g. B. napus or B. campestris cultivars with high or low glucosinolate contents, no preference in oviposition was observed (157), a finding similar to those for flea beetles (119) and some populations of Pieris brassicae (127).

Pathogens, competitors, and potential symbionts. Glucosinolates and their hydrolysis products have been tested in vitro against the crucifer pathogen Leptosphaeria maculans (stem canker) which attacks the green parts of crucifers (158). Allyl, 3-butenyl, 2-hydroxy-3-butenyl, indolyl and 1-methoxy-indolyl glucosinolates and the products generated by enzymatic hydrolysis were tested against an aggressive culture of the fungus. Individually, the thioglucosidase and glucosinolates had no effect on pathogen growth except slight reduction of growth upon exposure to allyl glucosinolate (158) at high concentrations that might be found in seeds, but would not be expected in leaves. Degradation of products of all glucosinolates except 5-vinyl-oxazolidine-2-thione, reduced fungal growth *in vitro*. The most toxic was the hydrolysis product of allyl glucosinolate at pH 7, presumably allyl isothiocyanate. This is consistent with findings on the leaf pathogen *Perenospora parasitica* (159) and early observations of the acute toxicity of allyl isothiocyanate (compared to others) on several pathogenic fungi (160). 3-butenyl isothiocyanate is also fungitoxic, whereas toxicity of the hydrolysis products of these alkenyl glucosinolates is reduced at pH 4, presumably reflecting lower toxicity of nitriles, a finding again consistent with observations on *P. parasitica* (159). Among the products of hydrolysis from indolyl glucosinolates, the indole-3-carbinol is more active in fungitoxic effects, although this effect could be due to differential solubility of hydrolysis products in the culture medium, rather than intrinsic differential toxicity (158). The observation that 2-hydroxy-3-butenyl hydrolysis products are not toxic to these pathogenic fungi suggests that goitrogenic glucosinolates, which are being bred out of crucifers because of their effects on livestock (4), confer no benefit in resistance to this pathogen.

The potential toxicity of the thioglucosidase-glucosinolate system *in situ* has been considered (161). Based on survival rates of the yeast seed pathogen *Nematospora sinecauda* in solutions of glucosinolates with and without thioglucosidase, it was concluded that seed infections could be arrested by fungal disruption of seed tissues and concomitant activation of the thioglucosidase-glucosinolate system (161). However, the individual components *in situ* did not prevent initial infection (161). Several potential hydrolysis products of glucosinolates in *Brassica juncea* were tested *in vitro*: allyl isothiocyanate was toxic at concentrations well within the range expected in seeds, but the corresponding nitrile was not toxic within this range. The effectiveness of glucosinolate hydrolysis in arresting invasions by the yeast was demonstrated by showing that seed homogenates of two low-glucosinolate cultivars of *B. campestris* were not toxic towards the yeast (161).

Products of glucosinolate hydrolysis sometimes provide substrates for reactions that affect pathogen-infected crucifers. Clubroot disease (*Plasmodiophora brassicae*), in which cell proliferation leads to the overgrowth symptoms and formation of clubs, has been proposed to result from *in situ* conversion of indolymethyl glucosinolate to the auxin precursor indole-3-acetonitrile (162). Experimental confirmation of this hypothesis was provided by the demonstration that *Brassica napus* tissues converted ^{14}C-indolylmethyl glucosinolate to the corresponding nitrile (163). Conversion of the ^{14}C-indole-3-acetonitrile to the auxin ^{14}C-indole acetic acid in *Brassica napus* is catalyzed by a nitrilase whose activity is higher in clubroot diseased plants (164). Autolysis of clubroot resistant cabbage tissue produces significantly less thiocyanate ion (presumably from hydrolysis of indolyl-3-methyl or *p*-hydroxybenzyl glucosinolates) than commercial cultivars (120 ug/g dry weight vs 204 ug/g dry weight) (165,166 cf. 167). However, no relationship between total glucosinolate levels and clubroot resistance occurs

(166-168). Resistance is partly mediated by glucosinolate composition; cabbages with lower levels of indole-3-methyl glucosinolate deprive the pathogen of its supply of auxin precursor.

Allelopathy. Various allelopathic effects have been attributed to crucifers; they have a reputation as poor companion plants. To some degree, this is a result of their growth characteristics, in which they often overtop other plants and exploit the growing space before more slowly growing plants become established (169). The infusion of glucosinolates into soil however, either from plant parts used as a cover crop and tilled under, or from roots of living plants, which leach or secrete intact glucosinolates into the growing medium (170) makes it worth inquiring about the effects of these compounds on other plants or plant symbionts (Vera, C.L.; McGregor, D.I.; Downey, R.K. Can. J. Plant Sci. 1988, in press). Effects of ionic thiocyanate (product from hydrolysis of indolylmethlyl glucosinolates and p-hydroxy- substituted glucosinolates) by growing cabbage, bean, and tobacco plants for 5 weeks in hydroponic solution with exogenous thiocyanate ion. Cabbage plants grew with up to 50 ppm thiocyanate ion; bean plants tolerated only 5 ppm and tobacco plants died at 5 ppm (171). With increasing levels of thiocyanate ion in the medium, leaf chlorosis increased in cabbage and beans; levels of thiocyanate ion increased in the leaf tissue of beans (in contrast to low accumulations of thiocyanate in leaves of cabbage), suggesting that the thiocyanate ion is an allelopathic agent (171). Other allelopathic effects of cruciferous plants may result indirectly from the effects of crucifers on vesicular-arbuscular mycorrhizal (VAM) fungi, which generally support symbiosis with crucifers poorly or not at all (172-175). Glucosinolate level in roots and susceptibility to penetration and establishment of symbiosis with Brassica cultivars was not correlated (173-175). VAM fungi germ tubes respond to the presence ofgrowth promoters (probably primary nutrients) that crucifer roots apparently exude at lower levels than compatible host species (173). While the growth of VAM fungi from spore germination to formation of appressoria (attachment points) to the roots looked identical near hosts and near crucifers, the process required more time near crucifers, and most of the VAM fungi growing with Brassica napus failed to produce arbuscles (174,175), the organ believed to be the site of exchange between VAM fungus and host plant. These findings were extended to wild crucifers as well as a range of Brassica cultivars, but no correlation between susceptibility to infection and glucosinolate levels or the types of glucosinolates found in seeds of the cultivars was found (175; Tommerup, I.C.; Sang, J.P. Univ. Western Australia, unpublished data). The involvement of hydrolysis products in this interaction is equivocal because the VAM fungi do not penetrate cell membranes. In normal hosts, the VAM fungus invaginates the cell membrane, and the intimate local association may persist for weeks; in highly incompatible interactions, however, this membrane disintegrates within an hour of the fungus penetrating the wall (176,177). Given the subcellular compartmentation of glucosinolates and thioglucosidases (85), it is possible these are involved. The critical tests have yet to be made. The observation that some species of crucifers, e.g. Raphanus raphanistrum, supported symbioses with four species of VAM fungi, of which three produced arbuscles, and two produced spores (175), is promising.

Literature Cited

1. Hegnauer, R. Phytochemistry 1986, 25, 1519-1535.
2. Rodman, J.E. In Phytochemistry and Angiosperm Phylogeny; Young, D.A.; Seigler, D.S., Eds.; Praeger Publ.: New York, 1981; p 43.
3. Fenwick, G.R.; Heaney, R.K., Mullin, W.J. CRC Critical Reviews in Food Science and Nutrition 1983, 18, 123-201.
4. Tookey, H.L.; VanEtten, C.H.; Daxenbickler, M.E. In "Toxic Constituents of Plant Foodstuffs", 2nd. ed.; Liener, I.E., Ed.; Academic: New York, 1980; p 103.
5. Larsen, P.O. In The Biochemistry of Plants; Stumpf, P.K.; Conn, E.E., Eds.; Academic: New York, 1981; Vol. 7, p 501.
6. Maheshwari, P.N.; Stanley, D.W.; van de Voot, F.R.; Gray, J.I. Can. Inst. Food Sci. Technol. J. 1980, 13, 28-32.
7. MacLeod, A.J.; Panesar, S.S., Gil, V. Phytochemistry 1981, 20, 977-980.
8. MacLeod, A.J.; Rossiter, J.T. Phytochemistry 1986, 25, 855-858.
9. Tookey, H.L. Can. J. Biochem. 1973, 51, 1654-1660.
10. Ohtsuru, M.; Hata, T. Biochim. Biophys. Acta 1979, 567, 384-391.
11. MacLeod, A.J.; Rossiter, J.T. Phytochemistry 1986, 25, 1047-1051.
12. Gil, V.; MacLeod, A.J. Phytochemistry 1980, 19, 2547-2551.
13. MacLeod, A.J.; Rossiter, J.T. Phytochemistry 1987, 25, 669-673.
14. Uda, Y.; Kurata, T.; Arakawa, N. Agric. Biol. Chem. 1986, 50, 2735-2740.
15. Uda, Y.; Kurata, T.; Arakawa, N. Agric. Biol. Chem. 1986, 50, 2741-2746.
16. Hasapis, X.; MacLeod, A.J. Phytochemistry 1982, 21, 559-563.
17. MacLeod, A.J.; Rossiter, J.T. Phytochemistry 1985, 24, 1895-1898.
18. Petroski, R.J.; Tookey, H.L. Phytochemistry 1982, 21, 1903-1905.
19. Petroski, R.J.; Kwolek, W.F. Phytochemistry 1985, 24, 213-216.
20. Benn, M. Pure & Appl. Chem. 1977, 49, 197-210.
21. Brocker, E.R.; Benn, M.H. Phytochemistry 1983, 22, 770-772.
22. Tookey, H.L.; Wolff, I.A. Can. J. Biochem. 1970, 48, 1024-1028.
23. Petroski, R.J. Plant Science 1986, 44, 85-88.
24. Daxenbichler, M.E.; Spencer, G.F.; Kleiman, R.; VanEtten, C.H.; Wolff, I.A. Anal. Biochem. 1970, 38, 373-382.
25. Gmelin, R.; Virtanen, A.I. Acta Chem. Scand. 1959, 13, 1464-1475.
26. Walker, N.J.; Gray, I.K. J. Agric. Food Chem. 1970, 18, 346-352.
27. Cole, R.A. Phytochemistry 1976, 15, 759-762.
28. Schluter, M.; Gmelin, R. Phytochemistry 1972, 1, 3427-3432.
29. Luthy, J.; Benn, M.H. Can. J. Chem. 1977, 55, 1028-1031.
30. Gil, V.; MacLeod, A.J. Phytochemistry 1980, 19, 2071-2076.

31. Hasapis, X.; MacLeod, A.J. Phytochemistry 1982, 21, 1009-1013.
32. Gil, V.; MacLeod, A.J. Phytochemistry 1980, 19, 1365-1368.
33. Bjorkman, R. In The Biology and Chemistry of the Cruciferae; Vaughan, A.J.; MacLeod, A.J.; Jones, B.M.G., Eds.; Academic London, 1976; p 191.
34. Bjorkman, R.; Janson, J.-C. Biochim. Biophys. Acta 1972, 276, 508-518.
35. Mithen, R.F.; Lewis, B.G.; Heaney, R.K.; Fenwick, G.R. Phytochemistry 1987, 26, 1969-1973.
36. Horn, P.J.; Vaughan, J.G. Phytochemistry 1983, 22, 465-470.
37. Gil, V.; MacLeod, A.J. Phytochemistry 1980, 19, 227-231.
38. Kaoulla, N.; MacLeod, A.J.; Gil, V. Phytochemistry 1980, 19, 1053-1056.
39. Gil, V.; MacLeod, A.J. Phytochemistry 1980, 19 1657-1660.
40. Prochazka, Z. Naturwissenschaften 1959, 46, 426.
41. MacLeod, A.J.; Panchasara, S.D. Phytochemistry 1983, 22, 705-709.
42. Gill, M.S.; MacLeod, A.J.; Moreau, M. Phytochemistry 1984, 23, 1937-1942.
43. Larsen, L.M.; Olsen, O.; Sorenson, H. Phytochemistry 1983, 22, 2314-2315.
44. Bjerg, G.; Fenwick, G.R.; Spinks, A.; Sorenson, H. Phytochemistry 1987, 26, 567-568.
45. Kjaer, A.; Malver, O. Phytochemistry 1979, 18, 1565.
46. Truscott, R.J.W.; Burke, D.G.; Minchinton, I.R. Biochem. Biophys. Res. Comm. 1982, 107, 1258-1264.
47. Truscott, R.J.W.; Minchinton, I.R.; Burke, D.G.; Sang, J.P. Biochem. Biophys. Res. Comm. 1987, 107, 1368-1375.
48. Truscott, R.J.W.; Minchinton, I.; Sang, J. J. Sci. Food Agric. 1983, 34, 247-254.
49. Truscott, R.J.W.; Johnstone, P.K.; Minchinton, I.R.; Sang, J.P. J. Agric. Food Chem. 1983, 31, 863-867.
50. Hanley, A.B.; Belton, P.S.; Fenwick, G.R.; Janes, N.F. Phytochemistry 1985, 24, 598-600.
51. Olsen, O.; Sorensen, H. Phytochemistry 1979, 18, 1547-1552.
52. Olsen, O.; Rasmussen, K.W.; Sorensen, H. Phytochemistry 1981, 20, 1857-1861.
53. Grob, K.; Matile, P. Phytochemistry 1980, 19, 1789-1793.
54. Carlson, D.G.; Daxenbichler, M.E.; VanEtten, C.H. J. Amer. Soc. Hort. Sci. 1985, 110, 634-638.
55. Hill, C.B.; Williams, P.H.; Carlson, D.G.; Tookey, H.L. J. Amer. Soc. Hort. Sci. 1987, 112, 309-313.
56. Carlson, D.G.; Daxenbickler, M.E.; Tookey, H.L.; Kwolek, W.F.; Hill, C.B.; Williams, P.H. J. Amer. Soc. Hort.Sci. 1987, 112, 179-183.
57. Carlson, D.G.; Daxenbichler, M.E.; VanEtten, C.H.; Kwolek, W.F.; Williams, P.H. J. Amer. Soc. Hort. Sci. 1987, 112, 173-178.
58. Sang, J.P.; Minchinton, I.R.; Johnstone, P.K.; Truscott, R.J.W. Can. J. Plant Sci. 1984, 64, 77-93.
59. Palmer, M.V.; Yeung, S.P.; Sang, J.P. J. Agric. Food Chem. 1987, 35, 262-265.
60. MacGibbon, D.B.; Allison, R.M. Phytochemistry 1970, 9, 541-544.

61. MacGibbon, D.B.; Allison, R.M. New Zealand J. Sci. 1971, 14, 134-140.
62. Phelan, J.R.; Vaughan, J.C. J. Exp. Bot. 1980, 31, 1425-1433.
63. Phelan, J.R.; Allen, A.; Vaughan, J.G. J. Exp. Bot. 1984, 35, 1558-1564.
64. Pihakaski, K.; Iversen, T.-H. J. Exp. Bot. 1976, 27, 242-258.
65. Iversen, T.-H.; Baggerud, C. Z. Pflanzenphysiol. 1980, 97, 399-407.
66. Iversen, T.-H.; Myhre, S.; Baggerud, C. Z. Pflanzenphysiol. 1983, 112, 391-401.
67. Pihakaski, K.; Pihakasi, S. J. Exp. Bot. 1978, 29, 335-345.
68. Reese, E.T.; Clapp, R.C.; Mandels, M. Arch. Biochem. Biophys. 1958, 75, 228-242.
69. Ohtsuru, M.; Tsuruo, I.; Hata, T. Agric. Biol. Chem. 1969, 33, 1309-1314.
70. Ohtsuru, M.; Tsuruo, I.; Hata, T. Agric. Biol. Chem. 1969, 33, 1315-1319.
71. Tani, N.; Ohtsuru, M.; Hata, T. Agric. Biol. Chem. 1974, 38, 1623-1630.
72. Oginsky, E.L.; Stein, A.E.; Greer, M.A. Proc. Soc. Exp. Biol. Med. 1965, 119, 360-364.
73. MacGibbon, D.B.; Beuzenberg, E.J. New Zealand J. Sci. 1978, 21, 389-392.
74. Weber, G., Oswald, S., Zoeller, U. Zeits. Pflanzenkrank Pflanzenschutz 1986, 93, 113-124.
75. Cole, R.A. Phytochemistry 1978, 17, 1563-1565.
76. VanEtten, C.H.; Daxenbichler, M.E. J. Agric. Food Chem. 1971, 19, 194-195.
77. Afsharypuor, S.; Lockwood, G.B. Plant Cell Rep. 1985, 4, 341-344.
78. Tang, C.-S. Phytochemistry 1973, 12, 769-773.
79. Tang, C.-S. Phytochemistry 1971, 10, 117-131.
80. Mahadevan, S. Annu. Rev. Plant Physiol. 1973, 24, 69-88.
81. Wallbank, B.E.; Wheatley, G.A. Phytochemistry 1976, 15, 673-766.
82. Cole, R.A.; Finch, S. Rep. natn. Veg. Res. Stn. 1978, 1977, 84.
83. Finch, S. In Insect-Plant Interactions; Miller, J.R.; Miller, T.A. Eds.; Springer-Verlag: New York, 1986, p 67.
84. Jorgensen, L.B. Nord. J. Bot. 1981, 1, 433-445.
85. Luthy, B.; Matile, P. Biochem. Physiol. Pflanzen 1984, 179, 5-12.
86. Bones, A.; Iversen, T.-H. Israel J. Bot. 1985, 34, 351-376.
87. Iversen, T.-H.; Baggerud, C.; Beisvag, T. Z. Pflanzenphysiol. 1979, 94, 143-154.
88. Matile, P. Biochem. Physiol. Pflanzen 1980, 175, 722-731.
89. Maheshwari, P.N.; Stanley, D.W., Beveridge, T.J.; van de Voort, F.R. J. Food Biochem. 1981, 5, 39-61.
90. Iversen, T.-H. Protoplasma 1970, 71, 451-466.
91. Jorgensen, L.B.; Behnke, H.-D.; Mabry, T.J. Planta 1977, 137, 215-224.
92. Behnke, H.-D.; Eschlbeck, G. Protoplasma 1978, 97, 351-363.
93. Grob, K.; Matile, P. Plant Sci. Lett. 1979, 14, 327-335.
94. Sukhija, P.S.; Loomba, A.; Ahuja, K.L.; Munshi, S.K. Plant Science 1985, 40, 1-6.

95. Tsuruo, I.; Hata, T. Agric. Biol. Chem. 1968, 32, 479-483.
96. Verschaffelt, E. Proc. Acad. Sci., Amsterdam 1910, 13, 536-542.
97. Nair, K.S.S.; McEwen, F.L. Can. Ent. 1976, 108, 1021-1030.
98. Feeny, P.; Paauwe, K.L.; Demong, N.J. Ann. Entomol. Soc. Amer. 1970, 63, 832-841.
99. Feeny, P. Ann. Missouri Bot. Gard. 1977, 64, 221-234.
100. Vincent, C.; Stewart, R.K. J. Chem. Ecol. 1984, 10, 33-39.
101. Hovanitz, W.; Chang, V.C.S.; Honch, G. J. Res. Lepid. 1963, 1, 249-259.
102. Petterson, J. Symp. Biol. Hung. 1976, 16, 203-208.
103. Petterson, J. Swedish J. Agric. Res. 1973, 3, 95-103.
104. Mitchell, N.D. Entomol. Exp. Appl. 1977, 22, 208-219.
105. Finch, S. Entomol. Exp. Appl. 1978, 24, 350-359.
106. Finch, S.; Skinner, G. Entomol. Exp. Appl. 1982, 31, 133-139.
107. Guerin, P.M.; Stadler, E. In Proc. 5th Symp. Insect-Plant Relationships, Visser, J.H.; Minks, A.K., Eds.; Cent. Agric. Publ. Doc.: Wageningen 1982, p 95.
108. Ellis, P.R.; Taylor, J.D.; Littlejohn, I.H. In Proc. 5th Int'l. Symp. Insect-Plant Relationships, Visser, J.H.; Minks, A.K., Eds.; Cent. Agric. Publ. Doc.: Wageningen 1982, p 131.
109. Thorsteinson, A.J. Can. J. Zool. 1953, 31, 52-72.
110. Nayar, J.K.; Thorsteinson, A.J. Can. J. Zool. 1963, 41, 923-929.
111. Schoonhoven, L.M. Proc. K. Ned. Akad. Wet. (C) 1967, 70, 556-568.
112. David, W.A.L.; Gardiner, B.O.C. Entomol. Exp. Appl. 1966, 9, 247-255.
113. Blau, P.A.; Feeny, P.; Contardo, L. Science 1978, 200, 1296-1298.
114. Wolfson, J.L. Environ. Entomol. 1982, 11, 207-213.
115. DeBoer, G.; Hanson, F.E. Entomol. Exp. Appl. 1984, 35, 177-193.
116. Wieczorek, H. J. Comp. Physiol. 1976, 106, 153-176.
117. Hicks, K.L. Ann. Entomol. Soc. Amer. 1974, 67, 261-264.
118. Nielsen, J.K. Entomol. Exp. Appl. 1978, 24, 41-54.
119. L.M.; Nielsen, J.K.; Ploger, A.; Sorensen, H. In Adv. Protection and Utilization of Cruciferous Crops; Sorenson, H., Ed.; Nijhoff/Junk: Dordrecht, 1985, p 230.
120. Mitchell, B.K.; Gregory, P. J. Comp. Physiol. 1981, 144, 495-501.
121. Sutcliffe, J.F.; Mitchell, B.K. J. Comp. Physiol. 1982, 146, 393-399.
122. Ma, W.C.; Schoonhoven, L.M. Entomol. Exp. Appl. 1973, 16, 343-357.
123. Stadler, E. Entomol. Exp. Appl. 1978, 24, 711-720.
124. Stadler, E. In Chemical Ecology of Insects; Bell, W.J.; Carde, R.T., Eds,; Sinauer: Sunderland, 1984, p 3.
125. Chew, F.S.; Robbins, R.K. Symp. R. Ent. Soc. Lond. 1984, 11, 65-79.
126. David, W.A.L.; Gardiner, B.O.C. Bull. Entomol. Res. 1962, 53, 91-109.
127. Rothschild, M. In Perspectives in Chemoreception and Behavior; Chapman, R.F.; Bernays, E.A.; Stoffalano, J.G., Eds.; Springer-Verlag: New York, 1986, p 175.

128. Rodman, J.E.; Chew, F.S. Biochem. Syst. Ecol. 1980, 8, 43-50.
129. Renwick, J.A.A.; Radke, C.D. Environ. Entomol. 1983, 12 446-450.
130. Slansky, F.; Feeny, P. Ecol. Monogr. 1977, 47, 209-228.
131. Wolfson, J.L. Entomol. Exp. Appl. 1980, 27, 223-232.
132. Louda, S.M.; Rodman, J.E. Biochem. Syst. Ecol. 1983, 11, 199-207.
133. Louda, S.M.; Farris, M.A.; Blua, M.J. J. Chem. Ecol. 1987, 13, 569-581.
134. Courtney, S.P.; Chew, F.S. Oecologia 1987, 71, 210-220.
135. Spencer, K.C. In Allelochemicals: Their Role in Agriculture, Forestry, and Ecology; Waller, G., Ed.; American Chemical Society Symp. 1987, 330, 275-288.
136. Hasapis, X.; MacLeod, A.J.; Moreau, M. Phytochemistry 1981, 20, 2355-2358.
137. Usher, B.F.; Feeny, P. Entomol. Exp. Appl. 1983, 34, 257-265.
138. Chew, F.S. Oecologia 1980, 46, 347-353.
139. Aplin, R.T.; d'Arcy Ward, R.; Rothschild, M. J. Entomol. (A) 1975, 50, 73-78.
140. Berenbaum, M. Amer. Natur. 1980, 115, 138-146.
141. Yu, S.J. J. Chem. Ecol. 1987, 13, 423-436.
142. Gould, F. Ecol. Entomol. 1984, 9, 29-34.
143. Yu, S.J. Pesticide Biochem. Physiol. 1983, 19, 330-336.
144. Ahmad, S.; Brattsten, L.B.; Mullin, C.A.; Yu, S.J. In Molecular Aspects of Insect-Plant Associations, Brattsten, L.B.; Ahmad, S., Eds.; Plenum Press: New York, 1986, p 73.
145. Brusewitz, G.; Cameron, B.D.; Chasseaud, L.F.; Gorler, K.; Hawkins, D.R.; Koch, H.; Mennicke, W.H. Biochem. J. 1977, 162, 99-107.
146. Bjorkman, R. Phytochemistry 1973, 12, 1585-1590.
147. Kawakishi, S.; Kaneko, T. Phytochemistry 1985, 24, 715-718.
148. Self, L.S.; Guthrie, F.E.; Hodgson, E. Nature 1964, 204, 300-301.
149. Self, L.S.; Guthrie, F.E.; Hodgson, E. J. Insect Physiol. 1964, 10, 907-914.
150. Brattsten, L.B. In Molecular Aspects of Insect-Plant Associations; Brattsten, L.B.; Ahmad, S., Eds.; Plenum Press: New York, 1986, p 211.
151. Berenbaum, M.R. In Molecular Aspects of Insect-Plant Associations; Brattsten, L.B.; Ahmad, S., Eds.; Plenum Press: New York, 1986, p 257.
152. Seo, S.T.; Tang, C.-S. J. Econ. Entomol. 1982, 75, 1132-1135.
153. Ahman, I. Entomol. Exp. Appl. 1985, 39, 247-253.
154. Ahman, I. J. Chem. Ecol. 1986, 12, 1481-1488.
155. Lichtenstein, E.P.; Strong, F.M.; Morgan, D.G. Agric. Food Chem. 1962, 10, 30-33.
156. Lichtenstein, E.P.; Morgan, D.G.; Mueller, C.H. Agric. Food Chem. 1964, 12, 158-161.
157. Ahman, I. Oikos 1985, 44, 191-194.
158. Mithen, R.F.; Lewis, B.G.; Fenwick, G.R. Trans. Brit. Mycol. Soc. 1986, 87, 433-440.
159. Greenhalgh, J.R.; Mitchell, N.D. New Phytol. 1976, 77, 391-398.

160. Walker, J.C.; Morell, S.; Foster, H.H. Amer. J. Bot. 1937, 24, 536-541.
161. Holley, R.A.; Jones, J.D. Can. J. Bot. 1985, 63, 521-526.
162. Butcher, D.N.; El-Tigani, S.; Ingram, D.S. Physiol. Plant Pathol. 1974, 4, 127-140.
163. Searle, L.M.; Chamberlain, K.; Rausch, T.; Butcher, D.N. J. Exp. Bot. 1982, 33, 935-942.
164. Rausch, T.; Butcher, D.N.; Hilgenberg, W. Physiol. Plant. 1981, 52, 467-470.
165. Chong, C.; Chiang, M.S.; Crete, R. HortScience 1981, 16, 663-664.
166. Chong, C.; Chiang, M.S.; Crete, R. Euphytica 1985, 34, 65-73.
167. Mullin, W.J.; Proudfoot, K.G.; Collins, M.J. Can. J. Plant Sci. 1980, 60, 605-612.
168. Hooker, W.; Walker, J.C.; Link, K.P. J. Agric. Sci. 1945, 70, 63-78.
169. Bell, D.T.; Muller, C.H.; Amer. Midl. Natur. 1973, 90, 277-299.
170. Elliott, M.C.; Stowe, B.B. Plant Physiol. 1971, 48, 498-503.
171. Ju, H.-Y.; Bible, B.B.; Chong, C. J. Chem. Ecol. 1983, 9, 1255-1262.
172. Medve, R.J. Amer. Midl. Natur. 1983, 109, 406-408.
173. Glenn, M.G.; Chew, F.S.; Williams, P.H. New Phytol. 1985, 99, 463-472.
174. Tommerup, I.C. New Phytol. 1984, 98, 497-502.
175. Tommerup, I.C. Proc. 6th North Amer. Conf. Mycorrhizae 1985, 330.
176. Maclean, D.J.; Sargent, J.A.; Tommerup, L.C.; Ingram, D.S. Nature 1974, 249, 186-187.
177. Maclean, D.J.; Tommerup, I.C. Physiol. Plant Path. 1979, 14, 291-312.

RECEIVED April 1, 1988

Chapter 13

Naturally Occurring Carbon–Phosphorus Compounds as Herbicides

Robert E. Hoagland

Southern Weed Science Laboratory, Agricultural Research Service, U.S. Department of Agriculture, Stoneville, MS 38776

Since the discovery in 1959 of the first naturally occurring compound containing a covalent C-P bond (2-aminoethylphosphonic acid, AEP), many phosphonate-related compounds have been identified in living systems. Some information on distribution, metabolic pathways and chemical properties exists, but no precise role for these compounds is known. In the early 1970's, the importance of some natural and synthetic phosphonates and their biological activity (e.g., antibiotics, herbicides) was recognized. The most successful synthetic phosphonate herbicide is glyphosate [*N*-(phosphonomethyl)glycine]. Bialaphos (L-2-amino-4-[(hydroxy)(methyl)phosphinoyl] butyryl-L-alanyl-L-alanine), isolated from *Streptomyces viridochromogenes*, was discovered and reported as an antibiotic (1972),and later found to be herbicidal. Phosalacine, a bialaphos analog, has been isolated from *Kitasatosporia phosalacinea*. Peptidases cleave amino acid moieties from the latter two compounds yielding the active component phosphinothricin whose site of action is inhibition of glutamine synthetase in plants. This compound is being marketed as the herbicide glufosinate. The biosynthetic pathway for bialaphos and the cloning and characterization of genes that code for these conversions have recently been elucidated.

Naturally occurring compounds with potential for use as herbicides have recently become of great interest to weed scientists and agricultural chemists. This interest may be attributed to the facts that natural phytotoxins (1) have a co-evolved species specificity, (2) often have diverse chemistries that could lead to completely novel synthetic herbicides, (3) may have low mammalian toxicity and pose less of a biohazard because they are specific and more biodegradable than many commercial herbicides, and (4) may also be less costly to register as herbicides by regulatory agencies such as EPA. Furthermore, microbial plant pathogens may be directly used to control weeds, and two microorganism products currently are commerically available for weed control (1). This has generated interest in toxins produced by these pathogens as pesticides. Many naturally occurring phytotoxic compounds have been found to be produced by higher plants, bacteria, fungi, and animals. An excellent review of herbicides from some of these natural sources has recently been published (2).

The focus of this chapter will be to examine some of the naturally occurring phosphonate compounds (i.e., bialaphos, phosalacine, and phosphinothricin) with respect to their natural sources, biochemistry, and biological properties. Special reference is given to their use and/or potential use as herbicides, including their relationship to some synthetic phosphonates that have been examined for biological activity.

Discovery of Carbon-Phosphorus Compounds in Nature

Phosphorus is required for the growth and reproduction of all plants and animals. Organic phosphorus compounds in the form of orthophosphates are involved in these many life processes and are contained in basic biochemical constituents (i.e., proteins, carbohydrates, nucleic acids, and phospholipids). They are also used in the generation of high energy bonds, provide buffering capacity, and the phosphate moiety imparts water solubility to these compounds (3). In spite of the multitude of compounds occurring as orthophosphate-oxygen compounds (esters, diesters, and phosphoric acid anhydrides), some organophosphate compounds occur in nature with carbon-phosphorus bonds; i.e., phosphonates. These compounds had been considered as possible naturally occurring compounds by Chavane (4) as early as 1947, but actual identification of the first phosphonate in a living system did not occur until 1959 when Horiguchi and Kandatsu (5) found 2-aminoethylphosphonic acid (AEP, Figure 1) in an amino acid hydrolysate from rumen protozoal lipid. Over the past 30 years, numerous reports have shown the presence of AEP and other phosphonates in a wide range of organisms, including bacteria, protozoa, coelenterates, molluscs, and unicellular plants.

Detection of Phosphonates

The chemistry of phosphonates has developed relatively slowly compared to that of phosphates. This is due to the fact that the carbon-phosphorus bond is difficult to synthesize and because of the vast amount of information on and distribution of phosphate and orthophosphate esters. Furthermore, early studies on natural phosphonates were slow due to the lack of sensitive methodology for detection of the carbon-phosphorus bond. These early studies consisted mostly of various chromatographic and isolation techniques. Early analytical methods for the detection of AEP in biological materials have been reviewed (6). The procedures usually involved ion-exchange chromatography for the separation of AEP followed by detection using ninhydrin. Various chromatographic modifications of these procedures have been expanded by several reseachers (7-10). More recently, gas-liquid chromatography and mass spectrometry have been employed for aminophosphonate determination (11-13). Use of these latter two techniques often requires derivitization (14-16). Isotachophoresis has also been used to analyze aminophosphonates (17). The use of sophisticated techniques such as NMR has made analysis of tissues and tissue extracts more rapid and sensitive. NMR has been used by a number of researchers for measuring phosphonates in tissue extracts (18-21), various biological fluids (22, 23), whole cells (24, 25), and soil (26).

Biosynthesis and Metabolism of AEP

Phosphonates can undergo various biochemical transformations which involve synthesis or catabolism of the carbon-phosphorus bond, or that proceed leaving the C-P bond intact. This latter category includes synthesis and degradation reactions of phosphonolipids, phosphonopeptides, other complex compounds, and N-methylation and transamination (27, 28). Before 1983, enzymatic conversion of phosphoenolpyruvate to 3-phosphonopyruvate via intramolecular phosphate-phosphonate rearrangement involving PO_3H_2 migration from oxygen to carbon was generally accepted (29-34). However, the enzyme responsible for this has not been isolated and characterized. In 1983 alkylphosphinous acids were determined as precursors of aminophosphonates (including phosphinothricin) and a novel pathway of C-P bond biosynthesis was proposed which involves a reduction of the phosphate to phosphite ester prior to rearrangement of phosphoenolpyruvate (35-37) (Figure 2). Reduced 3-phosphonopyruvate may then be oxidized to 3-phosphonopyruvate or directly undergo transformations leading to various alkylphosphonate compounds. Confirmation of this proposal requires isolation of reduced 3-phosphonopyruvate and the enzymes involved. Details on some of these biosynthetic transformations have recently been reported and will be presented later in the section concerning synthesis of bialaphos.

The use of labeled presursors of AEP in intact cells and cell-free extracts of *Tetrahymena* showed that AEP was formed from phosphonoenolpyruvate via 3-phosphonopyruvate, but the exact mechanism of 3-phosphonopyruvate conversion to AEP was unclear. Presently two pathways are proposed as summarized in Figure 3. One proposal suggests amination followed by decarboxylation of 3-phosphonoalanine (31). A second mechanism suggests decarboxylation as the first step followed by phosphonoacetaldehyde amination (33). Both pathways may be operative in some systems (33). Biosynthesis of phosphonolipds and N-methylated derivatives have been reviewed elsewhere (38, 39).

Details of catabolic routes of natural phosphonates have not been extensively examined except for AEP and 3-phosphonalanine. In bacteria AEP is degraded in a two-step process involving transamination whereby AEP donates its amino group to pyruvate and is converted to phosphonoacetaldehyde (Figure 4) (40-43). Recently, an enzyme responsible for this transamination step has been purified to homogeneity from *Pseudomonas aeruginosa* (44).

Occurrence and Distribution of Phosphonates

AEP and other phosphonates have been found in a wide diversity of living organisms; but these plant, animal, and microbial sources represent only a minute fraction of the total living organisms. The true extent of the distribution of these compounds is confounded by (1) the lack of complete surveys within and among species, (2) the use of insensitive techniques for detection, and (3) the

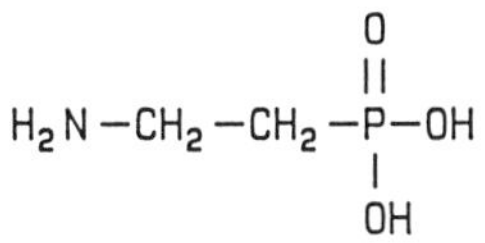

```
                  O
                  ||
H2N—CH2—CH2—P—OH
                  |
                  OH
```

Figure 1. Structure of 2-aminoethylphosphonate (AEP).

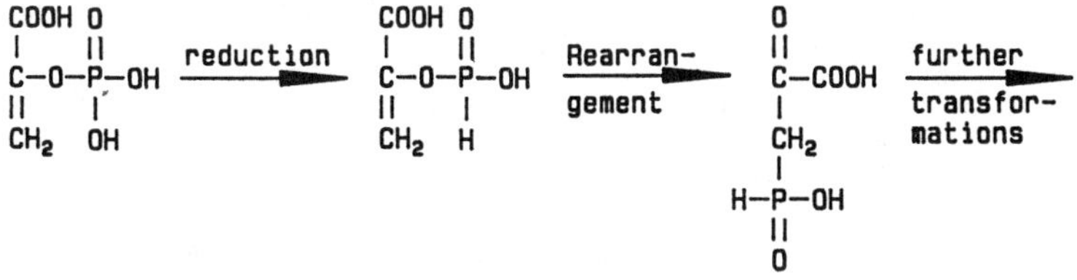

```
COOH O                        COOH O                           O
|    ||      reduction        |    ||       Rearran-           ||           further
C—O—P—OH   ──────────►    C—O—P—OH   ──────────►    C—COOH   ──────────►
||   |                        ||   |        gement             |            transfor-
CH2  OH                       CH2  H                           CH2          mations
                                                               |
                                                             H—P—OH
                                                               ||
                                                               O
```

Figure 2. Proposed pathway for C-P bond biosynthesis.

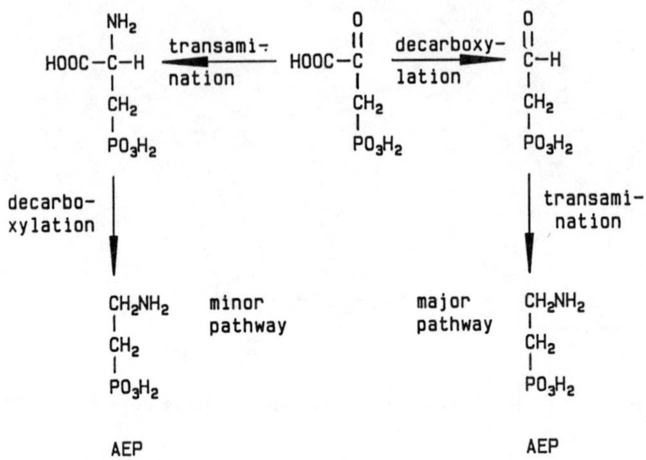

Figure 3. Biosynthesis of 2-aminoethylphosphonate (AEP).

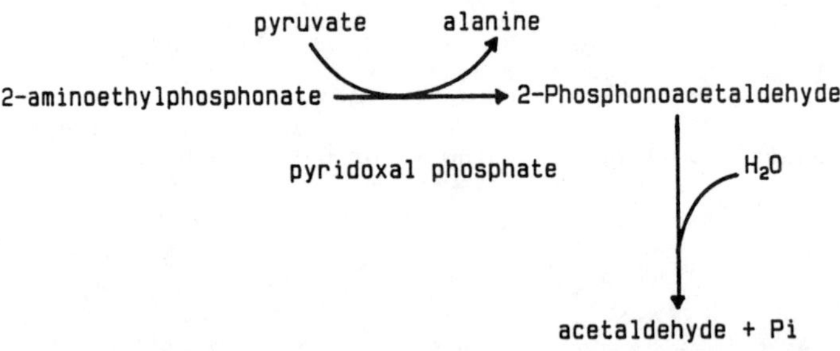

Figure 4. Degradation of 2-aminoethylphosphonate (AEP) by *Bacillus cereus*.

distribution of phosphonates via food chains to organisms lacking biosynthesis routes for these materials. Table I summarizes the date and source of isolation of various naturally-occurring phosphonate compounds. Compounds listed here are amino-phosphonates, but other phosphonates which are not substituted amines have been found. These include the antibiotics phosphomycin, phosphazomycin A, and phosphazomycin B; the phosphinous acid, 2-hydroxy-2-carboxy-ethylphosphinate; and oxophosphonates. These oxophosphonates have been shown to be intermediates in the biosynthesis of AEP and β-phosphonalanine (37).

Phosphonates usually represent a small portion of the total phosphorus content of individual species. For example, in fractions of several marine animals, phosphonate percentages ranged from 0.16 to 0.75% of total phosphorus (50). Aminophosphonic acids isolated from plankton represented 3% of the total phosphorus content (51). Phosphonate studies in bacteria suggest low levels; i.e., less than 1% of total phosphorus (7, 9, 52-55). Phosphonates in protozoa associated with the alimentary tract in ruminants are in the range of several percent of total phosphorus, but in other protozoa, such as *Euglena gracilis*, the level is 0.10% of total phosphorus (53) to 15% in *Tetralymena pyriformis* (14). *T. pyriformis* contains relatively high lipid concentrations in microsomes and cilia (56, 57); and due to its high phosphonate content and ease of laboratory culture, this organism has been used in many studies pertaining to the biochemistry of AEP and lipids containing AEP.

The phylum Coelenterata has been shown to contain many organisms with phosphonate compounds, some with levels up to 50% of total phosphorus (50). Phosphonate occurrence in other invertebrates is limited. AEP has, however, been detected in three Echinodermata [two species of starfish (50, 58) and a sea urchin (59), and in one species each of *Nemathelminthes* (53), *Spongia* (59)], and *Annelides* (59). Several arthropods (marine crustaceans) also contain AEP (8). The highest level of AEP in living tissue yet reported is in snail (*Helisomia* sp) eggs where nearly all phosphorus occurs as phosphonate phosphorus and 85% of this is AEP (24). The isolation of AEP from tissues of various ruminating chordates, including brain (60), milk (61), liver (62), bile (63), blood (64), and sperm (64) have suggested to some researchers that the aminophosphonates present are released by digestion of protozoa. Phosphonates have also been found in non-ruminating mammals, including rat (12), guinea pig (65), and man (66, 67). AEP has been detected in human tissues such as brain, liver, skeletal muscle, and heart in amounts of 0.01, 0.02, 0.04, and 0.06% of wet weight, respectively (11).

Phosphonate distribution in the plant kingdom is apparently limited to the lower fungi and unicellular plants, but as mentioned previously, surveys are incomplete. AEP has been detected in marine phytoplankton (68, 51) and in a fungus, *Pythium prolatum* (69).

Enzyme Inhibition by Various Phosphonate Compounds. The earliest report of enzyme inhibition by phosphonate compounds was the report of glutamine synthetase (GS) inhibition by phosphonic analogs of glutamic acid (70). Since then, numerous reports have indicated interactions of phosphonates with many different enzymes. Table II is a compilation of some enzymes which show *in vitro* inhibition by various substituted phosphonates. Since phosphonates inhibit a wide

Table I. Date and Source of First Isolation of Various Aminophosphonates

Chemical Name	Common Name	Date and Source of first isolation	Ref.
2-aminoethyl phosphonate	ciliatine (AEP)	rumen protozoa (1959)	5
2-carboxy-2-aminoethyl phosphonate	β-phosphonoalanine	*Zoanthus sociatus* *Tetrahymena pyriformes* (1964)	45
2-(*N*-methyl)2-aminoethyl phosphonate	-	*Anthopleura xantogrammica* (1967)	46
2-(*N*,*N*-dimethyl)2-aminoethyl phosphonate	-	*A. xantogrammica* (1967)	46
2-(*N*,*N*,*N*-trimethyl-2-aminoethyl phosphonate	-	*A. xantogrammica* (1967)	46
3-amino-3-carboxy-propyl (methylphosphonate)	phosphinothricin	*Streptomyces viridochromogenes* (1972)	47
2-amino-1-hydroxyethylphosphonate	-	*Acanthamoeba castellanii* (1973)	48
4-carboxy-4-amino-2-*t*-butenyl phosphonic acid	-	*Streptomyces plumbeus* (1976)	18
3-(*N*-acetyl-*N*-hydroxyamino)propyl phosphonate	antibiotic FR-900098	*Streptomyces rubellomurines* (1981)	49
3-(*N*-acyl-*N*-hydroxyamino)propyl phosphonate	fosmidomycin	*Streptomyces lavendulae* (1981)	49
3-(*N*-acetyl-*N*-hydroxyamino)-2-hycroxypropylphosphonate	antibiotic FR-33289	*Streptomyces rubellomurinus* subsp. *indigoferus* (1981)	49
3-(*N*-formyl-*N*-hydroxyamino)-1-*t*-propenylphosphonate	antibiotic FR-32863	*Streptomyces lavendulae* (1981)	49
2-hydroxyphenyl-1-aminoethyl phosphonate	-	*Actinomycetes* strain K-26 (1982) *Actinomadura spiculosospora*	27
2-aminoethylphosphinate	-	*Streptomyces hygroscopicus* (1983)	37
2-amino-2-carboxyethyl phosphonate	-	*Streptomyces hygroscopicus* (1983)	37
3-amino-3-carboxy-propyl(phosphonate)	-	*Streptomyces hygroscopicus* (1983)	37

Table II. *In vitro* Enzyme Inhibition by Various Aminophosphonates

Enzyme name and EC number	Ref.	Enzyme name and EC number	Ref.
Adenylosuccinase (4.3.2.2)	71	Carboxypeptidase A (3.4.2.1)	91, 92
Adenylosuccinase synthetase (6.3.4.4)	71	Carnosine synthetase (6.3.2.11)	93
D-Ala-D-Ala synthetase (6.3.2.4)	72, 73, 74	Cholinesterase (3.1.1.8)	94
Alanine dehydrogenase (1.4.1.1)	71	Chymotrypsin (3.4.21.1)	95
L-Alanine racemase (5.1.1.1)	75, 76 73	Cytidine deaminase (3.5.4.5)	96
Alkaline phosphatase (3.1.3.1)	77	5-Dehydroquinate (4.6.1.3)	83
D-Amino acid trans-aminase (2.6.1.21)	78	3-Deoxy-D-arabino-heptulosonate-7-phosphate synthase (4.1.2.15)	97
Angiotensis converting enzyme (3.4.15.1)	79, 80, 81, 82	Elastase (3.4.21.11)	95
Anthranilate synthase (4.1.3.27)	83	5-Enolpyruvylshikimate-3-phosphate synthase 5-phosphoshikimate-1-carboxy-vinyl-transferase (2.5.1.19)	98
Arginase (3.5.3.1)	71, 84	Ethanolamine phosphate-cytidyl transferase (2.7.7.14)	99
Arginine transamidinase (3.5.3.6)	71, 84	Glutamine synthetase (6.3.1.2)	70, 87, 100-104
Asparagine synthetase (6.3.1.1)	85	-Glutamylcysteine synthetase (6.3.2.2)	105
Asparagine synthetase (glutamine hydrolyzing, 6.3.5.4)	86	Isoleucyl-t-RNA synthetase (6.1.1.5)	106
Asparaginyl-t-RNA synthetase (6.1.1.12)	87	Leucine aminopeptidase (3.4.11.1, cytosolic)	107
Aspartase (4.3.1.1)	71	Leucine aminopeptidase (3.4.11.2, mircosomal)	107
Aspartate amino-transferase (2.6.1.1)	71	Leucyl-t-RNA synthetase (6.1.1.4)	106
Aspartate carbamoyl transferase (2.1.3.2)	88-90	Sphingosine 1-phosphate lyase (4.7.7.)	121
Methionyl-t-RNA synthetase (6.1.1.10)	108, 109, 110	Tyrosinase (1.10.3.1)	107
Ornithine carbamoyl transferase (2.1.3.3)	111-113	Tyrosine amino transferase (2.6.1.5)	122
Phenylalanyl-t-RNA synthetase (6.1.1.b)	114	Tyrosine decarboxylase (4.1.1.25)	122
Phospholipase C (3.1.4.3)	115	Tyrosyl-t-RNA synthetase (6.1.1.1)	122, 114
Pyruvate kinase (2.7.1.40)	116	UDP-NAMA-L-Ala synthetase (6.3.2.8)	72, 73
Rennin (3.4.4.15)	118-120	Valyl-t-RNA synthetase (6.1.1.9)	106-110, 114, 123-124
Serine transhydroxy-methylase (2.1.2.1)	120		

range of enzyme systems, it is highly probable that specific phosphonates can be found which will inhibit key plant enzymes, making them possible herbicide candidates. Some phosphonate compounds can serve as substrates for several enzymes. Readers are referred to a more comprehensive review for further information on enzyme phosphonate interactions (38).

Phosphonates as Herbicides, Plant Growth Regulators, and/or Fungicides

There are many reports in the scientific and patent literature describing phosphonates as herbicides, fungicides, and plant growth regulators. Table III is a compilation of some of these compounds and shows their wide diversity of chemical structure and biological activity. The most widely known and used of the synthetic phosphonates with agricultural applications is the herbicide glyphosate, *N*-(phosphonomethyl)glycine, which was introduced in 1971 (124). This compound combines high, broad-spectrum herbicidal activity with low mammalian toxicity and a short half-life in soils. The interest in glyphosate is demonstrated by the voluminous reports and patents dealing with its application and biochemical and physiological action. Also, its utility as a herbicide generated an enormous response in the synthesis of analogs. Perhaps up to a thousand glyphosate derivatives and analogs have been synthesized and screened for herbicidal activity (151). This stimulus of synthesis produced some compounds with herbicidal activity, but none with activity comparable to glyphosate. Biochemical and physiological studies on the effects of glyphosate have been examined by various workers. Figure 5 indicates some of the major enzyme systems tested for *in vitro* and *in vivo* effects of glyphosate. The mode of action of glyphosate is considered to be the inhibition of 5-enolpyruvylshikimate-3-phosphate synthase activity (168). There probably exist nearly 2,000 publications in the scientific literature on glyphosate and its activity in plants, soils, and bacterial systems. The reader is referred to reviews which cover the many aspects of the important research conducted on this herbicide (169, 170, 171). Glyphosate is not metabolized appreciably in plants, but some microorganisms degrade it rapidly, and the major metabolite is aminophosphonic acid. This compound is essentially non-toxic to fish and wildlife.

The next three most important compounds in Table III with regard to the present topic are phosphinothricin, bialaphos, and phosalacine which are naturally occurring substituted phosphonate compounds with potent herbicidal activity. Bialaphos and phosalacine are peptides containing a phosphinothricin moiety. Bialaphos has recently been marketed as a herbicide in Japan. This naturally occurring herbicidal compound is non-selective and controls several species of monocot and dicot weeds. It has a half-life of about 20-30 days in soil and has low toxicity to non-target organisms (126). Hoechst AG is developing chemically synthesized phosphinothricin as the herbicide HOE 39866, with the common name glufosinate (125). In contrast to the many literature references on glyphosate, relatively few reports exist on the biochemical, physiological, and applied aspects of bialaphos, phosalacine, and phosphinothricin. Most of these deal with the

Table III. Phosphonate Compounds with Herbicidal, Fungicidal, and Plant Growth Regulating Activity

Compound	Name or Remarks	Reference
Herbicides:		
$HOOC\text{-}CH_2\text{-}NH\text{-}CH_2\text{-}PO_3H_2$	glyphosate	69
$HOOC\text{-}CH(NH_2)\text{-}CH_2\text{-}CH_2\text{-}P(CH_3)O_2H$	phosphinothricin and glufosinate, (ammonium salt of phosphinothricin)	125
phosphinothricin-ala-ala	bialaphos	126
phosphinothricin-ala-leu	phosalacine	127
$CH_3\text{-}NH\text{-}CH_2\text{-}PO_3H_2$	post-emergence herbicide	128
$R\text{-}NH\text{-}CH_2\text{-}PO_3H_2$	herbicide	129
$(HOOC\text{-}CH_2\text{-}NH\text{-}CH_2)_2PO_2H$	post-emergence herbicide	130
$R'\text{-}NH\text{-}C(R)_2\text{-}PO(OR)_2$	R'=alky or acyl and R=alky or aryl (herbicide, defoliant)	131-136
$CH_3\text{-}CH(NH_2)\text{-}P(CH_3)O_2H$	and N-glycyl derivative	137
$R\text{-}C(PO_3H_2),NR'R''$	R=R'=H and R"=pyridyl herbicide	138-147
$Ch_3\text{-}CH(NH_2)CONH\text{-}CH(CH_3)PO_2H_2$	and other peptides, contact herbicides	148
$CH_3\text{-}CH_2\text{-}O\text{-}PO_2^-\text{-}CO\text{-}NH_2\ NH_4^+$	fosamine ammonium	172, 175

Continued on next page.

Table III. Continued.

Compound	Name or Remarks	Reference
$CH[(NHCH_2COOR)-PO_2H-O-C_6H_5]_2$	herbicide	149
$(CH_3)_2-Ph-N(CH_2-PO(OCH_3)_2-CO-CH_2Cl$	herbicide	150
Herbicide - Plant Growth Regulator:		
$HOOC-CH_2N(CH_2PO_3H_2)_2$		151-154
$H_2N-CH_2-PO_3H_2$		131, 155
$(H_2NCH_2)_2PO_2H$		131
$CH_3-CH_2CH_2-PO_3H-O-CH_2CH_3$	and analogs, PGR	156
$CH_3(CH_2)_3NH-Ph-PO(OCH_2CH_2CH_2CH_3)_2$		157-160
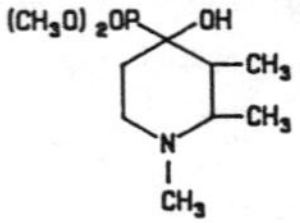 [structure: piperidine ring with $(CH_3O)_2OP$, OH, CH_3, CH_3, N–CH_3]		161
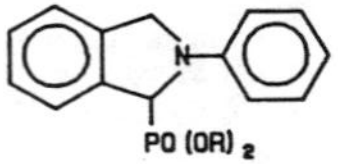 [structure: isoindoline with N-phenyl and $PO(OR)_2$]		162
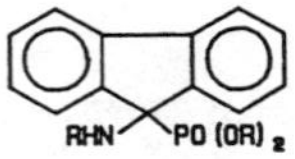 [structure: fluorene with RHN and $PO(OR)_2$]	maximum activity when R=n-butyl	163
Fungicides:		
$HOOC-CH(NH_2)-CH_2-PO_2HR$	R=methyl or ethyl	164, 165
$R'-NH-CR(CH_3)PO_2H_2$	R=H_3 or CH_3; R'=aryl	166
$R'_2NCH_2PO(OR)_2$	R=alkyl; R'=aklyl or cycloalkyl	167

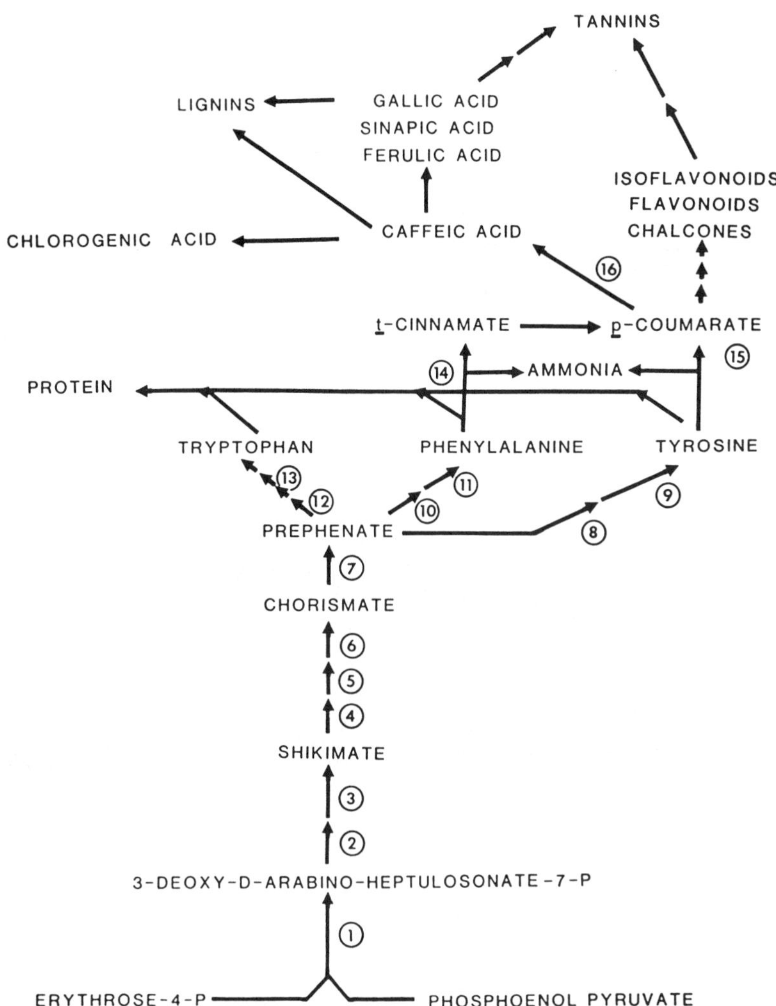

Figure 5. Schematic of various intermediates and products of phenolic metabolism and some of the enzymes which have been examined for effects of glyphosate. Enzymes: 1, 3-deoxy-2-oxo-D-arabinoheptulosate-7-phosphate synthase; 2, 5-dehydroquinate synthase; 3, shikimate dehydrogenase; 4, shikimate kinase; 5, 5-enolpyruvylshikimate-3-phosphate synthase; 6, chorismate synthase; 7, chorismate mutase; 8, prephenate dehydrogenase; 9, tyrosine aminotransferase; 10, prephenate dehydratase; 11, phenylalanine aminotransferase; 12, anthranilate synthase; 13, tryptophan synthase; 14, phenylalanine ammonia-lyase; 15, tyrosine ammonia-lyase; and 16, polyphenol oxidase. Reproduced from Ref. 169.

effects of phosphinothricin on GS since the compound was initially found to be a strong GS inhibitor and because it is an analog of another potent GS inhibitor, methionine sulfoximine. A brief review of the biochemical, physiological, and mode of action research on these naturally occurring phosphonate herbicides will be presented in a later section of this chapter.

Fosamine-ammonium, used primarily for brush control (172), is translocated in both susceptible and tolerant woody plants, but susceptible plants absorb and translocate significantly greater quantities of the herbicide than do tolerant species. It was proposed that the phytotoxicity of fosamine-ammonium is due to its action or the action of a metabolite at or above the point of contact (173). Mesquite growth was inhibited for up to three years following treatment with this compound. These studies indicate the herbicide inhibites protein and nucleic acid synthesis (174). Fosamine- ammonium is rapidly degraded by microorganisms. The half-life in soil is about 10 days, compared to about 2-3 weeks in greenhouse-grown apple seedlings (175).

Some phosphonates are used as commercial plant growth regulators. Ethephon (2-chloroethylphosphonic acid) is used to accelerate ripening, induce flowering, promote abscission, and to stimulate color (175). The action of this compound is via production of ethylene, a product of ethephon degradation. The glyphosate analog, glyphosine [*N*,*N*,-bis(phosphonomethyl)glycine] is used commercially as a suger cane ripener. At high levels it shows some herbicidal activity and causes growth cessation, chlorosis, and desication (155). This compound can reduce numbers of 70-S ribosomes and chloroplastic ribosomal RNA, but has no effect on photophosphorylation (152).

Other phosphonate plant growth regulators are propylphosphonic acid, ethylhydrogen propylphosphonate, and NIA 10637 (176). Ethyl propylphosphonate retards the growth of various woody species, and causes physical responses such as increased cold hardiness, seed germination induction, and affects flowering and fruiting. Several other aromatic phosphonates are reported to have plant growth regulator activity. Several phosphonates have potential as fungicides. Various organophosphonates (some of the organophosphonate class) have been developed over the past 15 years which have relatively low mammalian toxicity and environmental persistance compared to the organochlorine insecticides (177). Some of these compounds are presented and discussed elsewhere (178). There are apparently no reports of naturally occurring phosphonates with useful insecticidal activity.

Phosphinothricin, Bialaphos, and Phosalacine as Herbicides

Discovery and Early Development. Phosphinothricin was isolated from cultures of *Streptomyces viridochromogenes* (47) and from *S. hygroscopicus* as the phosphinothricylalanylalanine peptide (now called bialaphos) (179). This tripeptide was shown to be active as a bactericide against Gram-negative and Gram-positive bacteria and as a fungicide against the fungi *Botrytis cinerea*, *Rhizoctonia solani*, and *Piricularia oryzae* (180). The herbicidal activity of bialaphos was described in 1979 (181). Phosphinothricin was first synthesized in 1972 (47), and details and discussion of its synthesis are presented elsewhere (104). Herbicidal properties of phosphinothricin as a commerically synthesized ammonium salt were first reported in 1977 (182). Phosalacine is a phosphonate peptide analog of bialaphos differing only in that it possesses a C-terminal leucine residue instead of alanine. This compound was reported to be a herbicidal isolate from *Kitasatosporia phosalacinea*, an Actinomycete from soil (183). *Streptomyces* and *Actinomyces* belong to separate families within the same order (Actinomycetales) and produce very similar phosphonate peptides. More recently, phosalacine has been patented as a defoliant for hops (*Humulus lupulus*) (184).

General Properties and Herbicidal Activity. Bialaphos is active on the foliage of various weeds, including perennials (185). Its phytotoxicity is expressed more slowly than that of paraquat, but more rapidly that that of glyphosate. When five weeds with an LD_{50} range of <0.125 to 8.5 kg/ha were tested under greenhouse conditions (186), phosphinothricin was found to have an LD_{50} for sicklepod (*Cassia obtusifolia*) that is nearly 70-fold higher than for couchgrass (*Elymus repens*). Phosphinothricin, bialaphos, and phosalacine are all water soluble compounds that are absorbed by plants and microorganisms. Translocation is also rapid. Metabolism by peptidases in plants or in some microorganisms releases the active ingredient, phosphinothricin (Figure 6). Metabolism of bialaphos has also been reported in mice (187). After feeding, no parent compound was detected, but 3 and 4 metabolites were found in feces and urine, respectively, and the major metabolite was phosphinothricin. Metabolism of bialaphos occurs in the soil resulting in a half-life of about 20 to 30 days (126).

Other Biochemical Effects of Glufosinate and Related Compounds. Few reports are available on the effects of phosphinothricin on plant biochemical sites other than GS. Free IAA levels were lowered via increased conjugation and oxidation of IAA by glyphosate (188). Glufosinate was even more active than glyphosate while glyphosine and aminomethylphosphonic acid had lower activity. All of these compounds selectively reduced ethylene production.

Phenylalanine ammonia-lyase (PAL) activity has been shown to be increased dramatically over a several day time course by the phosphonate herbicide glyphosate (169, 189). Similar studies with glufosinate showed that this herbicide increased extractable PAL activity at 24 and 48 hours (about 2-fold above control levels) in both light- and dark-grown soybean plant tissue, but at 96 hours enzyme levels were reduced below control levels. The major metabolite of glyphosate, aminomethylphosphonate, and the glyphosate analog, glyphosine had little growth effect or effect on PAL activity (154). Tests of other phosphonates for herbicidal activity effects on PAL showed that after 48 hours, *N*-(phosphonomethyl)-2-aminoethanol, *N*-(phosphonomethyl)-3-aminopropanol, 3-phosphonopropionate, and 2-amino-4-phosphobutyrate did not significantly effect growth or extractable PAL activity in soybean seedlings (190). In this same study, however, phosphonoacetate was very toxic and caused a slight increase in extractable PAL activity levels while phosphonoformate and *N*-phosphonoacetyl-L-aspartate reduced growth and PAL by 25 and 30%, respectively.

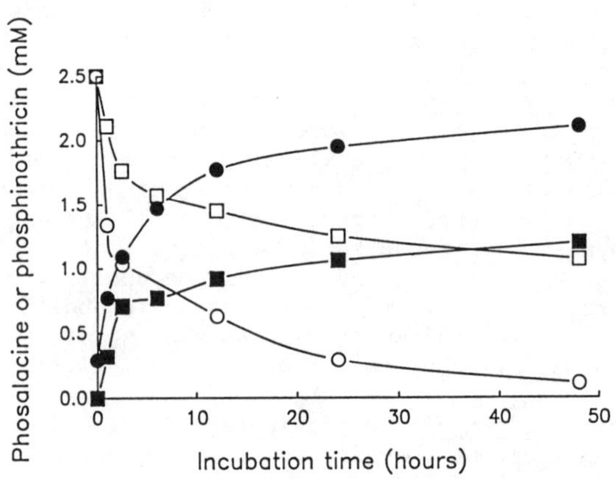

Figure 6. Formation of phosphinothricin (●, ■) from phosalacine (○, □) by cell free extracts from Bacillus subtilis (●, ○) and spinach leaves (■, □). Redrawn with permission from Ref. 183. Copyright 1984 Japanese Antibiotics Research Association.

Phosphinothricin can also increase extractable PAL activity on a fresh weight and a specific activity basis in soybean tissues (Hoagland, unpublished data). These effects on glyphosate, glufosinate, and phospinothricin are considered to be only secondary with regard to its mode of action which involves GS inhibition, as discussed below.

Inhibition of GS by Naturally Occurring Phosphonates. It has been known since 1959 that some phosphonic and phosphinic acid analogs of glutamic acid were inhibitory to glutamine synthetase (70). Since then, various other P-C containing analogs have been synthesized and some also inhibit this enzyme from mammalian (191-193) and plant sources (101, 104, 191, 194). Phosphinothricin was reported as an inhibitor of GS as early as 1972 (47). There are also known naturally occurring compounds other than those containing C-P bonds that exhibit potent inhibition of GS activity. Comparative structures of these natural products and substituted phosphonate analogs that are GS inhibitors are given in Figure 7. Oxetin, derived from *Streptomyces* sp., is a recently discovered GS inhibitor (195). Tabtoxinine-β-lactam, also an inhibitor of GS (196), is produced by *Pseudomonas tabaci* (197). The GS inhibitor L-methionine sulfoximine (198) is found in the bark of a tree, *Cnestis glabra* (199), but the inhibitory action of synthesized methionine sulfoximine has been known and utilized for years (20)). L-(*N*^{5}-phosphono)methionine-S-sulfoximine is a metabolite of the *N*-phosphono compound, L-(*N*^{5}-phosphono)methionine-S-sulfoximinyl-L-alanyl-L-alanine (200), another *Streptomyces* product (201). This latter compound should, however, not be confused with carbon-phosphono compounds and the difference is apparent when structures are compared. Glyphosate did not inhibit GS activity in an enzyme preparation from pea (*Pisum sativum* L.) (104). Phosphinothricin was found to be a more potent inhibitor of GS from bean (*Phaseolus vulgaris* L.) pod tissue and ovine brain than methionine sulfoxime (202), while glyphosate and 2-amino-4-phosphonobutyric acid at concentrtions of up to 1 mM had no inhibitory activity.

Glutamine synthetase in plants is the key enzyme of the GS/GOGAT (glutamine synthetase/glutamine:2-oxoglutarate aminotransferase) pathway and plays a crucial role in the assimilation/reassimilation of ammonia (203, 204). Analysis of two GS isozymes from various plant species show a wide range of ratios for the isozymes with similar K_i values for phosphinothricin, but no correlation with whole plant susceptibility to phosphinothricin (186). This suggests that whole plant susceptibility is not related to differences in the degree of enzyme inhibition. GS isozymes from wheat roots and leaves is strongly inhibited by phosphinothricin *in vitro*, however, the root enzyme was more strongly inhibited by the herbicide than was the chloroplast enzyme (194). The kinetics of inhibition of GS from pea leaves by phosphinothricin and methionine sulfoximine indicated an apparent K_i of 0.073 mM and 0.16 mM, respectively (205). Phosphinothricin also inhibits GS in bacteria (47) and algae (206). Kinetic studies have shown that the nature of phosphinothricin inhibition of GS activity is irreversible and competitive with respect to glutamate.

```
      O
      ||
CH3-P-CH2-CH2-CH-CO-NH-CH-CO-NH-CH-COOH
      |        |        |        |
      OH       NH2      CH3      CH3
      Bialaphos

      O
      ||
CH3-P-CH2-CH2-CH-COOH
      |        |
      OH       NH2
      Phosphinothricin

      O
      ||
CH3-S-CH2-CH2-CH-COOH
      ||       |
      NH       NH2
      Methionine sulfoximine

      O
      ||
CH3-S-CH2-CH2-CH-COOH
      ||       |
      N-PO3H2  NH2
      L-(N-phosphono)-methionine-S-sulfoximine

      O
      ||
HN-C
 |  |
H2C-C-CH2-CH2-CH-COOH
    |          |
    OH         NH2
      Tabtoxinine-β-lactam

 O-CH-COOH
 |  |
H2C-CH-NH2
      Oxetin
```

Figure 7. Structures of several naturally occurring compounds with potent *in vitro* inhibition of GS activity.

Glutamine Antagonism and Ammonia Toxicity. Bialaphos at concentrations up to 3 mM did not inhibit GS activity in plants of Japanese barnyard millet (*Echinochloa utilis*) but lowered extractable activity in treated shoots (207). Phosphinothricin inhibited GS both *in vitro* and extractable activity of treated shoots of this species. Glutamine content was reduced in bialaphos-treated shoots 48 hours after treatment, but exogenous glutamine did not ameleorate bialaphos phytotoxicity. Glutamine has been reported to antagonize bialaphos-caused growth inhibition on *Bacillus subtilis* and *Camellia japonica* L. pollen tubes (47, 208) and the inhibition of GS in *E. coli* by phosphinothricin (47). This protective effect of glutamine has also been demonstrated in other plant species (209). Ammonia has been shown to accumulate to toxic levels in plants treated with bialaphos or phosphinothricin (210-214). This was not totally unexpected since ammonia increases in plants treated with other GS inhibitors, such as methionine sulfoximine (215), and tabtoxin (216).

Site(s) of Action of Phosphinothricin . Inhibition of GS by phosphinothricin resulted in substantial light-dependent accumulation of ammonia, inhibition of photosynthesis, and eventual plant death (214). Because ammonia can occur from three major sources in the plant: (1) inorganic nitrogen assimilation, (2) catabolic or anabolic processes, and (3) photorespiration, these researchers also examined the influence of nitrate assimilation and photorespiration as causes for ammonia accumulation in plants treated with phosphinothricin. Photorespiration was responsible for about 60% of the ammonia formed. Since nitrate reduction was found to have a negligible influence on inhibition of photosynthesis at an early stage, about 40% of the increased ammonia was suggested to originate from catabolic or anabolic routes. Correlations between ammonia concentration and photosynthesis levels were similar regardless of nitrate addition, and CO_2 fixation under non-photorespiratory conditions remains largely intact even at high ammonia levels. Contrary to this, nitrogen fertilizers have been reported to enhance bialaphos efficacy, possibly due to increasing ammonia levels (217). However, the herbicide glufosinate lowered nitrate reductase levels in roots of light-grown soybeans (218).

To clarify the relationship of phosphiothricin inhibition of GS with the reduction of CO_2 fixation and to evaluate the role of glutamine, studies were conducted on protoplasts, chloroplasts, and whole leaves (209). Results suggested that glutamine depletion caused by phosphinothricin is the main cause of early photosynthesis inhibition and three senarios which may be operative in the inhibition of photosynthesis were proposed (Figure 8). Basically these involve (1) inhibition of protein biosynthesis; a lack of regeneration of Q_B protein involved in light dependent electron transport would lead to blockage of photosynthetic electron transport; an amino donor would not be present even for glyoxylate transamination; (2) toxic glyoxylate accumulation; glyoxylate is an inhibitor of RuDP carboxylase/oxygenase (219), and (3) Calvin cycle intermediate depletion; lack of GS (or other enzymes that prevent carbon flow into photorespiration by the oxygenase reaction) leads to a lack of RuDP for the Calvin cycle.

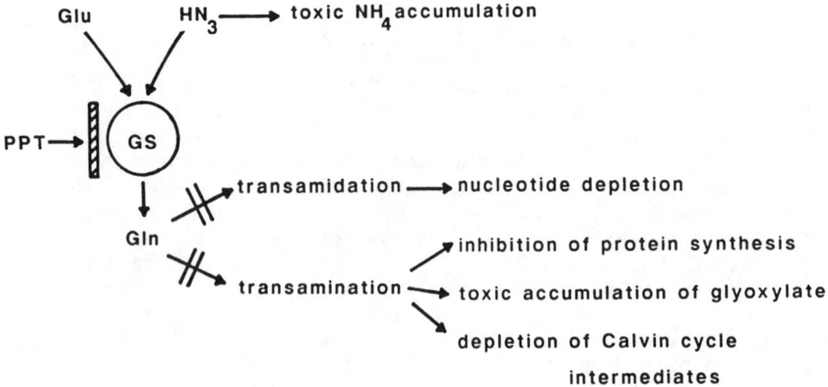

Figure 8. Possible sites of phosphinothricin mode of action. Redrawn with permission from Ref. 209. Copyright 1987 Verlag der Zeitschrift für Naturforschung.

Biotechnological and Biochemical Aspects of Bialaphos Production. The pathway for the biosynthesis of bialaphos has been determined utilizing C^{13}-labeled precursors, blocked mutants, metabolic inhibitors, and analysis of products accumulated and converted by a series of non-producing mutants of S. hygroscopicus (Figure 9). Very recently, experimentation has resulted in the isolation and manipulation of genes responsible for bialaphos biosynthesis in S. hygroscopicus (220). Using a plasmid vector, production genes for this herbicide were cloned from genomic DNA. Several plasmids were isolated which restored bialaphos productivity to mutants of S. hygroscopicus) that were blocked at different points in the biosynthetic pathway. A gene conferring resistance to bialaphos was also linked to the production genes. Mapping defined the location of these genes in a 16 kd cluster. How these genes are controlled; i.e., the transcriptional organization of the cluster and the involvement of regulatory genes remains to be determined.

Stereochemical Considerations. Absolute configuration of a bialaphos intermediate, 2-phosphinomethylmalic acid, has recently been found in the S configuration (221). This fact is important in the determination of the mechanism of transformation of this compound to dimethylphosphinothricin and the enzymes responsible for those metabolic steps.

The optical resolution of racemic phosphinothricin has been reported (222) and it was confirmed that the L form possesses high herbidical activity whereas the D-phosphinothricin had only very low phytotoxicity.

Potential of Naturally Occurring Phosphonates as Herbicides

One of the new strategies of herbicide discovery is the development of herbicide chemical classes based on products that occur naturally. Only a small percentage of higher plant species and microorganisms have been extensively examined for such compounds. Only a few natural products have been adequately examined for potential herbicidal activity (223). Naturally occurring phosphonate compounds are present in a variety of species and are probably in many others yet untested. Furthermore, some of these phosphonates that have been isolated and identified and those that have been synthesized have been shown to contain potent biological activity, including herbicidal and plant growth regulating properties.

Bialaphos is the first herbicide (and the first naturally occurring phosphonate compound) produced by fermentation technology to achieve commerical status. The active component of this compound, phosphinothricin, is of such chemistry that its synthesis was economically feasible and it could be produced as the herbicide glufosinate and marketed on a competetive basis with other synthetic herbicides.

Phosphonate compounds are now known to be widely distributed in nature and these compounds may exist in many living systems yet untested. This, coupled with the activity of various synthetic and naturally occurring phosphonates as enzyme inhibitors and the success of phosphinothricin, bialaphos, and phosalacine as herbicides suggests that increased screening programs coupled with

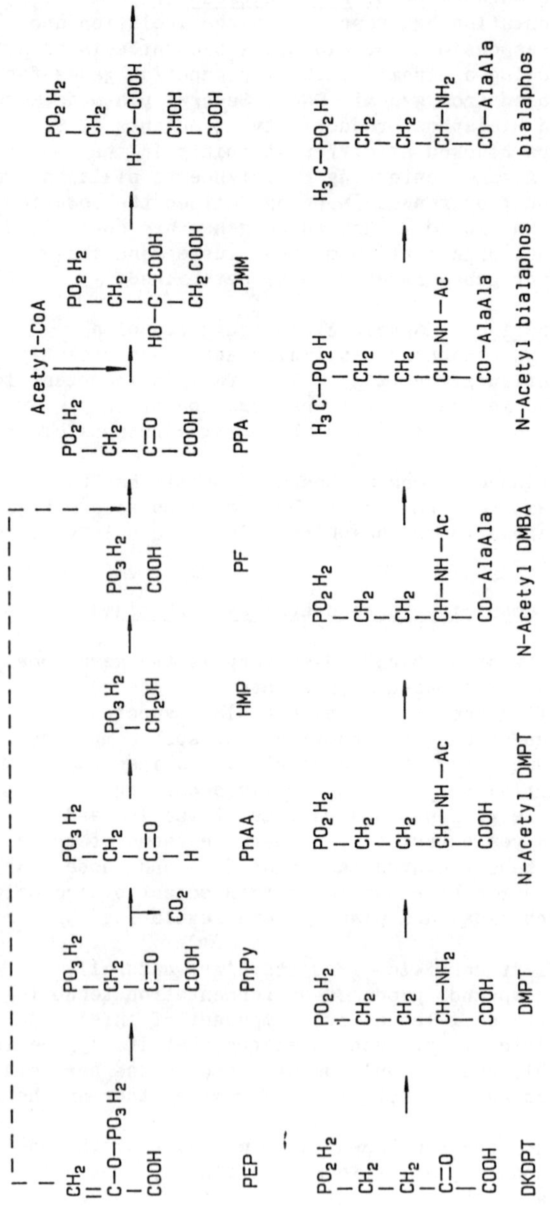

Figure 9. Biosynthetic pathway of bialaphos in *Streptomyces hygroscopicus*. PEP=phosphoenolpyruvate; PnPy=phosphonopyruvate; PnAA=phosphonoacetaldehyde; HMP=hydroxymethylphosphonate; PF=phosphonoformate; PPA=phosphinopyruvate; PMM=phosphinomethylmalate; DKDPT=deamino-α-keto-demethylphosphinothricin; DMPT=demethylphosphinothricin; DMBA=demethylbialaphos; AlaAla=alanylalanine. Redrawn with permission from Ref. 220. Copyright 1986 Springer Verlag.

traditional chemical synthesis backup and biotechnological approaches should lead to other novel phosphonate chemistries with useful herbicidal activity.

Literature Cited

1. Charudattan, R.; Walker, H.L. Biological Control of Weeds with Plant Pathogens; Wiley-Interscience: New York, 1982.
2. Duke, S.O. Rev. Weed Sci. 1986, 2, 15-44.
3. Katchman, B.J. In Phosphorus and Its Compounds; Wiley-Intersciene: New York, 1961, Vol. 2, Chapter 20.
4. Chavane, M.V. Compt. Rend. 1947, 224, 406-4 .
5. Horiguchi, M.; Kandatsu, M. Nature 1959, 184, 901-2.
6. Horiguchi, M. In Analytical Chemistry of Phosphorus Compounds; J. Wiley: New York, 1972; p. 703-24.
7. Czerkawski, J.W.; Faulds, C. J. Sci. Food Agr. 1974, 25, 45-55.
8. Tamari, M. Agr. Biol. Chem. 1979, 43, 651-2.
9. Dufva, G.S.; Bartley, E.E.; Arambel, M.J.; Galitzer, S.J.; Dayton, A.D. J. Animal Sci. 1982, 54, 837-40.
10. Maile, R.J.; Fischesser, J.G.; Anderson, R.J. J. Chromotog. 1977, 132, 366-8.
11. Alhadeff, J.A.; Daves, G.D. Biochim. Biophys. Acta 1971, 244, 211-3.
12. Alhadeff, J.A.; van Bruggen, J.T.; Daves, G. Biochim. Biophys. Acta 1972, 286, 103-6.
13. Karlsson, K.A. Biochim. Biophys. Res. Comm. 1970, 39, 847-51.
14. Harvey, D.J.; Horning, M.G. Mass Spectrometry 1974, 9, 955-69.
15. Huber, J.W. J. Chromatogr. 1978, 152, 220-3.
16. Rueppel, M.L.; Suba, L.; Marvel, J. J. Biomed. Mass Spectrometry 1976, 3, 28-31.
17. Fitzgerald, E.A. J. Chromatog. Sci. 1983, 21, 188-9.
18. Park, B.P.; Hirota, A.; Sakai, H. Agr. Biol. Chem. 1977, 41, 573-9.
19. Kowalik, J.; Azhmat, J.; Masterlerz, P. Phosphorus and Sulfur 1983, 18, 393-6.
20. DaSilveira, J.F.; Colli, W. Biochim. Biophys. Acta 1981, 644, 341-50.
21. Glonek, T.; Henderson, T.O.; Hilderbrand, R.L.; Myers, T.C. Science 1970, 169, 192-4.
22. Ostrovskii, D.N.; Sepetov, N.F.; Samilenko, A.A.; Sibeldina, L.A. Dokl. Acad. Nauk SSSR 1981, 256, 1010-3.
23. Ostrovskii, D.M.; Sepetov, N.F.; Reshetyak, O.I.; Sibeldina, L.A. Biokhimya 1980, 45, 517-25.
24. Miceli, V.M.; Henderson, T.O.; Myers, T.C. Science 1980, 209, 1245-7.
25. Deslauriers, R.; Ekeil, I.; Kroft, T.; Smith, C.P. Biochim. Biophys. Acta 1982, 721, 449-57.
26. Newman, R.H.; Tate, K.R. Comm. Soil Sci. Plant Anal. 1980, 11, 835-42.
27. Kasa, H.; Yamato, M.; Koguchi, T.; Okachi, R.; Kasai, M.; Shirahata, K.; Kawamoto, I.; Shuto, K.; Karasawa, A.; Deguchi, T.; Nakayama, K. Eur. Pat. Appl. 0.061.172, 1982.

28. Hilderbrand, R.L.; Henderson, T.O. In The Role of Phosphonates in Living Systems; CRC Press: Boca Raton, FL, 1983; Chapter 2.
29. Horiguchi, M.; Kittredge, J.S.; Roberts, E. Biochim. Biophys. Acta 1968, 165, 164-6.
30. Liang, C.R.; Rosenberg, H. Biochim. Biophys. Acta 1968, 156, 437-9.
31. Warren, W.A. Biochim. Biophys. Acta 1968, 156, 340-6.
32. Liang, C.R.; Rosenberg, H. Comp. Biochem. Physiol. 1968, 25, 673-81.
33. Horiguchi, M. Biochim. Biophys. Acta 1972, 261, 102-13.
34. Horiguchi, M.; Rosenberg, H. Biochim. Biophys. Acta 1975, 404, 333-40.
35. Seto, H.; Sasaki, T.; Imai, S.; Tsuruska, T.; Ogawa, H.; Satoh, A.; Inouye, S.; Niida, T.; Otake, N. J. Antibiot. 1983, 36, 96-8.
36. Seto, H.; Imai, S.; Tsuruoka, T.; Satoh, A.; Kajima, M.; Inouye, S.; Sasaki, T.; Otake, N. J. Antiobiot. 1982, 35, 1719-21.
37. Seto, H.; Imai, S.; Tsuruoka, T.; Ogawa, H.; Satoh, A.; Sasaki, T.; Otake, N. Biochem. Biophys. Res. Comm. 1983, 111, 1008-14.
38. Kafarski, P.; Mastalerz, P. Aminophosphonates. Natural Occurrence, Biochemistry, and Biological Properties Akademie-Industrie-Komplex, Inst. fur Wirkstofforschung: Berlin, 1984.
39. Hilderbrand, R.L. The Role of Phosphonates in Living Systems; CRC Press: Boca Raton, FL, 1983.
40. Cook, A.M.; Daughton, C.G.; Alexander, M. J. Bacteriol. 1978, 133, 85-90.
41. Lacoste, A.M.; Neuzil, E. Compt. Rend. Acad. Sci., Paris, Ser. D. 1969, 269, 254-7.
42. LaNauze, J.M.; Rosenberg, H. Biochim. Biophys. Acta 1968, 165, 438-47.
43. LaNauze, J.M.; Rosenberg, H.; Shaw, D.C. Biochim. Biophys. Acta 1970, 212, 332-50.
44. Dumora, C.; Lacoste, A.M.; Cassaigne, A. Eur. J. Biochem. 1983, 133, 119-25.
45. Kittredge, J.S.; Hughes, R.R. Biochemistry 1964, 3, 991-6.
46. Kittredge, J.S.; Isbell, A.F.; Hughes, R.R. Biochemistry 1967, 6, 289-95.
47. Bayer, E.; Gugel, K.H.; Hagele, M.; Hagenmaier, T.; Jessipov, S.; König, W.A.; Zähner, H. Helv. Chim Acta 1972, 55, 224-39.
48. Korn, E.D.; Dearborn, D.G.; Fales, H.M.; Sokoloski, E.A. J. Biol. Chem. 1973, 248, 2257-9.
49. Okuhara, M.; Goto, T. Drugs Exptl. Clin. Res. 1981, 7, 559-64.
50. Quin, L.D. Biochemistry 1965, 4, 324-30.
51. Kittredge, J.S.; Horiguchi, M.; Williams, P.M. Comp. Biochem. Physiol. 1969, 29, 859-63.
52. Steiner, S.; Conti, S.F.; Lester, R.L. J. Bacteriol. 1973, 116, 1199-1211.
53. Horiguchi, M.; J. Agr. Chem. Soc. 1966, 40, R25-R30.
54. Sarma, G.R.; Chandramaouli, V.; Venkitasubramanian, R.A. Biochim. Biophys. Acta 1970, 218, 561-3.

55. Ling, J.R.; Buttery, P.J. Br. J. Nutr. 1978, 39, 165-79.
56. Kennedy, K.E.; Thompson, G.A. Science 1970, 168, 989-91.
57. Thompson, G.A.,; Bambery, R.J.; Nozawa, J. Biochemistry 1971, 10, 4441-7.
58. Karlsson, K.A. Chem. Phys. Lipids 1979, 5, 6-43.
59. Quin, L.D.; Shelburne, A.F. J. Marine Res. 1969, 27, 73-84.
60. Shimizu, H.; Kakimoto, J.; Nakajima, T.; Kanazawa, A.; Sano, I. Nature 1965, 207, 1197-8.
61. Kandatsu, M.; Horiguchi, M. Agr. Biol. Chem. 1965 29, 781-2.
62. Hasegawa, S.; Tamari, M.; Kametaka, M. J. Biochem. 1976, 80, 531-5.
63. Tamari, M.; Ogawa, M.; Hasegawa, S.; Kametaka, M. Agr. Biol. Chem. 1976, 40, 2057-62.
64. Tamari, M.; Kametaka, M. Agri. Biol. Chem. 1980, 44, 1957-8.
65. Glonek, T.; Kopp, S.J.; Kot, E.; Pettegrew, J.W.; Harrison, W.H.; Cohen, M.M. J. Neurochem. 1982 39, 1210-9.
66. Baer, E.; Stanacev, N.Z. J. Biol. Chem. 1964, 239, 3209-14.
67. Barnard, M.; Canioni, P.; Cozzone, P.J. Biochimie 1983, 65, 449-75.
68. Baldwin, M.W.; Braven, J. J. Marine Biol. Assoc. 1968, 48, 603-8.
69. Baird, D.D.; Upchurch, R.P.; Homesley, W.B.; Franz, J.E. Proc. Northcentr. Weed Contr. Conf. 1971, 26, 64-8.
70. Mastalerz, P. Arch. Immun. Ter. Dosw. 1959, 7, 201-10.
71. Brand, L.M.; Lowenstein, J.M. Biochemistry 1978, 17, 201-10.
72. Atherton, R.F.; Hall, J.M.; Hassall, C.H.; Lambert, W.J.; Lloyd, P.S.; Ringrose, P.S.; Westmacott, D. Antimicrob. Ag. Chemother. 1982, 23, 571.
73. Lambert, M.P.; Neuhaus, F.C. J. Bactriol. 1972, 110, 978-87.
74. Lacoste, A.M.; Poulsen, M.; Cassaigne, A.; Neuzil, E. Current Microbiol. 1979, 2, 113-7.
75. Neuzil, E.; Cassaigne, A. Exp. Ann. Biochim. Med. 1980, 34, 165-210.
76. Atherton, F.R.; Hall, M.J.; Hassall, C.H.; Lambert, R.W.; Lloyd, W.J., Ringrose, R.S. Antiomicrobial Agents Chemother. 1979, 15, 696-705.
77. Landt, M; Baltz, S.C.; Butler, L.G. Biochemistry 1978, 17, 915-9.
78. Soper, T.S.; Manning, J.M. J. Biol. Chem. 1981, 256, 4263-8.
79. Petrillo, E.W.; Spitzmiller, E.R. Tetrahedron Lett. 1979, (51), 4929-30.
80. Thorsett, E.D.; Harris, E.E.; Peterson, E.R.; Greenlee, W.J.; Patchett, A.A.; Ulm, E.H.; Vassil, T.C. Proc. Nat. Acad. Sci, USA 1982, 79, 2176-80.
81. Galardy, R.E.; Knotoyiannidiou-Ostrem, V.; Kortylewicz, A.; Biochemistry 1983, 22, 1990-5.
82. Petrillo, E.W.; Karanewsky, D.S.; Spitzmiller, E.R.; Duggan, M.E. Phosphorus and Sulfur 1983, 18, 489.
83. Roisch, U.; Lingens, F. Hoppe-Seyler's Z. Physiol. Chem. 1980, 361, 1049-58.
84. Lacoste, A.M.; Cassaigne, A.; Neuzil, E. Compt. Rend. Acad. Sci. Paris, Ser. D. 1972, 275, 3009-12.
85. Jayaram, H.N.; Cooney, D.A. Cancer Treatment Reports 1979, 63, 1095-1108.

86. Pike, D.C.; Beevers, L. Biochim. Biophys. Acta 1982, 708, 203-9.
87. Lejczak, B.; Starzemska, H.; Mastalerz, P. Experientia 1981, 37, 461-2.
88. Hoogenraad, N.J. Arch. Biochem. Biophys. 1974, 161, 76-82.
89. Swyryd, E.A.; Seaver, S.S.; Stark, G.R. J. Biol. Chem. 1974, 249, 6945-50.
90. Collins, K.D.; Stark, G.R. J. Biol. Chem. 1971, 246, 6599-6605.
91. Jacobsen, N.E.; Bartlett, P.A. J. Amer. Chem. Soc., 1981, 103, 654-7.
92. Jacobsen, N.E. Bartlett, P.A. Amer Chem. Soc. Symp. Ser. 1981, 171, 221-4.
93. Seely, J.E.; Marshall, F.D. Life Sci. 1982, 30, 1763-8.
94. Petrenko, N.S.; Panasok, A.I.; Zbaritchenko, S.B.; Ivanova, Z.H.; Kim, T.V.; Suvalova, E.A. Fiziol. Aktiv. Veshchestva 1975, 7, 20-23.
95. Lamden, L.A.; Bartlett, P.A. Biochem. Biophys. Res. Comm. 1983, 112, 1085-90.
96. Ashley, G.W.; Bartlett, P.A. Biochem. Biophys. Res. Comm. 1982, 108, 1467-74.
97. Rubin, J.L.; Gaines, C.G.; Jensen, R.A. Plant Physiol. 1982, 70, 833-9.
98. Boocock, M.R.; Coggins, J.R. FEBS Lett. 1983, 154, 127-33.
99. Plantavid, M.; Maget-Dana, R.; Douste-Blazy, L. Biochimie 1975, 57, 951-7.
100. Meek, T.D.; Villafranca, J.J. Biochemistry, 1980, 19, 5513-9.
101. Wedler, F.C.; Horn, B.R. J. Biol. Chem., 1976, 251, 7530-8.
102. Wedler, F.C.; Horn, B.R.; Roby, W.G. Arch. Biochem. Biophys., 1980, 202, 482-90.
103. Wu, C. J. Biochem., 1977, 55, 332-9.
104. Maier, L.; Lea, P.J. Phosphorus and Sulphur, 1983, 17, 1-19.
105. Sekura, R.; Meister, A. J. Biol. Chem., 1977, 252, 2599-2605.
106. Neale, S. Chem.-Biol. Interactions, 1970, 2, 349-67.
107. Lejczak, B.; Makowiecka, E.; Kafarski, P.; Mastalerz, P. in (38).
108. Biryukov, A.I.; Ishmuratov, B.Kh.; Khomutov, R.M. FEBS Lett., 1978, 91, 249-52.
109. Khomutov, R.M.; Osipova, T.I.; Biryukov, A.I.; Ishmuratov, B.Kh. Bioorg. Khimia, 1979, 5, 56-63.
110. Biryukov, A.I.; Osipova, T.I.; Khomutov, R.M. FEBS Lett., 1978, 91, 246-8.
111. Hoogenraad, N.J. Arch. Biochem. Biophys., 1978, 188, 137-44.
112. Mori, M.; Aoyagi, K.; Tatibana, M.; Ishikawa, T.; Fshii, H. Biochem. Biophys. Res. Comm., 1977, 76, 900-4.
113. Pennick, M.; Gigot, D. J. Biol. Chem., 1979, 254, 6392-6.
114. Anderson, J.W.; Fowden, L. Chem.-Biol. Interact., 1970, 2, 53-5.
115. Rosenthal, A.F.; Pousada, M. Biochim. Biophys. Acta, 1968, 164, 226-37.
116. Izbicka-Dimitrijevic, E.; Masterlerz, P.; Kochman, M. Eur. J. Biochem., 1981, 114, 565-8.
117. Pfeiffer, F.R. U.S. Patent 3 681 480, 1972.
118. Turcotte, J.G.; Yu, C.S.; Lee, H.L.; Pavaranam, S.K.; Sen, S.; Smeby, R.R. J. Med. Chem., 1975, 18, 1184-90.

119. Pfeiffer, F.R.; Miao, C.K.; Itoka, S.C.; Weisbach, J.A. J. Med. Chem., 1972, 15, 58-60.
120. Schrich, L.V.; Diller, A. J. Biol. Chem., 1971, 246, 3961-6.
121. Stoffel, W.; Grol, M. Chem. Phys. Lipids, 1974, 13, 372-88.
122. Iron, A.; Ruart, M.; Duboy, J.P.; Beranger, M.; Cassaigne, A.; Neuzil, E. Biochem. Soc. Trans., 1981, 9, 246.
123. Ovens, S.L.; Bell, F.E. J. Biol. Chem., 1970, 245, 5515-23.
124. Southgate, C.B.; Dixon, H.B. Biochem. J., 1978, 175, 461-5.
125. Fischer, H.P.; Bellus, D. Pestic. Sci., 1983, 14, 334-46.
126. Mase, S. Jpn. Pestic. Inf., 1984, 45, 27-30.
127. Omura, S.; Hmotozawa, K.; Imamura, N.; Murata, M. J. Antibiot., 1984, 37, 939-40.
128. Maier, L. Ger. Patent 24 48 869, 1980.
129. Fredericho, P.M.; Summers, L.A. Z. Naturforsch., 1981, 36c, 242-5.
130. Maier, L. Phosphorus and Sulfur, 1983, 14, 295-322.
131. Birum, G.H. U.S. Patent 4 032 601, 1977.
132. Birum, G.H. U.S. Patent 4 036 913, 1977.
133. Issleib, K.; Hannirg, R.; Erfurt, G.; Dehre, H. GDR Patent
123 479, 1976.
134. Kojima, T.; Otsuka, Y. Japanese Patent 76 98 221, 1976; Chem. Abstr., 1977, 86, 106772.
135. Gunther, E.; Kochman, W.; Lottge, W.; Rothling, T.; Greuzburg, A. Ger. Patent 2 151 460, 1973.
136. Roy, B.; Devlin, J. British Patent 1 508 772, 1978.
137. Hoyle, W.; Vogel, R. Eur. Patent Appl. 10 067, 1980.
138. Okamoto, Y. Internatl. Congr. Phosphorus Compounds, Rabat, 1977, p. 649-52.
139. Suzuki, F.; Fujikawa, Y.; Yamamoto, S.; Mitzutani, H.; Funabashu, C.; Ohya, T.; Ikai, T.; Oguchi, T. German Patent 2 831 578, 1979.
140. Kawamura, Y.; Oya, T.; Igai, T.; Takematsu, T. Japanese Patent 80 43 054, 1980; Chem. Abst., 1980, 93, 63620.
141. Kawamura, Y.; Oya, T.; Igai, T.; Takematsu, T. Japanese Patent 80 43 055, 1980; Chem. Abst., 1980, 93, 63619.
142. Nissan Chem. Ind. Ltd. Japanese Patent 80 98 105, 1980; Chem. Abst., 1980, 93, 232711.
143. Nissan Chem. Ind. Ltd. Japanese Patent 80 98 194, 1980; Chem. Abst., 1980, 94, 4121.
144. Suzuki, F.; Fujikawa, Y.; Kamamoto, S.; Hizutani, M.; Oya, T.; Kawamura, Y. Japanese Patent 79 144 383, 1979; CA, 1980, 93, 26547.
145. Nissan Chem. Ind. Ltd. Japanese Patent 80 94 309, 1980; Chem. Abst., 1980, 93, 199240.
146. Nissan Chem. Ind. Ltd. Japanese Patent 80 89 293, 1980; Chem. Abst., 1980, 93, 181017.
147. Nissan Chem. Ind. Ltd. Japanese Patent 80 55 107, 1980; Chem. Abst., 1980, 93, 90191.
148. Pickles, W.; Baylis, E.K. Eurpean Patent 0 010 066, 1980.
149. Otsuka Chem. Co. Ltd. Japanese Patent 81 55 394, 1981; CA, 1981, 95, 204153.
150. Large, G.B. U.S. Patent 4 170 463, 1979.
151. Franz, J.E. In Advances of Pesticide Science, Vol. 2; Pergamon Press: Oxford, 1979; pp. 139-47.

152. Croft, S.M.; Arntzen, C.J.; Vanderhoef, L.N.; Zettinger, C.S. BBA, 1974, 335, 211-7.
153. Slovin, J.P.; Tobin, E.M. BBA, 1981, 637, 177-84.
154. Hoagland, R.E. Weed Sci., 1980, 28, 393-400.
155. Hartman, F.A. U.S. Patent 3 894 861, 1975.
156. Ratts, K.W. U.S. Patent 3 961 934, 1974.
157. Gunther, E.; Lottge, W.; Kochmann, W. GDR Patent 94 280, 1972.
158. Kramer, M.; Gunther, E.; Lottge, W.; Bech, R.; Kochmann, W. Environ. Qual. Saf. Suppl., 1976, 686; CA, 1976, 85, 11765.
159. Gunther, E.; Lottge, W. Ger. Patent 2 022 228, 1971.
160. Barth, P.; Gunther, E.; Kochmann, W.; Kramer, W.; Lottge, W.; Rothling, T.; Wolter, G. GDR Patent 191 142, 1976.
161. Azerbaev, I.N.; Tynibaev, B.D.; Abiyurov, B.D. USSR Patent 366 844, 1973.
162. Phillips, W.G. U.S. Patent 4 164 406, 1979.
163. Czerwinski, W.; Gancarz, R.; Przybylka, E.; Wieczorek, J.S. Acta Agrobot., 1982, 34, 253-60.
164. Ogawa, Y.; Yoshida, H.; Kondo, Y. Japanese Patent 74 13 123, 1974; Chem. Abst., 1974, 80, 121102.
165. Ogawa, Y.; Inouye, S.; Yamashina, H.; Niida, T. Sci. Rep. Meiji Seika Kaisha, 1976, 15, 22-8.
166. Takyama, S.; Sasaki, S.; Kimura, I. Japanese Patent 75 101 536, 1975; CA, 1975, 83, 189317.
167. Azerbaev, I.N.; Bosyakov, Yu.; Dzailavlov, S.D.; Bobrov, L.G.; Rogozhin, A.K. USSR Patent 557 579, 1978.
168. Amrhein, N.; Schab, J.; Steinrucken, H.L. Naturwiss, 1980, 67, 356-7.
169. Hoagland, R.E.; Duke, S.O. In Biochemical Responses Induced by Herbicides; Moreland, D.E.; St. John, J.B.; Hess, F.D., eds.; ACS Symposium Series No. 181; Amer. Chem. Soc.: Washington, DC, 1982; pp. 175-205.
170. Grossbard, E.; Atkinson, D., eds. The Herbicide Glyphosate; Butterworths: London, 1985.
171. Duke, S.O. In Herbicides: Chemistry, Degradation, and Mode of Action; Kearney, P.C.; Kaufman, D.D., eds.; Marcel Dekker Inc.: New York, 1988; pp. 1-70.
172. Zoebisch, O.C.; Rushing, T.T.; Barrier, G.E. Proc. Northeast Weed Sci. Soc., 1974, 28, 347-9.
173. Kitchen, L.M.; Rieck, C.E.; Witt, W.W. Weed Res., 1980, 20, 285-9.
174. Morey, P.R.; Dahl, B.E. Weed Sci., 1980, 28, 251-5.
175. Herbicide Handbook, WSSA: 5th Ed., Champaign, IL., 1983.
176. Dollwet, H.H.; Kumamoto, J.J. Plant Physiol., 1970, 46, 786-9.
177. Murphy, S.D. Pesticides in Toxicology: The Basic Science of Poisons, 2nd ed., Doull, J.; Klaasen, C.D.; & Amdur, M.O., eds., Macmillan, New York, 1980, chap. 16.
178. Hilderbrand, R.L. In The Role of Phosphonates in a Living System; CRC Press: Boca Raton, FL., 1983; chapter 6.
179. Kondo, Y.; Shomura, T.; Ogawa, Y.; Suzuki, T.; Moriyama, Ch.; Yoshida, J.; Inonye, Sh.; Niida, T. Sci. Rept. Meiji Seika Kaisha, 1973, 13, 34-41.
180. Meiji Seika Kaisha DOS 2 236 599 (1973, priority 28.7.1971).

181. Meiji Seika Kaisha, J 5 4092 628 (1979, priority 29.12.1977).
182. Hoescht AG, DOS 2 717 440, 1977.
183. Omura, S.; Murata, M.; Hanaki, H.; Hinotozawa, K.; Oiwa, R.; Tanaka, H. J. Antibiot. 1984, 37, 829-35.
184. Jap. Kokai Tokkyo Koho, Japanese Patent 61 176 505, 1986.
185. Tachibana, K.; Kaneko, K J. Pestic. Sci. 1986, 11, 297-304.
186. Ridley, S.M.; McNally, S.F. Plant Sci. Lett. 1985, 39, 31-6.
187. Suzuki, A.; Nishide, K.; Shimura, M.; Yamamoto, I. J. Pestic. Sci. 1987, 12, 105-7.
188. Lee, T.T.; Dumas, T.; Jevnikar, J.J. Pestic. Biochem. Physiol. 1983, 20, 354-9.
189. Duke, S.O.; Hoagland, R.E. in The Herbicide Glyphosate; Butterworths: London, 1980; Chapter 6.
190. Hoagland, R.E. Abstracts, 186th National Meeting, Amer. Chem. Soc.; 1983, Pestic. Div. Abst. No. 33.
191. Wedler, F.C.; Horn, B.R.; Roby, W.G. Arch. Biochem. Biophys. 1980, 202, 482-90.
192. Wu, C. Can. J. Biochem. 1977, 55, 332-9.
193. Lejezak, B.; Stargemsko, H.; Mastalez, P. Experientia 1981, 37, 461-2.
194. Manderscheid, R.; Wild, A. J. Plant Physiol. 1986, 123, 135-42.
195. Omura, S.; Murata, M.; Imamura, N.; Iwai, H.; Taneka, H.; Furusaki, A.; Matsumoto, T. J. Antibiot. 1984, 37, 1324-32.
196. Siden, S.L.; Durbin, R.D. Nature 1968, 219, 379-80.
197. Langston-Unkefer, P.L.; Macy, P.A.; Durbin, R.D. Plant Physiol. 1984, 76, 71-4.
198. Tate, S.S.; Meister, A. In The Enzymes of Glutamine Metabolism, Academic Press: New York; pp. 77-127.
199. Jeannoda, V.L.; Valeolalso, J.; Creppy, E.E.; Dorjeomer, G. Phytochemistry 1985, 24, 854-5.
200. Ronzio, R.A.; Rowe, W.B.; Meister, A. Biochemistry 1969, 8,1066-75.
201. Sekizawa, Y.; Takematsu, T. In Pesticide Chemistry, Human Welfare and the Environment, Vol. 2, Natural Products; Pergamon Press: Oxford, 1983; pp. 261-8.
202. Hoagland, R.E. Abstracts, Weed Sci. Soc. Amer. Abst. No. 168, 1987.
203. Tolbert, E.N. In The Biochemistry of Plants, Vol. 2 Academic Press: New York, 1980; pp. 488-523.
204. Wallsgrove, R.M.; Keys, A.J.; Bird, J.F.; Cornelius, M.J.; Lea, P.J.; Miflin, B.J. J. Exp. Bot. 1980, 31, 1005-7.
205. Leason, M.; Cunliffe, D.; Parkin, D.; Lea, P.J.; Miflin, B.J. Phytochemistry 1982, 21, 855-7.
206. Lea, P.J.; Joy, K.W.; Ramos, J.L.; Guerrero, M.G, Phytochemistry 1984, 23, 1-6.
207. Tachibana, K.; Watanabe, T.; Kekizawa, Y.; Takematsu, T. J. Pesti. Sci. 1986, 11, 27-31.
208. Tachibana, K.; Watanabe, T.; Suzuki, Y.; Sekizawa, Y. Sci. Rpt. Meji Seika Kaisha 1980, 19, 27-31.
209. Sauer, H.; Wild, A.; Ruhle, W. Z. Naturforsch. 1987, 42, 270-8.
210. Wild, A.; Manderscheid, R. Z. Naturforsch 1984, 39c, 500-4.
211. Wild, A.; Sauer, H.; Ruhle, W. Z. Naturforsch. 1987, 42c, 23-9.

212. Tachibana, K.; Watanage, T.; Sekizawam Y.; Takemastsu, T. J. Pest. Sci. 1986, 11, 33-7.
213. Tachibana, K.; Watanabe, T.; Sekizawa, Y.; Konnai, M.; Takematsu, T. Abst. 5th Int. Cong. Pestic. Chem.; Kyoto, IVa-19, 1982.
214. Wild, A.; Sauer, H.; Fuhle, W. Z. Naturforsch. 1987, 42, 263-69.
215. Platt, S.G.; Anthon, G.E. Plant Physiol. 1981, 67, 509-13.
216. Turner, J.G. Physiol. Plant Pathol. 1981, 19, 57-67.
217. Misato, T.; Yamaguchi, I. Outlook Agric. 1984, 13, 136-9.
218. Hoagland, R.E. Abstracts, 185th National Meeting Amer. Chem. Soc. Pestic. Div. Abst. No. 96.
219. Cook, C.M.; Tolbert, N.E. Plant Physiol. 1982, 69, Suppl. 52.
220. Murakami, T.; Anzai, H.; Imai, S.; Satoh, A.; Nagaska, K.; Thompson, C.J. Mol. Gen. Genet. 1986, 205, 42-50.
221. Shimatohno, K.; Seto, H.; Otake, N.; Imai, S.; Setoh, A. J. Antibiot. 1986, 39, 1356-9.
222. Goi, J.; Miyado, S.; Shomura, T.; Suzuki, A.; Niwa, T.; Yamada, Y. Japanese Patent 3 55 47 630, 1980.
223. Newman, E.I. Pestic. Sci. 1982, 13, 575-82.

RECEIVED August 23, 1988

Chapter 14

Phytochemical Inhibitors of Velvetleaf (*Abutilon theophrasti*) Germination as Models for New Biorational Herbicides

Richard G. Powell and Gayland F. Spencer

Northern Regional Research Center, Agricultural Research Service, U.S. Department of Agriculture, Peoria, IL 61604

Higher plants produce many biologically active natural products that adversely affect the growth and development of other organisms. Usually these toxins are considered to be defensive substances useful to the plant in discouraging or preventing attack from herbivores and microorganisms, or in protecting the plant from stress exerted by the environment and competing plant species. The inhibitory compounds normally occur within the plant as glycosides, or in some other inactive form, and they are mobilized rapidly when the plant is subjected to injury or other external stress. They may enter into the environment by volatilization or by leaching during the plant's life span, or they may be activated or modified by microbial action upon death of the plant. A number of examples demonstrating the variety of structural types of naturally occurring germination and growth inhibitors are discussed. Emphasis is placed on lesser known compounds found in seeds of uncultivated plants, those that are effective in inhibiting germination of velvetleaf (Abutilon theophrasti), and on phytotoxins that may serve as models for new biorational herbicides.

The growth inhibitory action of one plant upon another has long been recognized. Pliney the elder (23-79 A.D.) stated that "The shadow of the walnut tree is poison to all plants within its compass" (1). This type of phenomenon, the chemical interaction between plants, was named "allelopathy" by Molisch in 1937 (2) and was meant to describe both beneficial and detrimental biochemical interactions among all classes of plants, including microorganisms. Rice (3) defined the term as "Any direct or indirect harmful effects of one plant (including microorganisms) on another through the production of chemical compounds that escape into the environment." Rice has since expanded his definition to include beneficial interactions (4). The subject of allelopathy has attracted considerable interest as evidenced by recent comprehensive reviews (4-6). In its strictest sense, allelopathy refers only to those phenomena that actually occur in nature and does not apply to situations normally found in agriculture where human intervention by cover planting, herbicide application, or cultivating is employed. Thus, allelochemicals may be looked upon as naturally occurring plant growth regulators or as naturally occurring herbicides. Although the active principle (juglone) in walnut is now well known, a multitude of reported incidences of allelopathy remain where detailed characterization of the compound(s) responsible for the observed effects is needed.

Reports of allelopathic phenomena in nature have usually been made by persons largely untrained in natural products chemistry. Thus, many observers are not aware of the plethora of unique chemicals that are produced by plants and/or of the modern analytical and separation techniques that are available to isolate, identify and quantitate them. Duke (7) has reviewed naturally occurring compounds having potential as herbicides, and Cutler (8) recently outlined current methods useful for isolating, characterizing and screening natural products for herbicidal activity. In order to take advantage of the natural herbicide pool available from plants, structures of the various germination and growth inhibiting compounds need to be established and their effects on target weeds and on cultivated crops must be studied in detail.

Plant-derived chemicals appear to have potential for selective control of weeds in certain conventional and minimum tillage cropping systems and could serve well as models for totally new classes of biorational synthetic herbicides. The Environmental Protection Agency, under definition of pesticides as given by the Federal Insecticide Fungicide and Rodenticide Act, considers biorational pest control agents to include naturally occurring chemicals obtained from either plants or animals and chemicals synthesized by man, provided that they are identical to their naturally occurring counterparts in geometric and stereochemical structure (9). Thus, it has been suggested that registration of these "biorational" compounds could prove to be less expensive and time consuming than for other new herbicides. One can also anticipate that these natural products will rapidly be degraded in the environment, reducing problems with toxic residues and contamination of water supplies, and that many of them may be relatively specific in their action.

Previous investigations have focused primarily on microbially produced toxins as potential herbicides (7, 10) with relatively little attention being given to plant-derived natural products. In order to gain information as to the variety of growth and germination regulatory compounds present in plants, particularly in plant seeds of uncultivated species, and in the interest of eventually exploiting these phytochemicals for use as herbicides, the present search for natural inhibitors of velvetleaf germination was initiated. Bioassay procedures were as described, or slight modifications of those outlined, by Wolf, Spencer, and Kwolek (11) and this report summarizes recent findings in our laboratory.

Benzyl Isothiocyanate and Related Natural Products. Benzyl isothiocyanate (1) was isolated from mature papaya (*Carica papaya* L.) seeds and found to be a powerful inhibitor of imbibed velvetleaf (*Abutilon theophrasti* Medic.) seeds (11). Complete inhibition of germination occurred at a concentration of 6 x 10^{-4} M, and only 28% (compared to controls) of the seeds germinated after 4 days exposure to a 4 x 10^{-4} M treatment. Seedlings treated at the lower level were unhealthy in appearance and were shorter than controls. However, velvetleaf seeds were unaffected at 2 x 10^{-4} M, suggesting a critical concentration under which 1 is well tolerated.

Neither corn nor soybean germination was as susceptible to 1 as velvetleaf. Corn was unaffected at 1 x 10^{-3} M. Soybeans germinated (42%) at this level, however, the seedlings remained yellow in appearance (unlike the controls) and roots and shoots were considerably shorter than controls. It was apparent from these results that a sublethal dose of 1 alters the rate of germination and also totally inhibits germination of a certain percentage of seeds. At 4 x 10^{-4} M, 1 totally inhibited velvetleaf seedling growth and adverse effects on growth were noted at even lower concentrations.

Compound 1, normally derived enzymatically from benzyl glucosinolate, is found in both the fruit and seeds of papaya in significant amounts (12, 13), and may account for up to 2% of the seeds by weight (14). Other members of the family Cruciferae are known to contain benzyl isothiocyanate (15), and it also occurs in several species of the genus *Tropaeolum* (family Tropaeolaceae) (16). Related compounds, including allylisothiocyanate (2) and phenethylisothiocyanate (3), are known germination inhibitors (17) and 1 has also been reported to inhibit growth of papaya and mung bean seedlings (18). In addition, 1 has strong bactericidal, fungicidal, and insecticidal activities (19, 20) and alters the growth of various animal cells (21, 22).

A formulation of 1 in dry granules, with tung oil and calcareous loess, inhibited germination of several crop and weed seeds; however, rapid loss of activity was evident after a few days in moist sand (23). Sodium methyldithiocarbamate (metham), a synthetic soil fumigant available to farmers for many years, has been used to control pathogens, insects, and weeds in many horticultural crops (24). When applied to the soil, metham is degraded to methyl isothiocyanate (4) within a few hours depending on soil conditions and 4, in turn, is degraded to nontoxic products within several days. Methyl isothiocyanate (4) effected 100% control over velvetleaf germination in the laboratory at 4 x 10^{-4} M (Powell, R. G., and Spencer, G. F., unpublished data). Additional research on the many naturally occurring isothiocyanates and their glucosinolate precursors (25), and on synthetic analogs, could easily lead to more stable and effective herbicidal formulations.

Cryptocaryalactone and Other Pyranone Derivatives. Seeds of *Cryptocarya moschata* yielded two new germination inhibitors; (-)-cryptocaryalactone (5) and its deacetylated analog (6) (26). Applied at 4 x 10-3 M, 6 completely arrested germination of velvetleaf and decreased the germination rate of soybeans. Corn did not appear to be affected at this level. The acetylated derivative (5) was less effective giving only 50% inhibition of velvetleaf at 4×10^{-3} M. A related phenyltetrahydrofurano-2-pyrone (7), known both as altholactone and as goniothalenol, gave 100% inhibition at 5×10^{-3} M (Powell, R. G., and Spencer, G. F., unpublished data). Compound 7 occurs in the bark of Polyalthia species (27) and in the bark of Goniothalamus giganteus (Annonaceae) (28).

Substituted 2-pyranones are important natural products having diverse biological activities (29). Other reported germination inhibitors in this class include parasorbic acid (8) and psilotin (9). Compounds 5 and 6 are not as phytotoxic as benzyl isothiocyanate (1) but they do have an effect similar to that of 9 on turnips (30). Siegel reported that the activity of 9 could be reversed by addition of gibberellic acid (GA3) to the medium (30). Spencer, et al., have shown that GA3 has no effect on activity when applied simultaneously, to velvetleaf, with 6.

Benzoxazolinone (BOA) and 2,4-Dihydroxy-1,4-benzoxazin-3-one (DIBOA). Acanthus mollis seed proved to be an excellent source of glycoside 10. Enzymatic hydrolysis of 10 gives rise to 2,4-dihydroxy-1,4-benzoxazin-3-one (DIBOA, 11) and heating of 11 readily yields benzoxazolinone (BOA, 12) (31). Based on dry weight, 10 comprises almost 4% of the seed or 281 mmoles per kg. Compound 11 was 100% lethal to velvetleaf seed at 2×10^{-3} M. At lower concentrations, germination was delayed and seedling growth was slower than that of controls. Both the glucoside (10) and BOA (12) gave only slight germination and growth inhibition at 5×10^{-3} M.

These compounds (10-12), and related methoxy derivatives, are known constituents of corn, wheat, wild rye, giant reed, Job's tears, and a bamboo species (32-34); all members of the grass (Graminae) family. Apparently, only the parent glycosides occur in significant amounts in uninjured plants. These compounds had not been detected in seeds, or in dicotyledonous plants, before their discovery in bear's breech (A. mollis) although they have been reported in monocotyledonous seedlings as early as two or three days after germination (35, 36). Barnes, et al. (37) have found that 11 was by far the most active chemical in rye shoot tissue, in their assay using cress seed, and reported that it was present in high concentration. The benzoxazinones, as a group, may well be important growth regulators in grasses and contributors to allelopathy of many cereal crop residues, particularly in no-till situations.

Aiupova and coworkers (38) demonstrated that radicle elongation of cucumbers, oats, radishes and cabbage was sensitive to 12 and that doses required for 50% inhibition of elongation varied from 1.5 to 8.2 kg/ha. Germination was not affected at the doses given and, unfortunately, it was not possible to determine molar concentrations with the information provided. Other reported biological activities for these compounds include mycelial growth inhibition of Fusarium and Helminthosporum species by both 11 and 12 (39-41), toxicity and antifeedant activity of 11 to insects (35, 36) and anticonvulsant activity of 12 in small animals (42).

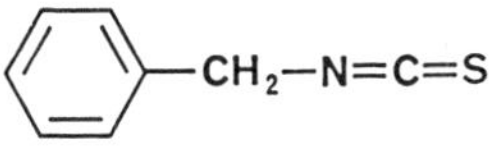

$C_6H_5-CH_2-N{=}C{=}S$

1

$CH_2{=}CH-CH_2-N{=}C{=}S$

2

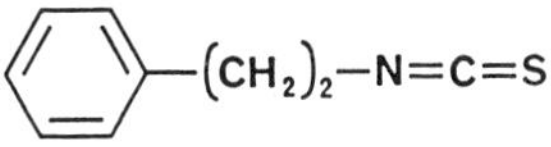

$C_6H_5-(CH_2)_2-N{=}C{=}S$

3

$CH_3-N{=}C{=}S$

4

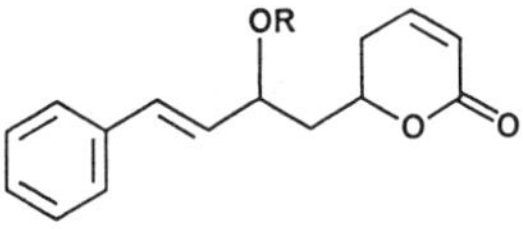

OR O O

5 R=Ac

6 R=H

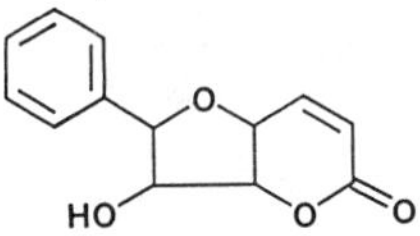

O HO O O

7

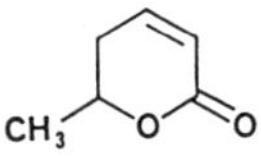

CH_3 O O

8

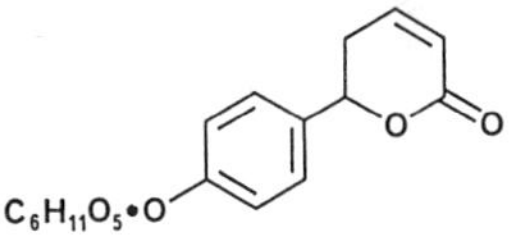

$C_6H_{11}O_5{\cdot}O$ O O

9

OH N O O $O{\cdot}C_6H_{11}O_5$ → OH N O O OH → NH O O

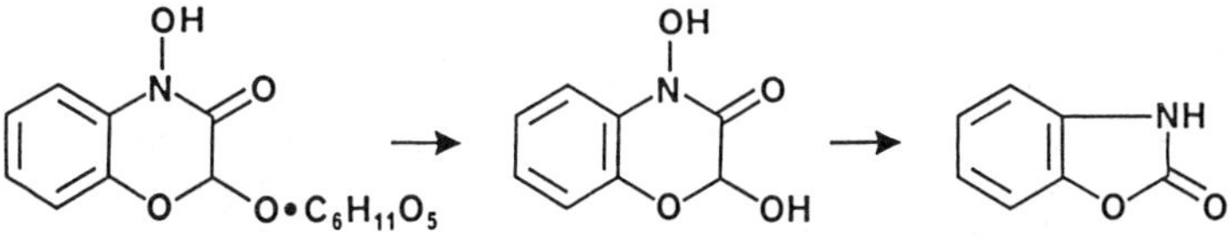

10 11 12

5,7-Dihydrochromone, a Flavanoid Decomposition Product. Flavanoids enjoy a reputation in allelopathy as inhibitors but bioassay data on specific compounds are not entirely convincing (3, 43). They are widespread in nature (44) and it would seem that, if they were indeed highly phytotoxic, their effects should be easily demonstrated. Decomposition reactions are quite complex, however, it has been clearly shown that dihydroxychromones can result from the enzymatic, peroxidative degradation of flavanoids (45, 46).

An antigermination factor present in Polygonum lapathifolium seeds was identified as 5,7-dihydroxychromone (13) by Spencer and Tjarks (47). Compound 13 had been isolated previously from P. persicaria seeds (48) and from peanut (Arachis hypogea) hulls (49), but, its antigermination properties had not been reported. In an earlier example of chemical inhibition from a Polygonum species (50) it was implied that the ubiquitous flavone luteolin (14a) was an allelopathic principle in the leaves. Spencer and Tjarks (47) tested both 13 and 14a in their velvetleaf germination bioassay system and found that luteolin (14a) and the related flavanoids, crysin (14b), apigenin (14c), kaempferol (14d), and quercitin (14e) were all less than one-tenth as effective as 13. Thus, it would appear that phytotoxicity commonly ascribed to flavanoids may well be primarily due to their decomposition products and that chromones structurally related to 13 deserve additional attention for their potential herbicidal activity.

Naturally Occurring Naphthoquinones. Juglone (15a), from the European walnut (Juglans regia) or from the North American walnut (J. nigra), is often cited as a classic example of an allelopathic chemical (1, 51). The bound phytotoxin has been identified as the 4-glucoside of 1,4,5-trihydroxynaphthalene which is converted to 15a on hydrolysis and oxidation. The characteristic brown staining of the hands caused by handling walnuts is attributed to the release of 15a. In walnut, the occurrence of 15a as a relatively non-toxic glycoside is strictly limited to green parts of the tree; it is not present in the ripe nuts. Compound 15a is water soluble and exerts its effect on other plants after being leached into the soil from stems and leaves. Juglone must persist in the soil around the tree for a reasonable length of time in order for it to be a useful ecological agent, however, information is lacking on the rate of turnover and on the concentrations at which it is effective in the natural environment.

Many other naphthoquinones are known to occur in nature (52), however, 15a has commonly been considered to be the only naphthoquinone phytotoxin produced by higher plants (3). Considering the historical and commercial importance of natural naphthoquinones, it is surprising that their herbicidal properties have not been examined more thoroughly in the literature. Spencer, et al., (53) pointed out that plumbagin (15b), isolated from an extract of Plumbago europaea, is at least as effective in inhibiting velvetleaf germination as is 15a. The few other reports concerning naphthoquinones include 5-hydroxy-2-methoxynaphthoquinone from Platycaria strobilacea as a lettuce germination inhibitor (54), the absence of phytotoxicity of lawsone (15c) when sprayed on young plants (55), and the isolation of several compounds inhibitory to lettuce from a shoot blight fungus (56). A few synthetic amino

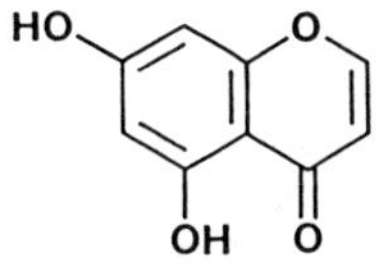

HO O OH O

13

HO O R3 R2 R1 OH O

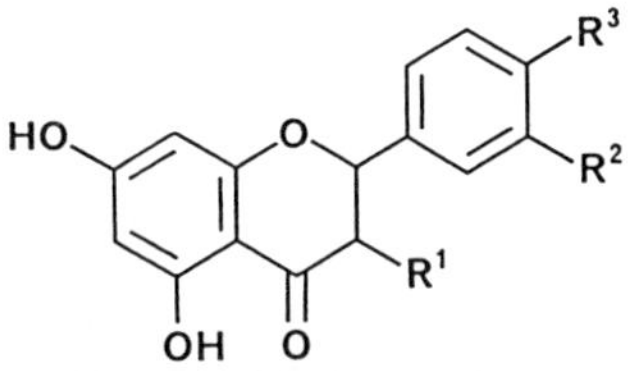

14a R^1=H, R^2=R^3=OH
14b R^1=R^2=R^3=H
14c R^1=R^2=H, R^3=OH
14d R^1=OH, R^2=H, R^3=OH
14e R^1=R^2=R^3=OH

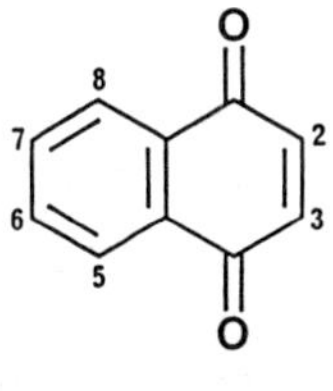

O 8 7 2 6 3 5 O

15

O O

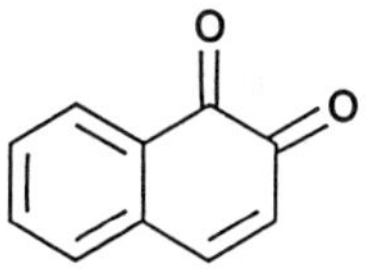

16

naphthoquinone derivatives have been evaluated as herbicides (57, 58) and were indicated to have potential in sugar beets and in rice.

Spencer et al., (53) compared the antigermination activity of some natural and synthetic naphthoquinones (15a-15r,16a,16b); their more important findings are summarized in Table I. Examination of the data provides some insight into the types of substitution affecting activity. One methyl group has nearly the same effect as a proton, in most cases, while methylation of an hydroxyl group yields divergent properties. This latter transformation decreased the potency of juglone (compare 15a with 15d) and markedly enhanced that of lawsone (15c vs 15e). In the latter instance one can infer that substitution serves to stabilize the molecule by prohibiting tautomerization. Juglone (15a), the natural allelopath, appears to be effective over a wide range of concentrations; whereas its isomer, lawsone (15c), is much less active at low concentrations and seems to have a rather steep dose/response relationship. Another set of isomers, plumbagin (15b) and 7-methyljuglone (15f), both natural products, give quite different responses. Of the compounds examined, those with natural origins tended to be phytotoxic at lower concentrations than those with no known natural source. It is important to note that, so far, only juglone has been demonstrated to be allelopathic in ordinary environmental situations. It is now apparent, however, that other naphthoquinones are phytotoxic and that these may have potential as growth regulators or as herbicides.

Miscellaneous Terpenes. Isolation of the velvetleaf antigermination factor in seeds of *Eryngium paniculatum* gave 17, (-)-2,4,4-trimethyl-3-formyl-2,5-cyclohexadienyl angelate, as the only active component (59). Ester 17 had been reported in roots of *Bupleurum gibraltaricum* (60), and several isomeric and analogous terpene aldehyde-esters are reported in other Umbelliferae (61-64). These compounds were simply biosynthetic curiosities until Spencer's observation (59) that one member of the series (17) had antigermination properties. Further studies concerning structure-activity relationships led to preparation and comparison of several analogs; results of these experiments are summarized in Table II. Compound 18 is active, albeit less active than 17. This difference may be due to the location of the ring-methyl group, the nature of the acyl group, or to stereochemistry (17 is optically active and 18 is racemic). The positional isomer of 18, compound 19, was inactive at all test concentrations.

The alcohol portion of 17 is very similar to 20, a precursor in the synthesis of strigol; strigol is a promoter for germination of witchweed (*Striga asiatia*) (65). The difference is only that of the 5,6-double bond. Activity of 20 was about the same as that of 18 (both compounds are racemic) and acetylation of 20, yielding 21, negated activity.

When subjected to even mild heat, 17 and 18 undergo elimination and rearrangement to give 2,3,6-trimethylbenzaldehyde (62). Although this compound was not available for bioassay, its isomer 26 was inactive, suggesting that the effects of 17 or 18 are not due to their degradation product. Results from compounds 22 and 23 demonstrate that activity is diminished by

Table I. The effect of selected naphthoquinones on velvetleaf germination

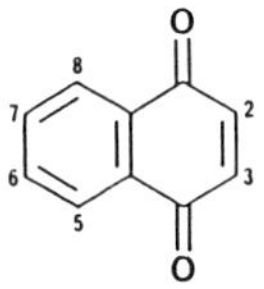

15 1,4-Naphthoquinones	Substitution	Concentration mM (% germination)[a]			
a Juglone	5-OH	0.4(100)	0.8(74)	1.0(41)	3.0(6)
b Plumbagin	2-CH_3, 5-OH	0.2(100)	0.5(66)	0.7(22)	0.8(12)
c Lawsone	2-OH	1.2(97)	1.8(71)	2.1(33)	2.5(10)
d 5-Methoxy	5-OCH_3	2.0(100)	3.0(84)	4.0(71)	6.0(5)
e 2-Methoxy	2-OCH_3	0.2(96)	0.4(70)	0.5(52)	0.6(10)
f 7-Methyljuglone	5-OH, 7-CH_3		4.0(100)	6.0(45)	8.0(5)
g Menadione	2-CH_3	0.7(92)	0.9(61)	1.0(47)	1.2(3)
h 2-Acetoxy	2-$OCOCH_3$		3.0(92)	4.0(51)	5.0(26)
i 2-Methyl-5-methoxy-	2-CH_3,5-OCH_3	1.0(98)	2.0(58)	4.0(27)	5.0(0)
j 1,4-Naphthoquinone	none	0.6(96)	0.9(69)	1.2(28)	2.5(3)
k Phthiocol	2-OH, 3-CH_3	1.0(93)	2.0(68)	4.0(18)	6.0(0)
16 1,2-Naphthoquinones					
a 1,2-Naphthoquinone	none			10.0(100)	
b 4-Methoxy	4-OCH_3	0.5(95)	0.8(75)	1.0(28)	2.0(5)

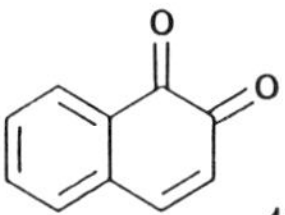

[a]Percent germination as compared to controls. Figures below 85 are significantly different from controls at the 95% level by the Chi-square 1-tailed test.

Table II. The effect of terpene aldehyde-esters and derivatives on velvetleaf germination

Compound	Concentration (mM)/Percent Germination[a]				
	2	4	6	8	10
17	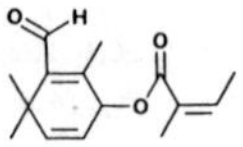 92	41	0		
18	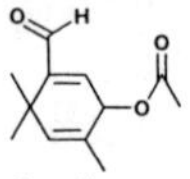	90	30	10	0
19					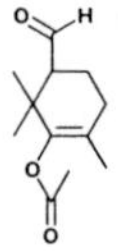 100
20	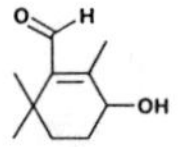	70	52	22	0
21					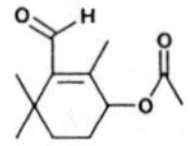 100
22			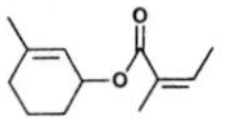 100	28	0
23	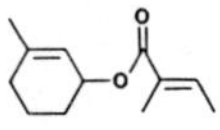	100	16	5	0
24					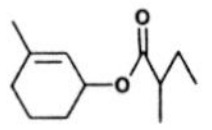 100
25					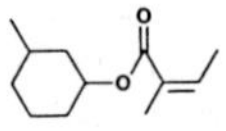 100
26					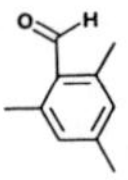 100

[a]Relative to controls; control rate = 36-40 seeds/40. Values below 85 significantly different from controls.

removal of the formyl group and that the orientation of the acyl double bond is not important. Moreover, 24 and 25 are inactive, an outcome suggesting that the cyclic double bond(s) is required.

Spencer, et al. (66), isolated several sesquiterpene inhibitors from Iva Axillaris seed. Of these, 27 and 28 (ilicic acid) were the most active. Effective concentrations for both, as inhibitors of velvetleaf germination, were near 4 x 10^{-3} M. Compound 27 is also known to occur in Parthenium tomentosum (67) and in Inula viscosa (68), while 28 has been isolated from Ambrosia ilicifolia (69) and from Inula graveolens (70).

Cinnamic Acid Derivatives. Investigation of germination inhibitors in star anise (Illicium verum) fruit led Wolf to significant observations concerning some common phenylpropanoids (71). The antigermination factor of star anise proved to be p-methoxycinnamaldehyde (29), a minor constituent previously unreported in this plant. Compound 29 is a flavor component of baked potatoes (72) and it occurs in many other plants including sweet basil (Ocimum basilicum) (73). In Sphaeranthus indicus (74) it may be present at levels as high as 7.4%. Essentially no germination of velvetleaf occurred after 4 days exposure to a 2.5 x 10^{-3} M dose of 29.

Wolf then compared six additional closely related compounds; o-methoxycinnamaldehyde (30), p-methoxycinnamic acid (31), cinnamic acid (32), cinnamaldehyde (33), cinnamyl alcohol (34) and anethole (35). It appeared that 30 was more active than 29 (Table III) however the data was not significant statistically. Of all the compounds tested, 33 was the most active. Addition of a methoxyl group reduces activity; compare 31 with 32 and 29 with 33. Cinnamyl alcohol (34) compares closely with the methoxy aldehydes and is more active than cinnamic acid (32). The major component of star anise oil (35) showed no effect on velvetleaf germination.

In reviewing the data it is clear that aldehydes and alcohols, among these phenylpropanoids, are more inhibitory than the corresponding acids. Relatively little data are available comparing these compounds in other biological systems; however, 32, 33 and 35 have all been reported to inhibit activities such as fungal growth and insect pupation. Compounds 33 and 35 are used in flavorings and in perfumes and reputedly have low toxicities in animals.

The Sesbanimides - Glutarimide Derivatives. Sesbania drummondii, S. punicea and S. vesicaria are legumes native to the Gulf Coastal Plains of the U.S.A. and all three species have long histories of toxicity to livestock (75). Studies of the antileukemic principles in S. drummondii (76, 77) and of the factor responsible for animal toxicity in S. punicea (78) culminated in isolation of sesbanimide A (36a) as the primary active compound in both species. Lesser amounts of sesbanimides B (37) and C (38) were also present in S. drummondii. Compound 36a occurs at a level of approximately 4.5 x 10^{-4}% in S. drummondii seed and its presence has also been confirmed in S. vesicaria seed by MS/MS (Powell, R. G., and Plattner, R. D., unpublished data). Suffness and Cordell have commented on the antitumor activity of 36a (79), and speculated

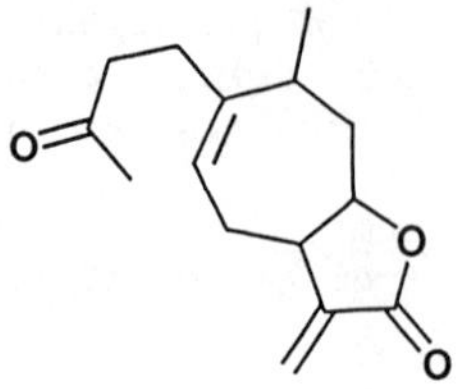

27

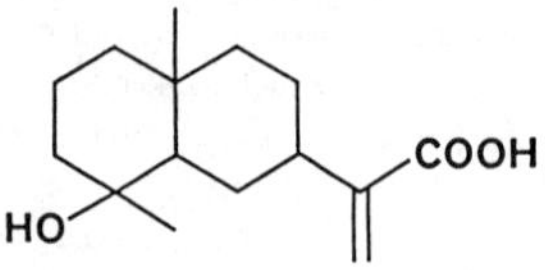

28

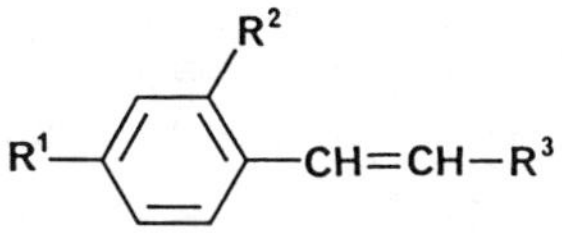

29 R^1=OCH_3, R^2=H, R^3=CHO
30 R^1=H, R^2=OCH_3, R^3=CHO
31 R^1=OCH_3, R^2=H, R^3=CO_2H
32 R^1=R^2=H, R^3=CO_2H
33 R^1=R^2=H, R^3=CHO
34 R^1=R^2=H, R^3=CH_2OH
35 R^1=OCH_3, R^2=H, R^3=CH_3

Table III. Velvetleaf germination results for selected cinnamyl compounds at 3 concentrations

Compound	% Germination, Relative to Controls		
	5 mM	2.5 mM	1 mM
29. *p*-methocycinnamaldehyde	1 a*	8 a	83 cd
30. *o*-methoxycinnamaldehyde	0 a	3 a	77 bc
31. *p*-methoxycinnamic acid	83 cd	83 cd	107 e
32. Cinnamic acid	5 a	77 bc	100 de
33. Cinnamaldehyde	0 a	0 a	57 b
34. Cinnamyl alcohol	0 a	0 a	73 bc
35. Anethole	95 d	100 de	98 de

*Values followed by the same letter are not statistically different at the 0.05 level by Chi-square l-tail

that the role of 36a in the seed may be to protect it from fungal attack. In solution, a solvent-dependent equilibrium exists between the tautomeric ring-closed hemiacetal (36a) and the ring-opened hydroxyketone (36b) forms of sesbanimide A.

Considerable structural resemblance is evident between the sesbanimides and glutarimide antibiotics produced by Streptomyces species such as cycloheximide (39) and streptimidone (40) (80). The sesbanimides (36-38) have been reported only in higher plants and there is currently no evidence, other than structural similarity, indicating that they could be of microbial origin.

Sesbanimide A (36a) and cycloheximide (39) give total inhibition of velvetleaf germination at the 8×10^{-4} M level (Powell, R. G., and Spencer, G. F., unpublished data). Compound 39 is extremely toxic to mammals, however, it has found limited agricultural use as a fungicide and as an abscission agent in citrus (81, 82). Abscission of oranges in Florida occurs at application rates of less than 0.11 kg/ha (0.1 lb/acre). Streptimidone (40) was reported to be effective as a pre-emergence herbicide against crab grass, morning glory and German millet at a soil spray rate of 20 lb/ acre (83). Further studies of the sesbanimides are needed in order to determine whether or not they also may be useful as plant growth regulatory or weed control agents.

Trewiasine and Related Ansamacrolides. The ansamacrolide antibiotics are a complex group of natural products that have received attention both for their structural complexity and for their broad range of biological activity (84). Representatives of this group include herbimycins A-C (41a-41c) (85-87), geldanamycin (41d) (84), and the ansamitocins (42a-42f) (88, 89). All of these compounds are microbial metabolites. Herbimycins A and B exhibit contact and selective activity, especially against dicotyledonous plants (86), and are more effective in the pre-emergence system rather than post-emergence. Geldanimycin (41d) did not entirely inhibit cress germination or radicle elongation even at relatively high test doses (90). Soon after emergence, however, radicles exposed to higher levels of 41d turned brown and began to disintegrate, indicating a post-germination effect.

The ansamitocins, produced by an organism thought to be a Nocardia species but later identified as Actinosynnema pretiosum (91), are simple C-3 esters of 42a (maytansinol or ansamitocin P-0). A series of more complex C-3 esters of maytansinol (maytansinoids), represented by 43a-43d, are found only in higher plants; typically in Maytenus and Putterlickia species (Celastraceae) (92, 93). The antitumor activity exhibited by maytansine (43) has led to clinical trials. Additional maytansinoids, including 44a-44c, 45a, 45b, are present in seeds of Trewia nudiflora (Euphorbiaceae), a tree native to parts of India (94, 95). Trewiasine (44a) gives greater than 50% inhibition of radicle elongation in velvetleaf at levels as low as 4×10^{-5} M but has no effect on germination up to 1×10^{-3} M (Powell, R. G., and Spencer, G. F., unpublished data). Growth inhibition at these levels is sufficient to warrant further studies to determine the suitability of using these or other ansamacrolides as herbicides.

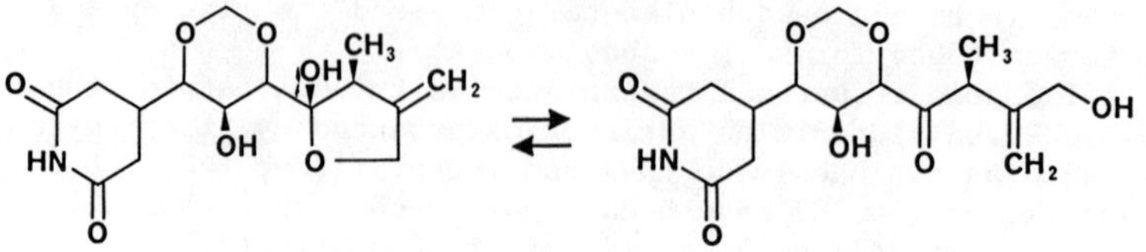

36a ⇄ 36b

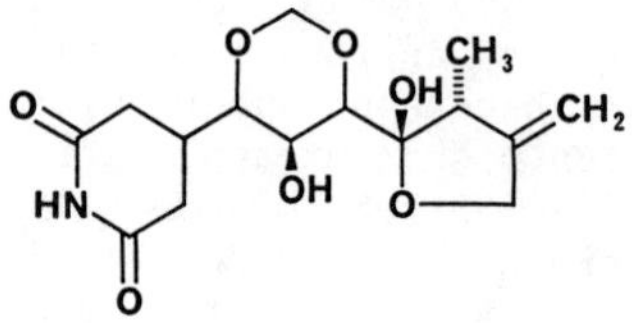

37

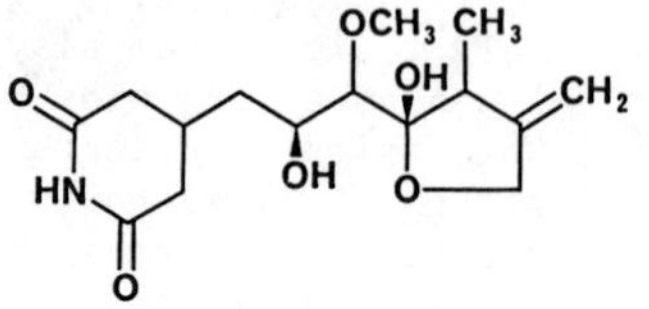

38

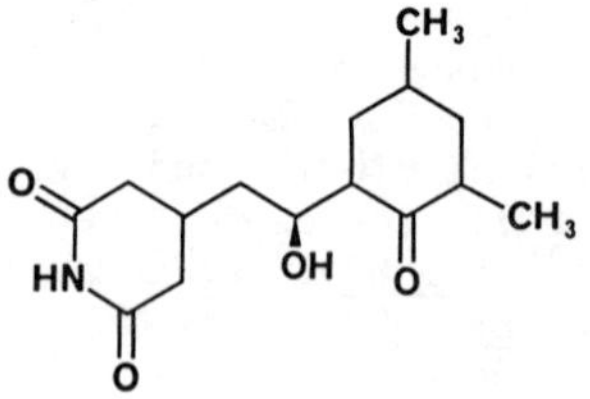

39

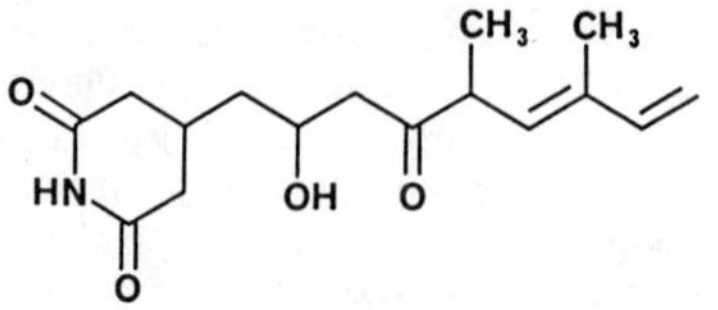

40

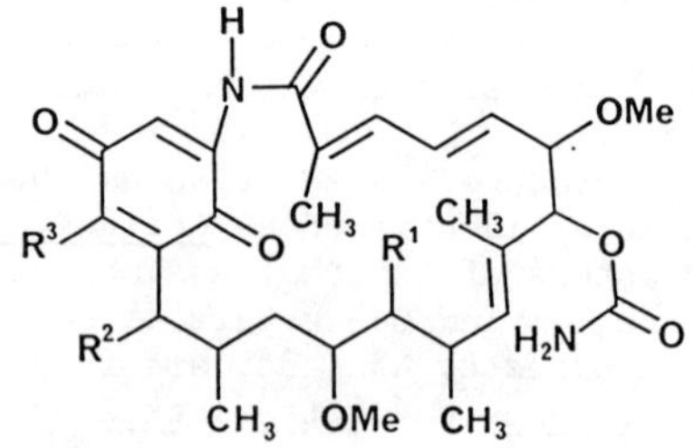

41a $R^1 = R^2 = OCH_3$, $R^3 = H$
41b $R^1 = OH$, $R^2 = R^3 = H$
41c $R^1 = OH$, $R^2 = OCH_3$, $R^3 = H$
41d $R^1 = OH$, $R^2 = H$, $R^3 = OCH_3$

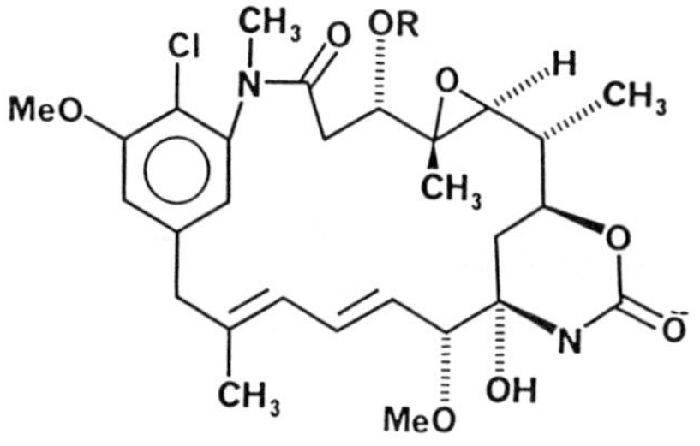

42a R = H
42b R = $COCH_3$
42c R = $COCH_2CH_3$
42d R = $COCH(CH_3)_2$
42e R = $COCH_2CH_2CH_3$
42f R = $COCH_2CH(CH_3)_2$

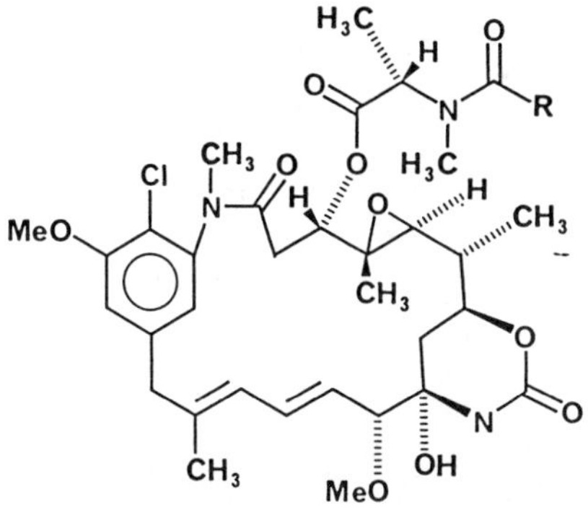

43a R = CH_3
43b R = CH_2CH_3
43c R = $CH(CH_3)_2$
43d R = $CH_2CH(CH_3)_2$

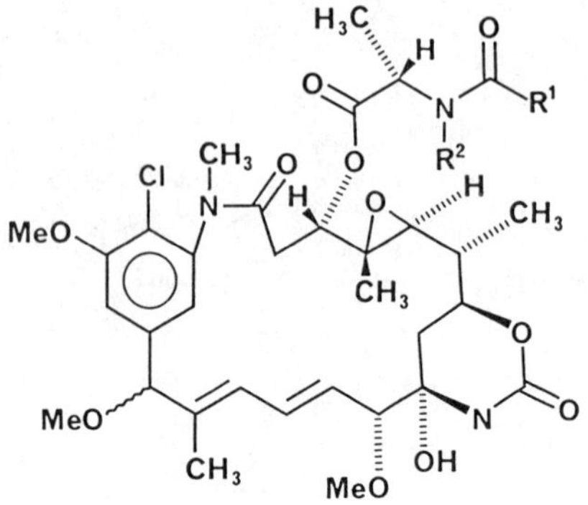

44a $R^1 = CH(CH_3)_2$, $R^2 = CH_3$
44b $R^1 = C(CH_3){=}CH_2$, $R^2 = CH_3$
44c $R^1 = CH(CH_3)_2$, $R^2 = H$

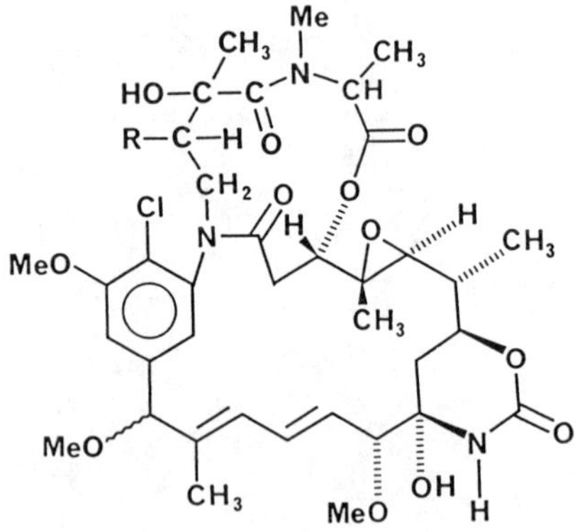

45a R = H
45b R = OH

Summary and Conclusions

Recent and continuing shifts toward reduced or no-till agriculture have increased the need for newer, more cost-effective and efficacious herbicides. Herbicides with new or novel mechanisms of action are desired as weed resistance to long-used herbicides develops. Use of selective herbicides has encouraged shifts in problem weeds to species more closely related to the associated crop; a situation requiring that future herbicides have even greater selectivity. As the need for new and novel herbicides is increasing, older methods for discovering useful herbicides, classical synthesis and screening, are yielding fewer returns. At the same time, licensing and regulation of new pesticides is increasingly difficult and expensive. Thus, studies of natural products are becoming more attractive to producers of agrichemicals, particularly as models for new structural types of active compounds.

This limited review provides evidence that higher plants themselves contain a wide variety of potentially useful growth and germination inhibitors. Compounds discussed markedly inhibit germination or growth of velvetleaf and, undoubtedly, will have similar effects on other plant species to a greater, or lesser, degree. Structures of these "biorational" materials range from quite simple, allyl isothiocyanate (2) for example, to very complex, for compounds such as the ansamacrolides. The simpler compounds and their structural analogs should be available by synthesis at prices competitive with present herbicides. Many of the compounds discussed may play a role in dormancy as most have been found as constituents of seeds. Others, such as the sesbanimides and the maytansinoids may be microbial metabolites that have been modified by plants for their own benefit. Utilization of tricothecene mycotoxins by the *Baccharis* plant for defensive purposes is a classical example of this type of allelopathic interaction (*96-98*).

It is estimated that less than 15% of the known species of higher plants have been examined for bioactive constituents (*99*) and few of the compounds identified have even been considered as potential herbicides. Certainly there is a need for expanded effort along these lines of research and there is every reason to believe that the many, as yet unexamined plant species, will yield numerous additional herbicidal compounds having novel structures and unexpected modes of action. Similar studies of toxic secondary plant metabolites for their fungicidal, insecticidal, and pharmaceutical activities are also warranted. Duke (*7*) has pointed out that the success of natural compounds as commercial herbicides will depend on several factors including the following: a, the position of regulatory agencies toward toxicological screening and licensing of natural products; b, the cost of production of these compounds; c, the efficacy and selectivity of the compounds in the field; and d, the success of industry in patenting natural compounds, many of which are known in the literature. The current position of regulatory agencies appears to favor toxicological screening and licensing of biorational products. Production costs of natural products should not differ substantially from those of synthetics having similar structures. Efficacy and selectivity of natural products in the field, in certain instances, may be superior to synthetics. Success in patenting compounds not yet in the literature will probably be the determining factor in whether or not natural products ultimately find widespread use as weed control agents.

Acknowledgments

Many thanks are due to the cooperation and technical assistance of L. W. Tjarks, D. Weisleder, R. D. Plattner, R. B. Wolf, R. Kleiman, T. S. Wilson, C. D. Williams and many others who assisted with portions of this work.

Literature Cited

1. Soderquist, C. J. J. Chem. Educ. 1973, 50, 782-83.
2. Molisch, H. "Der Einfluss einer Pflanze auf die andere-Allelopathie." Fischer, Jena, 1937.
3. Rice, E. L. Allelopathy; Academic Press: New York, 1974.
4. Rice, E. L. Allelopathy; 2nd Edition; Academic Press: New York, 1984.
5. Thompson, A. C., Ed.; The Chemistry of Allelopathy; ACS Symposium Series #268, American Chemical Society: Washington, D.C., 1985.
6. Putnam, A.; Tang, C.-S., Eds; The Science of Allelopathy; John Wiley & Sons, Inc.: New York, 1986.
7. Duke, S. O. Reviews of Weed Science. 1986, 2, 15-44.
8. Cutler, H. G. In The Science of Allelopathy; Putnam, A.; Tang, C.-S., Eds.; John Wiley and Sons, Inc. New York, 1986, p 147.
9. Upholt, W. M. In Handbook of Natural Pesticides: Methods; Mandava, N. B., Ed.; CRC Press: Boca Raton, Florida, 1985; Vol. I, p 273.
10. Duke, S. O. In The Science of Allelopathy; Putnam, A.; Tang, C.-S., Eds.; John Wiley & Sons, Inc: New York, 1986, p 287.
11. Wolf, R. B.; Spencer, G. F.; Kwolek, W. F. Weed Science. 1984, 32, 612-15.
12. Chan, H. T., Jr.; Heu, R. A.; Tang, C.-S.; Okazaki, E. N.; Ishizaki, S. M. J. Food Sci. 1978, 43, 255-56.
13. Tang, C.-S. Phytochemistry. 1971, 10, 117-21.
14. Tang, C.-S.; Syed, M. M; Hamilton, R. A. Phytochemistry. 1972, 11, 2531-33.
15. Daxenbichler, M. E.; VanEtten, C. H.; Brown, F. S; Jones, Q. J. Agric. Food Chem. 1964, 12, 127-130.
16. Kjaer, A.; Madsen, J. O; Maeda, Y. Phytochemistry. 1978, 17, 1285-87.
17. Putnam, A. R. Chem. Eng. News. 1983, 61, 33-34.
18. Syed, M. M. Phytotoxic compounds from Carica papaya L. and their fate in the soil; Diss. Abstr. Int. B36:4259; 1976.
19. Dem'yanenko, A. P.; Grabenko, A. D.; Danchenko, M. N; Rotmistrov, M. N. Fiziol. Aktiv. Veschestva. 1973, 5, 32-8; Chem. Abstr. 81:8684q.
20. Shcherbanovskii, L. R.; Nilov, G. I. Fitontsidy, Mater. Soveschch; 6th:109-113; 1972, Chem. Abstr. 78:53207q.
21. Horakova, K.; Kalafut, F. Acta Morphol. Acad. Sci. Hung. 1970, 18, 311-47.

22. Wattenberg, L. W. Cancer Res. 1981, 41, 2991-94.
23. Dale, J. E. Weed Science. 1986, 34, 325-27.
24. Teasdale J. R.; Taylorson, R. B. Weed Science. 1986, 34, 520-24.
25. VanEtten, C. H.; Tookey, H. L. In CRC Handbook of Naturally Occurring Food Toxicants; Rechcigl, M., Ed.; CRC Press: Boca Raton, FL, 1983 p 15.
26. Spencer, G. F.; England, R. E; Wolf, R. B. Phytochemistry. 1984, 23, 2499-2500.
27. Loder, J. W.; Nearn, R. H. Heterocycles. 1977, 7, 113-18.
28. El-Zayat, A. A. E.; Ferrigni, N. R.; McCloud, T. G.; McKenzie, A. T.; Byrn, S. R.; Cassady, J. M.; Chang, C.; McLaughlin, J. L. Tetrahedron Letters. 1985, 26, 955-56.
29. Adityachaudhury, N; Das, A. K. J. Sci. Ind. Res. 1979, 38, 265-77.
30. Siegel, S. M. Phytochemistry. 1976, 15, 566-67.
31. Wolf, R. B.; Spencer, G. F.; Plattner, R. D. J. Nat. Prod. 1985, 48, 59-63
32. Tipton, C. L.; Klun, J. A.; Husted, R. R; Pierson, M. D. Biochemistry. 1967, 6, 2866-70.
33. Tang, C.-S.; Chang, S. H.; Hoo, D.; Yanagihara, K. H. Phytochemistry. 1975, 14, 2077-79.
34. Zuniga, G. E.; Argandona, V. H.; Niemeyer, H. M.; Corcuera, L. J. Phytochemistry. 1983, 22, 2665-68.
35. Klun, J. A.; Robinson, J. F. J. Econ. Entomol. 1969, 62, 214-20.
36. Argondona, V. H.; Niemeyer, H. M.; Corcuera, L. J. Phytochemistry. 1981, 20, 673-76.
37. Barnes, J. P.; Putnam, A. R.; Burke, B. A.; Aasen, A. J. Phytochemistry. 1987, 26, 1385-90.
38. Aiupova, A. T.; Molchanov, L. V.; Kadryov, C. S.; Aliev, N. A.; Giasov, K.; Loi, N. P.; Tsoi, Z.; Umarov, A. A. Agrokhimiia. 1979, 10 107-11.
39. Virtanen, A. I.; Hietala, P. K. Acta Chem. Scand. 1955, 9, 1543-44.
40. Wahlroos, O.; Virtanen, A. I. Acta Chem. Scand. 1958, 12, 124-28.
41. Dawe, D. H. Diss. Abstr. Int. B34, 3031 (1973).
42. Orcutt, J. A.; Prytherch, J. P.; Konicov, M.; Michaelson, S. M. Arch. Int. Pharmacodyn. 1964, 152, 121-31.
43. Vickery, M. L.; Vickery, B. Secondary Plant Metabolism; University Park Press: Baltimore, 1981.
44. Mabry, T. J.; Harborn, J. M.; Mabry, H., Ed.; The Flavanoids; Academic Press: New York, 1975.
45. Barz, W. Degradation of flavanoids and isoflavanoids. In: Czygan, F. C., Editor, "Pigments in Plants"; Gustav Fischer Verlag, Stuttgart and New York, 1980, p 210.
46. Barz, W. and Hosel, W. Recent Adv. Phytochem. 1979, 12, 339-69.
47. Spencer, G. F. and Tjarks, L. W. J. Plant Growth Regul. 1985, 4, 177-80.
48. Romussi, G. and Ciarallo, G. Phytochemistry. 1974, 13, 2890-91.
49. Pendse, R., Rama Rao, A. V. and Venkataraman, K. Phytochemistry. 1973, 12, 2033-34.

50. Datta, S. C. and Chatterjee, A. K. Comp. Physiol. Ecol. 1980, 5, 54-9.
51. Harborne, J. B. "Introduction to Ecological Biochemistry"; Academic Press, London, 1982, p 210.
52. Thomson, R. H. "Naturally Occurring Quinones"; Academic Press, New York, 1957, p 55.
53. Spencer, G. F., Tjarks, L. W., England, R. E. and Seest, E. P. J. Nat. Prod. 1986, 49, 530-33.
54. Kondo, A. and Kawazu, K. Kogawa Daiguku Nogakubu Gakujutsu Hokoku. 1979, 30, 215-18.
55. Tripathi, R. D., Srivastova, H. S. and Dixit, S. N. Experientia. 1978, 34, 51-2.
56. Otomo, N., Soto, H. and Sakamura, S. Agric. Biol. Chem. 1983, 47, 1115-19.
57. Entwistle, I. D., Gilkerson, T. and Devlin, B. R. J. Herbicidal. Compositions Containing 1-4-Quinones or Derivatives Thereof. British Patent 1 534 275, 1978.
58. Yukinaga, T., Hideo, K. and Hide, O. Plant Growth Regulating Agents. Japanese Patent 76 09 807, 1976.
59. Spencer, G. F. J. Nat. Prod. 1986, 49, 924-26.
60. Bohlman, F., Zdero, C., and Grenz, M. Chem. Ber. 1975, 108, 2822-23.
61. Bohlman, F. and Zdero, C. Chem. Ber. 1969, 102, 2211-15.
62. Bohlman, F. and Zdero, C. Chem. Ber. 1971, 104, 1957-61.
63. Lemmich, J., Pederson, P. A. and Nielsen, B. E. Acta Chem. Scand. 1971, 25, 344-46.
64. Lemmich, J., Pederson, P. A., Good, M. S. and Nielsen, B. E. Acta Chem. Scand. 1971, 25, 995-99.
65. Pepperman, A. D. and Blanchard, E. in: "The Chemistry of Allelopathy,", Ed. by A. C. Thompson, American Chemical Society: Washington, DC, 1985, p 415.
66. Spencer, G. F., Wolf, R. B. and Weisleder, D. J. Nat. Prod. 1984, 47, 730-32.
67. Rodriguez, E., Yoshioku, H. and Mabry, T. J. Phytochemistry. 1971, 10, 1145-54.
68. Bohlmann, F., Czerson, H. and Schoneweiss, S. Chem. Ber. 1977, 100, 1330-34.
69. Herz, W., Chikamatsu, H. and Tether, L. R. J. Org. Chem. 1966, 31, 1632-34.
70. Fardella, G. Fitotherapia. 1979, 50, 3; Chem. Abstr., 1979, 91, 154332a.
71. Wolf, R. B. J. Nat. Prod. 1986, 49, 156-58.
72. Coleman, E. C., Ho, C. T. and Chang, S. S. J. Agric. Food Chem. 1981, 29, 42-8.
73. Pogany, D., Bell, C. L. and Kirch, E. R. Perfum. Essent. Oil Rec. 1968, 59, 858-65.
74. Nigam, S. S. and Rao, A. K. Riechst., Aromen., Koerperpflegem. 1968, 18, 169-70.
75. Kingsbury, J. M. "Poisonous Plants of the United States and Canada;" Prentice-Hall: Englewood Cliffs, New Jersey, 1964, p 353.

76. Powell, R. G., Smith, C. R., Jr. and Weisleder, D. J. Amer. Chem. Soc. 1983, 105, 3739-41.
77. Powell, R. G., Smith, C. R., Jr. and Weisleder, D. Phytochemistry. 1984, 23, 2789-96.
78. Gorst-Allman, C. P., Steyn, P. S. and Vleggaar, R. J. Chem. Soc. Perkin Trans. I. 1984. 1311-14.
79. Suffness, M. and Cordell, G. A. "Antitumor Alkaloids" in The Alkaloids, Chemistry and Pharmacology, A. Brossi, Ed.; Academic Press: Orlando, FL, 1985; p 18-21,
80. Johnson, F. "The Chemistry of Glutarimide Antibiotics" in Progress in the Chemistry of Organic Natural Products, W. Herz, H. Griesback and G. W. Kirby, Eds., Springer-Verlag: New York, 1971; p 140..
81. Nickell, L. G. In Encyclopedia of Chemical Technology; John Wiley & Sons: New York, 1982; Vol. 18, p 1.
82. Cooper, W. C. U. S. Patent 3 663 199, 1972.
83. Kohberger, D. L., Fisher, M. W., Galbraith, M. M., Hillegas, A. B., Thompson, P. E. and Ehrlich, J. Antib. & Chemo. 1960, 10, 9-16.
84. Rinehart, K. L., Jr. and Shield, L. S. Progress in the Chemistry of Organic Natural Products, 1976, 33, 231.
85. Omura, S., Iwai, Y., Takahashi,Y., Sadakane, N.,Nakagawa, A., Oiwa, H., Hasegawa, Y. and Ikai, T. J. Antibiotics. 1979, 32, 255-61.
86. Iwai, Y., Nakagawa, A., Sadakane, N., Omura, S., Oiwa, H., Matsumoto, S., Takahashi, M., Ikai, T. and Ochiai, Y. J. Antibiotics. 1980, 33, 1114-19.
87. Shibata, K., Satsumabayashi., S., Nakagawa, A. and Omura, S. J. Antibiotics. 1986, 39, 1630-33.
88. Higashide, E., Asai, M., Ootsu, K., Tanida, S., Kozai, Y., Hasegawa, T., Kishi, T., Sugino, Y. and Yoneda, M. Nature. 1977, 270, 721.
89. Asai, M., Mizuta, E., Izawa, M., Haibara, K. and Kishi, T. Tetrahedron. 1979, 35, 1079-85.
90. Heisey, R. M. and Putnam, A. R. J. Nat. Prod. 1986, 49, 859-65.
91. Hasegawa, T., Tanida, S., Hatano, K., Higashide, E. and Yoneda, M. Int. J. Syst. Bacteriol. 1983, 33, 314.
92. Smith, C. R., Jr. and Powell, R. G. "Chemistry and Pharmacology of Maytansinoid Alkaloids" in Alkaloids: Chemical and Biological Perspectives, Volume 2, W. S. Pelletier, Ed.; John Wiley & Sons: New York, 1984; p 149 .
93. Reider, P. J. and Roland, D. M. In The Alkaloids, Chemistry and Pharmacology; A. Brossi, Ed.; Vol. 23, p 71.
94. Powell, R. G., Weisleder, D. and Smith, C. R., Jr. J. Org. Chem. 1981, 46, 4398-4403.
95. Powell, R. G., Weisleder, D., Smith, C. R., Jr., Kozlowski, J. and Rohwedder, W. K. J. Am. Chem. Soc. 1982, 104, 4929-34.
96. Jarvis, B. B., Midiwo, J. O., Tuthill, D. and Bean, G. A. Science. 1981, 214, 460-62.
97. Jarvis, B. B., Pena, N. B., Rao, M. M., Comezoglu, N. S., Comezoglu, T. F. and Mandava, N. B. In The Chemistry of

Allelopathy; ACS Symposium Series #268, American Chemical Society: Washington, D.C., 1985, p 149.
98. Cutler, H. G. and Jarvis, B. J. Environmental and Experimental Biology. 1985, 25, 115-28.
99. Balandrin, M. F., Klocke, J. A., Wurtele, E. S. and Bollinger, W. H. Science. 1985, 228, 1154-60.

RECEIVED October 10, 1988

Chapter 15

Allelopathy in the Florida Scrub Community as a Model for Natural Herbicide Actions

Nikolaus H. Fischer[1], Nesrin Tanrisever[1], and G. Bruce Williamson[2]

[1]**Department of Chemistry, Louisiana State University, Baton Rouge, LA 70803–1804**

[2]**Department of Botany, Louisiana State University, Baton Rouge, LA 70803–1804**

We are investigating the hypothesis that allelochemicals released from plants of the Florida scrub community deter the invasion of fire-prone sandhill grasses. Bioassays guided our chemical isolations of phytotoxins from three scrub species. Constituents of the endemic scrub members, *Ceratiola ericoides*, *Conradina canescens* and *Calamintha ashei*, were tested for effects on the germination and radicle growth of lettuce (*Lactuca sativa*) and little bluestem (*Schizachyrium scoparium*), a native grass of the Florida sandhill community.

From *Ceratiola ericoides* we isolated two triterpenes, ursolic acid and erythrodiol, as well as flavanones, catechins, proanthocyanidins, a chalcone, and dihydrochalcones. Water leaf washes provided ursolic acid and a novel dihydrochalcone, ceratiolin, which exhibited no significant phytotoxic activity. However, ceratiolin spontaneously decomposed in water to hydrocinnamic acid, which showed considerable phytotoxic activity.

From *Conradina canescens* we isolated copious amounts of ursolic acid and a series of monoterpenes. Major monoterpenes constituents were 1,8-cineole, camphor, borneol, myrtenal, myrtenol, α-terpineol, and carvone. We tested camphor, myrtenal, borneol, and carvone for their phytotoxic activity on lettuce and bluestem and found strong effects on lettuce.

Calamintha ashei provided, besides ursolic acid and caryophyllene oxide, the monoterpenes menthofuran, epievodone, and calaminthone, all of which are phytotoxic. A saturated aqueous solution of epievodone had stimulatory effects on the germination of bluestem. Epievodone in a saturated aqueous solution of ursolic acid, a natural mild detergent, strongly inhibited the germination and growth of bluestem. Also, a mixture of epievodone, calaminthone, and caryophyllene oxide totally inhibited germination of bluestem but had only minor effects on lettuce.

0097–6156/88/0380–0233$06.00/0

Data obtained support our proposal that natural surfactants such as ursolic acid enhance the solubilization of allelopathic lipids via micellization and play a significant role in their transport to target seeds or seedlings. Two mechanisms of production and delivery of natural herbicides are emerging from these studies: (a) the formation of a highly phytotoxic derivative from a relatively non-toxic precursor, and (b) the production of natural surfactants that assist in transporting relatively insoluble phytotoxic lipids into the soil.

In Florida and the Southeastern Coastal Plain, the well drained sandy ridges of relict shorelines support two different plant communities. The more prevalent vegetation type is the sandhill, dominated by longleaf pine and oaks with a dense herbaceous cover of grasses and forbs. Throughout the sandhill vegetation are islands and strands of the scrub vegetation, which are dominated by a closed canopy of sand pine with a dense understory of evergreen oaks, and nearly devoid of herbaceous cover. The distinct species sets are segregated across abrupt ecotones with the less common scrub species regarded as endemics. Characteristic differences that distinguish the two vegetation types are summarized in Table I (1).

Table I. Contrast in scrub and sandhill vegetation types

	Scrub	Sandhill
Shrub layer:	Very dense	Open
Herbaceous layer:	Nearly none	Complete cover
Foliage phenology:	Evergreen	Deciduous
Surface litter:	Light	Heavy
Fire frequency:	20-50 years	3-8 years
Plant relative growth rates:	Slow	Fast
Age of plants at first reproduction:	Old	Young

In 1895, Nash (2) stated, "The scrub flora is entirely different from that of the high pine land (sandhill), hardly a single plant being common to both; in fact these two floras are natural enemies and appear to be constantly fighting each other."

The role of fire in maintaining the two communities has been proposed as an alternative to edaphic factors. In the absence of fire, scrub species will invade sandhill communities and proliferate (3,4). Observations after wildfires indicate that invading scrub species are killed by the repeated burning necessary to maintain healthy sandhill communities (4-6). Scrub regenerates successfully only with long-interval, catastrophic fires (7,8), but this fire-cycle does not explain the lack of sandhill species colonizing the scrub communities.

The two plant communities could be maintained as separate entities where frequent surface fires in the sandhill would preempt the success of scrub invaders, and during the long fire-free intervals in the scrub, allelopathy would preclude successful sandhill colonizers. Once established, the pattern of the communities would be maintained, though ecotones might expand and contract episodically with natural changes in fire frequency or habitat disruption. In order to test this allelopathy/fire hypothesis (1), our ecology and natural products chemistry groups initiated a cooperative investigation.

This paper reviews our findings of three early colonizers of the Florida scrub, *Ceratiola ericoides* (A. Gray) Heller (Empetraceae), *Calamintha ashei* (Weatherby) Shinner (Lamiaceae), and *Conradina canescens* Torr. & Gray (Lamiaceae). Preliminary results indicated that water-soluble allelotoxins released from the leaves of these scrub plants inhibit germination and growth of sandhill species (9). We present here a summary of our chemical and bioactivity data involving allelotoxins released from the above shrubs and their effects on the germination and growth of little bluestem (*Schizachyrium scoparium*) (Michx.) Nash (Poaceae). This native sandhill grass was used to test for potential benefit of allelopathy by scrub plants to deter the invasion of fire-prone sandhill grasses into the scrub. We also include evidence for the role of natural detergents such as ursolic acid as solubilization enhancers of bioactive lipids and discuss their function in the release and transport of allelopathic lipids in rain washes from an allelopathic source plant to a target species.

Materials and Methods

Bioassays. The preliminary bioassays were performed by D.R. Richardson (9) and involved application of test washes derived from leaves of *Ceratiola ericoides*, *Conradina canescens*, and *Calamintha ashei* to target seeds (*Schizachyrium scoparium* and *Lactuca sativa*) and subsequent monitoring of germination and growth (Table II). Once each month fresh leaves were collected from tagged test plants in the scrub community, soaked in distilled water at a ratio of 1 g of fresh leaf weight per 10 ml of water for 24 h under refrigeration at 8°C. Soaking permitted release of water soluble compounds from the leaf surface or through natural leaching as might occur during rainfall, but the refrigeration prevented decay of organic materials and release of internal products in whole leaves. The water solutions were decanted, and 5 ml of solution was added to sterile petri dishes (10 cm diameter) lined with one sheet of filter paper. Each dish contained thirty seeds of target plants with three or four replicate dishes per test wash. The test solutions were measured for pH and osmolarity and differed only slightly from water controls, so no adjustments were required. The petri dishes were kept in the dark until controls exhibited radicle growth of several cm, 4 days for lettuce and 15 days for little bluestem. After the dishes had been frozen to terminate growth, percent germination and radicle length of germinated seeds were recorded (9).

Chemical Analyses

Ceratiola ericoides. Volatiles from the hexane extract of *C. ericoides* were obtained by passing steam over the concentrate of the hexane extract (3 g) coated on the inside walls of a distillation column. The steam distillate (30 ml) was extracted with 10 ml of nanograde CH_2Cl_2, and the extract solution was subjected to GC-MS analysis.

The flavonoids 1-3, 5, 6 and a mixture of ursolic acid (UA, 25), and erythrodiol (26) used for the germination and radicle growth studies were isolated by previously described methods (10). Commercially available ursolic acid and hydrocinnamic acid (HCA) were also used in the bioassays. The bioassays of the flavanoids (1-3, 5, 6) involved preparation of a saturated aqueous solution of the natural triterpene mixture by sonication for 1 h of an excess of triterpene mixture in water and filtration. Saturated solutions of the flavonoids were prepared by adding 5 mg of each flavonoid to 22 ml of aqueous saturated triterpene solution and sonication of this mixture for 1 h. After filtration, a maximal concentration of 227 ppm of each compound was obtained. The results of the bioassays performed on lettuce and bluestem are summarized in Table III.

Florida scrub soil is acidic with pH levels as low as 4 in some areas (9). To test for possible effects of acid on ceratiolin, we carried out an acid treatment under reflux for 2 h with 4M HCl. ^1H NMR spectral analysis of the product mixture as well as GC-MS comparisons with standards showed that HCA (11) and its methyl ester were formed under these conditions.

Calamintha ashei and *Conradina canescens*. The isolations of the mono- and sesquiterpenes and ursolic acid from *C. ashei* used in the bioassay studies (11,12) as well as the methods for isolation and structure determination of the *C. canescens* terpenoids have been previously described (13,14).

Micellization studies of scrub plant leachates and their pure constituents by the Wolff method (15). Measurements of relative acridine fluorescence intensity were performed on a SLM 4800 spectrofluorimeter with an excitation slitwidth of 8 nm and an emission slitwidth of 2 nm. Emission wavelength was 425 nm and excitation wavelength was 360 nm or 395 nm, depending on the absorbance (11,13,14).

Results and Discussion

Biological activities of Ceratiola ericoides constituents. *Certiola ericoides* is a monotypic genus, one of only three genera of the family Empetraceae. An endemic of the Florida scrub community, it is locally dominant on disturbed sites and there is an absence of herbaceous growth around *Ceratiola* plants (9). Preliminary results from extensive monthly bioassays (Table II) showed that water-soluble litter extracts strongly inhibited germination and radicle growth of native grasses of the adjacent sandhill community; however, water soluble extracts of fresh leaves were only mildly allelopathic (9). Although chromatographic patterns of the flavonoids of Ceratiola had been previously carried out in connection with a biochemical systematic study of the Empetraceae (16), we performed an extensive re-investigation of this species in search for possible allelopathic constituents (10). Chemical analysis of the dichloromethane extract of ground aerial parts of C. ericoides yielded two known dihydrochalcones, angoletin (1) and

Table II. Effect of water washes of scrub species on Schizachyrium scoparium germination (G) and radicle length (RL)[1]

	Conradina canescens		Calamintha ashei		Ceratiola ericoides fresh plant		Ceratiola ericoides litter	
	G	RL	G	RL	G	RL	G	RL
March	74*	126*	66*	133	80	114	51*	28
April	112	82	93	76	103*	91	133	84*
May	83	79	85	72*	83	99	81	81
June	19*	50*	25*	44*	85	83	9*	79
July	52*	84*	20*	76*	55*	80	48*	93
August	104	91	84	72*	84	102	100	108
September	111	108	103*	104*	93	129*	85*	101*
October	56*	117	63*	83	74	108	70	95
November	161*	81*	123	83	147	84	95	91

[1]Numbers are germination as a percent of the control and radicle lengths as percent of control. An asterix indicates a significant difference ($p < 0.05$). (Data from ref. 9.)

2',6'-dihydroxy-4-methoxy-3',5'-dimethyldihydrochalcone (2), as well as 2',4'-dihydroxychalcone (6) and three known flavanones, 7-hydroxyflavanone, 8-methylpinocembrin (4), and 6,8-dimethylpinocembrin (3). Methanol extracts of ground leaves provided catechin (8), epicatechin (9) and epicatechin-(4β→8;2β→O→7)-epicatechin (A-2 dimer) (10). From water washes of freshly harvested leaves a novel dihydrochalcone, ceratiolin (7), was isolated (10). (See Figure I.)

Fresh C. ericoides leaves were odorless, but dried plant material emitted an odor much like that detected in the surroundings of living plants in the natural environment. Therefore, the atmosphere around dried leaves was flushed with nitrogen for eight hours, and the volatiles were trapped in CH_2Cl_2 at -70°C. These air samples contained angoletin (1), 2',6'-dihydroxy -4'-methoxy -3',5'-dimethyldihydrochalcone (2), 2',4'-dihydroxychalcone (6), and 6,8-dimethylpinocembrin (3) as major constituents. All four compounds were also found in the CH_2Cl_2 extract of C. ericoides (10).

Biological activities of the above Ceratiola flavonoids had not been previously reported except for assessments of the cytotoxic and antimicrobial potential of angoletin (1) and its benzyl derivatives, which exhibited no activity (17). For reasons that will be discussed later, five flavonoids (1-4,6) were tested in a saturated

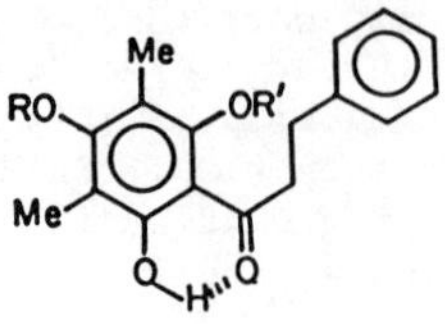

1, R = H, R′ = Me

2, R = Me, R′ = H

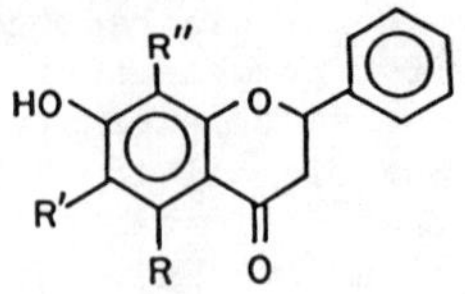

	R	R′	R″
3,	OH	Me	Me
4,	OH	H	Me
5,	H	H	H

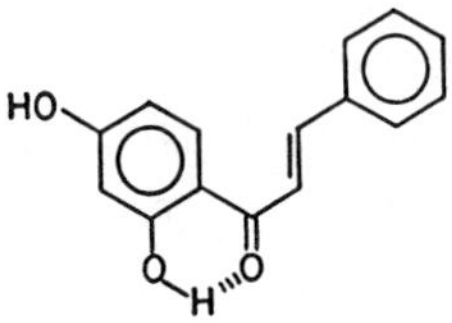

6

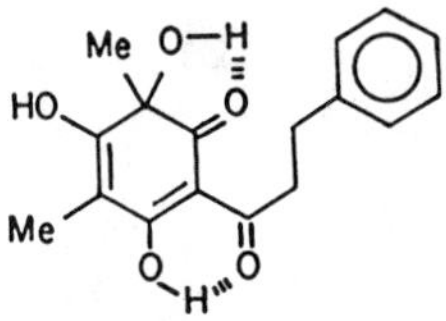

7

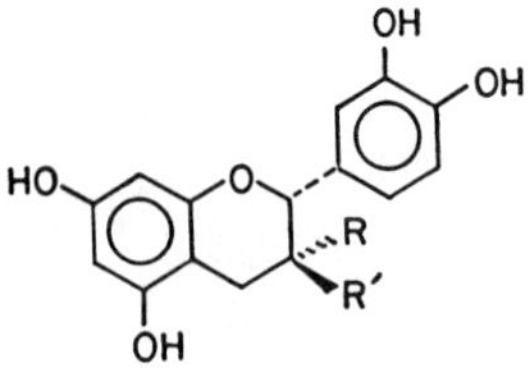

8, R = H, R′ = OH

9, R = OH, R′ = H

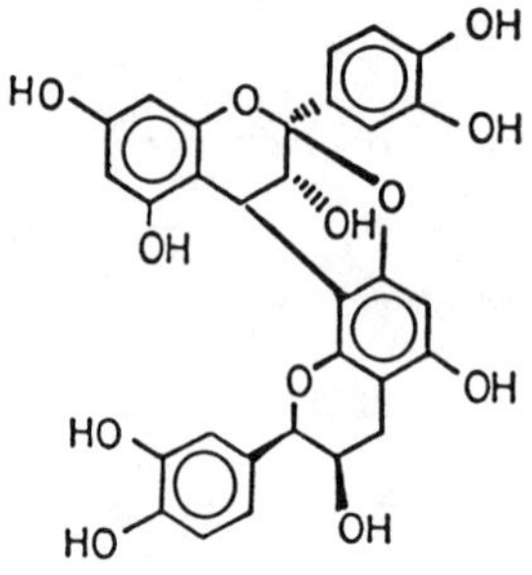

10

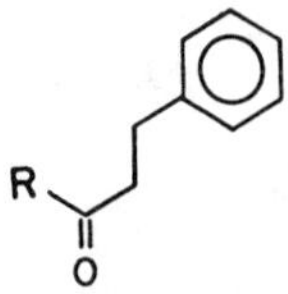

11, R = OH

12, R = CH_3

13, R = $-CH_2-(CO)_2-CH_3$

Figure I. Constituents of Ceratiola ericoides: angoletin (1), 2′,6′-dihydroxy-4-methoxy-3′,5′-dimethyldihydrochalcone (2), 6,8-dimethylpinocembrin (3), 8-methylpinocembrin (4), 5-desoxypinocembrin (5), 2′,4′-dihydroxychalcone (6), ceratiolin (7), catechin (8), epicatechin (9), epicatechin-(4β→8;2β→O→7)-epicatechin (A-2 dimer) (10) hydrocinnamic acid (11), 4-phenyl-2-butanone (12), and triketone (13).

aqueous solutions of an ursolic acid(UA)-erythrodiol mixture (ED) (Table III). The latter are two triterpenes (25 and 26) isolated from the the cutical wax of C. ericoides leaves.

We observed only minor inhibition of germination with lettuce and no significant effects with bluestem. Comparison of the activity on radicle growth of each compound in an aqueous solution of the triterpene mixture with the triterpene mixture control yielded opposite effects on L. sativa and S. scoparium. The compounds mildly stimulated growth of L. sativa while inhibitions as strong as 52% for 6,8-dimethylpinocembrin (3) were observed for S. scoparium radicle growth. When compared with water controls, which might be a better comparison for assessing the behavior of the natural system, only 6,8-dimethylpinocembrin showed a statistically significant inhibition (66%). Although the other flavonoids caused some decrease in the radicle lengths of S. scoparium test seedlings, none of these activities was statistically significant.

Table III. Effects of C. ericoides flavonoids on the germination and radicle length of L. sativa and S. scoparium dissolved in a saturated aqueous solution of an ursolic acid-erythrodiol mixture [1]

Compound[2]	L. sativa		S. scoparium	
	G	RL	G	RL
1	97 (97)	101 (90*)	86 (81)	66* (85)
2	100 (100)	120* (107)	94 (89)	70 (90)
6	88* (88*)	136* (121)	106 (100)	71 (91)
3	86* (86*)	115 (103)	97 (92)	52* (66*)
4	90* (90*)	132* (117)	100 (95)	57* (73)
UA-ED(25-26)	--- (100)	--- (89*)	--- (95)	--- (128)

[1]Germination (G) and radicle lengths (RL) are expressed as percentages of their controls. Each sample was compared to an ursolic acid-erythrodiol mixture (UA-ED) control. Data in parentheses indicate comparison to H_2O controls. An asterix indicates significant difference from the control at $p < 0.05$.
[2]Angoletin (1); 2',6'-dihydroxy-4'-methoxy-3',5'-dimethyl-dihydrochalcone (2); 2',4'-dihydroxychalcone (6); 6,8-dimethyl-pinocembrin (3); 8-methylpinocembrin (4).

Tannins inhibit bacteria (17) and their effects can be significant in allelopathic interactions when nitrifying bacteria are involved (18,19). Rice and Pancholy demonstrated that a 2 ppm concentration of condensed tannins added to soil totally inhibits oxidation of NH_4^+ within three weeks (20). This inhibition could indirectly influence the growth of plants with roots near the soil surface. Synergistic inhibitory effects of tannins were observed on Avena sativa coleoptile segments. Condensed tannin alone caused minor inhibitions, whereas in the presence of low concentrations of indoleacetic acid root segment elongation was completely blocked (21).

Catechin (8), epicatechin (9), and the A-2 dimer (10) at 62 and 620 ppm had minor to moderate effects on the germination of L. sativa, and only catechin at 62 ppm exhibited a significant germination inhibition (66% of control) on S. scoparium (Table IV). The growth inhibitions of L. sativa by the two catechins were low to moderate. The strongest inhibitory effects (64%) were exhibited by epicatechin at 620 ppm upon L. sativa, and significant growth promotion (136%) was found on S. scoparium at the 62 ppm level of epicatechin. In summary, the germination and growth inhibitions caused by the flavonoids, catechins, and the A-2 dimer (10) do not reach levels that were observed with plant leachates and chromatographic fractions of C. ericoides that contained mixtures of the above compounds.

Table IV. Effects of Ceratiola ericoides catechins on the germination (G) and radicle length (RL) of L. sativa and S. scoparium [1]

Compound	Conc. (ppm)	L. sativa G	L. sativa RL	S. scoparium G	S. scoparium RL
catechin (8)	62	95	87	66*	122
catechin (8)	620	97	92*	84	107
epicatechin (9)	62	94	75*	77	136*
epicatechin (9)	620	97	64*	73	113
A-2 dimer (10)	62	85	96	114	102
A-2 dimer (10)	620	92	70*	114	76

[1]Germination (G) and radicle lengths (RL) are expressed in percent of controls. An asterix indicates significant difference from control ($p < 0.05$).

Whole leaf water washes of C. ericoides obtained at ambient temperature were extracted with ethyl acetate-chloroform; the extracts provided, besides ursolic acid, one major compound, ceratiolin (7), which is a new and novel dihydrochalcone (10). Ceratiolin was suspected to play a significant role in the allelopathic activity of Ceratiola, since plant products on the leaf surface are more likely than are internal leaf constituents to be washed into the soil during rain. For this reason, ceratiolin was investigated further as a potential source for allelopathic activity against little bluestem. However, ceratiolin had no effect on the germination of S. scoparium at the 125 ppm level. In fact, it promoted the radicle growth of this target species (145%). These data were contrary to the results in the seasonal studies of water washes of Ceratiola leaves and, in particular, its litter (Table II). During the isolation and structural studies, we could not detect ceratiolin in the crude extracts of ground dried leaves of C. ericoides, but we did find it in the leaf water washes of fresh aerial parts and in the solutions from short-term whole plant dippings in methanol, in which it represented the major component (10). Also, the amount of ceratiolin extracted from dry plant material, which had been stored at room temperature for six months, was considerably lower than the yield from fresh plant material. This indicated that ceratiolin may be thermally unstable and underwent decomposition during storage of the plant material. Therefore, decomposition experiments were carried out with water solutions of pure ceratiolin. One solution was exposed to sunlight, and the other was kept in the dark. After 3 days hydrocinnamic acid (HCA) (11) was detected as a product in both experiments. The sample which had been kept in the dark also contained trace amounts of the triketone (13). Since the formation of these compounds was facile in water, these decomposition products must also form under natural conditions, especially in decaying *C. ericoides* litter.

A saturated aqueous solution of ceratiolin showed no inhibitory activity against the two test seeds (Table V). On the other hand, HCA exhibited high germination and growth inhibitions on both target seeds. At 63 ppm, HCA showed significant inhibitory activity on *S. scoparium* germination and growth, while equimolar amounts of ceratiolin (125 ppm) exhibited no activity on germination and stimulated growth. This difference may explain earlier findings that litter washes of *C. ericoides* showed considerable activity while fresh leaf washes were significantly less effective (Table II). Since soil under *C. ericoides* is acidic and the degradation of ceratiolin into HCA is promoted under acidic conditions, activity is likely to be greater in the sandy soil and decomposing litter than in fresh leaf rain washes. Low concentrations of HCA (2-10 ppm) have been detected in the soil under *Ceratiola*. Presently, quantitative analyses are in progress to determine seasonal

Table V. Effects of hydrocinnamic acid (HCA) and ceratiolin on germination and radicle growth of *Lactuca sativa* and *Schizachyrium scoparium* [1]

Compound	Conc. (ppm)	*L. sativa* G	*L. sativa* RL	*Schizachyrium scoparium* G	*Schizachyrium scoparium* RL
HCA (11)	1000	0*	---	0*	---
	500	0*	---	2*	17*
	250	4*	9*	50*	32*
	125	95	40*	68*	46*
	63	109	56*	74*	65
	31	100	72*	113	88
	16	104	76*	109	85
Ceratiolin(7)	125	102	83	102	145

[1]Germinations (G) and radicle lengths (RL) are in percent of the controls. An asterix indicates significant difference from the control ($p < 0.05$).

concentration changes of HCA and other decomposition products of ceratiolin. The triketone (13) has not yet been tested for its allelopathic potential due to insufficient amounts of material. Future bioassays will include compound 13 as well as the 4-phenyl-2-butanone (12), which was found in the volatiles of C. ericoides. The latter compound also appears to be a degradation product of ceratiolin, possibly being formed via acid cleavage of the triketone (13).

Biological activities of Calamintha ashei and Conradina canescens constituents. The mints Calamintha ashei and Conradina canescens are endemic to the Florida scrub community. Previously neither species had been investigated chemically. Water washes of fresh C. ashei and C. canescens leaves had significant inhibitory effects on S. scoparium germination and growth (Table II) (9). The cutical waxes contained copious amounts of ursolic acid(UA,25) (11). Further chemical analysis of the aerial parts of C. ashei provided the known monoterpene menthofuran (17) and the sesquiterpene caryophyllene epoxide. In addition, the new menthofurans, calaminthone (15) its desacetyl derivative (16), and epievodone (14) were isolated (11,12). The effects of the C. ashei constituents on the germination and growth of L. sativa and S. scoparium are summarized in Table VI. (See Figure II.)

Table VI. Effects of saturated aqueous solutions of Calamintha ashei constituents on the germination and radicle growth of S. scoparium and L. sativa [1]

		S. scoparium		L. sativa	
Compound(s)	conc. (ppm)	G	RL	G	RL
mixture of 14, 15, caryophyllene oxide	285	0*	--	97	61*
calaminthone (15)	250†	129	88	113	138*
epievodone(14)	250†	185*	93	96	131*
caryophyllene oxide	625†	128	91	104	66*
ursolic acid (UA)	---	86	136	104	93
epievodone + satd. aq. soln of UA	250†	57*	33*	97	81*
calaminthone (volatiles)		36*	91	---	---
epievodone (volatiles)		11*	77	---	---

[1]Numbers are germination (G) and radicle lengths (RL) as percent of water control solutions. An asterix indicates significant difference from the water control ($p < 0.05$).

†This concentration represents the maximal concentration based on the total amounts weighed. In all cases only part of each sample dissolved.

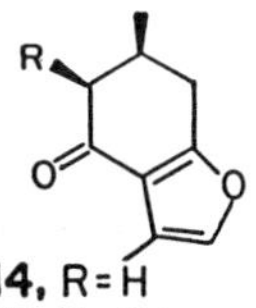

14, R=H
15, R=OAc
16, R=OH
17, R=H; 4-desoxy

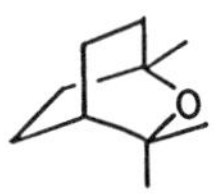

18

19, R, R' = O
20, R = H; R' = OH

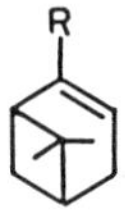

21, R = -CHO
22, R = $-CH_2OH$

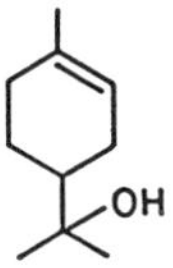

23

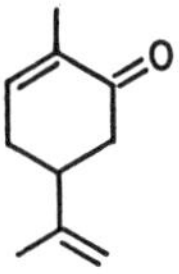

24

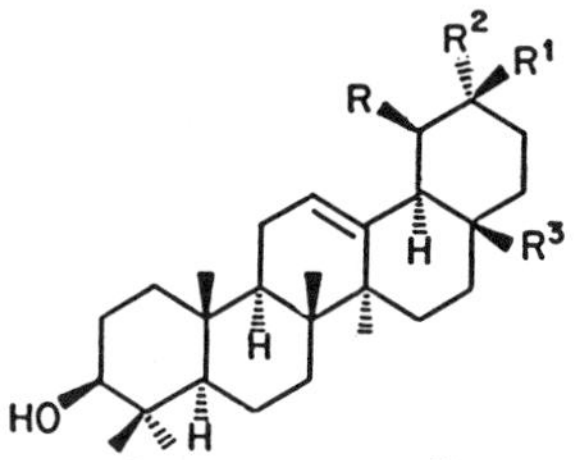

25, R = R^2 = Me, R^1 = H; R^3 = -COOH
26, R = H; R^1 = R^2 = Me; R^3 = $-CH_2-OH$

Figure II. Constituents of *Conradina canescens* and *Calamintha ashei*: epievodone (**14**), calaminthone (**15**), desacetylcalaminthone (**16**), menthofuran (**17**), 1,8-cineole (**18**), camphor (**19**), borneol (**20**), myrtenal (**21**), myrtenol (**22**), α-terpineol (**23**), carvone (**24**), ursolic acid (**25**), erythrodiol (**26**, isolated from *C. ericoides*).

A saturated aqueous solution of a chromatographic fraction containing calaminthone, epievodone, and caryophyllene oxide completely suppressed germination of *S. scoparium* but had no significant effects on *L. sativa* germination. In contrast, saturated aqueous solutions of the pure compounds were stimulatory to *S. scoparium* germination (185% of control for epievodone). Significant inhibitory effects were observed when epievodone was administered as a volatile or in a saturated aqueous solution of UA (**25**), which exhibited no significant effects when tested alone. Two facts are of interest in the above bioassay results: (a) The highly specific activity of a mixture of calaminthone, epievodone, and caryophyllene oxide toward little bluestem constrasts with insignificant germination inhibition on lettuce; (b) the strong inhibitory effects exhibited by the terpene mixture toward bluestem contrast with the germination promotions of bluestem caused by individual compounds. A similar dramatic synergistic phytotoxic effect by a mixture of camphor, pulegone, and borneol was previously observed by Asplund (22). Our interpretations of the synergistic property of UA will be discussed in the following section.

Water washes of fresh *Conradina canescens* leaves were strongly inhibitory due to the presence of a series of monoterpenes together with copious amounts of UA (**25**) (13). In a directed, bioassay-monitored search for the active fractions and their constituents of *C. canescens*, we analyzed a highly active fraction of the diethylether extract by GC-MS analysis (14). Several minor monoterpenes were present, but 1,8-cineole (**18**), camphor (**19**), borneol (**20**), myrtenal (**21**), myrtenol (**22**), α-terpineol (**23**), and carvone (**24**) were the major constituents. In the water leaf soak of fresh *C. canescens* leaves monoterpenes **18**,**19**,**20**, and **23** were present. All of these monoterpenes are known compounds, and several (*e.g.*, 1,8-cineole, camphor, and borneol) are potent plant germination and growth inhibitors (24).

Biological activities of four *C. canescens* monoterpenes on lettuce and of three on *S. scoparium* were determined (Table VII). In general, lettuce was considerably more sensitive to saturated aqueous solutions of the monoterpenes than was *Schizachyrium*, despite the fact that both species suffered complete germination inhibition when treated with a saturated aqueous solution of the highly active fraction of the diethyl ether extract.

Release mechanisms of allelopathic plant lipids. The Florida scrub community shares many characteristics (Table I) of the chaparral vegetation of Southern California (25-28), except the scrub consists of completely different species that grow in a sandy soil under different climatic conditions. In the California chaparral and other ecosystems two major release and transport mechanisms of allelopathic agents have been proposed (19). Water-soluble allelotoxins such as low molecular weight organic acids, phenolic acids, and plant phenolics can be washed of the leaf surface into the soil by fog drip and rain. Alternatively, allelotoxins can be volatilized from the plant leaf surface and carried to the ground to be absorbed by organic soil particles, seeds, or roots. Allelopathic plant products such as low molecular weight alcohols, aldehydes, carboxylic acids, ketones, and certain phenolics, as well as mono- and sesquiterpenes, could be released by this "volatility

Table VII. Major monoterpenes in the active fraction of a diethyl ether extract of <u>Conradina canescens</u>[1] and their effects on the germination and radicle growth of <u>Lactuca sativa</u> and <u>Schizachyrium scoparium</u>[2]

Compounds	L. sativa		S. scoparium	
	G	RL	G	RL
camphor (19)	37**	7**	95	100
borneol (20)	7**	8**	77	54*
myrtenal (21)	19**	8**	59*	112
carvone (24)	0**		--	--

[1]GC conditions in the GC-MS analysis were: 30m bonded silica capillary column; injection temp. 250°C; 60° for 1 min.; then 5°/min. to 210°. Retention times and mass spectra of all compounds given in the table were confirmed by correlation with standards.
[2]Germination (G) and radicle lengths (RL) are in percent of the distilled water controls:
*treatment was significantly different from the control at $p \leq 0.05$;
**treatment was significantly different from the control at $p \leq 0.01$.

mechanism". The above mechanisms exclude the wide host of non-volatile, water-insoluble plant lipids from participation in allelopathic actions. Furthermore, operation of the "volatility mechanism" in the Florida scrub may be doubted for a number of reasons: Firstly, in Florida regular winds would carry allelopathic volatiles away from the source plants, not depositing them in the immediate vicinity of the plant for allelopathic action. Secondly, the sandy soil in the scrub lacks the lipophilic organic matter necessary for absorption of the volatiles from the air. Thirdly, the relatively high temperature of the surface sands would prevent absorption of volatile lipids.

Water-solubilization of lipids can be facilitated by the involvement of natural detergents such as UA (25) through micellization (23). This detergent effect allows release and transport in rain washes of volatile and non-volatile hydrophobic plant products. Therefore, one aspect of our research on the Florida scrub community involved the analyses of leaf washes for the detection of micelles (1,11,13,14). We used the method by Wolff for detection of micellization and determination of critical micelle concentrations, because it has a wide applicability in micelle detection of anionic, cationic, and non-ionic tensides (15). This method is based on the fact that below the critical micelle concentration (CMC) the fluorescence of acridine is independent of surfactant concentration, while above the CMC, quantum yields decrease drastically with increasing surfactant concentration. This change is due to the behavior of acridine, which only fluoresces in protic solvents and is sufficiently hydrophobic to be located mainly within the hydrocarbon-like interior of the micelle. Preliminary tests with the synthetic tenside sodium dodecyl sulfate (SDS) and natural surfactants such as ursolic acid (25) and hydrocinnamic acid (11) demonstrated the usefulness of the Wolff method for our studies (Figure III). We unambiguously established that the highly inhibitory water leachates of Ceratiola ericoides, Calamintha ashei, and Conradina canescens (not shown) form micelles. All three species contain UA in the leaf waxes, thus providing a possible matrix for micellization. In each case the typical abrupt change in fluorescence yield of acridine at the critical micelle concentration was observed (Figure III).

We estimated the concentration of the leachate solution by assuming a reasonable average molecular weight of 300. The dramatic decrease in fluorescence intensity near 10^{-4} and 10^{-3} molar concentrations clearly indicated micellization of the leachate solution of all three scrub species. This result strongly suggested that UA acts as a micellar host for the bioactive terpenoids in Calamintha and Conradina water washes, most likely increasing their rate of solubilization in aqueous foliar leachates and/or aiding their transport to target seeds or plants. This effect is complementary to the "volatility" mechanism of allelopathic terpenes, which had been proposed by Muller (28). The "synergistic" effects of ursolic acid as exemplified by the bioassays described in Table VI are presently not understood. Experiments are in progress that will allow separation of the contributions due to possible micelle-mediated increases in concentration of allelopathic agents and/or to improved transport mechanisms that facilitate movement of allelopathic agents to and through membranes of target seeds.

In summary, ursolic acid appears to serve several important ecological functions in the three scrub species studied:

(a) Its fixative action for Calamintha and Conradina constituents reduces the volatilization loss of the allelopathic terpenes, which would otherwise be rapid.

(b) Its detergent effect enhances water-solubilization of non-polar terpenes and other volatile and nonvolatile lipids in rain washes.

(c) It facilitates transport of phytotoxins to and entry into the membranes of target seeds or seedlings.

Conclusions and Outlook

Detailed knowledge of allelopathic actions in natural plant communities can provide excellent models for new strategies in developing highly selective herbicides. The collaborative effort of ecologists and chemists in the present study has produced considerable information related to allelopathic interference in a natural plant community. The Florida scrub represents an ideal natural ecosystem for studies of this type, since the sandy soil of the scrub contains little organic matter. This absence simplifies considerably soil analyses for allelotoxins, which are frequently obscured by chemicals released from the organic matter in the soil. Furthermore, uncontrolled reversible and irreversible absorption of source plant allelochemicals by organic soil particles are also reduced in the sandy scrub soil.

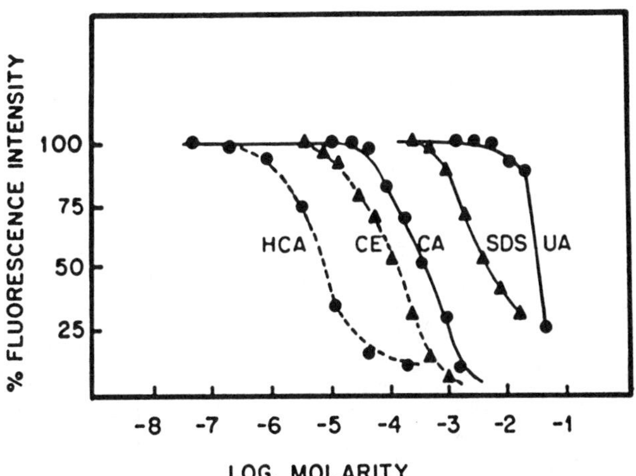

Figure III. Dependence of relative fluorescence intensity of 1.2×10^{-5} M acridine solution vs. concentration of leachates of Calamintha ashei (CA) and Ceratiola ericoides (CE) as well as aqueous solutions of hydrocinnamic acid (HCA), ursolic acid (UA) and sodium dodecyl sulfate (SDS).

Our studies in the Florida scrub have provided insight into new mechanisms of action that might also occur in other natural ecosystems. Chemical, microbial, or spontaneous decomposition of a nontoxic plant product into highly phytotoxic derivatives provides a possible mechanism for specific directed toxicity toward a target species. In *Ceratiola ericoides* the nontoxic ceratiolin represents an allelotoxin precursor, and its decomposition product HCA, which is mainly formed in the litter, acts as the specific allelotoxin for the sandhill target species. A promising practical aspect of this finding includes the synergistic allelopathic effects, exemplified by the dramatic increase in inhibition of bluestem germination caused by a mixture of volatile terpenes from *Calamintha ashei*. Further advances in the knowledge of bioactivity enhancements due to synergistic component mixtures could be of immense value in the development of natural biodegradable herbicides. The observation of highly selective phytotoxicity of *Calamintha* terpenes - *e.g.*, distinctly higher germination inhibition of bluestem than of lettuce - represents another useful guide in the development of selective herbicides (30).

Nonpolar cutical waxes and resins can possibly function as fixatives to prevent the loss of allelopathic volatiles by rapid evaporation. This property as a slow-release solvent for volatile lipids could be facilitated by a wide host of constituents in plant leaf waxes: triterpenes, long-chain fatty acids, hydrocarbons, alcohols, and esters, including fats. This fixative property, which is commonly exploited in the perfume industry to enhance long-term retention of active components, undoubtedly plays a significant role in the fixation of allelopathic plant volatiles.

Finally, the function of natural detergents such as triterpene acids and long-chain fatty acids as micelle-forming matrixes for volatile as well as nonvolatile bioactive lipids has distinct practical implications. Natural or synthetic surfactants enhance the solubilization of lipids in rain washes and thus facilitate the transport of allelopathic lipids into the soil to reach target seeds or seedlings.

Acknowledgments

This material is based upon work supported by the U.S. Department of Agriculture, Competitive Research Grants Program for Forest and Rangeland Renewable Resources under Agreement No. 85-FSTY-9-0139.

Literature Cited

1. Williamson, G. B.; Richardson, D. R.; Fischer, N. H. In *Frontiers of Allelochemical Research*; Rizvi, S.J.N., Ed.; Elsevier: Amsterdam, 1988 (in press).
2. Nash, G. V. *Bull. Torrey Bot. Club* 1895, *22*, 141-161.
3. Laessle, A. M. *Ecol. Monogr.* 1958, *28*, 361-387.
4. Veno, P. A. *Ecology* 1976, *57*, 498-508.
5. Williamson, G. B.; Black, E. M. *Nature* 1981, *293*, 643-644.
6. Hebb, E. A. *Southern J. Appl. Forestry* 1982, *6*, 144-147.
7. Harper, R. M. In *Florida Geological Survey*; Sixth Annual Report, 1914, pp.163-391.

8. Webber, H. J. Amer. J. Bot. 1935, 22, 344-361.
9. Richardson, D. R. Ph.D. Dissertation, University of South Florida, Tampa, 1985.
10. Tanrisever, N.; Fronczek, F. R.; Fischer, N. H.; Williamson, G. B. Phytochemistry 1987, 26, 175-179.
11. Tanrisever, N.; Fischer, N. H.; Williamson, G. B. Phytochemistry 1988, 27, 000-000. (in press).
12. Macias, F. A.; Fronczek, F. R.; Fischer, N. H. Phytochemistry, submitted.
13. Williamson, G. B.; Fischer, N. H.; Richardson, D. R.; De la Peña, A. J. Chem. Ecol., submitted.
14. De la Peña, A. Master's Thesis, Louisiana State University, Baton Rouge, 1985.
15. Wolff, T. J. Colloid and Interface Science 1981, 83, 658-660.
16. Moore, D. M.; Harborne, J. B. and Williams, C. A. Bot. J. Linn. Soc. 1970, 63, 277-283.
17. Hufford, C. D.; Oguntimein, B. O. Phytochemistry 1980, 19, 2036-2038.
18. Waage, S. K.; Hedin, P. A.; Grimley, E. Phytochemistry 1984, 23, 2795-2787.
19. Rice, E. L. Allelopathy; Second edition; Academic Press: New York, 1984.
20. Rice, E. L.; Pancholy, S. K. Amer. J. Bot. 1973, 60, 691-702.
21. Zinsmeister, H. D. Planta 1964, 61, 130-141.
22. Asplund, R. O. Weed Science 1969, 17, 454-455.
23. Turro, N. J.; Graetzel, M.; Braun, A. M. Angew. Chem., Int. Ed. Engl. 1980, 19, 675-696.
24. Fischer, N. H. In The Science of Allelopathy; Putnam, A. R. and Tang, C., Eds.; Wiley: New York, 1986, pp.203-218.
25. Muller, C. H.; Muller, W. H.; Haines, B. L. Science 1964, 143, 471-473.
26. Muller, C. H. Bull. Torrey Bot. Club 1966, 93, 332-351.
27. Muller, C. H.; Hanawalt, R. B.; McPherson, J. K. Bull. Torrey Bot. Club 1968, 95, 225-231.
28. Muller, C. H. Vegetacio 1969, 18, 348-357.
29. Muller, W. H.; Muller, C. H. Bull. Torrey Bot. Club 1964, 91, 327-330.
30. Duke, S. O. Rev. Weed Science 1986, 2, 15-44.

RECEIVED April 18, 1988

Chapter 16

Terpenoids as Models for New Agrochemicals

Stella D. Elakovich

Department of Chemistry, University of Southern Mississippi, Hattiesburg, MS 39406–5043

This chapter examines the evidence that some terpenoids serve as plant growth regulators (allelochemicals) as well as anti-herbivore agents. The plant origin and bioassay method of terpenoid allelopathic agents is explored. The role these same compounds play in plant-insect communication due to volatiles, and their involvement as insect pheromones is discussed. The agrochemical possibilities are far-reaching.

Terpenes and terpenoids are ubiquitous in the higher plant kingdom. Although the role of these compounds is still being debated, considerable evidence is accumulating that they have functions both as allelopathic agents and as anti-herbivore agents. As such, they possess the potential of being incorporated via genetic engineering methods into selected crop plants in order to both provide a defensive allelochemical edge over other plants, and also to minimize crop damage by insects. Many secondary compounds implicated in allelopathic interactions have also been reported to be involved in other protective or defensive roles for the plant (1). It is likely that many secondary compounds have more than one role in nature. This chapter will explore the role of terpenoids first as allelopathic agents--potential models for new herbicides--and then as important agents in plant-insect interactions.

Allelopathic Agents

In his classical paper titled "The Influence of One Plant on Another--Allelopathy", Hans Molisch in 1937 coined the term *allelopathy* to refer to biochemical interactions among all types of plants including microorganisms (2). Rice, in the second edition of his comprehensive monograph *Allelopathy* supports this early definition which includes both inhibitory and stimulatory interactions (3). Not surprisingly, the most commonly identified allelopathic compounds produced by higher plants are simple phenols

0097–6156/88/0380–0250$06.00/0

and cinnamic acid derivatives (3). As long ago as 1908, Schreiner and Reed (4) reported that vanillin, vanillic acid, and hydroquinone are commonly produced plant constituents inhibitory to seedling growth. This focusing on plant phenolic constituents as allelochemicals likely occurred for two main reasons: the ease of isolation and identification of phenolic compounds, and their high water solubility. Recently Fischer and Quijano (5) have proposed that compounds of low water solubility may act as allelochemicals via micelle formation, and they suggest a reevaluation of water-insoluble plant constituents as potential allelopathic agents.

Both mono- and sesquiterpenoids frequently occur in the steam volatile essential oils of higher plants. Monoterpenoids are 10-carbon compounds biogenetically derived from two isoprene units (6). Several hundred are known. Their function as germination and plant growth regulators was reviewed by Evenari (7) in 1949 and recently by Fischer (8). Sesquiterpenoids, 15-carbon compounds derived from three isoprene units (6), display wide structural variety and a dramatic increase in the number of known compounds--several thousand are known. The high degree of research activity in this area is exemplified by three reviews published in the past five years. Stevens (9) reviewed the biologically active sesquiterpene lactones, Fischer (8) reviewed the germination and plant growth regulatory functions of both mono- and sesquiterpenoids, and Elakovich (10) reviewed sesquiterpenes acting as phytoalexins and plant growth regulators. Since these are all quite recent reviews, this paper will only give a cursory review of work mentioned in them. Reports through late 1987 of mono- and sesquiterpenoids alleged to be allelopathic agents not included in these three recent reviews will be reviewed here in more detail.

Monoterpenes. Monoterpenes were implicated in allelopathy over 60 years ago when Sigmund (11) screened essential oils and pure monoterpenes and observed inhibition of seed germination and plant growth in wheat, rape and vetch. In his 1986 review, Fischer gives the structures of 14 monoterpenes that have been implicated in allelopathy (8). Ten of them, α- and β-pinene, camphene, camphor, 1,8-cineole, pulegone, borneol, limonene, α-phellandrene and *p*-cymene are common constituents of plant essential oils and have frequently been implicated in allelopathy. Muller and Muller (12) found camphene, camphor, cineole, dipentene, α- and β-pinene among the volatile inhibitors produced by *Salvia* species. Of these, camphor and cineole were the most toxic to root growth of test seedlings (13). Asplund (14) investigated the inhibitory nature of camphor, pulegone, borneol, cineole, limonene, α-phellandrene, *p*-cymene, α- and β-pinene toward radish seed germination and found that although compounds with a ketone group, camphor and pulegone, were more inhibitory than the others, all were inhibitory. Rather surprisingly, all stimulated germination at low concentrations.

The "bare zones", areas free of herbaceous vegetation, surrounding naturalized stands of *Eucalyptus camaldulensis* are due to the highly toxic cineole and less toxic pinene and α-phellandrane found among the terpenes produced by *E. camaldulensis* (15). This order of toxicity is reinforced by data from Weaver and Kish (16) who measured the effects of eight terpenes on cucumber and found even the smallest amounts tested of cineole, dipentene, citronellol, menthol and terpineol essentially eliminated both root and hypocotyl growth. Halligan (17) examined the relative activities of terpenoids from *Artemisia californica*, which like *Eucalyptus*, is characterized by bare zones under and adjacent to the thickets it forms. Of the five terpenoids found in major amounts, camphor was most toxic, followed by cineole.

Sour orange, *Citrus aurantium* L., also inhibits growth of herbaceous vegetation via an allelopathic mechanism (18-19). Among the volatile inhibitors identified are four terpenoids: α- and β-pinene, (±)-limonene, and citronellal. All of these compounds reduced the hypocotyl growth of *Amaranthus retroflexus*; all except the lower concentration tested of (+)-limonene also reduced radical length of *A. retroflexus*, and all except (-)-limonene inhibited the germination of *A. retroflexus* seeds. Inhibitory effects increased with increased inhibitor concentrations. α-Pinene was the most active of these inhibitors; this is the first report implicating citronellal in allelopathy.

The hexane extract of seeds from *Eryngium paniculatum* was found to inhibit germination of velvetleaf (*Abutilon theophrasti* Medic), a serious weed in corn and soybean fields (20). Fractionations monitored by bioassay gave the terpene aldehyde-ester (-)-2,2,2-trimethyl-3-formyl-2,5-cyclohexadienyl angelate (1) as the only active component. Structure-activity studies of (1) and nine analogs suggest that activity is diminished by removing the formyl group from the ring, the cyclic double bonds are required, and the orientation of the acyl double bond is not important.

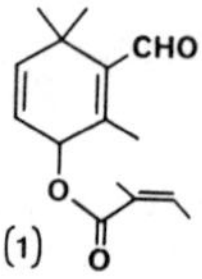

CHO

O

(1) O

Heisey and Delwiche (21) observed a highly toxic monoterpene alcohol as a major inhibitor present in *Trichostema lanceolatum* Benth. (vinegar weed). This alcohol, terpinen-4-ol, was less inhibitory than camphor, but almost twice as inhibitory as 1,8-cineole as measured by germination and radical growth inhibition of *Hordeum vulgare* (selected because of its genetic homogeneity) and *Bromus mollis* (selected because it is common to the California grassland).

In our investigations of allelopathic plants, we have examined the creeping perennial herb *Lippia nodiflora*, which is known for its rampant growth (22), and had earlier been reported to be a germination inhibitor (7). We found the essential oil inhibitory to lettuce seedlings, probably due in part to the presence of the terpenes β-pinene and *p*-cymene and the sesquiterpene α-caryophyllene identified in the complex essential oil. Investigation (23) of the essential oil of a second *Lippia* species, *Lippia adoensis*, revealed a simple mixture completely dominated by the terpene alcohol, linalool (81% of the total essential oil). The allelopathic monoterpenes 1,8-cineole and α-and β-pinene were also identified. Lettuce seedling bioassay of both the essential oil and of pure linalool showed similar activity. Both appear to enhance radical growth at low concentrations, although the enhancement is not statistically significant. At 400 ppm linalool causes a 42% reduction and *L. adoensis* leaf extract a 45% reduction in lettuce radical length.

Mechanisms of Allelopathic Inhibition by Mono-Terpenoids. There is a dearth of information available on the specific effects that allelochemicals produce. Muller and Muller (12) found camphor and cineole to be the most toxic among the volatile inhibitors produced by *Salvia* species. Indeed, cineole was used as the structural basis for the synthetic herbicide cinmethylin (24). Asplund (14) found terpenoids containing a ketone group, camphor and pulegone to be the most inhibitory among nine terperoids examined. Both Heisey and Delwiche (21) and Elakovich and Oguntimein (23) found monoterpene alcohols to be the major volatile inhibitors produced by *Trichostema lanceolatum* and *Lippia adoensis*, respectively. Certainly these results suggest that oxygenated terpenes are more highly inhibitory than terpene hydrocarbons. Einhellig (25) suggests that it is a rare exception when a single substance is responsible for allelopathy. He has shown (26) that both additive and synergistic inhibition may occur, but the mechanisms of this inhibition have not been determined.

Muller (27) has reviewed the possible mechanisms involved in allelochemical inhibition of herbs. When *Cucumis sativus* (cucumber) seedlings are exposed to *Salvia leucophylla* vapors, elongation of the hypocotyls is severely curtailed and growth is reduced to 25% of the control, indicating both stem and root cells are affected by the volatile terpenoids. Those cells which first develop in a germinating seed are more susceptible to these volatile terpenoids than are the cells of an older plant and cell elongation, cell division, and tissue maturation are all influenced (28). Somewhat in contrast is the work of Weaver and Klarich (29) who found that volatile substances, assumed to be monoterpenes, from *Artemisia tridentata* Nutt. suppressed seedling growth and the respiration rate of juvenile plant material, but elevated the respiration rate of mature leaves. They showed in laboratory tests that some terpenes raised and others lowered the respiration in wheat (*Triticum aestivum* L.). Volatiles of *S. leucophylla* essentially halted mitosis of *Allium cepa* root tips within 72 hrs of exposure (30). When lower concentrations of the terpenoids were used for treatments, the effect on chromosomes could be seen to be similar to that caused by colchicine treatment. The longer the *A. cepa* bulbs are exposed to terpenoids, or the more concentrated the terpenoids to which the cells are exposed, the greater are the chromosomal abnormalities. Increased cutin deposits in root epidermal cells, and increased lipid accumulation in root cortical cells also occur with increased exposure to *Salvia* terpenoids (31). These increased cutins and lipids may enhance the terpenoid inhibitory effects by making access of the allelochemicals into the cells easier. Einhellig (25) suggests that loss of membrane integrety by absorption of lipid soluble terpenes into the membranes is a likely starting point for their destructive action. Muller (27) suggests that even if susceptible plants do survive, they would be weaker and thus more likely to succumb to environmental stresses such as drought. It is thus possible to mitigate or intensify allelopathic effects by environmental factors.

Sesquiterpene Hydrocarbons and Their Derivatives. Recent reviews (8,10) of sesquiterpene hydrocarbons and their derivatives that have been implicated in allelopathy give structures for 20 compounds. Included are bergamotene, bisabolene, α-bulnesene, δ-cadinene, calamenene, β-caryophyllene, α-copaene, α-guayene, β-patchoulin, and β-selinene, all hydrocarbons frequently found in plant essential oils. Three sesquiterpene aldehydes, (-)-isobicyclogermacrenal, (-)-lepidozenal, and (+)-vitrenal are included along with the ketone zerumbone. The sesquiterpene alcohol, farnesol, its acetate, the related compound methyl farnesate, and the cadinene derivative, *epi*-khusinol acetate are all phytotoxic. Two inhibitory compounds having multi-functionality include the ent-2,3-*seco*-alloaromadendrane plagiochilin A and phomenone. Structure-activity studies of the latter suggest the epoxide function is necessary for growth inhibition, and acetylation is closely connected with the toxic properties (32).

Paul *et al.* (33) recently elucidated the structure of five new bioactive sesquiterpenoids from the green algae, *Caulerpa ashmeadii*. Four of these compounds, (2), (3), (4), and (5),

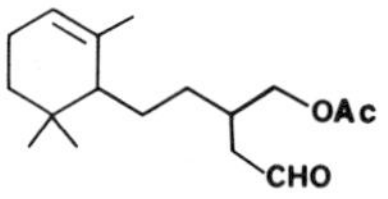

OAc CHO (2) CHO OAc (3) OAc OAc AcO (4) CHO CHO (5)

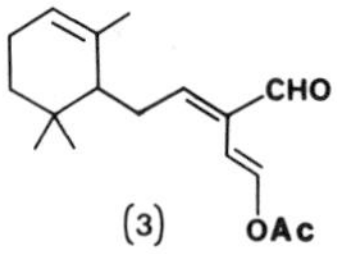

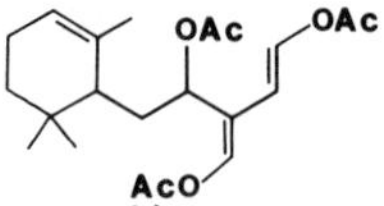

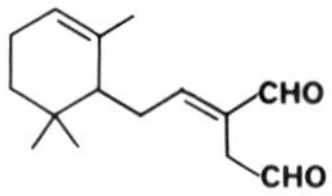

showed antimicrobial activity toward at least one marine bacterium and all were also toxic toward the damselfish, *Pomacentrus coeruleus*. The aldehyde and enol acetate functional groups present in these compounds can potentially react with proteins in a number of ways to alter or inactivate protein or enzyme function, supporting the hypothesis that these metabolites are used in chemical defense.

Seed germination bioassay of the ethyl acetate extract of dried, ground *Ambrosia peruviana* Willd., wild tansy, showed inhibitory activity (34). Five sesquiterpenes, four lactones of the pseudoguaianolide type and a diol were characterized from this extract. The diol structure was determined by spectral evidence to be (6), (+)-alloaromadendrane-4β,10α-diol. This is the first

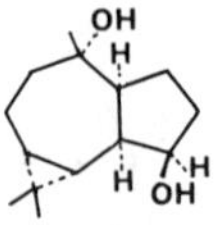

OH H H H OH (6)

reported sesquiterpene diol from *Ambrosia* species. It is antipodal to a compound isolated from the marine soft coral, *Sinularia mayi* and thus illustrates the frequently observed antipodal relationship between sesquiterpenes derived from marine and terrestrial organisms. Bioassay of isolated (6) showed it to cause reduction of cress roots and shoots at the three concentrations tested, but only the highest concentration inhibited lettuce seedling roots and shoots. At the two lower concentrations tested, lettuce roots and shoots were significantly stimulated.

Sesquiterpene Lactones. Of all of the terpenoid compounds, the sesquiterpene lactones possess the greatest variety of biological activities. More than 1000 of these compounds have been isolated, chiefly from the plant family Asteraceae (Compositae), but also from the Umbellifereae and Magnoliaceae (8). In 1976 Rodriguez, *et al.* (35) reviewed the biological activities and mechanism of action of these compounds. In a 1979 review, Fischer *et al.* (36) list 950 sesquiterpene lactones which had been characterized by 1977. They also discuss the biogenetic relationships of sesquiterpene lactones. The activities these compounds possess include antineoplastic agents, insect feeding deterrents, plant growth regulators, antimicrobial agents, schistosomicidal agents, vertebrate poisons and contact dermatitis in humans (8,10). Several structure-activity studies have led to the conclusion that plant growth regulation requires the presence of an exocyclic α,β-unsaturated lactone moiety (**7**) which can combine with sulfhydryl groups in key enzymes that control cell division (8,9,35).

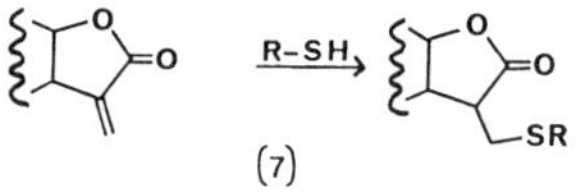

R-SH

SR

(7)

In his 1984 review, Stevens (9) gives structures of 8 sesquiterpene lactones which have allelopathic activity. Fischer (8), in his 1986 review, gives structures of 38 additional allelopathic sesquiterpene lactones. A recent report by Goldsby and Burke (34) gives structures for four sesquiterpene lactones of the pseudoguaianolide type isolated from the ethyl acetate extract of *Ambrosia peruviana*. Three of these, (**8**), (**9**), and (**10**), are structures not reported by Stevens (9) or Fischer (8). Of these, ambrosin (**8**) and psilostachyin B (**10**) were most active in preventing germination and root and shoot growth in lettuce and cress seedlings. Damsin (**9**) exhibited both stimulatory and inhibitory effects on seedling growth, depending on the concentration. It was also active against the fungal organism *Cladosporium herbarium*.

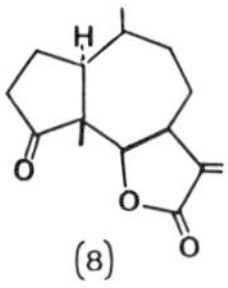

(8) 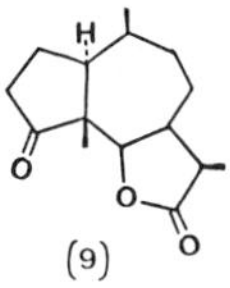 (9) 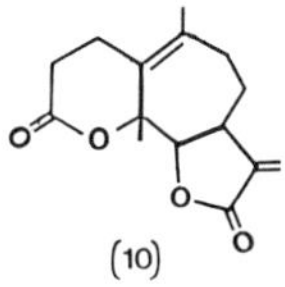 (10)

A glance at any current issue of a phytochemical journal reveals an ever increasing number of new sesquiterpene lactone structures isolated from higher plants. No doubt many of these, when subjected to growth inhibition assays, will also prove to be allelopathic.

Diterpenoids. Liverwarts contain oil bodies which are high in terpenoid content. Beyer *et al.* (37) isolated six new diterpenoids from the liverwart *Anastrophyllum minutum* (Schreb.) Schust. In addition, the known sesquiterpene hydrocarbons, anastreptene, β-barbatene, and bicyclogermacrene were isolated and identified. The six new diterpenoids, (11), (12), (13), (14), (15), and (16) represent a new diterpenoid type which the authors termed sphenolobane. The sphenolobane skeleton probably arises by the cyclization of geranylgeranyl pyrophosphate (37). The major compound, 3α,4α-epoxy-5α-acetoxy-18-hydroxysphenoloba-13E,16E-diene, (11), showed low growth inhibitory activity against rice seedlings, *Oryza sativa*. The inhibitory nature of compounds (12)-(16) was not tested.

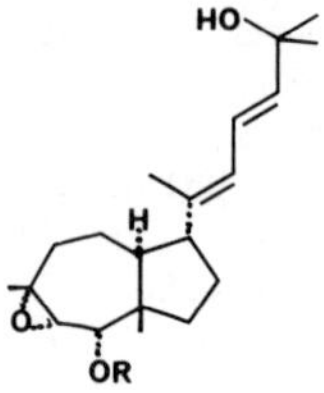

(11) R = $COCH_3$
(12) R = H
(13) 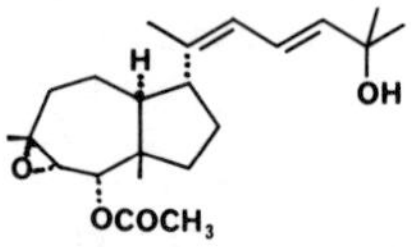
(14) 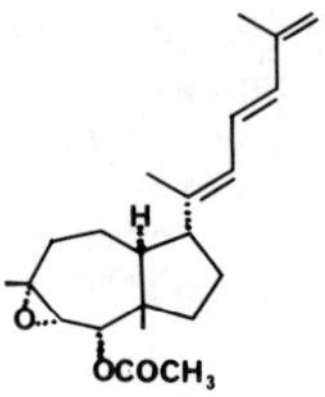
(15) R = $COCH_3$ 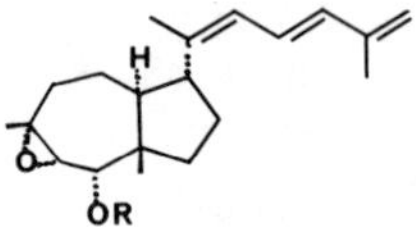
(16) R = H

Allelochemicals Involved in Plant-Insect Interactions

Allelochemicals can be used to minimize competition and invasion of weeds in crop fields especially by use of crop rotation schemes. Isolation and identification of specific allelochemicals has lagged behind the identification of allelopathic plants, largely due to the tedious, time-consuming nature of isolating large enough samples of pure compounds for a detailed structure study. Some of the isolated allelochemicals have shown potential for other uses, specifically as natural pesticides and growth regulators (38). Certainly these compounds have the potential of serving as models for useful new agrochemicals. As Lodhi *et al.* state (1), "It is clear that the involvement of biochemicals in an agroecosystem is a phenomenon that cannot be ignored, because these phytotoxins play a multipurpose role." Duke, in another chapter in this book, explores the potential of terpenoids derived from the genus *Artemisia* as pesticides. Powell and Spencer, in a separate chapter, explore the phytochemical inhibitors of *Arbutilon theophrasti* (velvetleaf) germination as models for new herbicides. The first part of this chapter has explored the role of terpenoids as allelopathic agents; the remainder of the chapter will examine the evidence that these same terpenoids are important agents in plant-insect interactions. Because of space constraints, the coverage will be representative, not exhaustive.

Olfactory responses toward plant volatiles. The great importance of terpenoids as agents of plant-insect communication probably results from their volatility combined with their great structural diversity which allows specificity (39). The olfactory responses toward monoterpene hydrocarbons, monoterpene alcohols, and sesquiterpene hydrocarbons isolated from pine pholem of *Hylobius abietis*, a weevil which is one of the most destructive pests of Nordic forest plantations, were measured by Selander *et al.* (40). The total steam distillate was either attractive or neutral to the weevils. A fraction of monoterpene hydrocarbons was slightly repellent even though two of the three major components (α-pinene and 3-carene) were clearly attractive. The third major component, limonene, was neutral, which suggests that the minor monoterpene hydrocarbons inhibit the attractant effects of the major components.

Those fractions containing oxygenated terpenes were attractive to the weevils as was a sesquiterpene hydrocarbon containing fraction. The only pure component tested, the terpene alcohol 4-terpineol was also attractive. In general, females were more attracted by the terpenes than were males, but concentration plays an important role. Similar olfactory responses have been observed in *Hylobius pales* Herbst. which were attracted to the monoterpenoids anethole, α-phellandrene, *d*-α-pinene, β-pinene, eugenol, terpineol, camphene, alloocimene, and 3-carene (41). The oxygenated terpenes and sesquiterpene hydrocarbons clearly play an important role in olfactory orientation.

Blust and Hopkins (42) examined the role of olfaction in the adaption of the specialist grasshopper, *Hypochlora alba* Dodge which feeds almost exclusively on *Artemisia ludoviciana*, Louisiana sagewort. Gereralist grasshoppers feed very little upon this plant which is rich in monoterpenes. Some 33 to 50% of these monoterpenes are one component, 1,8-cineole. The plant extract containing a mixture of monoterpenes elicited greater olfactory response than any of the 5 major monoterpene components tested individually. Of the individual monoterpenes tested, camphene, cineole, camphor, borneol and geraniol, borneol was the most stimulatory.

The soybean (*Glycine max*) genotype PI 227687 is resistant to insects, particularly to cabbage looper (*Trichoplusia ni*). *T. ni* thrives on other soybean commercial varieties such as Davis, but feeds less, grows more slowly, and shows poorer survival on resistant PI 227687 plants (43). Steam distillates from susceptible Davis cultivars attracted *T. ni* female adults, whereas steam distillates from resistant PI 227687 cultivars repelled them and were also more toxic to first-instar larvae. Since the primary selection of host plant is made by the adult *T. ni* moth in conjunction with selection of ovipositional sites, volatile plant chemicals (allelochemicals) are assumed to be important, and repellent compounds are the dominant volatile chemical messengers. Limited chemical differences in soybean plants can thus alter their acceptabilities to *T. ni* (43). Plant volatiles also contribute to rice resistance to the leafhopper *Nephotettix virescens* (44). When volatiles from a resistant rice cultivar were applied to a susceptible cultivar, leafhopper feeding was disrupted.

The monoterpene myrcene (**17**), although not yet implicated in allelopathy, is a common plant constituent and is present in the oleoresin of pine trees (*Pinus* spp.). Beetles of the genus *Ips* attack and colonize pine trees. They aggregate in response to the terpene alcohols ipsidienol (**18**) and ipsenol (**19**), and myrcene (**17**) can serve as the precursor of these pheromones (45). The stimulus for the biosyntheses of (**18**) and (**19**) in the *Ips* genus is associated with feeding, suggesting that aggregrating pheromones of *Ips* beetles are waste products from terpene metabolism. Bark beetles of the genus *Dendroctonus* metabolize α-pinene, a monoterpene which is ubiquitous in the oleoresin of *Pinus* spp., to the pheromones *trans*-verbenol (20) and verbenone (21) among other minor products (46). The same or similar systems of terpene metabolism are likely in different species within *Dendroctonus* and in closely related genera.

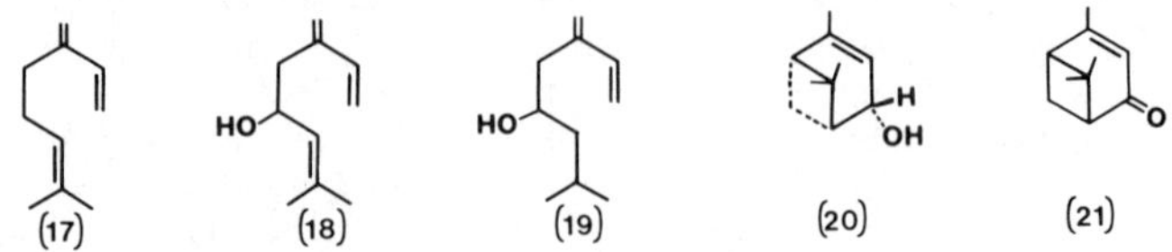

Insect mixed-function oxidases. Mixed-function oxidases (MFO, also known as microsomal cytochrome P-450 monooxygenase) are a group of widely distributed enzymes best known for their ability to degrade drugs, pesticides, and other compounds. Their primary function is to convert lipophilic compounds into more polar, more readily excreted hydrophilic metabolites. Brattsten *et al.* (47) have shown that MFO's of the southern armyworm, *Spodoptera eridania*, are induced by secondary plant substances added to its food, and the induction is rapid enough to provide this polyphagous insect with protection against these potentially offensive dietary factors. Eight terpenoids were evaluated for thier MFO activity inducing ability. The most potent inducers were the monoterpene hydrocarbons (+)-α-pinene and myrcene. Even a brief exposure to a small amount of these compounds appears to suffice to trigger an increase in MFO activity. Although insect MFO enzyme systems likely evolved for endogenous functions, their operation against hazardous secondary plant substances, such as the ubiquitous terpenoids, may be their major detoxification function.

Investigations with *Peridroma saucia* Hubner, the variegated cutworm, suggest that plant species differ in the degree to which they stimulate the MFO system, and thus an insect's ability to detoxify insecticides may depend on the terpenoids (and other compounds) produced by its host plant (48). Further studies in which fall armyworms, *Spodoptera frugiperda*, a generalist insect, were fed diets containing 0.2% of 13 monoterpenoids, 3 sesquiterpenes, 2 diterpenes, 5 triterpenes, and the tetraterpene carotene, showed the microsomes of these insects oxidized monoterpenes more favorably than other terpenoids, indicating a preference for the monoterpenes (49).

Conclusions

From this brief review it is clear that terpenoids play an increasingly important part in plant-plant interactions. As our knowledge of allelopathic compounds and their modes of action increase, we will be better able to apply this knowledge to solutions to problems in agriculture. Many of the same terpenoids which show allelopathic activity also possess activity as insect pheromones. Research in this area of insect-plant interactions may yield information of much value in the development of sophisticated methods of insect control as well as in the basic relations between insects and plants (48). Such studies will require extensive cooperation among chemists, biologists, and agricultural scientists, but will be rewarded by the development of new and better herbicides and plant growth hormones based on structures of allelochemicals, by useful crop rotation schemes, and by the development of genetically engineered cultivars which not only repel traditional insect pests, but also retard growth of unwanted plants. The agrochemical possibilities are far-reaching!

Literature cited

1. Lodhi, M.A.K., R. Bilal and K.A. Malik. 1987. Allelopathy in agroecosystems. *J. Chem. Ecol.* **13**, 1881-1891.
2. Molisch, H. 1937. Der Einfluss einer Pflanze auf die andere- Allelopathie. Fischer, Jena. 106 pp.
3. Rice, E.L. 1984. *Allelopathy*, 2nd Ed., Academic Press, Inc., New York.
4. Schreiner, O., and H.S. Reed. 1908. The toxic action of certain organic plant constituents. *Bot. Gaz.* (Chicago) *45*, 73-102.
5. Fischer, N.H., and L. Quijano. 1985. In A.C. Thompson (ed), *The Chemistry of Allelopathy*. American Chemical Society, Washington, D.C., pp. 133-147.
6. Robinson, T. 1980. *The Organic Constituents of Higher Plants*, 4th Ed., Cordus Press, North Amberst, MA, chapter 8.
7. Evenari, M. 1949. Germination inhibitors. *Bot. Rev.* **15**, 153.
8. Fischer, N.L. 1986. The function of mono and sesquiterpenes as plant germination and growth regulators. In A. Putnam and C.S. Tang (eds), *The Science of Allelopathy*, John Wiley and Sons, Inc., New York, pp. 203-218.
9. Stevens, K.L. 1984. Biological activity and chemistry of sesquiterpene lactones. In W.D. Nes, G. Fuller, and L.S. Tsai (eds), *Isopentenoids in Plants*, Dekker, New York, pp. 65-80.
10. Elakovich, S.D. 1987. Sesquiterpenes as phytoalexins and allelopathic agents. In G. Fuller and W.D. Nes (eds), *Ecology and Metabolism of Plant Lipids*, American Chemical Society, Washington, D.C., pp. 93-108.
11. Sigmund, W. 1924. Uber die Einwirkung von Stoffwech- selenprodukten auf die Pflanzen. III. Einwirkung N-frier pflanzlicher Stoffwechselendprodukte auf die Keimung von Samen (Aetherische Oele, Terpene u.a.). *Biochem. Z.* **146**, 389-419.
12. Muller, W.H., and C.H. Muller. 1964. Volatile growth inhibitors produced by *Salvia* species. *Bull. Torrey Bot. Club* **91**, 327-330.
13. Muller, C.H. 1965. Inhibitory terpenes volatilized from *Salvia* shrubs. *Bull. Torrey Bot. Club* **92**, 38-45.
14. Asplund, R.O. 1968. Monoterpenes: Relationship between structure and inhibition of germination. *Phytochemistry* **7**, 1995-1997.
15. del Moral, R., and C.H. Muller. 1970. The allelopathic effects of *Eucalyptus camaldulensis*. *Am. Midl. Nat.* **83**, 254-282.
16. Weaver, T. and L. Kish. 1982. Allelopathic potential of terpene secreting (aromatic) plants. *Proc. Mont. Acad. Sci.* **41**, 51-56.
17. Halligan, J.P. 1975. Toxic terpenes from *Artemisia californica*. *Ecology* **56**, 999-1003.

18. AlSaadawi, I.S. and A.J. AlRubeaa. 1985. Allelopathic effects of *Citrus aurantium* L. I. Vegetational patterning. *J. Chem. Ecol.* **11**, 1515-1525.
19. AlSaadawi, I.S., M.B. Arif, and A.J. AlRubeaa. 1985. Allelopathic effects of *Citrus aurantium* L. II. Isolation, characterization, and biological activities of phytotoxins. *J. Chem. Ecol.* **11**, 1527-1534.
20. Spencer, G.F. 1986. The effects of a terpene aldehyde-ester from *Eryngium paniculatum* and analogs on velvetleaf germination. *J. Nat. Prod.* **49**, 924-926.
21. Heisey, R.M. and C.C. Delwiche. 1984. Phytotoxic volatiles from *Trichostema lanceolatum* (Labiatae). *Amer. J. Bot.* **71**, 821-828.
22. Elakovich, S.D. and K.L. Stevens. 1985. Volatile constituents of *Lippia nodiflora*. *J. Nat. Prod.* **48**, 504-506.
23. Elakovich, S.D. and B.O. Oguntimein. 1987. The essential oil of *Lippia adoensis* leaves and flowers. *J. Nat. Prod.* **50**, 503-506.
24. Duke, S. O. 1987. Terpenoids from the genus *Artemisia* as potential pesticides. 194th ACS national meeting, August 30-September 4, 1987, New Orleans, LA. Agrochemicals abstract 104.
25. Einhellig, F.A. 1986. Mechanisms and modes of action of allelochemicals. In A.R. Putnam and C.-S. Tang (eds), *The Science of Allelopathy*. Wiley-Interscience, New York, pp. 171-188.
26. Einhellig, F.A. 1986. Interactions among allelopathic chemicals and other stress factors of the plant environment. In G.R. Waller (ed), *Symposium on Allelochemicals: Role in Agriculture, Forestry and Ecology*. American Chemical Society, Washington, D.C.
27. Muller, W.H. 1986. Allelochemical mechanisms in the inhibition of herbs. In A.R. Putman and C.-S. Tang (eds), *The Science of Allelopathy*. Wiley-Interscience, New York, pp. 189-199.
28. Muller, W.H. 1965. Volatile materials produced by *Salvia leucophylla*: effects on seedling growth and soil bacteria. *Bot. Gaz.* **126**, 195-200.
29. Weaver, T.W. and D. Klarich. 1977. Allelopathic effects of volatile substances from *Artemisia tridentata* Nutt. *Am. Midl. Nat.* **97**, 508-512.
30. Lorber, P. and W.H. Muller. 1980. Volatile growth inhibitors produced by *Salvia leucophylla*: effects on cytological activity in *Allium cepa*. *Comp. Physiol. Ecol.* **5**, 60-67.
31. Muller, W.H. and R. Hauge. 1967. Volatile growth inhibitors produced by *Salvia leucophylla*: effects on seedling anatomy. *Bull. Torrey Bot. Club* **94**, 182-191.
32. Capasso, R., N.J. Iacobellis, A. Bottalico and G. Randazzo. 1984. Structure-toxicity relationships of the eremophilane phomenone and PR-toxin. *Phytochemistry* **23**, 2781-2784.
33. Paul, V.J., M.M. Littler, D.S. Littler and W. Fenical. 1987. Evidence for chemical defense in tropical green alga *Caulerpa ashmeadii* (Caulerpaceae: Chlorophyta). *J. Chem. Ecol.* 13, 1171-1185.

34. Goldsby, G. and B.A. Burke. 1987. Sesquiterpene lactones and a sesquiterpene diol from Jamaican *Ambrosia peruviana*. *Phytochemistry* **26**, 1059-1063.

35. Rodriguez, E., G.H.N. Towers, and J.C. Mitchell. 1976. Biological activities of sesquiterpene lactones. *Phytochemistry* **15**, 1573-1580.

36. Fischer, N.H., E.J. Oliver, and H.D. Fischer. 1979. *The Biogenesis and Chemistry of Sesquiterpene Lactones*. In *Progress in the Chemistry of Organic Natural Products*, Vol **38**, Springer, New York, pp. 47-390.

37. Beyer, J., H. Becker, M. Toyota and Y. Asakawa. 1987. Diterpenoids with a novel skeleton from the liverwart *Anastrophyllum minutum*. *Phytochemistry* **26**, 1085-1089.

38. Putnam, A.R. and W.B. Duke. 1978. Allelopathy in agroecosystems. *Ann. Rev. Phytopathol.* **16**, 431-451.

39. Whittaker, R.H. and P.P. Feeny. 1971. Allelochemics: chemical interactions between species. *Science* **171**, 757-770.

40. Selander, J., P. Kalo, E. Kangus and V. Perttunen. 1974. Olfactory behaviour of *Hylobius abietes* L. (Col., Curculionidae). I. Response to several terpenoid fractions isolated from Scots pine phloem. *Ann. Ent. Fenn.* **40**, 108-115.

41. Thomas, H.A. and G.D. Hertel. 1969. Responses of the pales weevil to natural and synthetic host attractants. *J. Econ. Ent.* **62**, 383-386.

42. Blust, M.H. and T.L. Hopkins. 1987. Olfactory responses of a specialist and a generalist grasshopper to volatiles of *Artemisia ludoviciana* Nutt. (Asteraceae). *J. Chem. Ecol.* **13**, 1893-1902.

43. Khan, Z.R., A. Ciepiela, and D.M. Norris. 1987. Behavioral and physiological responses of cabbage looper, *Trichoplusia ni* (Hubner), to steam distillates from resistant versus susceptible soybean plants. *J. Chem. Ecol. 13*, 1903-1915.

44. Khan, Z.R. and R.C. Saxena. 1985. Effect of steam distillate extract of a resistant rice variety on feeding behavior of *Nephotettix virescens* (Homoptera: Cicadellidae). *J. Econ. Entomol.* **78**, 562-566.

45. Hughes, P.R. 1974. Myrcene: a precursor of pheromones in *Ips* beetles. *J. Insect. Physiol.* **20**, 1271-1275.

46. Hughes, P.R. 1975. Pheromones of *Dendroctonus*: origin of α-pinene oxidation products present in emergent adults. *J. Insect. Physiol.* **21**, 687-691.

47. Brattsten, L.B., C.F. Wilkinson and T. Eisner. 1977. Herbivore-plant interactions: mixed-function oxidases and secondary plant substances. *Science* **196**, 1349-1352.

48. Yu, S.J., R.E. Berry, and L.C. Terriere. 1979. Host plant stimulation of detoxifying enzymes in a phytophagous insect. *Pestic. Biochem. Physiol.* **12**, 280-284.

49. Yu, S.J. 1987. Microsomal oxidation of allelochemicals in generalist (*Spodoptera frugiperda*) and semispecialist (*Anticarsia gemmatalis*) insect. *J. Chem. Ecol.* **13**, 423-436.

RECEIVED April 12, 1988

Chapter 17

Chemical and Anatomical Response in *Gossypium* spp. Challenged by *Verticillium dahliae*

Robert D. Stipanovic, Marshall E. Mace, David W. Altman, and Alois A. Bell

Southern Crops Research Laboratory, Agricultural Research Service, U.S. Department of Agriculture, College Station, TX 77841

Verticillium dahliae, a fungal pathogen of plants, evokes anatomical and chemical responses in resistant cotton plants. Tyloses are formed which block infected xylem vessels and confine the pathogen, and the synthesis of fungitoxic terpene phytoalexins is induced in paravascular cells. The primary terpenes identified to date include desoxyhemigossypol (dHG), hemigossypol (HG), gossypol, and their methyl ether derivatives. Of these terpenes, dHG is the most toxic to a nondefoliating strain of *V. dahliae*. Terpene synthesis is also elicited in cotton cell suspension cultures by heat-killed conidia of *V. dahliae*. Gossypol, HG and the methyl ether derivatives of HG and dHG have been isolated from such cell cultures. Embryogenic callus cultures of *Gossypium hirsutum* have greatly reduced embryo production when heat-killed conidia are added to the agar medium. In the presence of citrate, in vitro synthesis of these terpenoids is inhibited.

An understanding of the fundamental mechanisms of pest resistance is crucial for strategically enhancing plant production. Plants synthesize chemicals (phytoalexins) which are toxic to pathogenic organisms as one defense method. Knowledge of the structure, pathway of biosynthesis, enzymes and genes controlling the biosynthesis of phytoalexins should assist development of a more resistant plant.

Work in our laboratory has concentrated on phytoalexins that are associated with the challenge of an important fungal pathogen, *Verticillium dahliae*, to *Gossypium* (cotton). We have characterized the phytoalexins and the correlated anatomical response of plants to infection with this organism. The following pages report current work in this area. Our goal is to enhance plant defense strategies, particularly as they relate to the role of phytoalexins.

Disease Description

Verticillium wilt is caused by the soil-borne fungus *V. dahliae*. First detected in 1927 (1), the disease occurs in major cotton production areas. The loss to the 1986 U.S. cotton crop from Verticillium wilt was estimated at 200,000 bales (2). Additional losses due to poorer fiber quality are substantial, but difficult to document.

More than 90% of commercial cottons are cultivars of *Gossypium hirsutum* L. (3). *Gossypium hirsutum* generally can incur significant losses due to Verticillium wilt, especially when incited by the defoliating strains of *V. dahliae*. *Gossypium barbadense* cultivars exhibit some of the highest levels of resistance to Verticillium wilt (4). Several studies of the anatomical and biochemical factors associated with *Verticillium* pathogenesis and on resistance to Verticillium wilt in many host plants have been published (5-7). Because of the susceptibility of *G. hirsutum* cultivars to Verticillium wilt, improved understanding of disease processes is especially important for the development of control procedures.

Pathogenesis of Verticillium wilt depends upon penetration of the pathogen's hyphae through the root cortex and into the xylem vessels. In susceptible plants these hyphae continue to grow in the vessels, and conidia of the pathogen spread systemically throughout the vascular system. Thus, an optimal plant protective mechanism must defend against penetration, survival and growth of hyphae, and distribution of conidia.

Plant Defense Mechanisms

The cotton plant's defense mechanisms may be divided into two broad categories: anatomical and chemical. Anatomical defenses include the protective root cortex and the induced formation of tyloses in xylem vessels. Defense chemicals include those that are preformed or present in healthy noninfected plants and those induced (e.g., phytoalexins) by the plant pathogen or its products. Herein, we will limit the discussion to our work on *V. dahliae*-induced tyloses and gossypol-related phytoalexins. Other compounds, such as the flavonoids that are involved in *V. dahliae* resistance in cotton leaves (8), will not be discussed.

Plant Response

Tylose Formation. Pathogenesis by fungal wilt pathogens must allow entry of the pathogen into xylem vessels and systemic spread of conidia in the vascular system. Occlusion of xylem vessels induced by the pathogen or its metabolites constitutes the initital, critical host resistance response (5,9). Tyloses constitute the major occlusive process in *V. dahliae*-infected cotton (9). Tyloses are outgrowths from paravascular parenchyma cells into the lumen of the xylem vessel. In *V. dahliae*-infected vessels, multiple tyloses form and occlude cells.

Extensive tylosis occurs in both susceptible and resistant cottons infected with *V. dahliae*. The release of secondary conidia of *V. dahliae* and the occlusion of vessels in infected stem xylem of Verticillium wilt-resistant *G. barbadense* cv. Seabrook Sea Island, 12B2 (SBSI) and wilt-susceptible *G. hirsutum* cv. Rowden have been compared (9). These studies showed that the vessels of SBSI became occluded before secondary conidia were released into the xylem fluid. Occlusion in Rowden occurred after the release of secondary conidia, allowing systemic distribution of *V. dahliae* and severe wilt to ensue.

Phytoalexin formation. The terpenoid phytoalexins are first detected histochemically in xylem at 18 and 24 hrs after inoculation with *V. dahliae* in the wilt-resistant SBSI and wilt-susceptible Rowden cultivars, respectively (9). The structure and proposed sequence of biosynthesis of these terpenoids are shown in Fig. 1. Gossypol (**G**), a dimeric sesquiterpene, is biosynthesized from cis-trans-farnesyl pyrophosphate or its equivalent (10,11). The biosynthetic precursors of **G**, desoxyhemigossypol (**dHG**), hemigossypol (**HG**) and their 6-methyl ether derivatives (**dMHG**) and (**MHG**), have been isolated from *V. dahliae* infected *G. barbadense* stele tissue (12,13). These terpenoids normally absent from the boll, cambium, and xylem are rapidly synthesized in these tissues following inoculation with *V. dahliae* (14). The resistant variety of *G. barbadense* produces toxic concentrations of terpenoids in inoculated xylem vessels 24 to 48 hours sooner than susceptible varieties (15-17).

When susceptible Rowden was compared to resistant SBSI 14 days after inoculation with *V. dahliae*, over 50% of the first three internodes of Rowden were infected with mycelium, while only about 10% of those from SBSI were infected (18). Rowden was severely wilted while SBSI was symptomless. Of the terpenoid aldehydes present, **HG** and **MHG** accounted for over 85% of the total, with **G** and its methyl ether derivatives, **MG** and **DMG**, comprising the remainder.

Two to 4 days after inoculation, SBSI contained higher terpenoid concentrations than Rowden; however, 14 days after inoculation the opposite is true (15). The higher concentration in Rowden after 14 days is ascribed to extensive secondary colonization in Rowden. Thus, Rowden, with more infected tissue, produces more phytoalexins than the resistant SBSI in which infection is contained.

The higher concentrations of methylated terpenoids in the resistant SBSI as compared to Rowden (Fig. 2) lead us to speculate that methylation increased the toxicity of the terpenes. However, experiments to date with a non-defoliating strain of *V. dahliae* indicate that this is not true (19). **HG** and **dHG** are more toxic to *V. dahliae* mycelia than their methyl ether derivatives (**MHG** and **dMHG**) (Fig. 3); similarly, they are better inhibitors of *V. dahliae* conidial germination (Fig. 4). In SBSI stele tissue 2 days after inoculation, **dHG** is present at concentrations sufficient to inhibit more than 95% of the germination of *V. dahliae* conidia (Table I; Fig. 4). If the other phytoalexins act in an additive fashion, then a lethal dose for conidia is present 2 days after inoculation.

Fig. 1. Terpenoid phytoalexins formed in *Gossypium* spp. (dHG = desoxyhemigossypol, dMHG = desoxyhemigossypol-6-methyl ether, HG = hemigossypol, MHG = hemigossypol-6-methyl ether, G = gossypol, MG = gossypol-6-methyl ether, DMG = gossypol-6,6'-dimethyl ether).

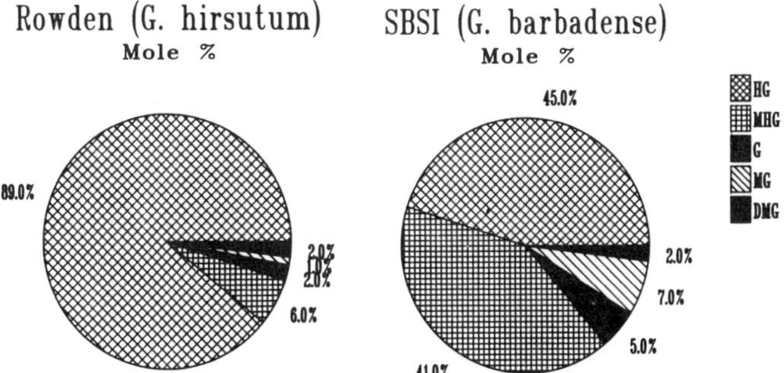

Fig. 2. Mole percent of terpenoid aldehydes in cotton stele tissue 14 days after stem puncture inoculation with *Verticillium dahliae* (HG = hemigossypol, MHG = hemigossypol-6-methyl ether, G = gossypol, MG = gossypol-6-methyl ether, DMG = gossypol- 6,6'-dimethyl ether).

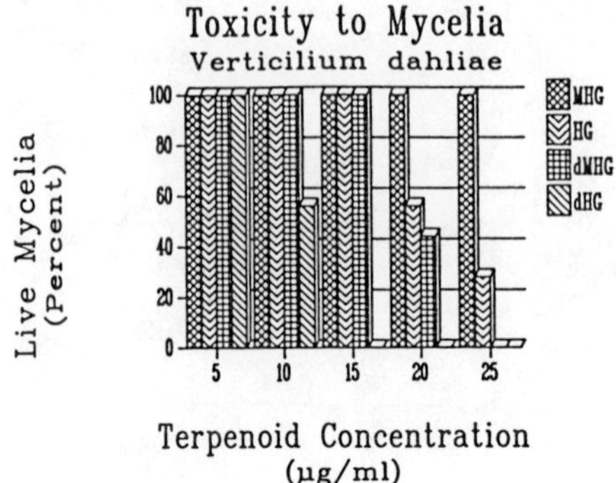

Fig. 3. Toxicity of cotton phytoalexins (MHG = hemigossypol-6-methyl ether, HG = hemigossypol, dMHG = desoxyhemigossypol-6-methyl ether, dHG = desoxyhemigossypol) to *Verticillium dahliae* mycelia.

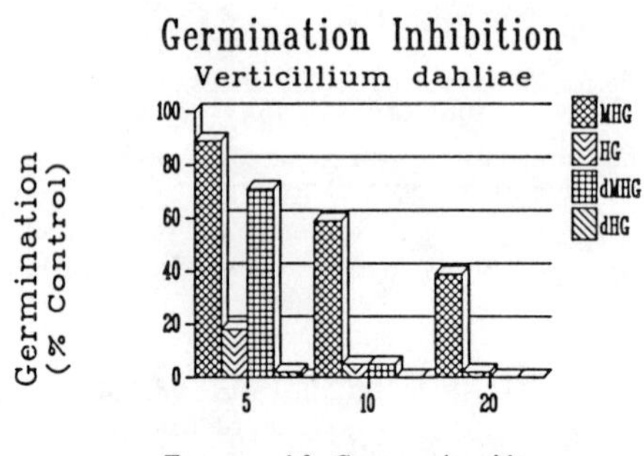

Fig. 4. Inhibition of *Verticillium dahliae* conidial germination by phytoalexins (MHG = hemigossypol-6-methyl ether, HG = hemigossypol, dMHG = desoxyhemigossypol-6-methyl ether, dHG = desoxyhemigossypol).

Table I. Terpenoid concentration in *Gossypium barbadense* cv. Seabrook Sea Island stele tissue 2 and 10 days after inoculation with *Verticillium dahliae*

Terpenoid	Concentration[a] 2 days	10 days
Desoxyhemigossypol	5.3 (0.8)[b]	25.0 (1.7)
Hemigossypol	10.1 (0.4)	25.7 (2.0)
Desoxyhemigossypol methyl ether	8.7 (0.5)	56.5 (2.6)
Hemigossypol methyl ether	18.7 (1.7)	79.2 (4.0)

[a] μg/ml water from stele tissue of the hypocotyl and internodes one to three.
[b] Mean of three experiments; standard deviation in parentheses.

Table II. Mean water solubility of terpenoid phytoalexins at pH 6.3

Terpenoid	Solubility (μg/ml buffer)[a]
Desoxyhemigossypol	50.2 (2.8)[b]
Hemigossypol	4.3 (0.2)
Desoxyhemigossypol methyl ether	2.9 (0.3)
Hemigossypol methyl ether	2.0 (0.1)

[a] Solubility in 0.15M potassium phosphate buffer, 24°C.
[b] Mean of three replications; standard deviation in parentheses.

For the terpenoid aldehydes to be effective, it is assumed they must be soluble in the fluid within infected xylem vessels. In water solubility studies (19), only dHG was found to have sufficient water solubility (Table II) to reach a fungicidal concentration at the pH of xylem fluid (pH 6.3). However, solubilizing agents might be present in this fluid, augmenting the solubility of the other terpenoids. But, dHG alone fits all the criteria required to act as a fungicide: i) accumulation to a concentration that kills all mycelia and conidia, ii) sufficient solubility in pH 6.3 water to attain fungicidal concentrations, and iii) location at the site of infection (18,19, and Mace M.E., USDA-ARS, unpublished data). These results relate only to the nondefoliating strain of *V. dahliae* studied. Experiments are being conducted to determine if these conclusions can be extrapolated to more virulent strains of *V. dahliae*.

Elicitation of Phytoalexins in Cell Culture

Induction of secondary natural products has been a significant subspecialty in plant cell culture leading to major commercialization efforts worldwide. Induction can occur with several kinds of culture manipulation, but elicitation has been defined as phytoalexin-induced synthesis triggered by substances isolated from microorganisms (20).

Elicitation of terpenoid phytoalexins has been reported for several genera including *Gossypium*. Cotton phytoalexin accumulation following fungal challenge in tissue culture has important research applications as a model system (21).

In 1981 the capacity of *G. hirsutum* cell suspensions to form G and HG in unstressed suspension cultures was reported (22). Elevated levels of terpenoid elicitation were subsequently demonstrated with heat-killed *V. dahliae* conidia (23, 24). Later studies showed that membrane changes, as monitored by fluorescent molecular probes, correlate with phytoalexin formation in cotton cell suspension cultures (25). Heinstein identified G and HG in his elicitation experiments with *G. arboreum* cv. Nanking (24). MHG and dMHG also were identified in elicitation experiments with this *G. arboreum* accession and *G. hirsutum* cultivars (23). *Gossypium arboreum* cell suspensions when challenged with 0.1% (w/v) cellulase (*Aspergillus niger*) failed to elicit terpenoid synthesis (Table III). This observation is contrary to the reported sesquiterpenoid phytoalexin elicitation by cellulase in *Nicotiana tabacum* (27). Thus, cotton elicitation appears to depend on a specific cell surface interaction rather than on a general cell wall attack.

In an embryogenic callus culture of *G. hirsutum* cv. Coker 310, a linear correlation ($r = -0.87$; $P < 0.01$) between the elicitor concentration and embryo production was observed (23). This suggests the potential for a cellular selection system for increased disease resistance as was done with alfalfa challenged by *Fusarium oxysporum* (28).

Phytoalexin elicitation in *G. arboreum* suspension culture has been observed when a non-pathogen, *Saccharomyces cerevisiae*, was used as the elicitor (24). Similary, *Aspergillus flavus* induces terpenoid synthesis in ovule cultures of *G. hirsutum* (Altman, D.W.; Stipanovic, R.D.; Mellon, J., USDA-ARS, unpublished data).

Table III. Mean response of *Gossypium arboreum* suspensions to sesquiterpenoid aldehyde elicitation [a]

Treatment	Fresh weight (g)	Gossypol equivalents[b]
Control	28.8	0.06
Verticillium dahliae	26.8	0.17
Cellulase	35.5	0.08
$LSD_{0.05}$	4.2	0.06

[a]Incubation for 72 hrs at pH 6.0.
[b]Determined by the aniline method of Smith (26) on a % dry weight basis.

Verticillium dahliae elicitation of terpenoid synthesis from *G. arboreum* suspension cultures has been arrested by addition of citrate (25). Concentrations of 2 mM gave 50% inhibition. A similar response also was observed for glyceolin elicitation with *V. dahliae* in soybean cell suspensions (25).

Effects of Citrate on Terpenoid Synthesis in Excised Roots

Prompted by the data on citrate activity (25), we have studied its effect on phytoalexin production in excised cotton roots with and without chilling. Four-day-old seedlings were cut at the transition zone and the roots were incubated at 28°C in petri dishes (2.5 gm root/dish) containing a medium composed of sucrose (0.5%), sodium acetate (0.1%) and 0.015 M PIPES buffer (pH 6.5), with and without 4.5 mM sodium citrate. After 12, 24, 36, and 48 hrs, the roots were removed and frozen at -135°C. A second set of four-day-old seedlings were held at 10°C for 4 days. The roots were excised, incubated at the time intervals above, and frozen at -135°C. The pH of all solutions was recorded.

After the roots were thawed, the terpenes were extracted and the extracts subjected to HPLC analysis. G was the major terpenoid produced in all systems. Cold treatment was not an effective method for inducing synthesis in this experiment (Fig. 5). However, the experiment demonstrated that citrate was a strong inhibitor of sesquiterpenoid production. Citrate did affect the pH of the medium. After 36 hrs the pH had risen about one pH unit over the control which was virtually unchanged. Experiments using seedlings as above with no buffer, or with PIPES, phosphate, or citrate buffer were also conducted (Table IV). All of the buffers suppressed terpenoid aldehyde synthesis, but citrate did so most effectively. In the presence of citrate, phytoalexin synthesis is effectively blocked. The role of pH is uncertain. The effects of pH stand in contrast to observations from soybean tissue culture systems in which citrate, but not di- and tricarboxylic hydroxy acids, inhibited phytoalexin formation (25).

Table IV. Effects of buffers on media pH and production of terpenoid aldehydes in cotton roots 48 hours after excision

Buffer	pH[a]	Gossypol equivalents[b]
None	5.80 (0.36)[c]	76.8 (15.7)
PIPES	7.25 (0.06)	22.1 (4.0)
Phosphate	7.35 (0.06)	15.0 (2.2)
Citrate	7.58 (0.15)	9.5 (0.6)

[a]Initial pH for all treatments was 6.5.
[b]Terpenoid aldehyde concentrations (ppm) fresh weight determined by the reaction with phloroglucinol (28).
[c]Mean of four replications; standard deviation given in parentheses.

Conclusions

Phytoalexins, the toxic chemicals produced by the cotton plant in response to disease, are an essential component in the defense response. Our studies in vivo, although concentrating on the interaction between a nondefoliating isolate of *V. dahliae* and

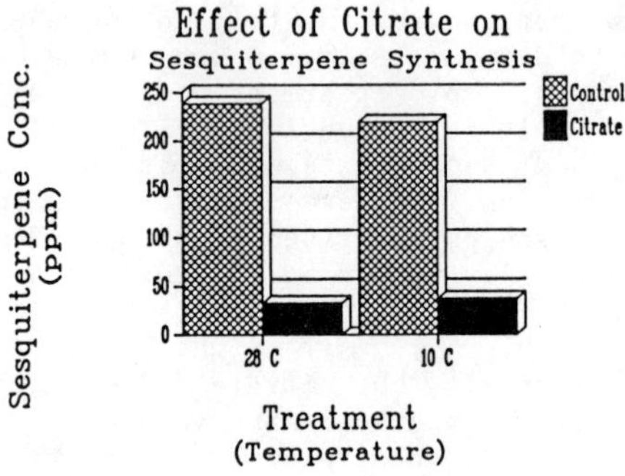

Fig. 5. Effect of citrate on the synthesis of phytoalexins in excised cotton roots (concentration expressed as ppm of sesquiterpenes).

cotton, indicate that resistant cotton plants respond by the rapid formation of tyloses which occlude infected vessels and halt the spread of infection. This is followed by a rapid increase in phytoalexin formation to concentrations that kill conidia and mycelia. The quick response to the infection, and the quality of the phytoalexins may be the difference between a resistant and a susceptible plant. Of the several phytoalexins formed, only **dHG** is sufficiently water soluble to accumulate to fungitoxic levels in fluids of infected tissue. If solubilizing agents are not involved, **dHG** appears to be the critical compound involved in the phytoalexin response. The elicitation of sesquiterpenoid phytoalexins in tissue culture by both pathogenic and nonpathogenic organisms indicates that this is a cell surface phenomenon that could be a useful research model.

LITERATURE CITED

1. Shapovalow, M.; Rudolph, B.A. Plant Dis. Rptr. 1930, 14, 9-10.
2. Gazaway, W. Beltwide Cotton Prod. Res. Conf. Proc. 1987, 5.
3. Lee, J.A. In Cotton: Kohel, R.J.; Lewis, C.F., Eds.; Amer. Soc. Agronomy: Madison, WI, 1984; pp. 1-25.
4. Wilhelm, S. In Fungal Wilt Diseases of Plants: Mace, M. E.; Bell, A.A.; Beckman, C.H., Eds.; Academic: New York, 1981; pp. 308-315.
5. Beckman, C.H.; Talboys, P.W. In Fungal Wilt Diseases of Plants: Mace, M.E.; Bell, A.A.; Beckman, C.H., Eds.; Academic: New York, 1981; pp 487-521.
6. Bell, A.A.; Mace, M.E. In Fungal Wilt Diseases of Plants: Mace, M.E.; Bell, A.A.; Beckman, C.H., Eds.; Academic: New York, 1981; pp 431-86.
7. Pegg, G.F. In Fungal Wilt Diseases of Plants: Mace, M.E.; Bell, A.A.; Beckman, C.H., Eds.; Academic: New York, 1981; pp 193-253.
8. Howell, C.R.; Bell, A.A.; Stipanovic, R.D. Physiol. Plant Path. 1976, 8, 181-188.
9. Mace, M.E. Physiol. Plant Pathol. 1978, 12, 1-11.
10. Masciadri, R.; Angst, W.; Arigoni, D. J. Chem. Soc., Chem. Comm. 1985, 1573-1574.
11. Stipanovic, R.D.; Stoessl, A.; Stothers, J.B.; Altman, D.W.; Bell, A.A.; Heinstein, P. J. Chem. Soc., Chem. Comm. 1986, 100-102.
12. Bell, A.A.; Stipanovic, R.D.; Howell, C.R.; Fryxell, P.A. Phytochem. 1975, 14, 225-231.
13. Stipanovic, R.D.; Bell, A.A.; Howell, C.R. Phytochem. 1975, 14, 1809-1811.
14. Bell, A.A.; Stipanovic, R.D. Mycopathol. 1978, 65, 91-106.
15. Bell, A.A. Phytopath. 1969, 59, 1119-1127.
16. Gaibullaev, I.K.; Avazkhodzhaev, M.K.; Egamberdiev, A.E. Genetika. 1975, 11, 34-36.
17. Zaki, A.I.; Keen, N.T.; Erwin, D.C. Phytopath. 1972, 62, 1402-1406.
18. Mace, M.E.; Bell, A.A.; Beckman, C.H. Can. J. Bot. 1976, 54, 2095-2099.
19. Mace, M.E.; Stipanovic, R.D.; Bell, A.A. Physiol. Plant Path. 1985, 26, 209-218.
20. Mantell, S.H.; Smith, H. Semin. Ser. Soc. Exp. Biol. 1983, 18, 75-108.

21. Heinstein, P.F. J. Nat. Prod. 1985, 48:1-9.
22. Heinstein, P.; El-Shagi, H. J. Nat. Prod. 1981, 44, 1-6.
23. Altman, D.W.; Stipanovic, R.D.; Mitten, D.M.; Heinstein, P. In Vitro Cell. Develop. Biol. 1985, 21, 659-664.
24. Heinstein, P. J. Nat. Prod. 1985, 48, 907-915.
25. Apostol, I.; Low, P.S.; Heinstein, P.; Stipanovic, R.D.; Altman, D.W. Plant Physiol. 1987, 84, 1276-1280.
26. Smith, F.H. J. Amer. Oil Chem. Soc. 1966, 44, 267-269.
27. Watson, D.G.; Rycroft, D.S.; Freer, I.M.; Brooks, C.J.W. Phytochem. 1985, 24, 2195-2200.
28. Hartman, C.L.; McCoy, T.J.; Knous, T.R. Plant Sci. Lett. 1984, 34, 183-194.
29. Bell, A.A. Phytopath. 1967, 57, 759-764.

RECEIVED April 22, 1988

Chapter 18

Mitotic Disrupters from Higher Plants

Effects on Plant Cells

Kevin C. Vaughn and Martin A. Vaughan

Southern Weed Science Laboratory, Agricultural Research Service, U.S. Department of Agriculture, Stoneville, MS 38776

Plant compounds that disrupt mitosis have been well-studied for their effects on animal cells but their effects on plant cells have been relatively poorly described. Most of the mitotic disrupters (e.g. podophyllotoxin, vinblastine, trewiasine) cause effects similar to the well known disrupter, colchicine. Squashes of these treated roots reveal numerous cells in prometaphase. At the electron microscopic level, the chromosomes appear condensed but no microtubules are associated with the chromosomes so that movement to the cell poles is impossible. Nuclear membranes reform around the chromosomes, resulting in oddly shaped, lobed nuclei. Cell elongation is also affected because of disruption to the cortical microtubules that are involved in determining cell shape, resulting in isodiametric cells in the zone of elongation. The combination of cell division and elongation inhibition results in a distinctly swollen club-shaped root. These affects are all due to a direct interaction of the disrupter with microtubules, by preventing further polymerization of tubulin into microtubules. Caffeine appears to affect only telophase by disrupting cell plate formation, probably due to failure of the vesicles to fuse to produce a normal cell plate. Caffeine may also have a direct effect on tubulin, like colchicine. Taxol actually promotes tubulin polymerization into microtubules and stabilization of the microtubules, possibly through interaction with microtubule-associated proteins. Although none of these mitotic disrupters have been utilized in agriculture, a number of these compounds are similar to existing herbicides and, if chemically modified or formulated, may be useful as herbicides.

Chemicals of plant origin, such as colchicine, vinblastine/ vincristine, and podophyllotoxin, have long been used as tools to study mitosis because of their ability to disrupt this cellular process. It was, in part, this anti-mitotic effect that produced intense research on the effect of these compounds on animal cell mitosis in pursuit of potential anti-tumor agents. Surprisingly, these compounds have been used relatively little with plants, with the exception of the induction of polyploidy with colchicine. Some of the reasons for this lack of interest may be due to the relative insensitivity of plants to these compounds (1,2). Generally 100-1,000 X increase of the compound is required to elicit the same level of response in plants that's obtained with animals.

Colchicine-induced disruption of mitosis in plant cells has been thoroughly described (3) although the effects, on plant cells, of other plant derived mitotic disrupters such as vinblastine/ vincristine (4, 5, 6), taxol (7, 8, 9), podophyllotoxin (6), and caffeine (6, 10, 11, 12) have not been as extensively studied. Other plant-derived anti-tumor compounds such as maytansine and trewiasine, have been little studied for their effects on plant cells.

Many compounds can affect processes such that the cells do not enter mitosis but remain in the G (gap) or S (synthesis) stages of the cell cycle (13). Most of these effects are secondary effects of these compounds, the result of inhibition of some other metabolic process required for the entrance into mitosis. This report will be limited to those plant compounds which disrupt mitosis, including cytokinesis, by directly or indirectly interfering with stages of mitosis *per se*.

In most cases, these mitotic disrupters interfere with the cellular structures known as microtubules. Microtubles are unbranched, hollow, cylinders composed of protein subunits, tubulin, and present in all eukaryotic cells (14). The microtubule is about 25 nm in outer diameter and is composed of 13 subunits when viewed in cross section. Immunofluorescence microscopy and transmission electron microscopy of protoplast ghosts have revealed microtubules as long as 20 μm in some plant cells (15).

Microtubules are associated with maintenance of structure through the cytoskeleton and cellular movement through the spindle and flagellar apparatus. The association of microtubules to cellular structure and movement is closely tied to the ability of microtubules to undergo dynamic assembly and disassembly at appropriate times and places in the cell. Microtubule assembly occurs by nucleation of non-identical tubulin subunits that exist in the cytoplasm as a free pool of α and β tubulin, each at about 55kD. The tubulin heterodimes form a heterodimer which are added at the assembly end of the "growing" microtubule. Recent data indicate that small oligomers of the tubulin heterodimers are added to the growing microtubule end, rather than individual subunits (16). Disassembly occurs primarily at the end of the microtubule opposite the assembly end. This dynamic process is referred to as "treadmilling". Although this is one of many models to describe the assembly of microtubules, there are other theories that explain microtubule assembly as well. Other cellular factors apparently influencing the rate and extent of microtubule polymerization and depolymerization are divalent metal cations, such as calcium, and microtubule-associated proteins (MAPs). At present, no MAPs have been identified in association with plant microtubules (17) and mixtures of only α and β tubulin can be assembled into microtubules *in vitro* (18, 19). Thus, there is no *a priori* reason that plant cell microtubules "must" include these non-tubulin proteins to assemble into microtubules.

Many functions in the plant cell are controlled by microtubules and these functions are defined by the various configurations of microtubules. These configurations are most easily observed in plant cells by observation of squashes of softened cells that have been incubated with antibodies to tubulin protein and then with fluorescently tagged secondary antibodies. This allows for a three-dimensional display of all the microtubules in the cell. Cortical microtubules occur close to the cell wall along the plasma membrane (Fig. 1A). It is believed that these microtubules are involved in orienting cellulose microfibrils during cell wall synthesis and in the orientation of new microfibrils that occurs in the cell wall during cell elongation (20). Spindle and kinetochore microtubules (Fig. 1C) are involved in the movement of chromosomes during mitosis. Phragmoplast microtubules (Fig. 1D) are thought to be involved in the movement and arrangement of Golgi-derived vesicles at the forming cell plate. Another microtubule configuration, the preprophase band of microtubules (Fig. 1B), is a ring of microtubules that appears before the cell is about to enter mitosis. The preprophase band appears to and was originally thought to predict the location of the area where the cell plate would form following the separation of daughter nuclei at telophase (21). However, compounds that disrupt latter stages of mitosis can alter the location of the new phragmoplast (22). It has been proposed that the preprophase band may serve as a stabilized tubulin pool for use in microtubule ., configurations required in the subsequent mitosis (23).

Although there are many similarities between plant and animal tubulin there are some striking differences. Electrophoretic separation of tubulin in certain gel systems reveals that plant tubulins run in the reverse electrophoretic mobility as animal tubulins (19, 24) (Fig. 2). That is, the order of migration is α, β in animal cells and β, α in plant cells. Antibodies to tubulin from many animal sources can cross react with plant tubulin, indicating that many antigenic sites on the tubulin molecules are conserved. Monoclonal antibodies to both α and β tubulin recognize a sequence near the carboxy terminus of both proteins (25) and these antibodies recognize tubulin from protists, plants and animals (e.g. 26, 27). Data reported below on the sensitivity differences between mitotic disrupter effects on plants and animals indicate that many recognition sites for these disrupters must differ for plant and animal tubulin, however.

Specific Compounds and Groups of Compounds

Colchicine. Colchicine (Fig. 3) is the most well studied and widely used of the plant alkaloids whose mode of action is to disrupt tubulin. This tropolone derivative is a three ring

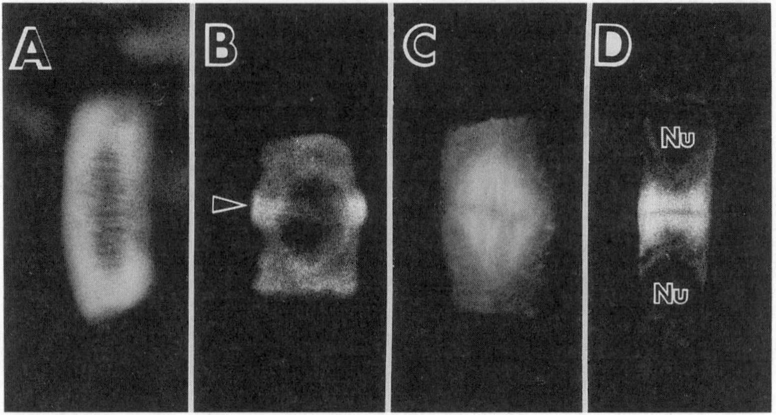

Figure 1. Immunofluorescence microscopy using antibodies to tubulin and secondary antibody labelled with fluorescein on stabilized onion root cells reveal different microtubule confirmations. A. Cortical microtubules at the cell periphery. B. A hoop like ring of microtubules (arrow) is characteristic of the preprophase band. C. A cell in the process of mitosis has distinct spindle microtubules that radiate from the poles of the cell and attach to the chromosomes at the cell plate. D. A band of short phragmoplast microtubules appears at the cell plate. The two newly formed nuclei (Nu) stand out in negative relief. All X 400.

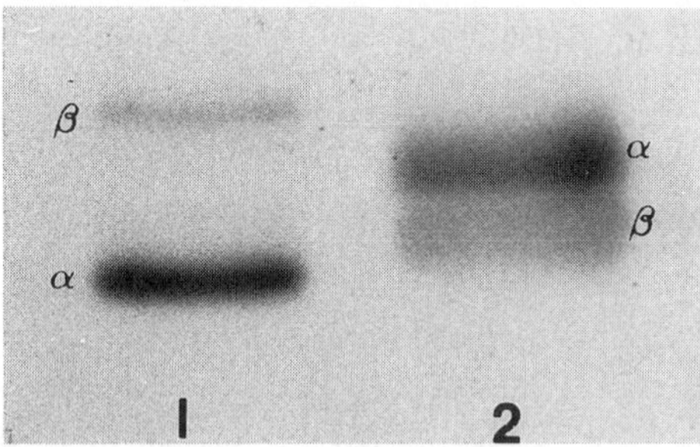

Figure 2. Western blot of carrot root and bovine brain extracts probed with rabbit antisera to sea urchin tubulin illustrate the difference in electrophoretic mobility between tubulin from these two sources: the α, β order in animal cells is reversed to β, α in plant cells. This antisera recognizes the α subunit better than the β. Confirmation of this was achieved by incubating duplicate blots in monoclonal anti-α and β tubulins.

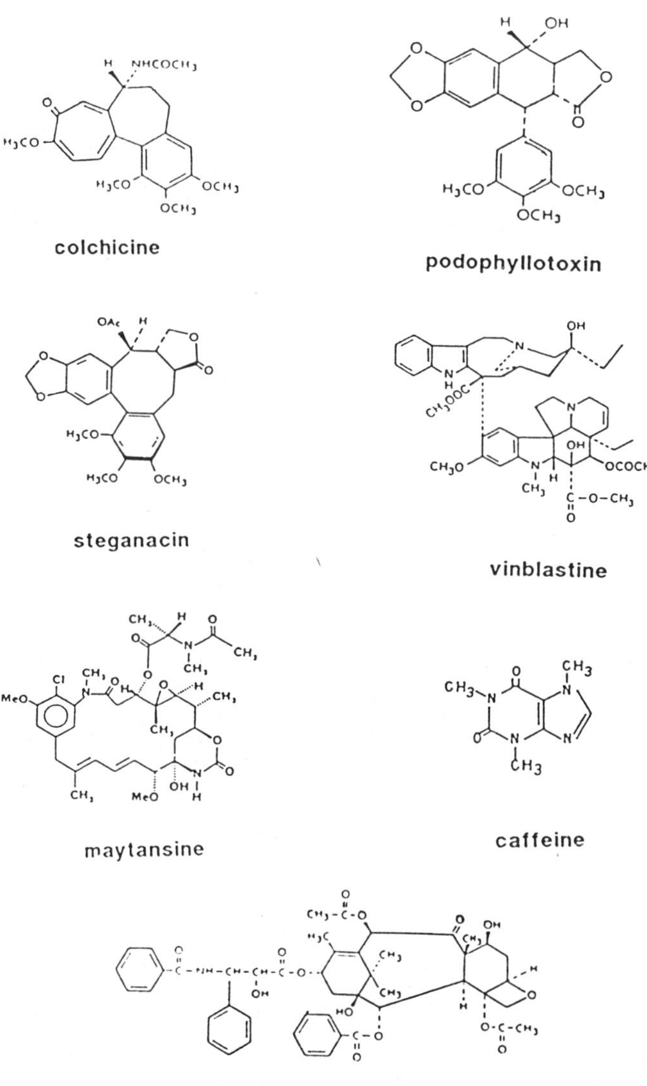

Figure 3. Structures of the mitotic disrupters described in this report.

structure compound isolated from bulbs of Colchicum autumnale, autumn crocus, as well as from other members of this family and related groups (1). Plants which produce colchicine are extremely resistant to the anti-mitotic effects of this compound, tolerating levels 100-1,000 times those that produce mitotic arrest in sensitive plant species (28, 29, 30). The molecular mechanism for this resistance is not known.

Colchicine was first recognized by Pernice in 1889 after the observation of abnormal mitotic figures in the the intestine of a dog who had died after eating Colchicum bulbs. Levan (31) described the mitotic abnormalities caused by colchicine treatment of Allium roots as colchicine or "c-mitosis", a normal mitotic cycle up to prometaphase followed by the random scattering of condensed chromatid pairs ("c-metaphase") (Fig. 4A). The mitosis progresses no further, though the chromatids separate ("c-anaphase"), and the nucleus reforms resulting in a "4c single nucleus" ("c-telophase"). The number of cells in mitosis in a root meristem appears to increase following colchicine treatment due to "arrest" at prometaphase and an increase in the time that the cell is stalled at this stage.

At the electron microscope level, the chromosomes appear highly condensed but unlike control cells at prometaphase, no microtubules are associated with the chromosomes. After the frustrated attempt at mitosis, the nuclear envelope reforms around the chromosomes, resulting in a lobed nucleus (Fig. 4B) or, less frequently, a multiple nucleus if the chromosomes were widely separated prior to nuclear reformation. All forms of microtubules appear to be effected by colchicine treatment. Because of the loss of both spindle and cortical microtubules, cells at the root tip neither divide nor elongate. This results in a conspicuously swollen (club or spear-shaped) root morphology.

Colchicine completely inhibits the in vitro assembly of tubulin into microtubules from both animal (32) and plant (2, 33) sources. The quantity of colchicine required to inhibit animal tubulin polymerization into microtubules is far below a 1:1 stoichiometry between tubulin and colchicine concentrations. Thus, models that have sought to explain the mechanism of action of this alkaloid have involved a "capping" concept to explain the non-stoichiometric effect of colchicine. These models (34) envision an interaction of tubulin dimers or oligomers and colchicine where, following binding to the dimer, the colchicine-tubulin complex is added to the assembly end of the growing microtubule, capping the microtubule so that no further tubulin dimers could be added. As mentioned above, microtubules are dynamic structures that undergo "treadmilling": assembly, or addition of tubulin dimers at one end and disassembly, or removal of tubulin dimers from the opposite end. Colchicine, by blocking the assembly end of the microtubule, causes an eventual loss of microtubule dimers through the unhindered disassembly end (35). There is evidence that colchicine-tubulin complex binds to the disassembly end as well as the assembly end as exchange of dimers is in dynamic equilibrium at both assembly and disassembly ends (36).

A photoaffinity analog of colchicine (37) has been used to probe the tubulin subunit (α or β) to which colchicine binds in its

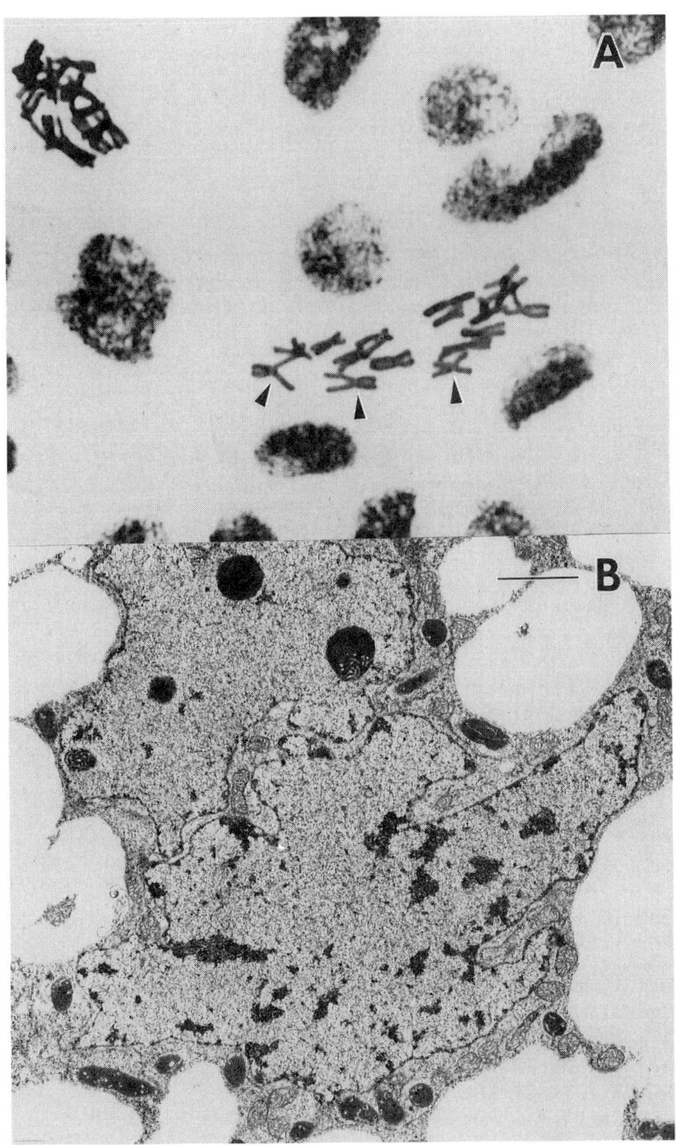

Figure 4. Effects of colchichicine (1mM) on plant cells. A. Characteristic "C-mitosis" revealing "ski-pairs" of chromosomes (arrows) after treatment of onion roots. X 400. B. Extensively lobed nucleus after a incomplete mitosis, the nuclear membrane reforms around the chromosomes that are unable to move to the poles. Bar=1.0 μm.

inhibition of microtubule polymerization. Only the α subunit was bound by this analog, indicating that colchicine binding is primarily associated with this subunit.

Plant tubulin is relatively much less sensitive to colchicine than animal tubulin (1) and a number of mechanisms for this possible general resistance have been suggested (3). Recent work, utilizing purified tubulin from higher plants, has indicated that plant tubulin has a much lower affinity for colchicine than animal tubulin (2, 33). These data indicate that the differences between plant and animal responses to colchicine is due to the tubulin *per se* rather than uptake, translocation, and metabolism of colchicine.

Other Compounds that bind at the "Colchicine Site" of Tubulin

Podophyllotoxin (Fig. 3), a lignan, isolated from the rhizomes and roots of the may apple (*Podophyllum peltatum*) has a long history as a folk remedy and was studied extensively as an anti-tumor agent (38). Podophyllotoxin, like colchicine, disrupts mitosis at prometaphase in animal (39) as well as in plant cells (5, 6, 40) (Fig. 5A) although few detailed descriptions are available. Like colchicine, podophyllotoxin is much more effective in disrupting mitosis in animal cells than in plant cells (1).

Podophyllotoxin inhibits colchicine binding to tubulin from animal sources (41) indicating that the binding site on the tubulin molecule is the same for colchicine and podophyllotoxin. However, tropolone, which inhibits colchicine binding, does not effect podophyllotoxin binding to tubulin (42). These data indicate that colchicine and podophyllotoxin have two binding sites, one of which is shared (42). Colchicine binding to tubulin from plant sources is unaffected by podophyllotoxin, indicating that the shared binding site on animal tubulin is not present or is not shared in plant tubulin.

Structurally, podophyllotoxin is quite different from colchicine with the exception of a common trimethoxy ring. This may explain binding to the same or shared site as colchicine on animal tubulin (43) but does little to explain why it apparently does not bind to the same site on plant tubulin. This is especially unusual considering the similar *in vivo* effects of podophyllotoxin in plant and animal cells and the similarity of these effects to that of colchicine (e.g. Fig. 5A).

Steganacin (Fig. 3), a lactone from *Stegnotaenia araliacea*, has anti-mitotic properties in animal cells similar to that of colchicine and has been shown to compete with colchicine for the same binding site on the tubulin molcule (44). Like podophyllotoxin, steganacin shares the trimethoxy ring with colchicine and produces mitotic arrest and inhibits microtubule polymerization in a manner similar to that of colchicine and podophyllotoxin. No studies of the effect of this compound have been conducted on plants to our knowledge.

"Vinca Alkaloids" and Maytansinoids

The so-called "Vinca alkaloids" are compounds with potent anti-tumor activity and are widely used as chemotherapeutic agents for

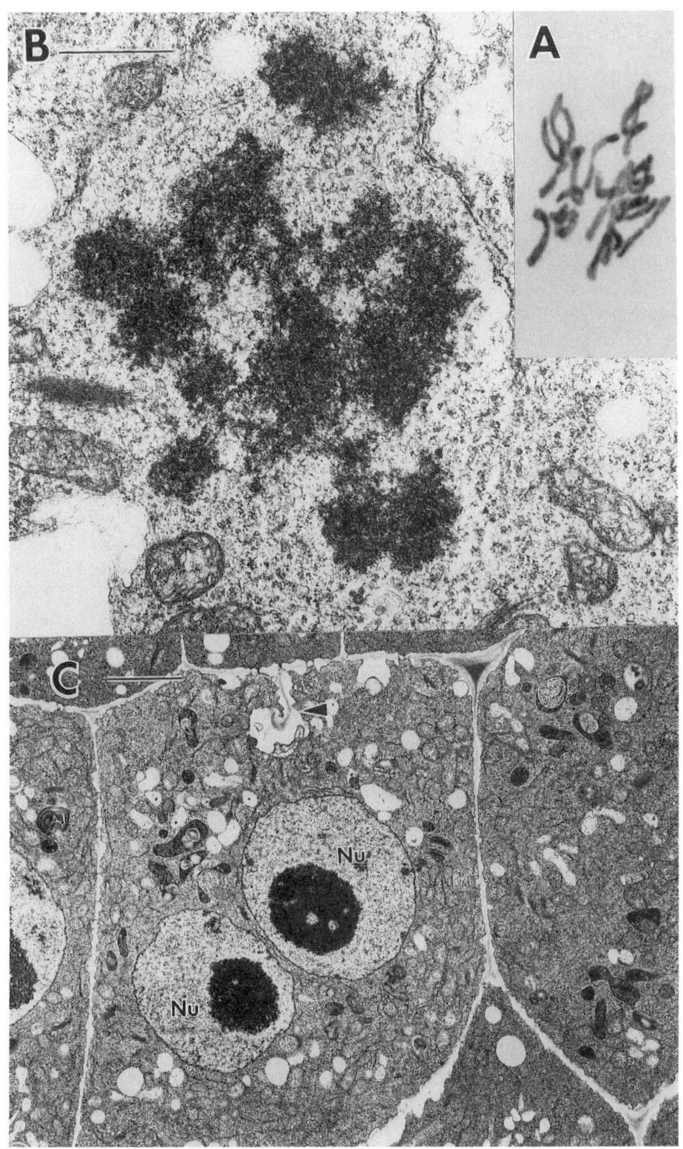

Figure 5. A. Light micrograph arrested prometaphase after treatment of onion roots with 0.1 mM podophyllotoxin. X 400. B. Electron micrograph of an arrested prometaphase after treatment of Catharanthus roseus with roots 0.1 mM vincristine. The chromosomes are condensed but no microtubules are associated with them. m= mitochondrion. Bar= 1.0 µm. C. Apparent binucleate cell after treatment with 0.1 mM vincristine. Note the patial cell wall (arrow). Nu= nucleus. Bar= 2.0 µm.

the treatment of Hodgkins' disease and leukemia (45). The most studied of these alkaloids are vinblastine (Fig. 3) and vincristine, both derived from _Catharanthus roseus_ (=_Vinca rosea_). Because of the importance of these compounds in chemotherapy, many studies have been undertaken to enhance the synthesis, recovery, and detection of these compounds, including breeding programs to enhance the levels of these useful alkaloids (46).

At low concentrations, both vinblastine and vincristine induce the same kinds of mitotic effects as colchicine in both plants and animals (Fig. 5B) (6, 47, 48). Other effects of these alkaloids are the formation of multipolar divisions in plant cells (5, 49) as well as a mitodepressive effect (50). In animal cells, high concentrations of both of these compounds induce the formation of paracrystalline arrays of tubulin but these do not occur even after treatment with 1 mM vinblastine in plant cells (5, 47). Because the _in vivo_ sensitivity of plant cells to vinblastine is much lower than that of animals (1), it has been suggested that plant tubulin might have only the low affinity binding sites rather than the high affinity sites of animal tubulin. Although vinblastine can stabilize the binding of colchicine to tubulin, it apparently does not bind to the same sites on the tubulin molecule (51). Use of a photoaffinity analog of vinblastine reveals that the vinblastine binds to both α and β tubulin (52). These data are consistent with the report of two binding sites per tubulin molecule by Ludueña _et al_. (48) using fluorescence inhibition as a measure of binding.

Kramers and Stebbings (4) reported that _C. roseus_, the source of both vinblastine and vincristine, is resistant to vinblastine. In a reinvestigation of this observation, we (47) found that although this species is resistant at up to 10^{-4} M vinblastine and 10^{-5} M vincristine, higher concentrations of each compound caused mitotic irregularities typical of that found in sensitive plants: arrested prometaphase and lobed nuclei (Fig. 5C). Neither vinblastine nor vincristine are highly soluble and it is likely that, in the original report of Kramers and Stebbings (4), the lack of effect at higher drug concentrations was not noted because the drug was not in solution. We have found that dissolving the Vinca alkaloids in dimethyl sulfoxide is necessary to obtain true solutions of these compounds at higher concentrations and the dimethyl sulfoxide itself had no effects at the concentrations utilized.

Another large group of mitotic disrupters are the maytansinoids which include maytansine (Fig. 3), isolated from _Mayteneus_, and trewiasine (Fig 3), isolated from _Trewia_ (53). These compounds are structurally quite dissimilar from the Vinca alkaloids but Ludueña _et al_. (48) have shown that they bind to sites that overlap the binding site of the Vinca alkaloids. Likewise, they bind in a molar ratio of two molecules of maytansinoid: one molecule of tubulin. Despite the overlapping binding sites of the Vinca alkaloids and the maytansinoids, the maytansinoids do not induce the formation of paracrystalline arrays of tubulin nor do they stabilize colchicine binding as do the Vinca alkaloids (48).

Unlike many of the other plant-derived mitotic disrupters, the maytansinoids are remarkably effective at disrupting mitosis in plant cells as well as in animal cells. Their effects are very similar to colchicine and other disrupters of this class producing arrested prometaphase figures and lobed, reformed nuclei (Fig. 6). Occasionally, a few remnant kinetochore microtubules are noted in the arrested prometaphase figures but this is the only distinction between the effects of the maytansinoids and that of colchicine. Large numbers of mitotic arrests occur even after treatment with concentrations of maytansinoids as low as 10^{-8} M. Generally 10^{-4}-10^{-3} M solutions of colchicine are required for similar levels of mitotic disruption (e.g. 4,6). C. roseus, the plant that is tolerant to all but very high concentrations (10^{-3}-10^{-4} M) of vinblastine and vincristine (47) is nevertheless equally or more sensitive to both maytansine and trewiasine as onion, a vinblastine-sensitive species (Vaughn, Vaughan, and Ludueña, in preparation). The dramatic effects of the maytansinoids at low concentrations indicate that they may be valuable tools to disrupt the mitotic process without the danger of secondary or non-specific effects noted with the normally high concentrations of compounds required to disrupt plant microtubules.

Taxol

Taxol (Fig. 3) is a diterpenoid isolated from the western yew (Taxus brevifolia) and other species of Taxus (54, 55) and, like the other plant alkaloids discussed above has shown anti-tumor activity (54). However, unlike the other mitotic disrupters, this agent actually promotes polymerization of tubulin into microtubules by lowering the critical concentration of tubulin required for polymerization rather than promoting depolymerization (56, 57). The actual molecular mechanism for this stabilizing effect is unknown.

The original light microscopic studies suggested that taxol works in the same manner as colchicine since taxol treatment also produces arrested prometaphases. Electron and immunofluorescence microscopic studies indicated that abundant microtubules were found both at their normal sites at the kinetochore as well as at other sites in the cell where microtubules are not normally found, or at least not found in abundance (Fig. 7A). In many cases, the microtubules are associated, either forming doublets or extensions that link them to each other or to other cellular structures (57). Foisner and Wiche (58) have found that taxol not only stimulates the association of tubulin but also that of the microtubule associated proteins (MAPs). Incubation of mixtures of tubulin and MAPs in the presence of taxol in an in vitro microtubule polymerization system produced microtubules with extensions that cross-linked the microtubules, similar to the extensions observed in vivo. Antibody labelling of these microtubules with anti-MAPs identified these extensions as containing MAPs (58).

The presence of MAPs in higher plant cells has not yet been established (17), although calmodulin often forms an association with plant microtubules as determined by immunofluorescence microscopy (59). Thus, if taxol does cause crosslinking of microtubules by MAP interaction, the effects of taxol in vivo on plant cells would indicate that plants also must have MAPs.

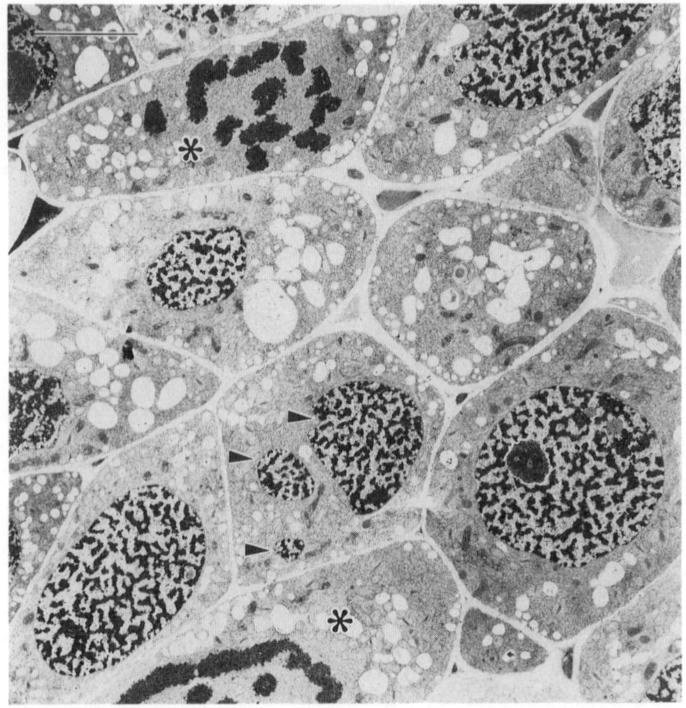

Figure 6. Electron micrograph of onion root tip treated with 1μM trewiasine. Two cells arrested prometaphase (*) and a lobed reformed nucleus in which three pieces are seen in this thin section (arrows). Bar= 5.0 μm.

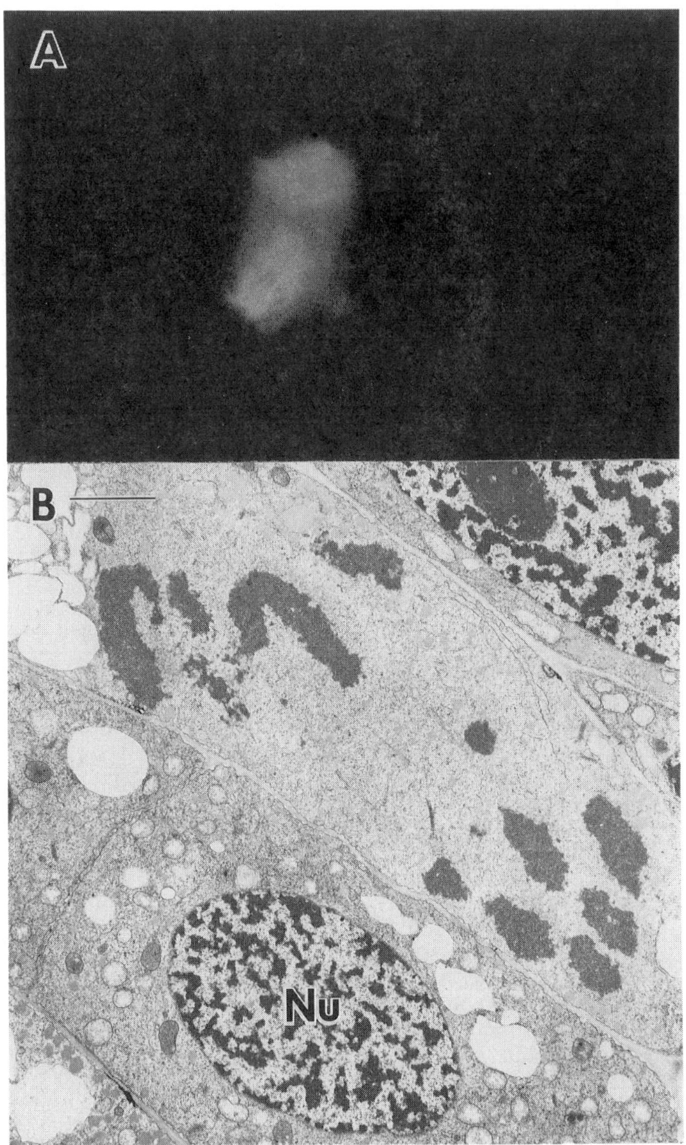

Figure 7. Effects of 1 μM taxol on goosegrass (A) and onion (B). Immunofluorescence micrograph of the goosegrass containing two spindle- like formations, oriented in different directions are noted (arrows). In this electron micrograph of onion, mitotis apparently proceeds normally as evidenced by the cell in mitosis and the normally formed nucleus (Nu). A=X400; in B bar=2.0μM.

Plant microtubules have been particularly difficult to assemble *in vitro*. Morejohn and colleagues (60, 61) took advantage of the unique properties and availability of taxol to polymerize it *in vitro*. Agents used by other investigators to isolate and polymerize plant tubulin (glycerol or DMSO at high molarity) allowed a limited polymerization of tubulin into microtubules (60), but the yield of microtubules is much lower than with taxol.

Although no careful comparison of plant and animal cells has been undertaken, the doses of taxol used to elicit effects in animal cells (57) and in plant cells (7, 8, 9) are similar, indicating similar affinities for tubulin. Interestingly, not all plants are affected to the same degree by taxol treatment. Onion, a species on which many microtubule disrupter studies have been preformed, shows virtually no change in microtubule orientation or other mitotic irregularities even after 24 h of treatment in 10 μM taxol (Fig. 7B) even though other species show large increases in the mitotic index and abnormally oriented mitotic figures. There are several possibilities that explain these data. Substances such as taxol, present in low concentrations in the cell, may be used by the cell to stabilize the microtubules during the various conformations that are required to maintain a proper cell cycle. Onion may have its tubulin stabilized by a cellular agent similar to taxol, rendering it less sensitive to an external stabilizer. The presence of a stabilizer may explain the relative insensitivity of onion to a number of mitotic disrupter herbicides, such as dimethyltetrachloroterephthalate (DCPA) (67).

Caffeine and Related Methylxanthines

Caffeine (Fig. 3) is an oxypurine contained in fruit of the coffee plant (*Coffea arabica*). Unlike the other alkaloids which disrupt mitosis, the methylxanthines, of which caffeine is a member, affect the completion of cytokinesis at telophase and not the formation the mitotic spindle at prometaphase. The precise mechanism by which caffeine produces this effect is not known. There is some evidence, although relatively small when compared to the other mitotic inhibitors, that caffeine acts directly on tubulin. Other evidence implicates an effect of caffeine on calcium concentrations and vesicle fusion at the cell plate. As more information is compiled on the molecular processes involved in cell plate formation, the molecular site of action of caffeine will no doubt be elucidated as well.

Treatment of root meristem cells with caffeine results in normal mitosis but apparent failure of cytokinesis produces large numbers of binucleate cells and cells with incomplete cell plates (6, 10, 12, 63). Vesicles derived from the Golgi apparatus apparently move toward the cell plate but the vesicles do not fuse so that an incomplete, reticulate cell plate results (Fig. 6). In extreme cases, no cell plate is formed and binucleate cells result (Fig. 8).

Caffeine is a potent inhibitor of cAMP phosphodiesterase in animal cells (64) and it was assumed that the caffeine effect on plant ctyokinesis was due to elevated levels of cAMP and decreased levels of calcium affecting the form and function of membrane

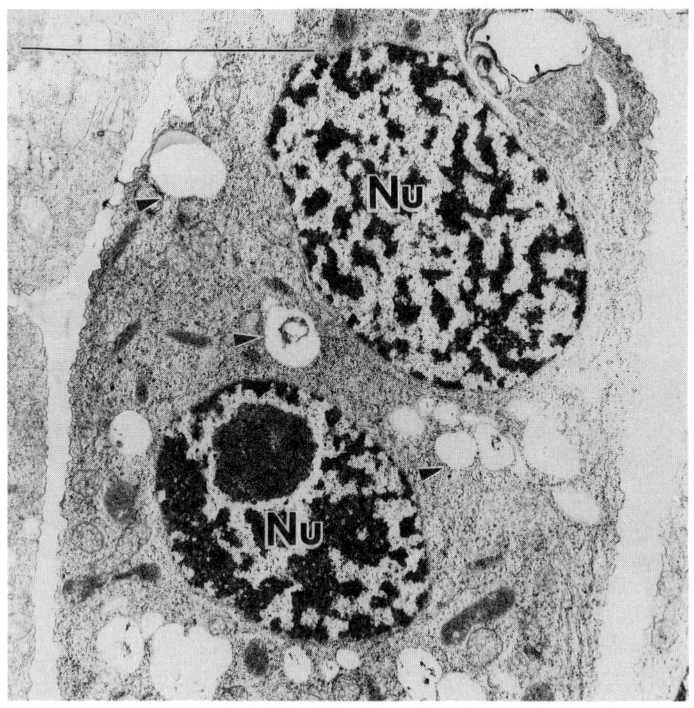

Figure 8. After 0.1 mM caffeine treatment the cells are binucleate. Vesicles (arrows) of abnormal size and content are found separating the two nuclei (Nu). Bar=5.0μm.

vesicles at the cell plate (10). However, the overwhelming evidence indicates that cAMP does not exist in higher plants (65). A role for calcium (10, 11, 12) and possibly magnesium (11) has been demonstrated for cytokinesis in plant cells suggesting an effect of caffeine on calcium levels at the cell plate. Other mitotic disrupters can cause calcium inbalance (66). However, it is now believed that, although the effects are real, they are not the primary cause of mitotic disruption.

Effects of mitotic disrupters are generally determined in fixed specimens and the sequence of events are often implied based upon static electron micrographs. Bonsignore and Hepler (12) observed living, dividing stamen hairs of _Tradescantia_ through mitosis after treatment with caffeine through the use of Nomarski differential interference microscopy. Quite susprisingly, they observed that an apparently normal cell plate was formed but that the plate was subsequently disturbed. This may be due to the failure of the vesicles to fuse properly so that a structure that appears in the microscope as a wall is laid down and later disperses as the phragmoplast microtubules disappear. Because phragmoplast microtubules disappear from the center of the cell first, the images of partial cell plates present only at the edge of the cell (10) may be explained.

A direct interaction between tubulin and caffeine and related alkaloids has also been noted in a number of studies. In the green alga _Chlamydomonas_ presumptive tubulin mutants that are resistant to colchicine exhibit heightened sensitivity to caffeine (67). _Chlamydomonas_ flagella are a convenient experimental system for studying effects of compounds on microtubules. The caffeine analog, isobutylmethylxanthine (IBMX), caused flagellar resorption, and, like other microtubule disrupters (68), caused changes in the mRNA levels for flagellar proteins such as tubulin (69). Moreover, Morejohn (unpublished) has found that IBMX inhibits the _in vitro_ polymerization of tubulin into microtubules, although the levels required were higher than that for other mitotic disrupters. Supporting these _in vitro_ data are the _in vivo_ studies of Juniper and Lawton (70). These workers observed that cortical microtubules of primary cell wall were relatively caffeine-insensitive but cold-sensitive whereas the cortical microtubules of the secondary cell wall were caffeine-sensitive but cold-insensitive. These data were the first to indicate that there may be differences between microtubules of different cell types, at least in their sensitivity to disruption. Vaughn _et al_. (6) observed that the dinitroaniline-resistant biotype of _Eleusine indica_ (with an altered tubulin composition) was much more sensitive to caffeine and related alkaloids than the susceptible biotype but, because the resistant biotype made abnormal cell walls, it could not be determined whether this was a differential caffeine effect on tubulin or an exacerbation of an already abnormal situation.

Comparison of Herbicides with Plant Derived Mitotic Disrupters

In general, plant-derived microtubule disrupters are much more effective on animal than on plant cells (1), with the exception of the maytansinoids and taxol as described above. What are the characteristics that mark an effective mitotic disrupter for plant cells and those that are effective for animal cells? Usually effective animal cell mitotic disrupters are those which have at least a fair solubility in water and are obtained from plants (or share a site of action with a natural plant compound). At the other extreme, effective plant cell mitotic disrupters generally have a low water solubility and are chemically-derived (Table 1). Compounds that affect both plants and animals, including the compounds propham and chlorpropham, are formulated for both herbicidal and insecticidal use, and are generally intermediate in water solubility.

Part of this division may reflect the needs for these compounds rather than specific microtubule sites that are attacked by each disrupter. For example, the dinitroaniline herbicides are highly insoluble compounds that are incorporated into the soil prior to crop planting. Crop plants that are tolerant of these herbicides are generally large seeded, lipid-rich plants that can sequester the herbicide in lipid bodies (71) away from the site of action, the microtubule. Sensitive weed species, chiefly the grasses, are small seeded and have much lower levels of lipid and are unable to sequester the herbicide. The solubility characteristics of this group of herbicides renders some selectivity, allowing them to be useful herbicides. Thus, the selective herbicidal properties of a compound may be the reason for the low solubility of many of the plant-specific disrupters, rather than chemical differences related to plant-animal differences in their tubulin.

Potential Use of Plant-Derived Microtubule Disrupters in Agriculture

This symposium's purpose is to highlight the potential of natural products as agricultural agents. At present, none of the plant derived mitotic disrupters are used to solve crop problems but there is a potential for some uses.

Taxol has been shown to protect animal cells from subsequent treatment with mitotic disrupters such as colchicine and maytansine (34). Because a number of herbicides are mitotic disrupters, we (72) investigated whether taxol could protect susceptible *Eleusine indica* (goosegrass) from subsequent treatment with the dinitroaliline herbicides trifluralin or oryzalin. Treatment of the susceptible biotype with sublethal levels of taxol for 24 h and subsequent treatment with 10 μM trifluralin or 1 μM oryalin resulted in no abnormal mitoses. Grasses are very sensitive to most mitotic disrupter herbicides and treatment of grass seedlings with a protectant such as taxol would allow selective weed control of weed grasses in crop grasses, such as Johnsongrass (*Sorghum halapense*) control in sorghum (*Sorghum bicolor*), an especially difficult herbicide problem. An effective method of implementation might involve simply treating dry seeds with a solution of taxol or a taxol derivative and then planting the treated seed into soil incorporated with dinitroaniline herbicides. Of course, the scarcity of taxol at present makes such a proposal seem quite ridiculous. The development of an inexpensive synthesis of a

Table 1. Grouping of mitotic disrupters based upon differential sensitivities of plants and animal microtubules to these compounds

Group	Characteristics
I Much more effective on animals	
colchicine	
vinblastine	Mainly plant-derived, at least moderately water soluble
podophyllotoxin	
griseofulvin	
oncodazole	
II Effective both on plants and animals	
taxol	Obtained from both chemical and plant
maytansinoids	
chlorpropham	sources; solubilities generally less than group I but greater than group III
propham	
III Effective only on plants	
Dinitroanilines	Mainly lipid soluble; maximum water solubility usually
Pronamide	∿ 2ppm without organic solvents added; none are plant-derived
Amiprophosmethyl	

compound with the properties of taxol is something for synthetic organic chemists to consider, however.

Another potential approach involves chemical modification of the disrupter by adding other substituent groups that would render them less toxic to humans, and increase the chance for selectivity to and within groups of plants. An example will make this more clear. Paraquat is a non-selective herbicide that acts by accepting electrons from the primary acceptor of photosystem I. A parquat radical dication results which reacts with molecular oxygen to generate superoxide and other toxic oxygen species such as hydroxyl radical and peroxide (73). Some years ago ICI introduced a bis-carbamoylmethyl derivative of paraquat called morfamquat. Unlike paraquat, this compound was very effective in controlling broad-leaved weeds but was very ineffective in controlling monocots. Although the exact mechanism for this discrimination is not known (74), the most believable explanation for this discrimination is that dicots are able to hydrolyze the carbomyl group from morfamquat, yielding a molecule that is much easier to move across membranes in the cell. Further decarboxylation would give rise to the parent compound paraquat. By using similar modified forms of plant derived mitotic disrupters, it may be possible to render these disrupters non-toxic to man, increase their efficiency at entry into the plant by increasing their lipophilicity, and perhaps achieve the selectivity desired of a herbicide.

Acknowledgments

Thanks are extended to R. H. Jones and P. Wolfe for technical assistance. The NIH kindly provided taxol for our studies. R. Powell and R. Ludueña provided generous samples of trewiasine and maytansine, respectively. This work was supported in part by a USDA Competitive Grant (86-CRCR-1-1933) to K. C. V.

Literature Cited

1. Deysson, G. In Microtubules and Microtubule Inhibitors; Borgers, M., deBrabander, M., eds. North Holland: Amsterdam, 1975; pp 427-451.
2. Morejohn, L. C.; Bureau, T. E., Tocchi, L. P.; Fosket, D. E. Proc. Natl. Acad. USA 1984, 81, 1440-1444.
3. Hart, J. W.; Sabnis, D. O. Curr. Adv. Plant Sci. 1976, 20, 1095-1104.
4. Kramers, M. R.; Stebbings, H. Chromosoma 1977, 61, 277-287.
5. Hillmann, G.; Ruthman, A. Planta 1982, 155, 124-132.
6. Vaughn, K. C.; Marks, M. D.; Weeks, D. P. Plant Physiol. 1987, 83, 956-964.
7. Bajer, A. S.; Cypher, C., Mole-Bajer, J., Howard, H. M. Proc. Natl. Acad. Sci. USA 1982, 79, 6569-6573.
8. Mole-Bajer, J.; Bajer, A. S. J. Cell Biol. 1983, 96, 527-540.
9. Weerdenburg, C.; Falconer, M. M.; Setterfield, G.; Seagull, R. W. Cell Motil. Cytoskel. 1986, 6, 469-478.
10. Paul, D. C.; Goff, C. W. Exp. Cell Res. 1973, 78, 399-413
11. Becerra, J.; Lopez-Saez, J. F. Exp. Cell Res. 1978, 111, 301-308

12. Bonsignore, C. L.; Hepler, P. K. Protoplasma 1985, 129, 28-35.
13. Hess, F. D. In Biochemical Responses Induced by Herbicides; Moreland, D. E., St. John, J. B.; Hess, F. D., Eds. ACS Symposium Series 181; American Chemical Society: Washington, DC, 1982; pp 207-230.
14. Little, M.; Ludueña, R. F.; Langford, G. M.; Asnes, C. F.; Farrell, K., J. Mol. Biol. 1981, 149, 95-107.
15. Van der Valk, P., Rennie, P. J.; Connally, T. A.; Fowke, L. C. Protoplasma 1980, 105, 27-43.
16. Mandelkow, E.; Mandelkow, E.M.; Bordas, J. Ann. N.Y. Acad Sci. 1986, 466, 650-655.
17. Tiwari, S. C.; Wick, S. M.; Williamson, R. E., Gunning, B. E. S. J. Cell Biol. 1984, 99, 63s-69s.
18. Morejohn, L. C.; Fosket, D. E. Nature 1982, 297, 426-428.
19. Dawson, P. J.; Lloyd, C. W. EMBO J. 1985, 4, 2451-2455.
20. Lloyd, C. W. Int. Rev. Cytol. 1984, 86, 1-51.
21. Pickett-Heaps, J. D. J. Cell Sci. 1969, 4, 397-420.
22. Clayton, L.; Lloyd, C. W. Eur. J. Cell Biol. 1984, 34, 248-253.
23. Wick, S. M.; Duniec, J. J. Cell Biol. 1983, 97, 235-243.
24. Hussey, P. J.; Gull, K. FEBS Lett. 1985, 181, 113-118.
25. Serrano, L.; Wandosell, F.; Avila, T. Anal. Biochem. 1986, 159, 253-259.
26. Gunning, B. E. S.; Wick, S. M. J. Cell Sci. Suppl. 1985, 2, 157-179.
27. Simmonds, D. H.; Seagull, R. W.; Setterfield, G. J. Histochem. Cytochem. 1985, 33, 345-352.
28. Cornman, I. Biol. Bull. 1942, 81, 297-298.
29. Levan, A.; Steinegger, E. Hereditas 1947, 33, 552-556.
30. Santos Diaz, J. D.; Mesquitz, J. F. Bol. Soc. Brot. Ser. 2 1978, 52, 221-240.
31. Levan, A. Hereditas 1938, 24, 471-486.
32. Borisy, G. G.; Olmstead, T. B.; Klugman, R. A. Proc. Natl. Acad. Sci. USA 1972, 69, 2890-2894.
33. Morejohn, L. C.; Bureau, T. E.; Tocchi, L. P.; Fosket, D. E. Planta 1987, 170, 230-241.
34. Schiff, P. B.; Horwitz, S. B. In Molecular Actions and Targets for Cancer Chemotherapeutic Agents; Academic Press: New York, 1981, pp 483-507.
35. Cabral, F. R.; Brady, R. C.; Schibler, M. T. Ann. N.Y. Acad Sci.; 1986, 466, 745-756.
36. Wilson, L.; Meza, I. J. Cell Biol., 1973, 58, 709-720.
37. Williams, R. F., Mumford, C. L.; Williams, G. A.; Floyd, L. J.; Aivaliotis, M. J.; Martinez, R. A.; Robinson, A. K.; Barnes, L. D. J. Biol. Chem. 1985, 260, 13794-13802.
38. Kelly, M. G.; Hartwell, J. L. J. Natl. Cancer Inst. 1954, 14, 967-1010.
39. Krishnan, A.; Hsu, D. J. Cell Biol. 1971, 48, 407-409.
40. Sullivan, B. J.; Wechsler, H. I. Science 1947, 105, 433-435.
41. Wilson, L.; Bryan, T. Adv. Cell Mol. Biol. 1975, 3, 21-72.
42. Cortesse, F.; Bhattacharyya, B.; Wolff, T. J. Biol. Chem. 1977, 252, 1134-1140.
43. Bhattacharyya, B.; Wolff, T. Proc. Natl. Acad. Sci. USA 1976, 73, 2375-2378.

44. Schiff, P. B.; Kende, A. S.; Horwitz, S. B. Biochem. Biophys. Res. Commun. 1978, 85, 737-746.
45. Creasy, W. A. Hand. Exper. Pharm. 1975, 38, 670-694.
46. Langone, T. T.; D'Onofrio, M. R.; van Vunakis, H. Anal. Biochem. 1979, 95, 214-221.
47. Vaughn, K. C.; Vaughan, M. A. Bot. Soc. Am. Abst. 1987.
48. Ludueña, R. F.; Anderson, W. H.; Prasad, V.; Jordan, M. A.; Ferrighi, K. C.; Roach, M. C.; Horowitz, P. M.; Murphy, D. B.; Fellous, A. Ann. N.Y. Acad. Sci. 1986, 466, 718-732.
49. Segawa, M.; Kondo, K. Experientia 1978, 34, 996-999.
50. Deysson, G.; Truhaut, R. C. R. Acad Sci. (Paris) 1963, 257, 3512- 3515.
51. Wilson, L.; Anderson, K.; Creswell, K. J. Cell Biol. 1974, 63, 373a.
52. Safa, A. R.; Hamel, E.; Felsted, R. L. Biochem. 1987, 26, 97-102.
53. Powell, R. G.; Weisleder, D.; Smith, C. R. Jr. J. Org. Chem. 1981, 46, 4398-4403.
54. Wani, M. C.; Taylor, H. L.; Wall, M. E.; Goggon, P.; McPhail, A. T. J. Am. Chem. Soc. 1971, 93, 2325-2327.
55. Miller, R. W.; Powell, R. G.; Smith, C. R. Jr. J. Org. Chem. 1981, 46, 1469-1474.
56. Schiff, P. B.; Fant, T.; Horwitz, S. B. Nature 1979, 277, 665-667.
57. Manfredi, J. J.; Horwitz, S. B. Pharmac. Ther. 1984, 25, 83-125.
58. Foisner, R.; Wiche, G. J. Ultrast. Res. 1985, 93, 33-41.
59. Wick, S. M.; Mato, S.; Duniec, T. Protoplasma 1985, 126, 198-206.
60. Morejohn, L. C.; Fosket, D. E. Nature 1982, 297, 426-428.
61. Morejohn, L. C.; Fosket, D. E. J. Cell Biol. 1984, 99, 141-147.
62. Chong, C.T.; Smith, D. Weed Sci. 1972, 20, 155-159.
63. Dela Pena, A.; Puertas, M. J.; Merino, F. Chromosoma 1981, 83, 241-248.
64. Rasmussen, H.; Tenenhouse, A. Biochem. 1968, 59, 1364-1370.
65. Amrhein, N. Planta 1974, 118, 241-258.
66. Hertel, H.; Quader, H.; Robinson, D. G.; Marme, D. Planta 1980, 149, 336-340.
67. Warr, J. R.; Flanagan, D.; Quinn, D. Exp. Cell Res. 1978, 111, 37-46.
68. Collis, P. S.; Weeks, D. P. Science 1978, 202, 440-442.
69. Lefebvre, P. A.; Nordstrom, S. A.; Moulder, J. E.; Rosenbaum, J. L. J. Cell Biol. 1978, 78, 8-27.
70. Juniper, B. E.; Lawton, J. R. Planta 1979, 145, 411-416.
71. Hilton, J. L.; Christiansen, M. N. Weed Sci. 1972, 20, 290-294.
72. Vaughan, M. A.; Vaughn, K. C. Plant Physiol. 1987, 83s, 107.
73. Vaughn, K. C.; Duke, S. O. Plant Cell Environ. 1983, 6, 13-20.
74. Brian, R. C. Pestic. Sci. 1972, 3, 409-414.

RECEIVED April 26, 1988

Chapter 19

Camptothecin and Other Plant Growth Regulators in Higher Plants with Antitumor Activity

J. George Buta and Andrzej Kalinski

Plant Hormone Laboratory and Nematology Laboratory, Beltsville Agricultural Research Center, Agricultural Research Service, U.S. Department of Agriculture, Beltsville, MD 20705

Camptothecin and other plant growth regulating compounds isolated from plants having antitumor activity are discussed concerning agricultural applications. These alkaloids and lactones have predominantly growth inhibiting activity and were isolated from plant material collected for the National Cancer Institute anticancer screening program. Unique selective growth regulating activity of the alkaloid, camptothecin, was found on several species of mono- and dicotyledonous plants. Studies on the mechanism of action indicated camptothecin inhibits the activity of barley DNA topoisomerase I.

Several years ago an investigation was begun in the Plant Hormone Laboratory (USDA, Beltsville, MD) to find new types of plant growth regulating chemicals. One source for these types of active compounds was a small portion of the extracts of higher plants that had been collected by the USDA for the National Cancer Institute (NCI) program to screen for new anticancer agents. Some 114,000 extracts from 3,394 species of plants as of January 1981 were screened by the NCI. The primary screening of extracts has involved an in vitro test for cytotoxicity in the KB cell culture and an in vivo test against L-1210 leukemia LE in mice. A recent publication of this research (1) summarized testing of antitumor plant-derived compounds in advanced development which include the following types of compounds: the alkaloids, maytansine from Maytenus spp., indicine-N-Oxide from Heliotropium indicum, homoharringtonine from Cephalotaxus harringtonia, ellipticine from Ochrosia elliptica, and lactones: bruceantin from Brucea antidysinterica, taxol from Taxus spp., and 4-beta-hydroxywithanolide E from Withania spp. Other compounds discussed but not in the advanced testing stage of the NCI program included: epoxides, psorospermin from Psorospermum febrifugum, phyllanthoside from Phyllanthus braziliensis, lactones, baccharin from Baccharus megapotamica, 6-senecioyloxychaparrinone from Simaba multiflora, eriofertopin from Eriophyllum confertiflorum, tripdiolide from Tripterygium wilfordii, and the phenol, fagarinone from Fagara macrophylla (2, 3).

Detection of Plant Growth Regulating Activity

Detection of growth regulating activity by extracts of higher plants from NCI was done by tobacco and bean bioassay. These plants were grown in the greenhouse and *Nicotiana tabacum* cv. Xanthi-nc plants were treated at 60 days after planting while *Phaseolus vulgaris* cv. Greenpod was treated at 14 days after planting. The ethanolic extracts were evaporated to dryness and the remaining residue of each extract was made into a 5% lanolin suspension. These lanolin suspensions were applied to the upper three axillary buds of Xanthi tobacco after decapitation and as a ring on the stem below the meristem of the Greenpod beans. A minimum of three plants were used per treatment and measurements taken on the inhibition of elongation of the meristematic portions of the plants. At least 50% inhibition of growth for at least one week was considered necessary for activity to be worthy of further examination. Other morphological effects could also be observed. More than seven hundred extracts of plants from the NCI collection that had shown activity in the tumor inhibition assays were tested and a partial listing of the active extracts are presented in Table I.

Identification of Plant Growth Regulators

Several studies have been done on the plant growth regulating activity of a small number of the plants indicated in Table I. Allelopathic effects have been reported for extracts of *Baccharis megapotamica* (*4*) and growth inhibition of beans, corn and tobacco resulted after treatment with a number of the macrocyclic tricothecenes related to baccharin which have been isolated from *Baccharis* sp. (*5*). Maytansine obtained from *Maytenus serrata* inhibited growth in tobacco callus and rice seedling bioassays (*6*). Parthenin, a sesquiterpenic lactone isolated from *Parthenium hysterophorus*, has also been reported to be allelopathic (*4*). The seed germination inhibitor detected in *Pinus pinea* was identified as abscisic acid, a plant hormone (*7*) and brassinolides (steroidal plant growth regulators) were isolated and identified in *P. thunbergii* (*8*). Aqueous extracts of leaves of *Polygonium orientale* inhibited seed germination and seedling growth of mustard, rice and lettuce and the active principles were determined to be phenolic, possibly flavone glycosides (*9*).

Based on the bioassay data (Table I) several plant extracts causing complete inhibition of axillary bud growth in tobacco were selected for fractionation in order to isolate the plant growth regulating compounds. The isolation procedures were generally a combination of solvent partitioning followed by chromatography using the above-mentioned bioassays to determine the course of the purification. Each isolation procedure had to be altered for the chemistry of the particular active compounds being isolated.

Cephalotaxus harringtonia. An ethanolic extract of seeds of *C. harringtonia* was phytotoxic when applied to axillary buds of *N. tabacum*. The harringtonine alkaloids identified as tumor inhibitors (*10*) were found to have minimal plant growth regulating activity in the bud growth bioassay. Fractionation by gel permeation chromatography and subsequent HPLC with accompanying bioassays led to the isolation of a tropone-lactone, harringtonolide (*11*) (Figure 1).

TABLE I. Plants from NCI Collection whose extracts cause plant growth inhibition in the Nicotiana bioassay

Species	
Alectryon subcinereum Radlk	Leonti nepetaefolia R.Br.
Baccharis cordifolia	Lepidotrichillia volkensii
Barsama abyssinica	Liatris provincialis Godfrey
Brucea antidysinterica	Linum perenne L.
Camptotheca acuminata	Maytenus senegalensis
Casuarina cunninghamiana Mig.	Parthenium hysterophorus
Caylusea abyssinica	Pedilonthus tithymailoides Poit
Cephalotaxus harringtonia	Paeonia suffruticosa Andr.
Cladonia subtenuis Evans	Phyllanthus brasiliensis
Crassocephalum mannii	Phyllostachys bambusoides
Cyphostemma kilimandscharicum	Pinus australis
Dinocarpus longa Lour.	Poinciana gillesii
Erythrina sp.	Polygonum coccineum
Euonymus atropurpureus Jacq.	Ricinus communis
Euphorbia bicolor Engelm. and Gary	Sesbania punicea
Fagara macrophylla	Solidago canadensis spp.
Gnidia subcordaxa Engl.	Sophora viciifolia Hance
Grindelia microcephala DC	Stephania abyssinica
Horkelica fusca Lindl.	Taxus media
Hymenoxys odorata DC	Terminalia superba
Hypoestes verticilliaris	Toddalia asiatica
Ichtyomethia grandiflora	Verbesina greenmanii Urb.

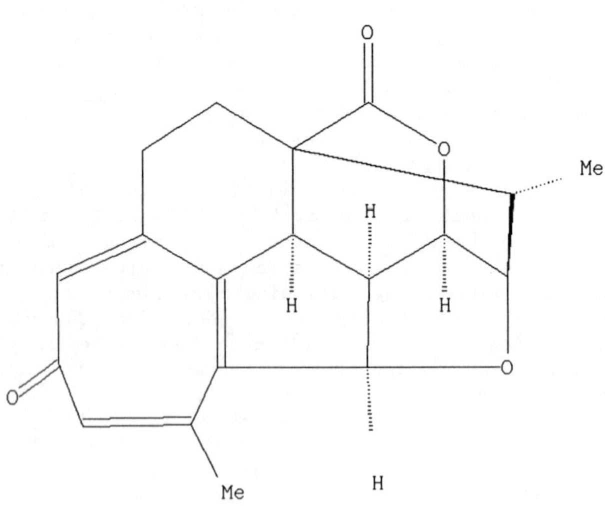

Figure 1. Harringtonolide.

Sesbania drumondii. An ethanolic extract of seed pods of S. drumondii was phytotoxic in the N. tabacum axillary bud bioassay. Fractionation on TLC using silica gel with methanol-methylene chloride (1:9) and bioassay of the fractions led to the isolation of an active fraction containing linoleic acid which can cause the growth regulating activity observed (12). The antitumor compound, sesbanimide, has apparently not been tested as a plant growth regulator (13).

Camptotheca acuminata. An ethanolic extract of stem-wood of C. acuminata completely inhibited axillary bud growth of tobacco but had no effect on the growth in the bean assay. Fractionation of the extract initially by TLC using silica gel with $Me_2CO-CHCl_3$ 1:1 allowed the isolation of a group of related alkaloids, principally camptothecin (Figure 2), as the active compounds (14). Camptothecin had been characterized as a potent antitumor compound that showed promise early in the NCI program but was later not advanced in the clinical studies (15). Camptothecin was applied to tobacco, bean and corn plants as emulsified sprays. Inhibition of tobacco bud and corn shoot growth but not bean meristems was found (14). This apparent selectivity of inhibition observed in the initial plant growth regulating assay studies was in contrast to non-selective phytotoxicity observed with applications of the natural products, harringtonolide and linoleic acid, discussed earlier.

The majority of the most recent efforts spent in this study of plant growth regulating natural products has been directed toward understanding more about camptothecin and its possible agricultural applications. Camptothecin was found to be very effective as an inhibitor of the sprouting of potatoes when applied as a 5×10^{-4}M spray. Weight loss of the tubers in storage was reduced and, even after 4 months at 15 C, no symptoms of internal sprouting or other deleterious effects were evident (16). Similarly, applications of the compound were found to increase the storage life of radishes (17).

Camptothecin (50-500 μM) generally slowed the rate of germination of seeds of crop plants tested and inhibited seedling growth of lettuce and radish but stimulated that of watermelon (15). These results prompted an expanded study of seed germination and seedling growth of a broader spectrum of plants as to the effect of camptothecin treatments.

Isolation of Camptothecin. For a large scale preparation, camptothecin was extracted from the stemwood of Camptotheca acuminata Dec. with hot EtOH. After removal of the solvent, the residue was dissolved in $EtOAc-CHCl_3$ (1:3) and fractionated on a Bio Beads S-X1 gel permeation column with the same solvent. The purity of the preparation was examined by reverse phase HPLC with MeOH: 3% HOAc (1:4). The camptothecin isolated contained 8% of the 10-methoxycamptothecin homologue and this mixture caused the same biological activity as camptothecin itself. Therefore the camptothecin alkaloids were used without further purification. Camptothecin was converted to the K salt form with 1N KOH and 25, 50, 100, 250 and 500 uM solutions were prepared.

Seedling Growth Assays With Camptothecin. Eight monocot and nine dicot seed species were tested for effects of camptothecin on germination and seedling growth (Table II). Small seeds were sown

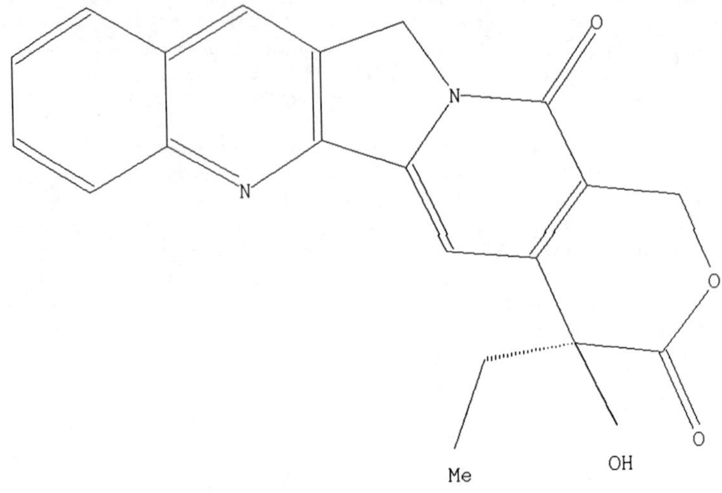

Figure 2. Camptothecin.

Table II. Seed species treated with Camptothecin

Common Name	Class	Species
	Monocot	
Annual ryegrass		*Lolium multiflorum* Lam.
Barley		*Hordeum vulgare* L.
Corn		*Zea mays* L.
Kentucky bluegrass		*Poa pratensis* L.
Perennial ryegrass		*Lolium perenne* L.
Sorghum		*Sorghum bicolor* L.
Tall fescue		*Festuca arundinacea* Schreb.
Wheat		*Triticum aestivum* L.
	Dicot	
Lettuce		*Lactuca sativa* L.
Mung bean		*Phaseolus aureus* L.
Mustard		*Brassica campestris* L.
Pumpkin		*Cucurbita pepo* L.
Radish		*Raphanus sativus* L.
Soybean		*Glycine max* L.
Tobacco		*Nicotiana tabacum* L.
Tomato		*Lycopersicon esculentum* Mill.

at rates of 16 per Petri dish quadrant and treated with 1.0 ml of camptothecin solution. Larger seeds were sown at rates between 5 and 10 per quadrant and treated with 1.5 to 2.0 ml of camptothecin solution. A minimum of 3 replicates were used for each treatment including controls. All seeds were germinated in a growth chamber at 20-21°C with a photoperiod of 16 hr light-8 hr darkness. The number of days from sowing to measurement ranged from 3 days for radish to 14 days for Kentucky bluegrass. No inhibition of total germination was found for the camptothecin solutions tested. The effects of camptothecin on seedling growth are shown in Figures 3 and 4.

Generally for all species tested, seedling growth was inhibited with highest concentrations of camptothecin (500 µM). However, different degrees of growth inhibition in monocots were found after treatment with the lower concentrations of camptothecin. Corn and sorghum were less inhibited than barley and wheat. Kentucky bluegrass and perennial ryegrass were less inhibited than tall fescue and annual ryegrass. Similarly among the dicot curcubits, watermelon seedling growth was stimulated by low concentrations of camptothecin while that of pumpkin was inhibited. Lettuce was highly inhibited by these concentrations of camptothecin while the very small tobacco seed was not much inhibited. An alteration of pattern of growth occurred with mustard where hypocotyl growth was increased while root growth was inhibited.

Mechanisms of Action of Camptothecin. Some efforts have been made to determine how camptothecin may function in affecting plant growth. A histological examination of axillary buds of tobacco inhibited by camptothecin indicated a selective inhibition of young developing vascular tissues (18). Examination of electron micrographs of inhibited potato sprouts indicated that camptothecin inhibited sprouting by inducing structural changes in vascular tissues and interfering with cell division in meristematic portions of the sprouts (16).

Interactions of camptothecin with plant hormones: gibberellin (GA_3) a growth stimulator; abscisic acid (ABA), a growth inhibitor, and the cytokinin zeatin (Z), a cell division promoter were examined. Camptothecin inhibited the GA_3-induced dark-germination of lettuce and hypocotyl elongation of seedlings as did ABA. However, in contrast to the reversal of ABA-induced inhibition by Z, inhibition of GA_3-induced dark-germination of lettuce seed by camptothecin could not be overcome by Z. Since the inhibition of germination of lettuce seeds is more pronounced at high temperatures (up to 35C) the experiments involving camptothecin were repeated. When Z was added under these conditions, the inhibition of germination by camptothecin was partially overcome (15). These experiments indicated a possible role for camptothecin involving cell division processes in plants.

Many studies of camptothecin as an antitumor compound have been done over the last fifteen years. Camptothecin has potent antitumor activity in a wide range of mammalian tumors (20).

The compound caused reversible DNA fragmentation and inhibited nucleic acid synthesis in mammalian neoplastic cells (21, 22). Because of the potential clinical applications for camptotecin, much research was done to understand the mechanism of action of the compound at the cellular level.

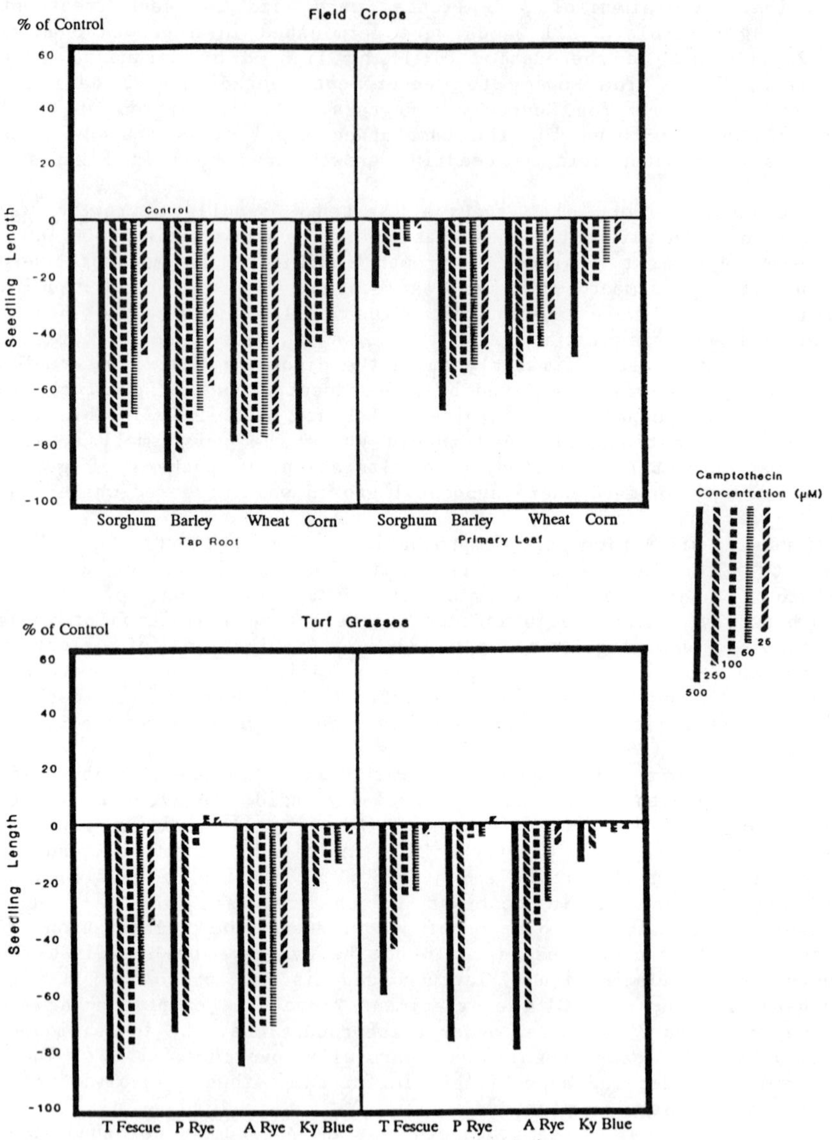

Figure 3. Effects of camptothecin on seedling growth of monocots.

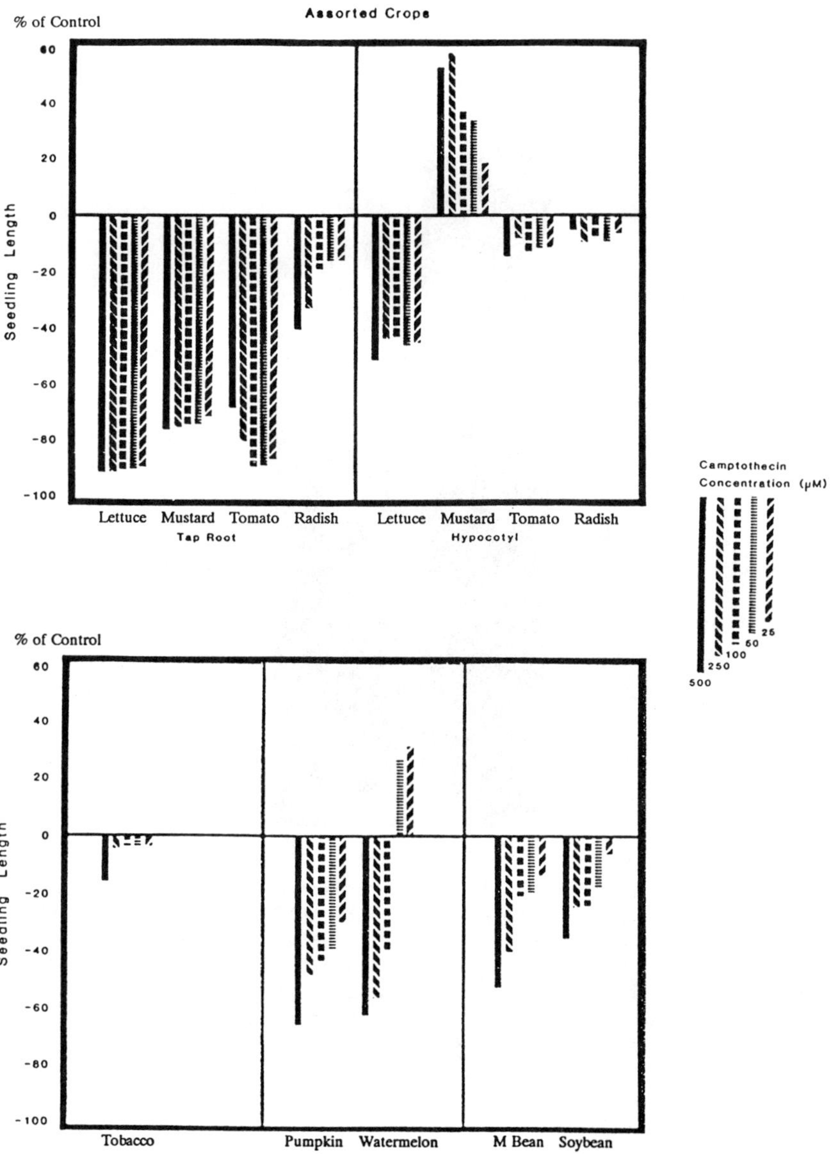

Figure 4. Effects of camptothecin on seedling growth of dicots.

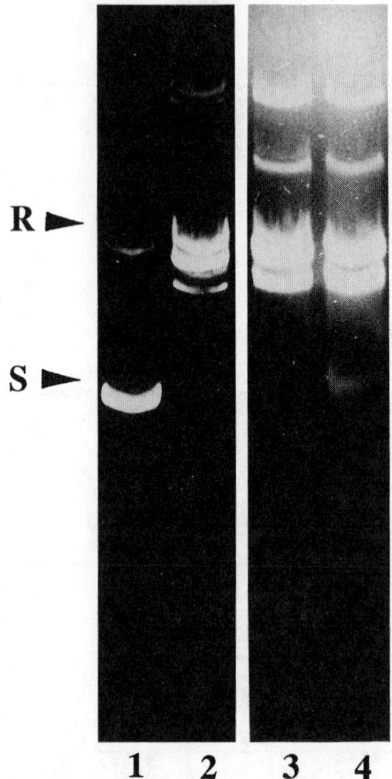

Figure 5. Effect of camptothecin on barley DNA topoisomerase I. (1) pBR322 DNA, no enzymne added. (2) pBR322 DNA + barley DNA topoisomerase. (3) pBR322 DNA + barley topoisomerase + camptothecin, 133 μM. (4) pBR322 DNA + barley topoisomerase + camptothecin, 66 μM. S is supercoiled pBR322 DNA. R is the linear form of pBR322 DNA. Reaction conditions for the agarose gel described in (25).

During the past 3 years several investigations have been performed on the influence of camptothecin on DNA topoisomerases and much progress has been made in understanding the possible cell target sites of this compound. In cells, the duplex DNA has a tertiary structure and this topological arrangement of the DNA can dictate biological functions such as transcription, replication, recombination and repair. DNA topoisomerases are enzymes that regulate the topological state of DNA by breaking and rejoining the DNA phosphodiester bonds by a mechanism involving either a single strand (type I topoisomerase) or a double strand (type II topoisomerase) cleavage (23).

Camptothecin (5 µM) was found to inhibit the relaxation of pBR322 DNA by purified calf thymus DNA topoisomerase I. A detailed analysis of the effect of camptothecin on DNA topoisomerase I *in vitro* showed that the compound induced single strand DNA cleavage. Precisely how camptothecin acts on DNA topoisomerase is not known. However, camptothecin did not affect DNA topoisomerase II activity (24).

The effect of camptothecin on plant topoisomerases has recently been examined. Using a partially purified enzyme from barley seeds (25), a strong inhibition of the relaxation of supercoiled pBR322 DNA by the barley DNA enzyme was observed with 66 µM camptothecin (26) (Figure 5). These studies, *in vitro*, using a partially purified DNA topoisomerase I led to the conclusion that this class of DNA enzymes could be the cellular targets of camptothecin in plants.

The findings concerning the unique DNA regulatory mechanism of camptothecin, as well as possible agricultural applications, illustrate best these studies on the discovery of new and interesting chemicals that are present in higher plants with potential uses as plant growth regulators in agriculture. The major portion of plants having antitumor activity and activity in the *Nicotiana* bioassay have not been studied further and opportunities for finding other new types of plant growth regulators remain attractive.

Acknowledgment

The assistance of David W. Spaulding with seedling growth analysis is appreciated.

Literature Cited

1. Suffness, M.; Douros, J. *J. Nat. Prod.* 1982. 45, 1-14.
2. Douros, J.; Suffness, M. *Cancer Treatment Rev.* 1981. 8, 63-87.
3. Douros, J.; Suffness, M. *Recent Results Cancer Res.* 1981. 76, 153-175.
4. Jarvis, B. B.; Pena, N. B.; Rao, M. M.; Comezoglu, N. S.; Comezoglu, T. F.; Mandava, N. B. "Chemistry of Allelopathy" ACS Sym. Ser. 1985. 268, 149-160.
5. Cutler, H. G.; Jarvis, B. B. *Environ. Exp. Bot.* 1985. 25, 115-128.
6. Komoda, Y.; Isogai, Y. *Sci. Pap. Coll. Gen. Educ. Univ. Tokyo* 1978. 28, 129-134.
7. Martinez-Honduvilla, C. J.; Santos-Ruiz, A. *Planta* 1978. 141, 141-144.

8. Yokota, T.; Baba, J.; Arima, M.; Morita, M.; Takahashi, N. Tennen Yuki Kogobutsu Toronkai Koen Yoshishu 1983. 26, 70-77.
9. Datta, S. C.; Chatterjee, A. K.; Comp. Physiol. Ecol. 1980. 5, 54-59.
10. Smith, C. R.; Mikolajczak, K. L.; Powell, R. G. Med. Chem. Acad. 1980. 16, 391-416.
11. Buta, J. G.; Flippen, J. L.; Lusby, W. R. J. Org. Chem. 1978. 43, 1002-1003.
12. Buta, J. G. J. Nat. Prod. 1983. 46, 755.
13. Powell, R. G.; Smith, C. R. Lloydia 1981. 44, 86-90.
14. Buta, J. G.; Worley, J. F. J. Agr. Food Chem. 1976. 24, 1085-1086.
15. Wall, M. E.; Wani, M. C. Med. Chem. Acad. 1980. 16, 417-436.
16. Tao, K. L. J.; Buta, J. G. Plant Growth Reg. 1986. 4, 219-226.
17. Wang, C. Y.; Buta, J. G.; Moline, H. E.; Hruschka, H. W. J. Am Hort. Sci. 1980. 105, 120-124.
18. Wang, C. Y.; Buta, J. G.; Hruschka, H. W. Hort. Sci. 1980. 15, 72-73.
19. Worley, J. F.; Spaulding, D. W.; Buta, J. G. Tob. Sci. 1979. 23, 43-44.
20. Gallo, R. C.; Whang-Peng, J.; Adamson, R. H. J. Natl. Cancer Inst. 1971. 46, 789-795.
21. Horwitz, M. S.; Horwitz, S. B. Biochem. Biophys. Res. Commun. 1971. 45, 723-727.
22. Spataro, A.; Kessel, D. Biochem. Biophys. Res. Commun. 1972. 48, 643-648.
23. Wang, J. C. Annu. Rev. Biochem. 1985. 54, 665-697.
24. Hsiang, Y.-H.; Hertzberg, R.; Hecht, S.; Liu, L. F. J. Biol. Chem. 1985. 260, 14873-14878.
25. Kalinski, A.; Chandra, G. R.; Muthukrishnan, S. Plant Physiol. 1987, S111.
26. Kalinski, A.; Buta, J. G.; Unpublished results, 1987.

RECEIVED April 1, 1988

Chapter 20

Natural Products in the Search for New Agrochemicals

John H. Cardellina, II

Natural Products Laboratory, Department of Chemistry, Montana State University, Bozeman, MT 59717

An investigation of the hypothesis that marine algae and invertebrates produce growth regulators and feeding deterrents which might be active against terrestrial plants and insects has been expanded to include studies of terrestrial plants, marine and terrestrial microorganisms and marine invertebrates in a search for leads to new agrochemical agents. New field (brine shrimp toxicity) and laboratory (phytotoxicity) screens have been incorporated and novel, bioactive compounds have been isolated. Results of this young project clearly point to marine organisms and plant pathogens of weeds as very promising sources of unique chemical entities to deal with agricultural problems.

Farmers annually endure significant crop losses from the combined effects of plant pathogenic bacteria and fungi, insects and weeds. Ranchers endure the analogous problem of weeds encroaching on quality grazing areas. This problem has been aggravated by the phenomenal escalation in cost, in dollars and man-hours, of developing new, more effective chemical remedies for these agricultural problems. Growing concern over the long-term ecological effects of synthetic agrochemicals not only has led to increased costs and development time, but has also resulted in a relatively conservative approach to new product development by the agrochemical industry. Many recently introduced products are merely analogs of compounds already approved for marketing; thus, the discovery of truly novel chemical structures with agrochemical activity would have a dramatic impact on agriculture over the next several decades.

0097–6156/88/0380–0305$06.00/0

The International Union of Pure and Applied Chemistry (IUPAC) has sponsored an International Conference on Chemistry and World Food Supplies (CHEMRAWN II) to address the challenge of increasing world food production two to four fold in the next several decades. Such an increase is deemed necessary to accommodate the needs of the eight billion world inhabitants expected by 2015 (1).

One of the conference objectives was to pinpoint areas for research and development with the greatest potential for meeting world food needs. Among the recommendations set forth by the future action committee at this conference were:

1) high priority for the development of new chemical methods for control of microbial and insect pests;

2) increased focus on the identification and application of new plant growth regulators;

3) emphasis on chemical/biological control of weeds by means of allelochemicals; and

4) formulation of programs to attract and train larger numbers of highly qualified scientists for agriculture-related scientific disciplines.

In a more recent outlook (2) on future agricultural research in the United States, these ideas were reiterated and amplified. Some of the problems cited for focus of future research included:

1) improving human health and nutrition through improvements in quality, quantity and safety of foods;

2) developing new scientific tools;

3) defending plants chemically;

4) controlling plant pathogens;

5) exploiting natural insecticides; and

6) training agriculture-oriented scientists.

Natural products are an attractive source of potential leads to new agrochemicals, not only for the diversity and novelty of chemical structures produced by living organisms, but also for the potential specificity of biological action and the greatly reduced likelihood of harmful bioaccumulation and/or soil and ground water residues. To have reasonable prospects for commercial development, a natural product must be available in abundance from the natural source or amenable to large scale, cost effective synthesis. Alternatively, the natural product might provide a template or assemblage of key functionalities, conformation and stereochemistry which are responsible for the observed activity and can be synthesized in simpler analogs with retention of activity.

With these concepts in mind and with the notion that the marine biosphere might be an untapped reservoir of agrochemically potent compounds, we embarked several years ago on an effort to evaluate the potential of marine natural products as agrochemical agents (3). This operation has since been expanded to include studies of microorganisms and terrestrial plants.

Materials and Methods

This effort utilizes a battery of assays in screening for antimicrobial, plant growth and insect control activity.

The antimicrobial assay consists of the common impregnated disc assay against three fungi (*Rhizoctonia solani*, *Phythium ultimum* and *Helminthosporium sativum*) and two representatives each from the Gram-positive (*Bacillus cereus* and *Corynebacterium michiganensis*) and Gram-negative (*Xanthomonas campestris* and *Pseudomonas aeruginosa*) bacteria. The plant growth and insecticidal assays were described in a previous report (3). The nicked leaf phytotoxicity (4) and brine shrimp toxicity (5) screens have been adapted from literature procedures. The phytotoxicity screen includes tests against johnsongrass (*Sorghum halapense*), spotted knapweed (*Centaurea maculosa*) and leafy spurge (*Euphorbia esula*).

Results

Collection and Screening. Marine specimens were collected in waters near Victoria, British Columbia, Seattle, Washington, and Bermuda. Terrestrial plants were obtained from a number of sites in southwestern Montana. Microbial cultures were obtained from marine invertebrates in Bermuda and infected weeds in Montana. Table I summarizes the screening results on extracts prepared from our collections to date. Water soluble extracts are differentiated from organic solubles; this distinction reveals interesting contrasts, as does a comparison of activities among different classes of organisms.

Some trends can be readily discerned from the data in Table I. Sponges are clearly the most attractive candidates for study at the moment, with the highest levels of activity in all screens. The organic solubles of all organisms exhibit higher levels of insecticidal and antimicrobial activity, while phytotoxicity activity is slightly favored in the aqueous extracts. Brine shrimp toxicity has shown a far better correlation to cytotoxicity (in our anticancer program) than to insecticidal activity.

Antimicrobials. Research in this area has thus far uncovered three very different structural types. The gorgonian genus Briareum is best known as a source of highly functionalized diterpenes (6-9), but the modest antibacterial activity in B. polyanthes from Bermuda was traced to bissetone, **1**, a unique, unprecedented pyranone (10). Bissetone inhibits the growth of Xanthomonas campestris at a moderately high dose (0.6 mg/disc), but its apparent biogenetic precursor, the ketal **2** (11), may be more active. Tests with a limited supply of the unstable **2** will be undertaken as soon as final details of the structure elucidation are clarified.

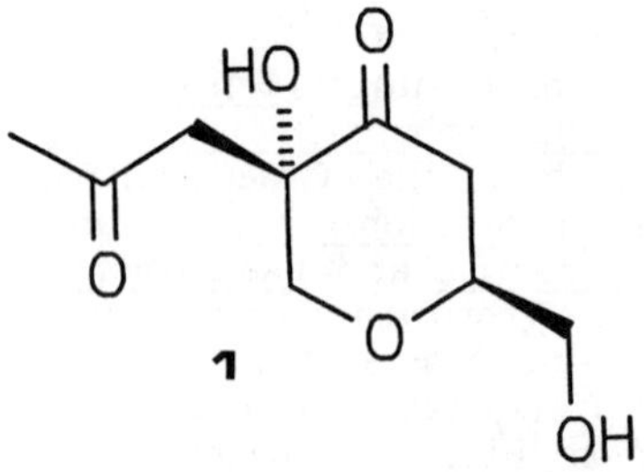

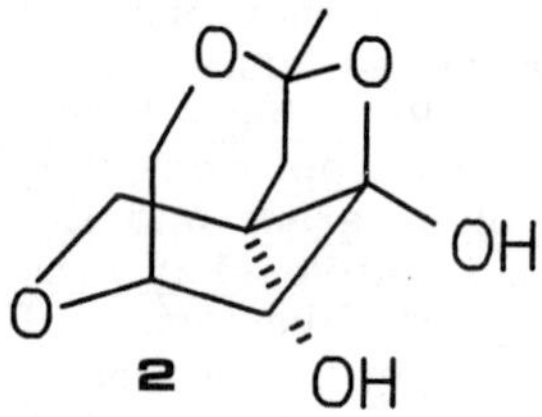

The tunicate Eudistoma olivaceum, collected in Bermuda, has yielded two families of novel ß-carbolines, **3-6** and **7-9** (12). Kobayashi et al. (13) have also isolated **3-6** and reported antimicrobial activity in this group. Quantities of **7-9** sufficient for bioassay were not available, but a synthesis of **9** is underway; once prepared, it will be scrutinized in our in-house assay systems.

Lomatium dissectum, an umbelliferous plant from southwestern Montana, contains substantial quantities of the tetronic acids **10**, which are fairly potent antimicrobials in vitro (14) and highly ichthyotoxic (15). The tetronic acids and the derived monoacetates were quite inhibitory to Gram-positive bacteria, particularly Corynebacterium michiganensis, and fungi, notably Rhizoctonia solani, at doses of 12-50 μg/disk. A variety of analogs or structural modifications of **10** have recently been prepared for biological testing.

Table I. Summary, Agrochemical Screening Program

Organism Type	Extract	AM[a]	PG[b]	PH[c]	BST[d]	I[e]
Terrestrial Plants						
	organic	5/16	1/1	-	2/6	0/1
	aqueous	6/21	2/2	-	0/9	-
Algae						
	organic	3/12	0/1	-	1/6	-
	aqueous	1/16	3/3	-	0/7	0/1
Sponges						
	organic	9/24	-	3/3	7/15	6/14
	aqueous	5/26	8/8	5/7	3/19	1/14
Tunicates						
	organic	6/14	-	1/6	4/7	1/2
	aqueous	1/15	4/4	2/3	0/11	0/4
Other Invertebrates						
	organic	9/12	-	1/2	2/8	2/3
	aqueous	3/15	1/2	0/2	0/8	0/2
Microorganisms		17/102	-	21/70	20/69	-

[a]antimicrobial
[b]plant growth
[c]phytotoxicity
[d]brine shrimp toxicity
[e]insecticidal
[f]indicates none tested thus far

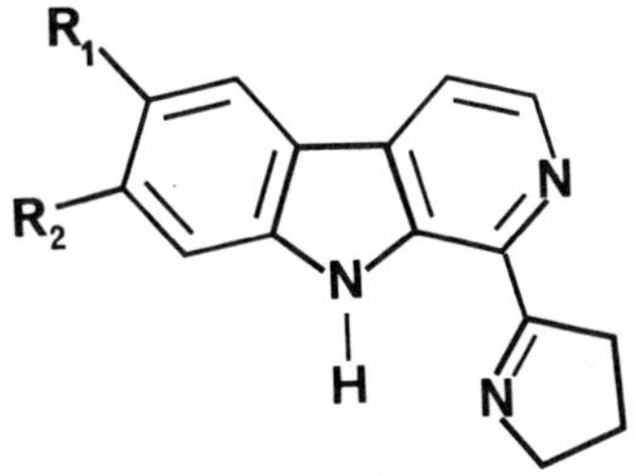

3 R_1 = H; R_2 = Br
4 Br H
5 H H
6 OH Br

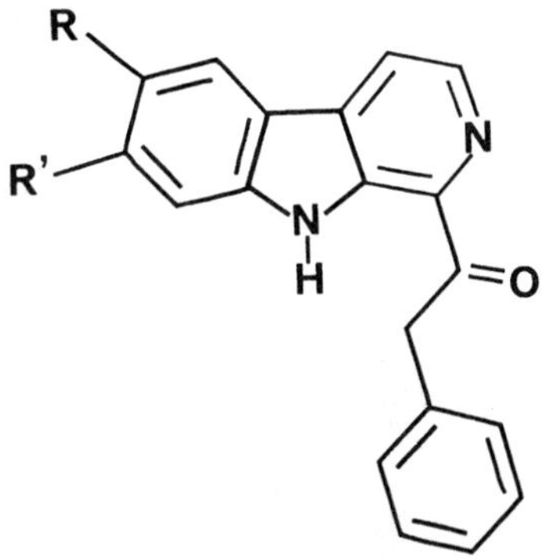

7 R = H, R' = Br
8 R = Br, R' = H
9 R = R' = H

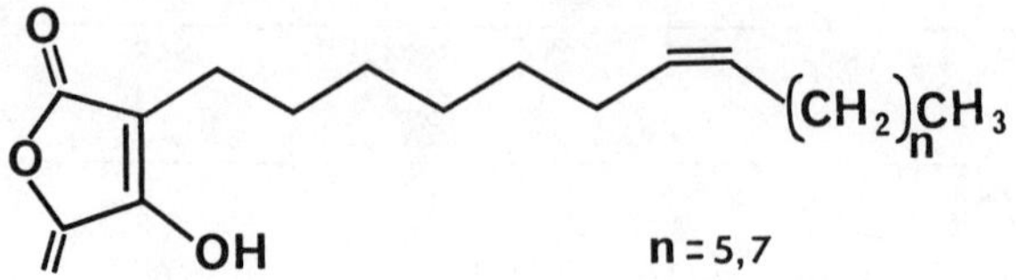

O OH $(CH_2)_nCH_3$ n = 5,7

10

Insecticides. We have now isolated and identified a total of twelve briaran diterpenes from the coelenterates Briareum polyanthes (7-9) and Ptilosarcus gurneyi (16), ten of them novel structures. Five of these compounds have been tested against the tobacco hornworm, Manduca sexta; the preliminary conclusions drawn from those studies are that an electrophilic site in the cyclohexane ring contributes significantly to toxicity (e.g. **11**, **12**), while the briaran class in toto precipitates reductions in weight gain (e.g. **13-15**). A recent recollection of B. polyanthes for reisolation and structure verification of **2**, the presumed precursor of bissetone (**1**), has now made substantial quantities of **11** and **13** available for structure modification and subsequent SAR studies.

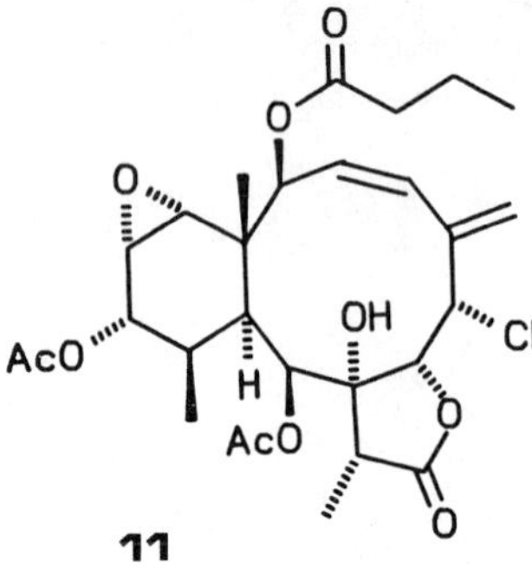

11

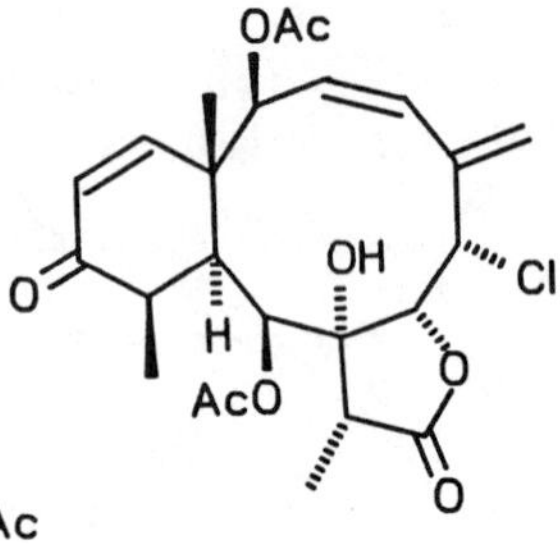

12

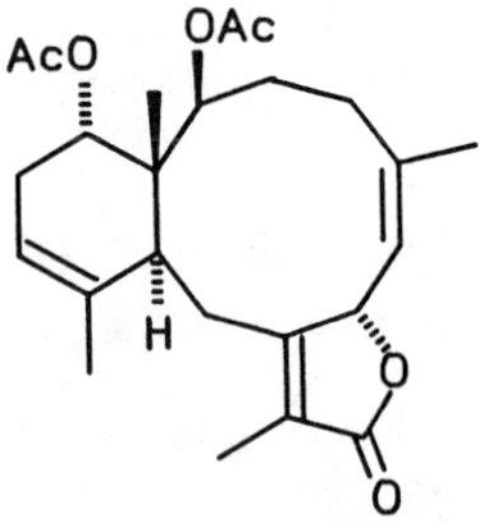

13

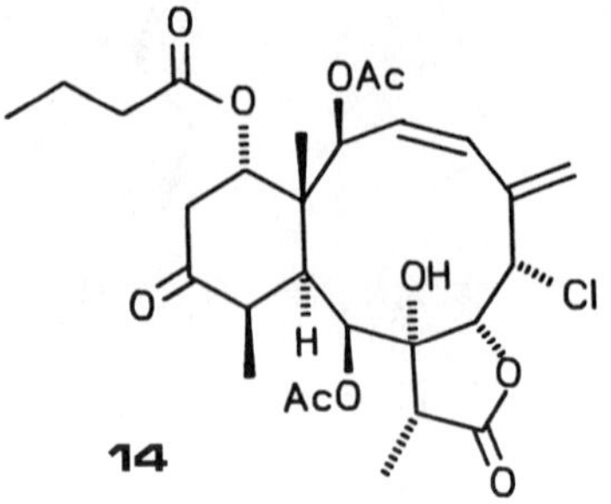

14

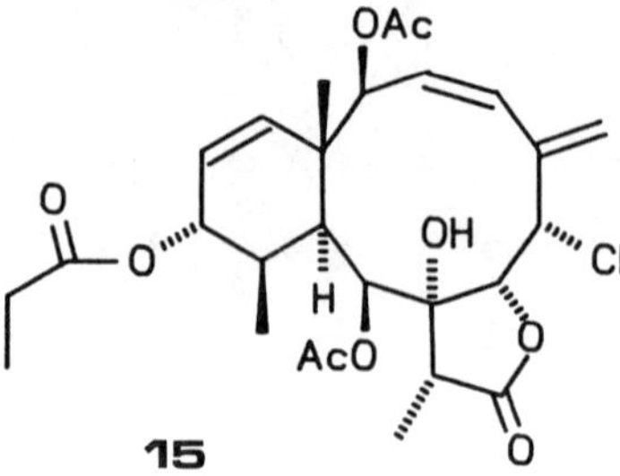

15

Our pursuit of insecticidal activity in the sponge Ulosa ruetzleri led to a pair of phosphonate esters, the creatinine derivative **16** and the hydantoin **17** (17). While **16** is inactive toward M. sexta at 250 ppm, the hydantoin phosphate **17** has an LD_{50} <10 ppm. The creatinine derivative **16** and analogs (**18, 19**) of **17** have been synthesized (17); testing of the hydantoin analogs is underway. Since hydantoins are known to exhibit CNS and anaesthetic activity (18), **17-19** may prove unattractive as leads for new insecticides because of their potential impact on mammalian species. Conversely, they represent a structurally desirable bioactive natural product type, since they are amenable to synthesis and, perhaps, scale up.

Herbicides and Plant Growth Regulators. Extracts of the calcareous green alga Neomeris annulata exhibited phytotoxicity, cytotoxicity and toxicity to brine shrimp, but no activity against the tobacco hornworm. We (19) have isolated from this species the first green algal brominated sesquiterpenes, **20-22**. Of the three, only **22**, neomeranol, is phytotoxic to johnsongrass (Sorghum halapense); the derived ketone **23** is not phytotoxic, however.

Extensive field studies led to our discovery, isolation and culture of a fungal pathogen of spotted knapweed, Centaurea maculosa (20), a critical weed pest on the rangelands of Montana and other northwestern states. The fungus, a variety of Alternaria alternata, produces both host-specific and non-specific phytotoxins. We have isolated and identified three classes of compounds with phytotoxic activity from the fungus--tenuazonic acid, **24**, a well characterized, broad-spectrum phytotoxin (21), a series of perylenequinones and the simple, yet previously unknown diketopiperazine of L-proline and L-tyrosine, **25**. While **24** was toxic to almost all test plants, **25** was active only against C. maculosa out of fifteen weed and crop plants used in the nicked leaf test (4). We have synthesized the diketopiperazine; although it is not exceptionally potent, the specificity, simplicity of structure and unlikelihood of negative impact on mammals and the environment make it an exceedingly attractive lead. Two of the perylenequinones, **26** and **27**, are also phytotoxic (22).

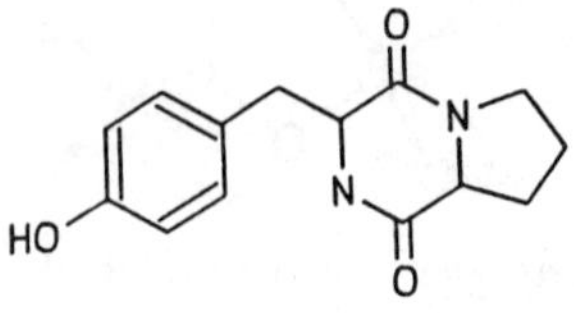

25 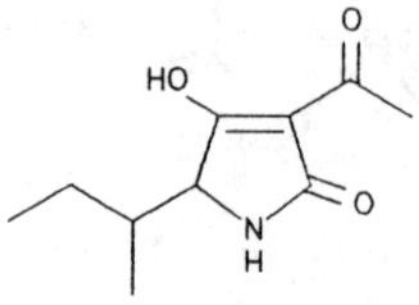 **24**

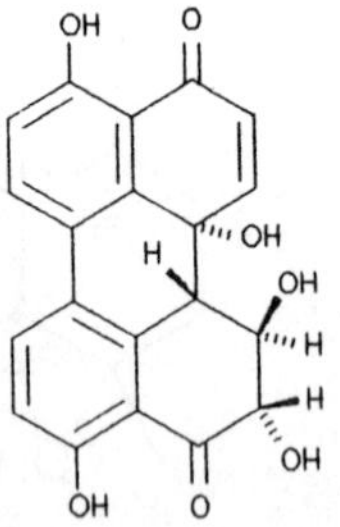

26 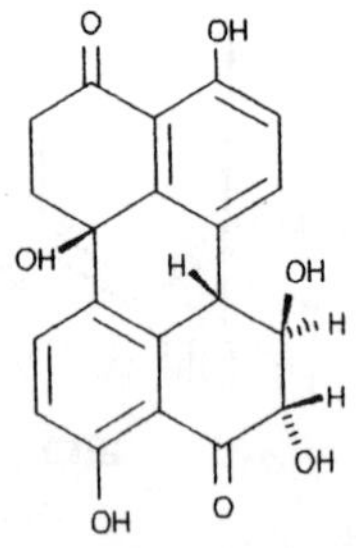 **27**

A new lead just under development is a potent plant growth inhibitor (lettuce assay) from the water solubles of the sponge Spheciospongia othella. These same extracts contain exceptionally high levels of nickel, which appears to be bound by peptides of molecular weight ca. 1400 (23). The plant growth inhibitor is not the nickel complex and is of lower molecular weight, according to size exclusion chromatography experiments.

Discussion and Conclusions

Our continued discovery of agrochemically active compounds from natural sources certainly substantiates the notion that new herbicides, insecticides and antimicrobials can be found not only in terrestrial plants, but in marine organisms and microbes as well. Our screening data for marine microbes suggest that a potential bonanza of new, very active compounds might be discovered there. Our work on knapweed pathogens and Strobel's pioneering work on weed pathogens as sources of phytotoxins (4) would seem to mandate additional attention to those organisms.

Other research groups have begun to consider the agrochemical potential of marine invertebrates. Ireland has recently reported insecticidal activity in an indole derivative (24) and a depsipeptide (25). Crews (26) has isolated the same depsipeptide and found it to have some molluscicidal activity.

Natural products seem more compelling than ever as leads to agrochemicals. The expansion of focus to include microbes and marine organisms, in addition to plants, coupled with recent strides in separation methodology and non-destructive instrumental techniques for structure elucidation, promise to make this field of endeavor an exciting area in the years to come.

Acknowledgments

I thank B. C. VanWagenen for the insecticidal assays, A. C. Stierle for the phytotoxicity assays, R. Dillman for the antimicrobial assays, D. E. Barnekow for brine shrimp assays and M. F. Raub for the plant growth assays. This work was supported by Sea Grant, Department of Commerce, the Montana Agricultural Experiment Station and Rohm and Haas Company. I thank Drs. Colin Swithenbank, Zev Lidert, Jon Mynderse, Stephen Wratten and Gary Strobel for their insight and counsel.

Literature Cited

1. Krieger, J. Chem. & Eng. News, 8 Aug 1983, 23-26.

2. "Research for Tomorrow", Crowley, J. J., ed.; U. S. Department of Agriculture, 1986.

3. Cardellina, J. H., II; Raub, M. F.; VanWagenen, B. C. In "Allelochemicals: Role in Agriculture and Forestry", Waller, G. R., ed.; ACS Symposium Series, No. 330, Washington, D. C., 1987, 562-571.

4. Suguwara, F.; Strobel, G. A.; Fisher, L. E.; Van Duyne, G. D.; Clardy, J. Proc. Natl. Acad. Sci. USA, 1985, **82**, 8291-8294.

5. Meyer, B. N.; Ferrigni, N. R.; Putnam, J. E.; Jacobsen, L. B.; Nichols, D. E.; McLaughlin, J. L. Planta Med., 1982, **45**, 31-34.

6. Burks, J. E.; van der Helm, D.; Chang, C. Y.; Ciereszko, L. S. Acta Crystallogr., 1977, **B33**, 704-709.

7. Grode, S. H.; James, T. R., Jr.; Cardellina, J. H., II Tetrahedron Lett., 1983, **24**, 691-694.

8. Grode, S. H.; James, T. R., Jr.; Cardellina, J. H., II; Onan, K. D. J. Org. Chem., 1983, **48**, 5203-5207.

9. Cardellina, J. H., II; James, T. R., Jr.; Chen, M. M. H.; Clardy, J. J. Org. Chem., 1984, **49**, 3398-3399.

10. Cardellina, J. H., II; Hendrickson, R. L.; Manfredi, K. P.; Strobel, S. A.; Clardy, J. Tetrahedron Lett., 1987, **28**, 727-730.

11. Cardellina, J. H., II; Hendrickson, R. L.; Manfredi, K. P.; Martin, G. E., manuscript in preparation.

12. Kinzer, K. F.; Cardellina, J. H., II Tetrahedron Lett., 1987, **28**, 925-926.

13. Kobayashi, J.; Harbour, G. C.; Gilmore, J.; Rinehart, K. L., Jr. J. Am. Chem. Soc., 1984, **106**, 1526-1528.

14. VanWagenen, B. C.; Cardellina, J. H., II Tetrahedron, 1986, **42**, 1117-1122.

15. VanWagenen, B. C.; Huddleston, J.; Cardellina, J. H., II J. Nat. Prod., in press.

16. Hendrickson, R. L.; Cardellina, J. H., II Tetrahedron, 1986, **42**, 6565-6570.

17. VanWagenen, B. C.; Cardellina, J. H., II, manuscript in preparation.

18. Lis, A. W.; Bijan, R. Physiol. Chem. Phys. 1970, **2**, 293-299.

19. Barnekow, D. E.; Cardellina, J. H., II; Martin, G. E., manuscript in preparation.

20. Stierle, A. C.; Cardellina, J. H., II; Strobel, G. A., Proc. Natl. Acad. Sci. USA, submitted.

21. Nishimura, S.; Kohmoto, K. Annu. Rev. Phytopathol., 1983, **21**, 87-116.

22. Stierle, A. C.; Cardellina, J. H., II; Strobel, G. A., J. Nat. Prod., submitted.

23. West, R. R.; Cardellina, J. H., II, unpublished data.

24. Moriarity, R. M.; Roll, D. M.; Ku, Y.; Nelson, C.; Ireland, C. M. Tetrahedron Lett., 1987, **28**, 749-752.

25. Zabriskie, T. M.; Klocke, J. A.; Ireland, C. M.; Marcus, A. H.; Molinski, T. F.; Faulkner, D. J.; Xu, C.; Clardy, J. J. Am. Chem. Soc., 1986, **108**, 3123-3124.

26. Crews, P.; Manes, L. V.; Boehler, M. Tetrahedron Lett., 1986, **27**, 2797-2800.

RECEIVED June 10, 1988

NATURAL PRODUCTS FROM INSECTS OR AFFECTING INSECTS

Chapter 21

Terpenoids from the Genus *Artemisia* as Potential Pesticides

Stephen O. Duke[1], Rex N. Paul, Jr.[1], and S. Mark Lee[2,3]

[1]Southern Weed Science Laboratory, Agricultural Research Service, U.S. Department of Agriculture, Stoneville, MS 38776
[2]Native Plants, Inc., 417 Wakara Way, Salt Lake City, UT 84108

Members of the higher plant genus *Artemisia* (Compositae) produce a myriad of terpenoid compounds in their glandular trichomes. Many of these compounds, especially the sesquiterpenoid lactones, are biologically active as fungicides, herbicides, antimicrobials, insecticides, and insect antifeedants. These compounds offer significant potential as sources of new pesticides. Two commercial examples of pesticides derived from or chemically related to these naturally-occuring terpenoids are toxaphene (an insecticide and herbicide) and cinmethylin (a herbicide). An antimalarial sesquiterpenoid from annual wormwood (*Artemisia annua*), artemisinin, has recently been found to be a potent phytotoxin. This compound and other *Artemisia*-derived terpenoids are discussed as examples of the considerations in screening natural compounds for use as pesticides.

Secondary products from higher plants represent an enormous diversity of biologically active compounds that have just begun to be exploited as pesticides or as sources of new pesticide chemistries (1,2). Most of the few examples of pesticide development from plant-derived natural products are from terpenoids that can be found in the genus *Artemisia* or closely related genera. The genus *Artemisia* consists of over 400 species and many species of this genus have long been known to produce compounds with potent biological activity. *Artemisia* includes the wormwoods, the sagebrushes, the mugworts, the sageworts, wormseed, and tarragon. Throughout history, this genus has been the source of folk medicines, spices, flavorings, and insect repellents. Currently, the pharmaceutical, food science, and fragrance industries are intensively studying the terpenoids of

[3]Current address: California Department of Food and Agriculture, Chemistry Laboratory Services, 3292 Meadowview Road, Sacramento, CA 95832

0097–6156/88/0380–0318$06.00/0

Artemisia. However, little effort has been expended in determining the potential of the terpenoids of *Artemisia* as pesticides. Recently, a very limited review of drugs, insecticides, and other agents from *Artemisia* was published (3). The present review surveys the terpenoids from *Artemisia*, their biological activities, and their potential for pesticide development.

Biological Significance of Terpenoids of *Artemisia*

Biosynthetic site. The surfaces of the leaves, flower parts, stems, and other aerial portions of the shoots of virtually all species of *Artemisia* investigated are covered with capitate glands filled with resinous oils (Fig. 1a)(*e.g.*, 4-9). In some species, up to 35% of the mature leaf surface is covered with these glands which contain most of the monoterpenes and virtually all of the sesquiterpene lactones of the leaf (4). These glands originate early during leaf development as a basal stalk originating from the epidermis (Figure 1b). As the basal stalk cells mature, the cuticle that covers the entire basal stalk splits away from the cell walls of the stalk cells (Figure 1c) and the subcuticular space fills with terpenoids. The resulting bilobed sac (Figure 1d) obscures the basal stalk cells which are likely to be the source of the terpenoids in the sac. As glands mature on older leaves, the sacs rupture (Fig. 1e), releasing resinous oils over the surface of the leaf.

Based on their biological activity, the function of the terpenoids found in these glands is almost certainly protective. Since many of these compounds are (10), or are likely to be, phytotoxic to the producing plant, sequestration of the compounds in glands allows for concentration of high levels of these toxins to form a chemical barrier to potential herbivores and pathogens without poisoning the plant.

In addition to glands, there is some evidence that terpenoids also accumulate in specialized cells of roots and shoots of some *Artemisia* species. For instance, roots of tarragon (*A. dracunculus* L.) contain resin-containing canals formed from cells in the cortex that accumulate high concentrations of terpenoids (11). There is some evidence to indicate that the terpenoid compositions of the oils from glands and resin canals are different (12).

Terpenoids produced by *Artemisia*. There is tremendous diversity in the terpenoids of *Artemisia* so that an encyclopedic listing such as that provided by Ahmad *et al.* (13) is not possible in a limited review such as this. However, Table I provides a partial listing of some of the terpenoids of three species of *Artemisia*. Many compounds were unidentified in the studies from which the data in Table I were obtained. The compounds in Table I are mostly monoterpenoids (10 carbon). Many of the more biologically active terpenoids from *Artemisia* are sesquiterpenoids (15 carbon) or triterpenoids (30 carbon). The sesquiterpenoid lactones are especially active and some species of *Artemisia* contain up to 3.5 % of their dry

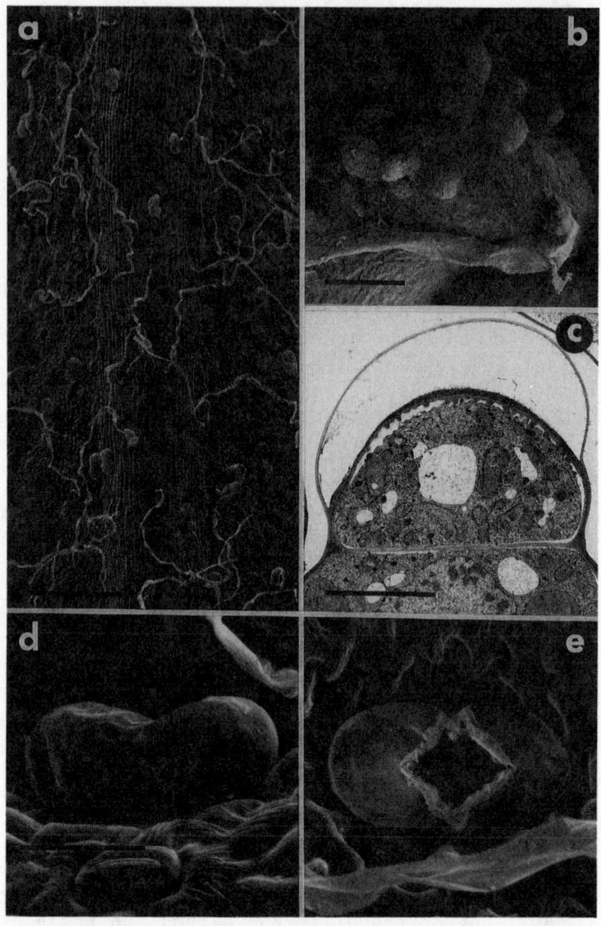

Figure 1. a. Scanning electron micrograph of a mature leaf of annual wormwood (*A. annua*). b. The apical meristem of an annual wormwood plant, including leaf primordia. Note basal cells of glands forming early during leaf morphogenesis. c. Tramsmission electron micrograph of cuticle separating from the cells of the basal stalk. d. Scanning electron micrograph of a mature capitate gland on the leaf of an annual wormwood plant. e. Capitate gland after rupture. Bars = 200, 25, 5, and 25 μm in a, b, c, and d, respectively. Magnification for e is equal to that for d.

weight as sesquiterpenoid lactones (4). There are biologically active diterpenoids (20 carbon) that are potent phytoalexins (17), however, few have been identified in *Artemisia*.

Table I. A Partial Listing of Terpenoids Identified from Essential Oils from Three *Artemisia* Species as Determined by Gas Chromatography[a]

Terpenoid	*A. vulgaris*	*A. douglasiana*	*A. nilagiraca*
α-pinene	+	+	+
camphene	+	+	+
β-pinene	+	+	+
sabinene	-	+	+
α-thujene	-	-	+
Δ-carene	-	+	-
myrcene	+	+	+
α-terpinene	+	-	+
p-cymene	+	+	+
1,8-cineole	+	+	+
γ-terpinene	-	+	+
limonene	+	-	+
terpinolene	-	-	+
ocimene	-	+	+
artemisia ketone	+	+	-
thujone	+	+	+
linalool	+	+	+
α-gurjunene	-	-	+
thujyl alcohol	+	-	-
linalyl acetate	-	-	+
artemisia alcohol	+	+	+
ambrosial	-	+	-
camphor	+	+	+
sabinene hydrate	-	-	+
isoborneol	+	+	+
caryophyllene	-	+	+
pinocarvacrol	-	-	+
phellandral	-	-	+
borneol	+	+	+
terpinen-4-ol	+	+	+
α-terpineol	-	+	+
isobornyl and bornyl acetate	+	-	+
β-bisabolene	-	-	+
δ-cadinene	-	-	+
carveol	-	-	+
isopinocamphone	-	-	+
d-pulegone	-	-	+
α-farnesene	-	-	+
nerolidol	-	-	+
eudesmol	-	-	+

[a]from references 14-16.

Chemical ecology of *Artemisia* terpenoids. Some of the best examples of chemical defenses by plants are in the genus *Artemisia*. For instance, some of the earliest documented cases of allelopathy were those of *A. absinthium* (18,19), *A. vulgaris* (20), and *A. californica* (21). The volatile terpenoids from these species strongly inhibit the growth of other plant species in their vicinity. More recently, Friedman (22) found *A. herba-alba* to also strongly inhibit the growth of plant competitors. The strong insect repellent, antimicrobial, flavoring (spice), and antihelminthic properties of many of the terpenoids from *Artemisia* indicate that these compounds provide strong protection against pathogens, nematodes, and herbivores. Although little literature exists documenting and confirming these functions in nature, considerable circumstantial evidence with isolated compounds exists to support this view.

Terpenoids from *Artemisia* with Known Pesticidal Properties

Two extensive reviews of the biological properties of terpenoids and sesquiterpenoids from plants were recently published (7,23). Many of the compounds discussed in these reviews are constituents of *Artemisia*.

Phytotoxins. Numerous terpenoids from *Artemisia* are strong phytotoxins (Table II). Asplund (27) found several terpenoids to be more powerful germination inhibitors than HCN. Only 5.8 μM HCN was required to inhibit germination 50 %, whereas 1.5 and 3.3 μM pulegone and camphor were equally effective. Two commercial herbicides are based on the chemistries of monoterpenes from Table II.

Although developed as an insecticide, chlorinated camphene (sold as Toxaphene or Camphechlor) was used as a herbicide in legume crops (*e.g.*,29) for about 10 years before it was removed from sale. The commercial product consisted of a mixture of all (>200) chlorinated forms of the molecule. The active herbicidal principal was never identified. The product was generally applied at about 2.2 kg/ha for control of sicklepod (*Cassia obtusifolia*) in soybeans. As discussed below under insecticides, identification of the phytotoxic principle may have resulted in a new product with superior selectivity and fewer toxicological problems.

The new herbicide cinmethylin (30,31) is a close analog of 1,8-cineole. It is a potent growth inhibitor with activity similar to

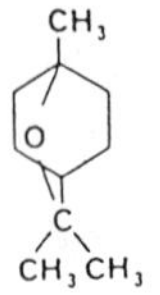

CH_3 / O / C / CH_3 CH_3

1,8-Cineole

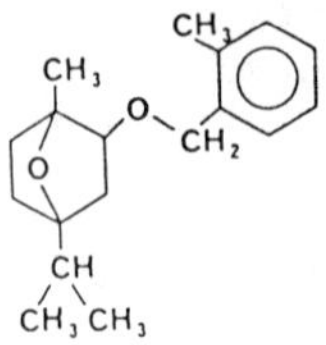

CH_3 / O / CH_2 / CH_3 / O / CH / CH_3 CH_3

Cinmethylin

artemisinin (10). Cinmethylin appears to act by stopping entry of cells into mitosis (32) - the type of effect one would expect from a metabolic inhibitor.

As discussed in more detail below, the sesquiterpenoid lactones are especially biologically active. This subject has recently been reviewed comprehensively by Fisher (33).

Table II Herbicidal Properties of Terpenoids of *Artemisia*

Terpenoid	Conc. (μM)	Assay	Effect (% inhibition)	Ref.
Achillin	100	Root growth of cucumber	35	(24)
Artemisinin	33	Root growth of lettuce, pigweed, purslane, and velvetleaf	50	(10)
Borneol	470	Lettuce seed germination	50	(25)
	sat. at.[a]	Madia sativa germination	100	(26)
Camphor	180	Lettuce seed germination	50	(24)
	3.3	Radish seed germination	50	(27)
	sat. at.	Madia sativa germination	100	(26)
	sat. at.	Root growth of barley	90	(28)
1,8-Cineole	78	Radish seed germination	50	(27)
	sat. at.	Madia sativa germinaton	100	(26)
	sat. at.	Root growth of barley	25	(28)
p-Cymene	51	Radish seed germination	50	(27)
Limonene	45	Radish seed germination	50	(27)
Pinene	30	Radish seed germination	50	(27)
Pulegone	1.5	Radish seed germination	50	(27)
Terpinen-4-ol	140	Lettuce seed germination	50	(25)
	sat. at.	Root growth of barley	45	(28)
Thujone	22	Lettuce seed germination	50	(25)
Viscidulin-C	100	Root growth of cucumber	30	(24)

[a]sat. at. = saturated atmosphere

Insecticidal and antifeedant compounds. A tremendous amount of literature exists on the role of terpenoids in plant-insect interactions. The role of terpenoids produced by plant epidermal glands and trichomes in insect resistance was recently reviewed in detail (34). The present review will deal with only a few representative examples and those that relate most closely to *Artemisia* terpenoids.

The most important group of insecticides derived from natural products are the pyrethroids. This insecticide class was derived from the chemistries of insecticidally active terpenoids from various *Chrysanthemum* species (35). These compounds include pyrethrins, cinerins, and jasmolins. The genera *Chrysanthemum* and *Artemisia* are both members of the Compositae and the capitate glandular development in the two genera are almost identical (36).

From the paucity of literature, very little effort apparently has been made to screen *Artemisia* for insecticides and insect antifeedants. Chemical extracts of *A. cana* cause high mosquito larvae mortality in ponds (37), however, the active principle has not been identified.

Chlorinated camphor (toxaphene) was used for a number of years as an insecticide and at one time it topped all other insecticides in sale in the U.S. (35) before its use was limited by the Environmental Protection Agency. Several specific forms of chlorinated camphor were found to be much more highly insecticidal than others (38,39). The approach of halogenating a terpenoid, producing a mixture of many compounds, and then separating and characterizing the active principals could be a valid strategy for discovering new pesticides.

Several terpenoids found in *Artemisia* have been associated with insect repellent or antifeedant activities of plant extracts. Numerous terpenoids of *A. vulgaris* were found to be mosquito repellents (14). The most potent repellent was terpenen-4-ol. Several monoterpenoids found in *Artemisia*, such as α-pinene (40), have insect repellent properties toward other insect species. However, some of these more volatile compounds also act as insect attractants (41). Sesquiterpenoids are known for their bitter taste and, thus might be expected to act as antifeedants. For instance, the sesquiterpenoid caryophyllene has aphid repellent activity (42). The epoxide of caryophyllene inhibits *Heliothis virescens* larvae growth by about 60% when incorporated at about 5 mM in their diet (43). Several terpenoids of *A. capillaris*, including caryophyllene, ar-curcumine, bornyl acetate, and γ-terpinene, had antifeedant activity against cabbage butterfly larvae (44).

Antimicrobial and antifungal compounds. Saturated atmospheres of many volatile terpenoids inhibit fungal growth (*e.g.*, Table III). Virtually all published studies of the antifungal activities

Table III Antifungal Properties of Terpenoids of *Artemisia*

Terpenoid	Conc.	Assay	Effect (% inhibition)	Ref.
Carene	sat. at.[a]	Boletus variegatus growth	86	(45)
		Ceratocytis pilifera growth	64	(46)
p-Cymene	sat. at.	Trichoderma virida growth	90	(47)
Limonene	sat. at.	Lenzites saepiaria growth	95	(47)
		B. variegatus growth	85	(45)
		T. virida growth	83	(47)
Myrcene	sat. at.	C. pilifera growth	68	(46)
		T. virida growth	81	(47)
Pinene	sat. at.	B. variegatus growth	86	(45)
		C. pilifera growth	72	(46)
		T. virida growth	84	(47)
Terpinolene	sat. at.	B. variegatus growth	77	(45)
		Rhizopogon roseolus growth	65	(45)

[a]sat. at. = saturated atmosphere

of these compounds are on fungi that are saprophytic on wood or, as pathogens or mycorhizae, infect woody plants that produce the compounds. One might expect these compounds to be more effective against fungal pathogens which did not coevolve with a host plant species that produces large amounts of terpenoids.

Nematicidal activity. Extracts of *A. siversiana* were reported to cause high mortality of *Meloidogyne incognita* and *Rotylenchulus reniformis* (48). However, the active principle was not identified.

Artemisinin as an Example of a Potential Pesticide from Wormwood

Artemisinin is a sesquiterpenoid lactone endoperoxide (Figure 2) found in high concentrations in annual wormwood (*A. annua*) (49). It is the active principle in preparations that have long been used in Chinese folk medicine as an antimalarial drug (50). Now there is great interest by pharmacologists in using it in modern medicine. Our laboratory (10) and another (51) recently found this compound to be a potent plant growth inhibitor with potential as a herbicide.

Artemisinin reduced growth of the root of lettuce and several weed species by about 50% at 33 µM (Table II). Similar results were obtained by Stevens and Merrill (52) on lettuce root with several plant-derived sesquiterpenoid lactones. They found 80 ppm (ca. 30 µM) centaurepensin, repin, and solstitiolide to reduce growth of lettuce roots by about 50%. Acroptilin was less effective as a growth inhibitor. They found subphytotoxic levels of these compounds to stimulate growth, whereas we found very low levels of artemisinin to stimulate germination.

Fischer and Quijano (53) found very little effect of 12 sesquiterpenoid lactones on germination of 12 plant species. Few herbicides significantly influence germination, whereas many of them inhibit growth. Therefore, their results provide little evidence of the potential of these compounds as herbicides.

Our objectives in the original work described here were to further determine structure-activity relationships for artemisinin analogues and to provide additional information on the mechanism of action of artemisinin as a phytotoxin.

Materials and Methods. The lettuce seedling bioassay was conducted as before (10). High purity epimers of artemisinin ether (arteether) were synthesized. Artemisitene and artemisinin degradation products were purified by HPLC.

Effects of Arteethers, Artemisitene, and Breakdown Products of Artemisinin on Lettuce Seedling Growth. Some artemisinin ethers are more active against some strains of *Plasmodium falciparum* than artemisinin (54). However, as a phytotoxin against lettuce, neither the alpha nor the beta epimer of arteether (Fig. 2) was more active than artemisinin (Figure 3).

Many sesquiterpenoids from plants have either a α-methylene-γ-lactone moiety or a cyclopentenone group (33). Either of these will readily react with thiol groups to form a

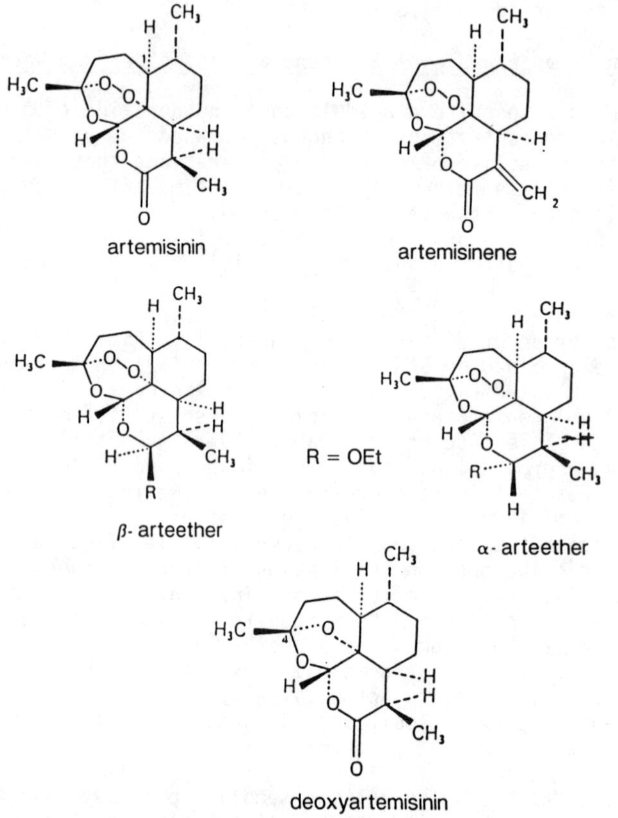

artemisinin

artemisinene

R = OEt

β- arteether

α- arteether

deoxyartemisinin

Figure 2. Structures of artemisinin and related compounds tested for herbicidal activity.

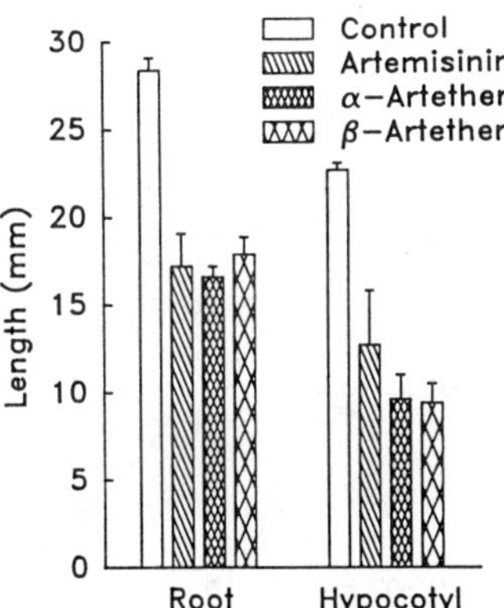

Figure 3. Comparison of the effects of the two epimers of artemether with artemisinin on the growth of roots and hypocotyls of lettuce. All chemicals were at 33 μM.

covalent linkage. If the thiol group is on an enzyme, this reaction could inhibit or inactivate the enzyme. Theoretically, a reaction of the thiol group of free cysteine should remove the biological activity of such compounds. In fact, cysteine has been successfully used to antidote the toxic effects of sesquiterpenoids on mammals (55). We tested the effect of cysteine on the phytotoxcity of artemisinin. At 3.3 mM, it completely prevented the growth-inhibiting effect of 33 μM artemisinin (Fig. 4). In experiments in which the seeds were imbibed for one day in either water, 3.3 mM cysteine, or 33 μM artemisinin and then transferred for two days to one of these three treatments (9 different treatments), it was found that: (1) artemisinin has little effect on seedling growth when administered after a day of imbibition and germination without the presence of artemisinin; (2) the seeds must be in the presence of artemisinin for more than one day in order for significant inhibition of hypocotyl growth to occur; and (3) cysteine has little effect on the inhibitory effect of artemisinin when applied after the artemisinin (Fig. 5).

These data indicated that cysteine might act directly with artemisinin. Two possibilities for a direct interaction were examined. The first was that the phytotoxicity of artemisinin might be due to contaminating artemisitene, which has the α-methylene-γ-lactone moiety. Artemisitene is a natural constituent of *A. annua* with 4 to 5 fold less activity against *Plasmodium* than artemisinin (56). We found artemisitene and artemisinin to have virtually the same growth-inhibiting activities (data not shown). Thus, the effects of our artemisinin preparations on plant growth were not due to contamination with highly active artemisitene.

The second possibility is that the endoperoxide of artemisinin is reduced to the epoxide (deoxyartemisinin) by cysteine. We tested the phytotoxicity of deoxyartemisinin and four thermal degradation products of artemisinin. At 50 μM, only one degradation product (an unknown) had any activity and its activity was no greater than that of artemisinin (data not shown). These data suggest that the activity of artemisinin is lost by decomposition. Artemisinin can thermally degrade in solution (57). However, GC/MS analysis of 33 μM artemisinin at 25°C for 5 days revealed that no degradation took place. Furthermore, inclusion of 3.3 mM cysteine had no effect on artemisinin concentration during the 5 day period. Therefore, the reduction of artemisinin activity by cysteine is not due to reduction of the endoperoxide to form deoxyartemisinin and the mode of action of artemisinin is unlikely to have anything to do with direct interactions with sulfhydryl groups of proteins.

To summarize, our recent studies have found equal or less phytotoxic activity with α- or β-arteethers, artemisitene, or the thermal decomposition products of artemisinin than with artemisinin. Furthermore, these results confirm earlier structure-activity conclusions (10), that the endoperoxide is a requirement for phytotoxicity.

Considerations in Development of Terpenoids as Pesticides.

As discussed in detail previously (1,2,58), the development of a

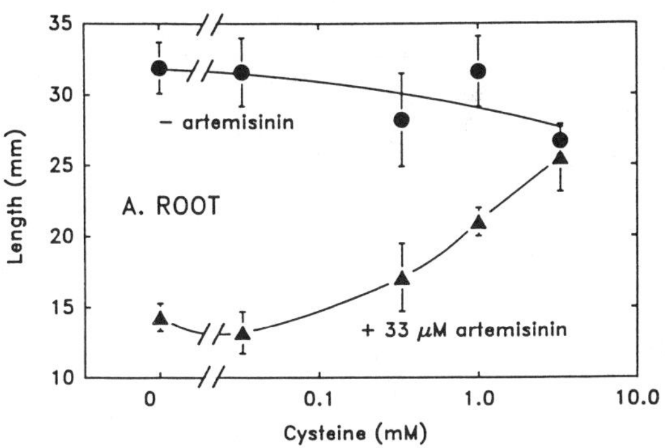

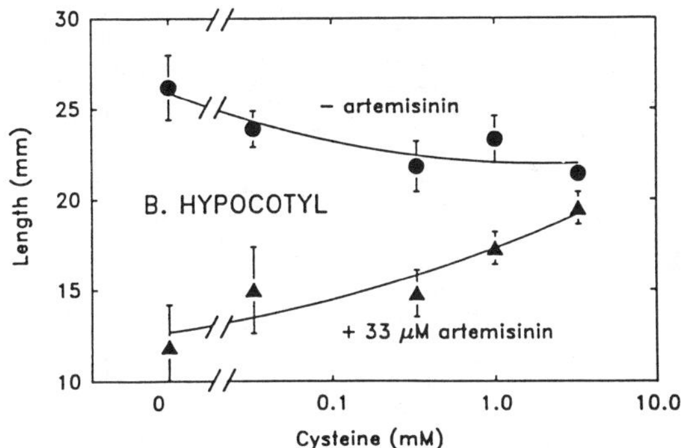

Figure 4. Effects of different concentrations of cysteine on the growth-inhibiting effects of 33 μM artemisinin on growth of lettuce seedlings after 3 days. The lettuce seeds were germinated for 3 days in solutions containing artemisinin with or without cysteine.

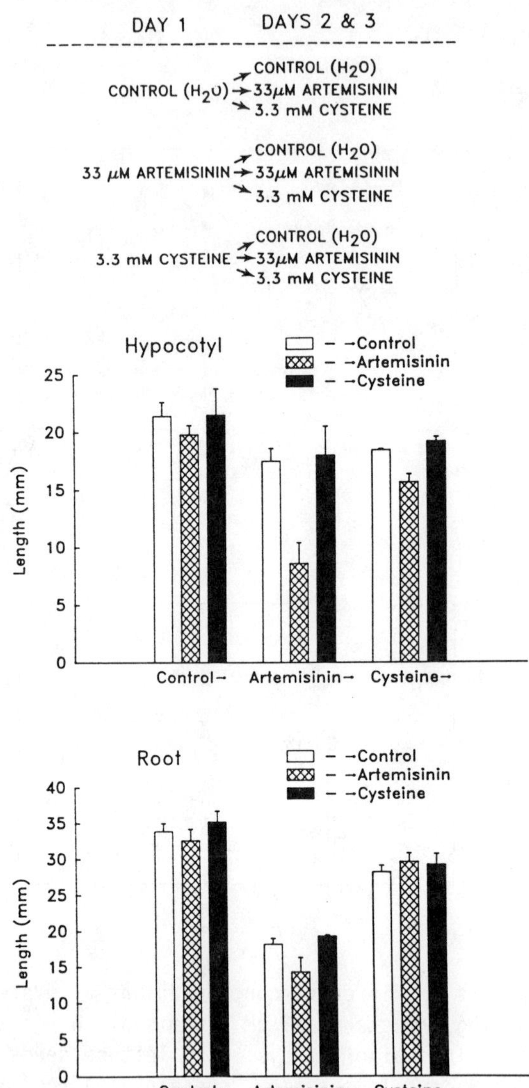

Figure 5. Effects of imbibing lettuce seeds for 1 day in water, 33 μM artemisinin, or 3.3 mM cysteine and then transferring to one of these treatments for 2 days of subsequent growth (see inset for experimental design).

pesticide from a natural product is a tremendously complicated venture. As with any pesticide, the efficacy, selectivity, potential market niche, and environmental properties must be considered. For instance, one consideration is the stability and persistence of a potential pesticide in the environment. It must persist long enough to be effective but not so long that it causes long term residue problems. Picman (59) recently found that the sesquiterpenoid lactone isoalantolactone disappeared from soil in 90 days. Its initial rate of disappearance was greater in organic than in mineral soils. This type of evaluation does not differ between synthetic and natural product-derived compounds.

With natural products, however, the beginning phase and the final phases of development are more complicated than traditional pesticide discovery strategies. In synthetic pesticide discovery programs, there is relatively little problem in conducting structure-activity studies. However, structure determination of a natural product can be a difficult task and chemical synthesis of the natural product and its analogues can be a major undertaking. Extremely small yields often make microbioassays desirable.

Later, the cost of chemical synthesis of a natural product with otherwise commercially attractive features may be prohibitive. However, biosynthesis may be a reasonable alternative. One biosynthetically derived herbicide, the microbial product bialophos, has already been marketed (2). With a plant product, such as a terpenoid, biosynthesis of the compound by intact plants in the field or greenhouse is highly unlikely to be a viable method of production unless the compound is highly efficaceous at low application rates and the production can be manipulated genetically or with growth regulators. The production of terpenoids can often be dramatically increased by certain types of stress. For instance, sublethal levels of the herbicide acifluorfen induce greatly increased biosynthesis of certain terpenoids in higher plants (60). Growth regulators are being explored to increase production of artemisinin by *A. annua*. Daminozide and chlormequat had little or no effect on production (61), whereas acifluorfen increased production by at least twofold (62).

An alternative to field production of the compound is production with tissue culture or cell cultures. Recent advancements in enhancing production of secondary plant products in plant tissue cultures and immobilized plant cell systems might be employed to make this choice viable. Already, efforts are being made to enhance production of artemisinin in *A. annua* shoot cultures (63). In these studies, sterol synthesis inhibitors greatly increased synthesis of artemisinin. Advancements in biotechnology and molecular biology may tip the balance toward biosynthesis of useful natural products in tissue or cell cultures.

Conclusions

Pharmacognosy has recognized the tremendous potential of the terpenoids of the genus *Artemisia* as a source of new pharmaceuticals. The biological activity of these compounds is not restricted to mammalian pathogens and, thus, is also a virtually untapped reservoir of potential pesticides and chemical bases for new pesticides. Two pesticides, chlorinated camphene and cinmethylin, have been developed from two of the simpler terpenes occuring in *Artemisia* as well as many other plant species. The biological activity of the large numbers of sesquiterpenoids of *Artemisia* indicate that they may be a lucrative source of pesticides. Synthesis of analogues and structure activity studies with these compounds might expand their potential usefulness.

Acknowledgments

We thank Ed Croom, Jr. and Hala N. Elsohly of the University of Mississippi for providing samples of arteether. The technical assistance of Al Lane was invaluable in the experiments reported here.

Literature Cited

1. Duke, S. O. Rev. Weed Sci. 1986, 2, 15-44.
2. Duke, S. O.; Lydon, J. Weed Technol. 1987, 1, 122-128.
3. Sherif, A.; Hall, R. G.; El-Amamy, M. Med. Hypotheses 1987, 23, 187-93.
4. Kelsey, R. G.; Shafizadeh, F. Biochem. Syst. Ecol. 1980, 8, 371-7.
5. Slone, J. H.; Kelsey, R. G. Amer. J. Bot. 1985, 72, 1445-51.
6. Cappelletti, E. M.; Caniato, R.; Appendino, G. Biochem. Syst. Ecol. 1986, 14, 183-99.
7. Kelsey, R. G.; Reynolds, G. W.; Rodriguez, E. In Biology and Chemistry of Trichomes; Rodriguez, E.; Healey, P. L.; Mehta, I., Eds.; Plenum: New York, 1984; pp 187-241.
8. Ascensao, L.; Pais, M. S. S. Bot. Gaz. 1987, 148, 221-7.
9. Eisner, T.; Eisner, M.; Meinwald, J. J. Chem. Ecol. 1987, 13, 943-6.
10. Duke, S. O.; Vaughn, K. C.; Croom, E. M., Jr.; Elsohly, H. N. Weed Sci. 1987, 35, 499-505.
11. Lassanyi, Z.; Pal, A. Acta Agron. Acad. Sci. Hung. 1984, 33, 377-83.
12. Lassanyi, Z.; Stieber, G. Acta Agron. Acad. Sci. Hung. 1976, 25, 269-80.
13. Ahmad, Y.; Hasan, S. K.; Sherwani, N. K. J. Chem. Soc. Pak. 1986, 8, 277-96.
14. Hwang, Y.-S.; Wu, K.-H.; Kumamoto, J.; Axelrod, H.; Mulla, M. S. J. Chem. Ecol. 1985, 11, 1297-1306.
15. Scora, R. W.; Kumanoto, J.; Horner, P. F.; Hollenberg, J. L. J. Nat. Prod. 1984, 47, 279-84.
16. Uniyal, G. C.; Singh, A. K.; Shah, N. C.; Naqvi, A. A. Planta Med. 1985, 51, 457-8.
17. Harborne, J. ACS Symp. Ser. 1986, 296, 22-35.
18. Funke, G. L. Blumea 1943, 5, 281-93.
19. Bode, H. R. Planta 1940, 30, 567-89.
20. Mann, H. H.; Barnes, T. W. Ann. Appl. Biol. 1945, 32, 15-22.
21. Muller, C. H.; Muller, W. H.; Haines, B. L. Science 1964, 143, 471-3.
22. Friedman, J. ACS Symp. Ser. 1986, 330, 53-68.

23. Ivie, G. W.; Witzel, D. A. in Handbook of Natural Toxins Vol. 1 Plant and Fungal Toxins; Keeler, R. F.; Tu, A. T., Eds.; Marcel Dekker: New York, 1983; pp. 543-84.
24. McCahon, C. B.; Kelsey, R. G.; Sheridan, R. P.; Shafizadeh, F. Bull. Torrey Bot. Club 1973, 100, 23-8.
25. Reynolds, T. Ann. Bot. 1987, 60, 215-23.
26. Halligan, J. P. Ecology 1975, 56, 999-1003.
27. Asplund, R. O. Phytochemistry 1968, 7, 1995-7.
28. Heisey, R. M.; Delwiche, C. C. Amer. J. Bot. 1984, 71, 821-8.
29. Sherman, M. E.; Thompson, L., Jr.; Wilkinson, R. E. Weed Sci. 1983, 31, 622-7.
30. May, J. W. Weeds Today 1984, 15(4), 7-8.
31. May, J. W.; Goss, J. R.; Moncorge, J. M.; Murphy, M. W. Proc. Brit. Crop Protect. Conf. 1985, 12, 265-70.
32. El-Deek, M. H.; Hess, F. M. Weed Sci. 1986, 34, 684-8.
33. Fischer, N. In The Science of Allelopathy; Putnam, A. R.; Tang, C.-S., Eds.; Wiley-Interscience, New York, 1986; Chapter 12.
34. Stipanovic, R. D. ACS Symp. Ser. 1983, 208, 69-100.
35. Green, M. B.; Hartley, G. S.; West, T. F. Chemicals for Crop Improvement and Pest Management, 3rd Edit.; Pergamon Press: Oxford, 1987; pp. 105-17.
36. Peterson, R. L.; Vermeer, J. In Biology and Chemistry of Plant Trichomes; Rodriguez, R.; Healey, P. L.; Mehta, I., Eds.; Plenum: New York, 1984; Chapter 4.
37. Sherif, A.; Hall, R. Proc. Calif. Mosq. Vect. Control 1984, 52, 76-9.
38. Casida, J. E.; Lawrence, L. J. Environ. Health Perspect. 1985, 61, 123-32.
39. Saleh, M. A.; Casida, J. E. In Advances in Pesticide Science, Part 3; Geissbühler, H., Ed.; Pergamon Press: Oxford, 1979; pp. 562-6.
40. Jacobson, M. Annu. Rev. Ent. 1966, 11, 403-22.
41. Brattsten, L. B. ACS Symp. Ser. 1983, 208, 173-95.
42. Gregory, P.; Tingey, W. M.; Ave, D. A.; Bouthyette, P. Y. ACS Symp. Ser. 1986, 296, 160-7.
43. Stipanovic, R. D.; Williams, H. J.; Smith, L. A. ACS Symp. Ser. 1986, 296, 79-94.
44. Yano, K. J. Agric. Food Chem. 1987, 35, 889-91.
45. Melin, E.; Krupa, S. Physiol. Plant. 1971, 25, 337-40.
46. Cobb, F. W., Jr.; Krstic, M.; Zavarin, E.; Barber, H.W., Jr. Phytopath. 1968, 58, 1327-35.
47. DeGroot, R.C. Mycologia 1972, 64, 863-70.
48. Mahmood, I.; Masood, A.; Saxena, S.; Husain, I. Acta Bot. India 1979, 7, 129-32.
49. Liersch, R.; Soicke, H.; Stehr, C.; Tüllner, H.-U. Planta Med. 1986, 52, 387-90.
50. Klayman, D. L. Science 1985, 228, 1049-55.
51. Chen, P. K.; Leather, G. R.; Klayman, D. L. Plant Physiol. 1987, 83 Suppl., Abstract 406.
52. Stevens, K. L.; Merrill, G. B. ACS Symp. Ser. 1985, 268, 83-98.
53. Fischer, N. H.; Quijano, L. ACS Symp. Ser. 1985, 268, 133-47.
54. Luo, X.-D.; Shen, C.-C. Med. Res. Rev. 1987, 7, 29-52.
55. Herz, W. In Effects of Poisonous Plants on Livestock; Keeler, R. F.; van Kampen, K. R.; James, L. F., Eds.; Academic Press: New York, 1978; pp. 487-97.

56. Acton, N.; Klayman, D. L. Planta Med. 1987, 53, 266-8.
57. Lin, A. J.; Klayman, D. L.; Hoch, J. M.; Silverton, J. V.; George, C. F. J. Org. Chem. 1985, 50, 4504-8.
58. Duke, S. O. In The Science of Allelopathy; Putnam, A. R.; Tang, C.-S., Eds.; Wiley-Interscience, New York, 1986; Chapter 17.
59. Picman, A. K. Biochem. Syst. Ecol. 1987, 15, 361-4.
60. Kömives, T.; Casida, J. E. J. Agric. Food Chem. 1983, 31, 751-5.
61. Liersch, R.; Soicke, H.; Stehr, C.; Tüllner, H.-U. Planta Med. 1986, 52, 387-90.
62. Bryson, C.T.; Croom, E. M., Jr.; Proc. Miss. Acad. Sci. 1988, 33, 5.
63. Kudakasseril, G. J.; Lam, L.; Staba, E. J. Planta Med. 1987, 53, 280-4.

RECEIVED April 12, 1988

Chapter 22

Chemistry and Biological Activity of Acylnornicotines from *Nicotiana repandae*

R. F. Severson[1], R. F. Arrendale[1], Horace G. Cutler[1], D. Jones[2], V. A. Sisson[3], and M. G. Stephenson[4]

[1]Richard B. Russell Research Center, Agricultural Research Service, U.S. Department of Agriculture, Athens, GA 30613
[2]Department of Entomology, University of Kentucky, Lexington, KY 40546
[3]Crops Research Laboratory, Agricultural Research Service, U.S. Department of Agriculture, Oxford, NC 27565
[4]Coastal Plain Experiment Station, Agricultural Research Service, U.S. Department of Agriculture, Tifton, GA 31793

The Nicotiana species, N. repanda, N. stocktonii, and N. nesophila, produce cuticular components that are toxic to tobacco hornworms [Manduca sexta(L.)]. Analyses of the methylene chloride extract of the leaf surface components showed that the major constituents were the duvatrienediols, hydrocarbons, and a series of hydroxyacylnornicotines (HOacylNN), containing 3-hydroxy-aliphatic acid moieties (C_{13}-C_{15}), with normal, iso-, and anteiso-methyl-branched chains. The major HOacylNN was 1′-(3-hydroxy-12-methyltridecanoyl)nornicotine. A series of C_{12}-C_{13} normal and methyl-branched chain acylnornicotines (acylNN) were also identified as minor components. The HOacylNN-acylNN fraction was isolated from the methylene chloride cuticular extract by solvent partitioning and Sephadex LH-20/$CHCl_3$ column chromatography. The HOacylNN isomers were isolated by preparative reverse-phase HPLC. The HOacylNN-acylNN fraction was toxic to tobacco hornworms, decreased the growth of wheat coleoptiles, and had antibiotic activity.

Many of the genus Nicotiana species have leaf trichomes that produce components with a wide range of biological activities (1-10). Goodspeed (11) classified Nicotiana leaf trichomes into five categories--simple (Figure 1), branched, glandular-headed (Figure 2), those with specialized cells, and hydathodes. The glanded trichomes may produce observable exudates (Figure 3), which along with non-trichome-produced cuticular components can be

0097–6156/88/0380–0335$08.00/0

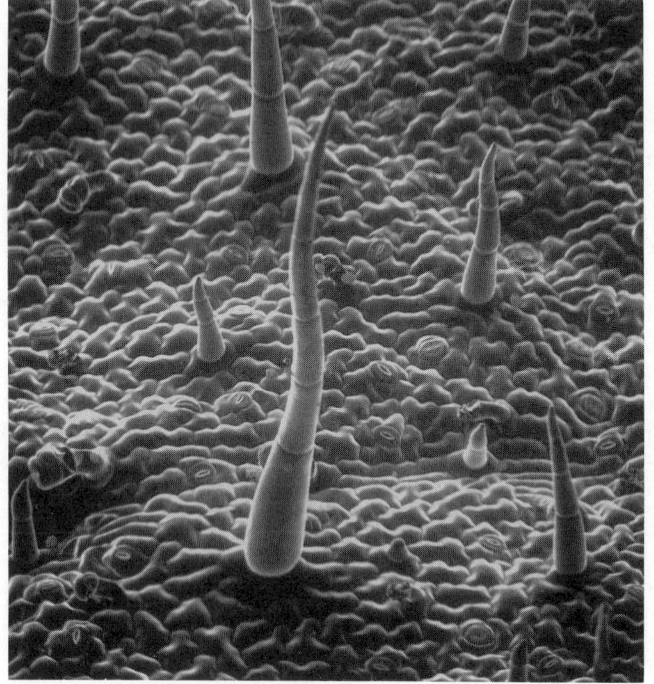

Figure 1. Electron micrograph of simple trichome.

quantitatively extracted by dipping the leaves into methylene chloride. These extracts can be quantified by glass capillary gas chromatography (GC-2) (12). The cuticular extracts of all Nicotiana species contain a homologous series of non-trichome-produced saturated hydrocarbons (generally C_{25}-C_{36}) with normal, iso-, and/or anteiso-branched isomers (Figure 4) (1,12-14). The trichome

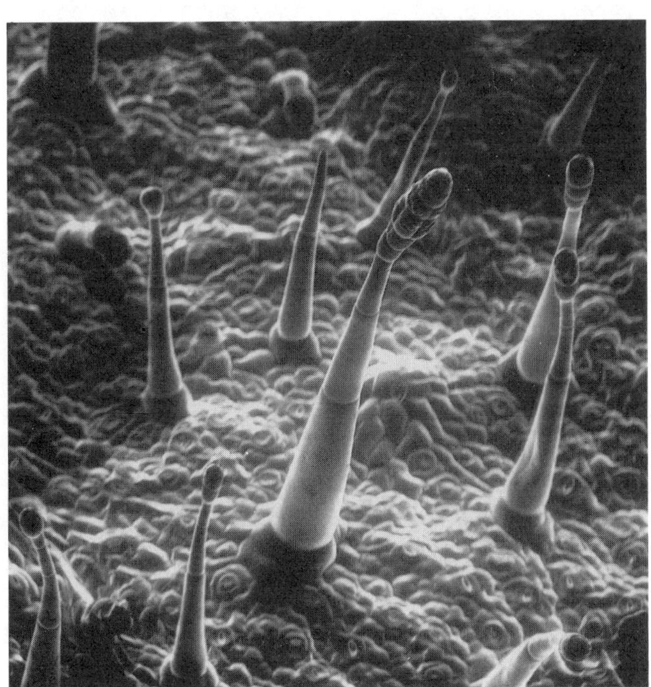

Figure 2. Electron micrograph of glanded trichome.

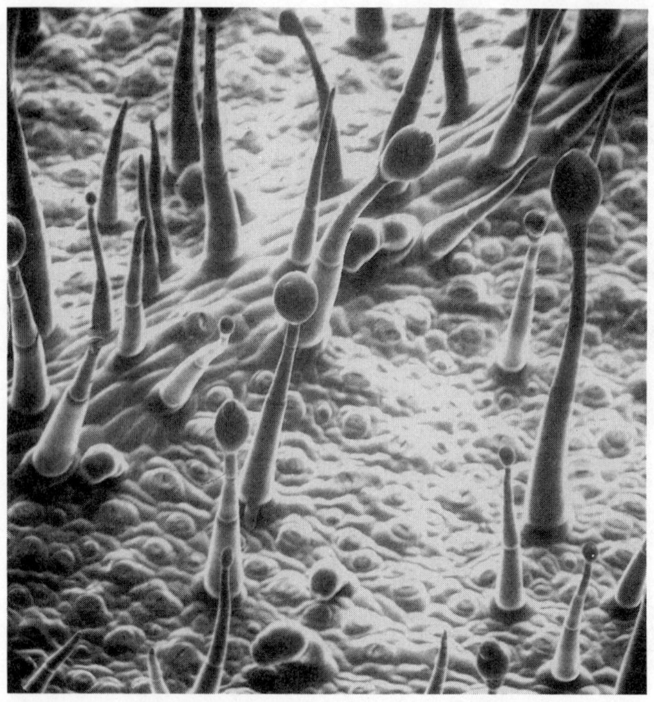

Figure 3. Electron micrograph of glanded trichome with exudates.

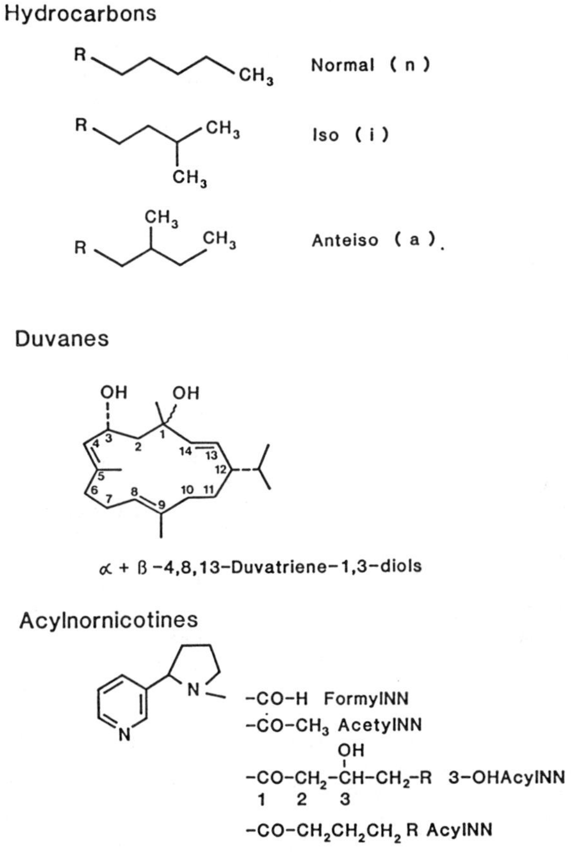
Hydrocarbons
R
CH3
Normal (n)
R
CH3
CH3
Iso (i)
CH3
R
CH3
Anteiso (a).
Duvanes
OH
OH
α + β -4,8,13-Duvatriene-1,3-diols
Acylnornicotines
N
N
-CO-H FormylNN
-CO-CH3 AcetylNN
OH
-CO-CH2-CH-CH2-R 3-OHAcylNN
1 2 3
-CO-CH2CH2CH2 R AcylNN

Figure 4. Major cuticular components of Nicotiana section Repandae.

exudates may contain duvane diterpenes, labdane diterpenes, sucrose and glucose esters, and/or a homologous series of fatty alcohols and/or wax esters (1,2,13). Many of these compounds inhibit the growth of etiolated wheat coleoptiles (1,2,7-9) and have antibiotic (1,2,9) and fungitoxic (6) properties. Among these, the duvatrienediols (Figure 4) are potent ovipositional stimulants for the tobacco budworm [Heliothis virescens(F.)] (10).

In 1970, Thurston and coworkers (15) reported that the Nicotiana species of the section Repandae (N. nesophila, N. repanda, and N. stocktonii) were resistant to tobacco hornworm [Manduca sexta (L.)] damage. They postulated that it was due to toxic compounds in the trichome exudates. This was confirmed by Jones et al. (16) and Huesing and Jones (17). We have recently isolated a fraction from the cuticular extracts of Repandae, which contains of a series of C_{12}-C_{13} acylnornicotines (acylNN) and C_{12}-C_{16} 3-hydroxyacylnornicotines (HOacylNN) and have shown that this fraction is highly toxic to the tobacco hornworm (18,19). Zado and Jones (20) demonstrated that the terminal biosynthesis of these compounds occurs in the trichomes. In this report we will discuss our methodology to isolate, characterize, and quantify these compounds and their insecticidal, antibiotic, and plant growth regulating activity.

Materials and Methods

The three Repandae species were grown under conditions used for flue-cured tobacco at the Tobacco Research Laboratory, Oxford, North Carolina and at the University of Georgia Coastal Plain Experiment Station, Bowen Farm, Tifton, Georgia. When plants were in full-flower, young leaves (about 30 g) were extracted by dipping into methylene chloride, contained in 8-oz wide-mouth bottles, as described by Severson et al. (12). Leaf area was measured by a Li-Cor Model LI-3000 area meter. After solvent removal under nitrogen, the components in the cuticular extract were treated with a 1:1 mixture of N,O-bis(trimethylsilyl)trifluoroacetamide (BSTFA)-dimethylformamide (DMF) for 30 min at 76°C to derivatize the hydroxylated components to trimethylsilyl (TMS) ethers. The cuticular components were then analyzed by capillary gas chromatography (GC-2) on a 0.3 mm i.d. x 25 m thin film (about 0.1 µm) SE-54 fused silica column (21). GC conditions were: oven temperature program of 150-182°C at 8°/min, 182-198°C at 4°/min, 198-280°C at 3°/min, and 5 min hold at 280°C, 35 cm/sec H_2 flow rate, 100 mL/min split, injection port temperature 250°C and flame ionization detector temperature, 310°C. (See Figure 5 for resulting GC-2 chromatograms of cuticular extracts of N. repanda, N. stocktonii, and N. nesophila.) Larger quantities of cuticular extracts, needed for isolation and characterization studies, were obtained by dipping whole plant tops into methylene chloride, as described previously (5). The isolation scheme used to obtain a HOacylNN-acylNN isolate is shown in Figure 6. About 4 kg of whole, green plant material yielded a methylene chloride cuticular extract residue of about 2.0 g, which was partitioned between 150 mL each of hexane and 80% MeOH-$H_2$0. The MeOH-$H_2$0 fraction was washed with 50 mL hexane and the combined hexane extracts were extracted with 80% MeOH-$H_2$0 (2 x 50 mL). The MeOH-$H_2$0 extracts were combined and

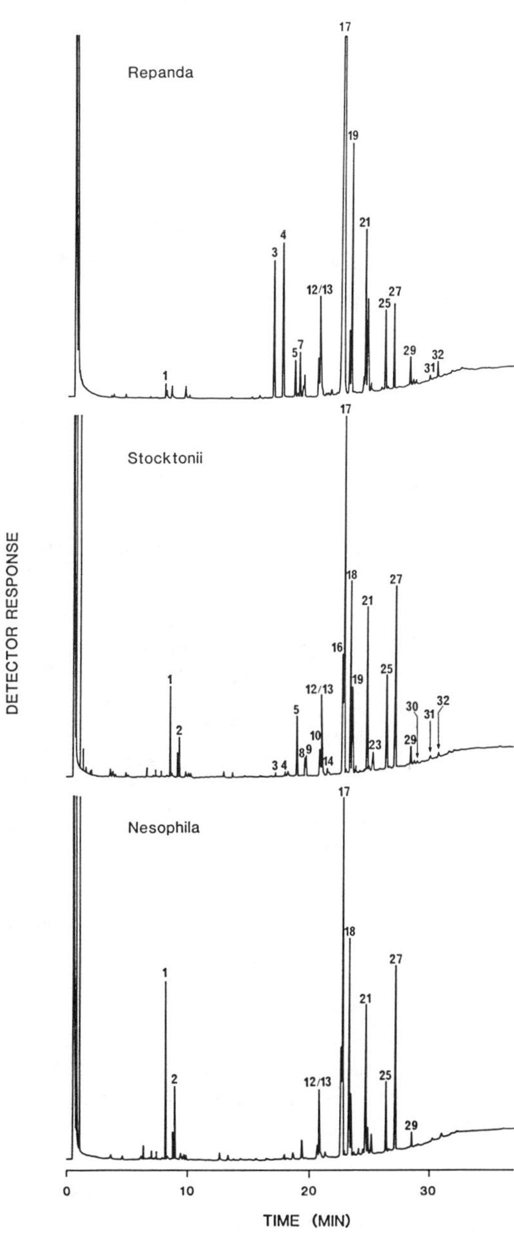

Figure 5. Gas chromatograms of the silylated cuticular components of Nicotiana section Repandae. (Peak identifications given in Table I.)

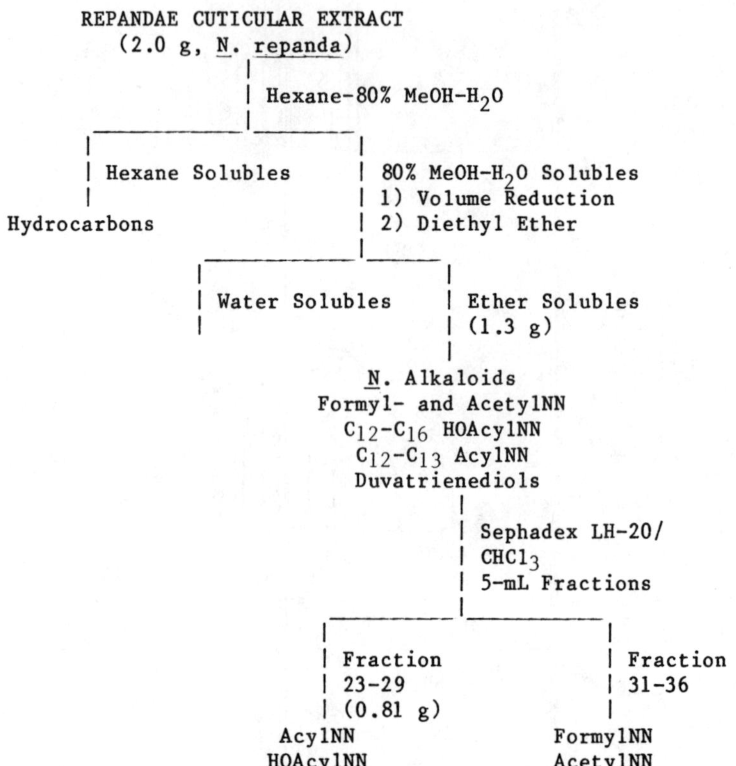

Figure 6. Scheme for the isolation of the acylNN-HOacylNN from the cuticular extract of *Nicotiana* section Repandae.

reduced to about 75 mL on a roto-evaporator. Distilled H_2O (50 mL) was added and the residue was extracted with 150 mL of diethyl ether (containing 2% EtOH). After washing with H_2O (3 x 75 mL) and drying over Na_2SO_4, the ether solubles from the MeOH-H_2O extract were taken to dryness (about 1.3 g), dissolved in 2 mL of $CHCl_3$ (0.75% EtOH), and chromatographed on Sephadex LH-20 (two 109 cm x 1.25 cm i.d. Chromatronix columns in series, $CHCl_3$ flow at 2 mL/min). Elution was monitored by UV absorption at 254 nm and 5-mL fractions were collected. Fractions 23-29 were combined to yield 810 mg of acylNN-HOacylNN isolate (98+% by GC-2, see Figure 7 for isolate GC-2 profiles).

The C_{12}, C_{13}, and C_{14} homologs of the HOacylNN were isolated using low pressure reverse-phase C_{18} liquid chromatography. About 200 mg of the LH-20 Fractions 23-29 (acylNN-HOacylNN isolate from N. nesophila), dissolved in 1 mL of acetonitrile, were chromatographed on a reverse-phase C_{18} column (109 cm x 1.25 cm Chromatronix column; 100 cm bed of C_{18} packing from Prep-Pak 500, Waters Associates), following a solvent gradient of 1:1 acetonitrile:H_2O to 9:1 acetonitrile:H_2O over 6 hr at 2 mL/min. Elution was monitored at 254 nm and 5-mL fractions were collected (Figure 8). Fractions of similar composition, as determined by GC-2, were combined, acetonitrile removed, and residue extracted with methylene chloride to yield: Fractions 48-57 , 12 mg of C_{12}-HOacylNN (90%); Fractions 62-67, 8 mg of C_{13}-HOacylNN (85+%); Fraction 69-72, 35 mg of C_{14}-HOacylNN (97+%, 96% iso-C_{14}-HOacylNN); and Fraction 73-88, 103 mg of C_{14}-HOacylNN (99+%). The rechromatography of combined fractions from N. nesophila yielded still more refined isolates; GC-2 profiles are shown in Figure 9. These fractions were tested for biological activity and were used to prepare HOacylNN esters.

The (C_2) acetyl-, (C_4) butanoyl-, (C_6) hexanoyl-, (C_8) octanoyl-, (C_{10}) decanoyl-, (C_{12}) dodecanoyl-, (C_{14}) tetradecanoyl-, and (C_{16}) hexadecanoyl-acylnornicotines were prepared by reacting the corresponding acid chloride with nornicotine (NN) isolated from flue-cured N. tabacum variety TI 1112. About twice the molar equivalent of the acid chloride was added to 30 mg of NN in 0.5 mL each of $CHCl_3$ and pyridine, in a 8-mL screw cap test tube, cooled in an ice bath. When the addition was complete, the mixture was heated for 15 min at 40°C, poured into a separatory funnel which contained 25 mL of saturated Na_2CO_3, and extracted with 1:1 ether:benzene (3 x 25 mL) ($CHCl_3$ for C_2-C_6 compounds). The organic extract was washed with H_2O until neutral, reduced in volume, and the acylNN fraction was isolated by low pressure reverse-phase C_{18} as described above. The 3-acetate, benzoate, and 3,5-dinitrobenzoate derivatives of the iso-C_{14}-HOacylNN were prepared and isolated as above. Component purity was determined by GC-2, high-pressure liquid chromatography, and mass spectrometry (22).

The acylNN-HOacylNN Sephadex LH-20 isolate, i-C_{14}-HOacylNN, and C_{14}-acylNN were bioassayed by topical application on first instar tobacco hornworm larvae. The HOacylNN isolates, i-C_{14}-HOacylNN esters and synthetic acylNN were bioassayed using etiolated wheat coleoptile and gram positive bacteria [B. subtilis (+), B. cereus (+), and M. thermosphactum (+), and gram negative bacteria, E. cloacae (-)]. Wheat seedlings (Triticum aestivwas L., cv Wakeland) were grown in the dark for four days on moist sand at 22°C, then

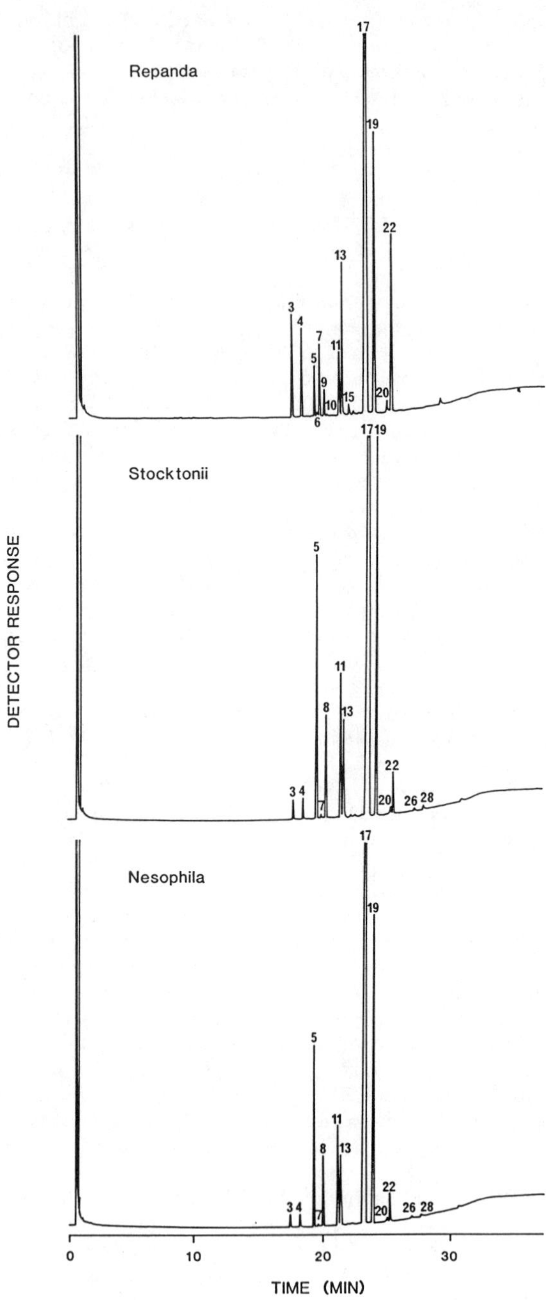

Figure 7. Capillary gas chromatograms of the silylated acylNN-HOacylNN isolates from *Nicotiana* section Repandae.

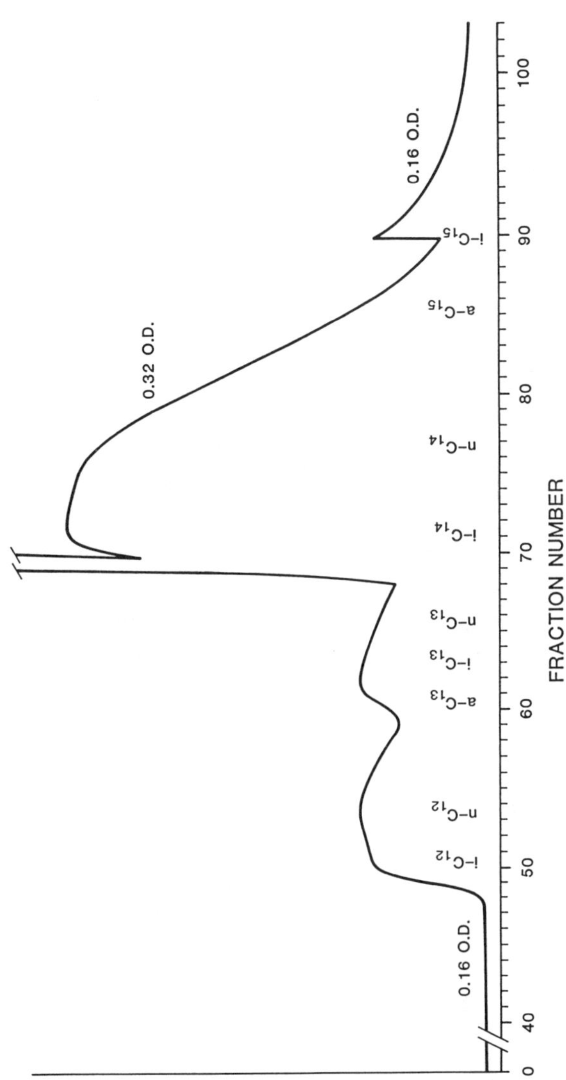

Figure 8. Preparative reverse-phase chromatographic separation of the HOacylNN isolate from N. nesophila.

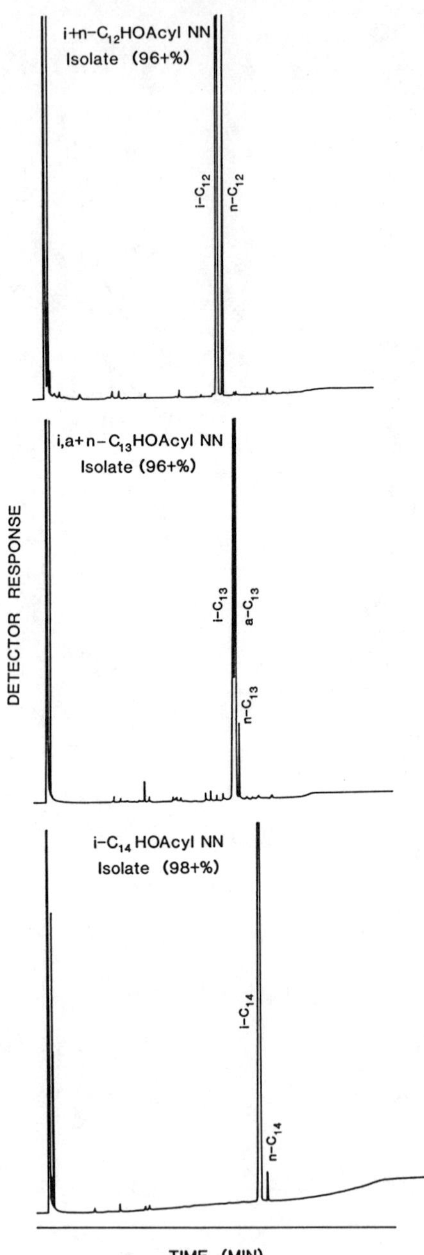

Figure 9. Capillary gas chromatogram of the silylated C_{18} reverse-phase HOacylNN isolates from N. nesophila.

harvested. The apical 2 mm were discarded and the next 4 mm were cut and retained for bioassay (23). Ten 4-mm sections were placed into test tubes containing dilutions of the compounds to be tested at 10^{-3}, 10^{-4}, 10^{-5} and 10^{-6} M formulated in acetone and 2% sucrose-phosphate citrate buffer (8). Assays were incubated for 18 hours at 22°C in the dark in a roller tube apparatus (0.25 cpm). Coleoptile sections were placed into a photographic enlarger to give a 3X image, and the length of each coleoptile was recorded. All assays were duplicated and the data were statistically analyzed (24). For bacterial assays, bacteria were heavily seeded on DST-Oxoid agar and 4-mm diameter discs, impregnated with the compound to be tested at 50, 250, and 500 µg per disc, were placed on the agar. Plates were incubated at 37°C overnight and the diameter of the growth inhibition zone was measured.

Results and Discussion

Extraction and Quantitation of Acylated Nornicotine

The cuticular components of the *Nicotiana* species in the section Repandae are readily extracted by dipping the fresh green leaf into methylene chloride. After removal of the extraction solvent and conversion of hydroxylated components to silyl ethers, the cuticular components can be separated and quantitated by capillary gas chromatography. The resulting chromatograms of *N. repanda*, *N. nesophila*, and *N. stocktonii* are shown in Figure 5 and identified in Table I. Extraction efficiency for the major cuticular components of the Repandae is given in Table II and their levels are given in Table III. Consistent with previous data, the trichome-produced components, the HOacylNN and duvatrienediols, are more efficiently extracted than the true, waxy leaf surface components (the aliphatic hydrocarbons) (12). *N. repanda* produced the highest levels of acylNN and HOacylNN components. Similar levels of typical *Nicotiana* hydrocarbons were found in all species and *N. stocktonii* had the highest levels of the duvatrienediols, the tobacco budworm ovipositional stimulants (5,10).

Fractionation and Characterization of Cuticular Components

As shown in Figure 6, the Repandae cuticular extracts were separated into polarity classes by solvent partitioning between hexane and 80%MeOH-H_2O. The typical C_{25}-C_{35} *Nicotiana* hydrocarbons (1,5,10) were the major constituents of the hexane fraction (greater than 97+% by GC-2). Major components in the MeOH-H_2O soluble fraction were the acyl-HOacylNN. The *Nicotiana* alkaloids (nicotine, nornicotine, anabasine, anatabine) and C_1 and C_2 acylNN, formylNN and acetylNN, and duvatrienediols, were minor constituents in the polar fraction and were readily characterized by GC-2/MS and GC-2 retention data (12,25,28).

The multicomponent MeOH-H_2O soluble fraction was further fractionated by Sephadex LH-20 chromatography, which resolved the major acylNN-HOacylNN fraction from duvatrienediols and the normal nicotine alkaloids, along with formylNN and acetylNN. The GC-2 chromatograms of the resulting acylNN-HOacylNN isolates are shown in

Table I. Major Components in the Cuticulae of N. Section Repandae

Peak No.	Component
1	α-4,8,13-Duvatriene-1,4-diol (α-diol)
2	β-4,8,13-Duvatriene-1,4-diol (β-diol)
3	1′-(10-Methylhendecanoyl)nornicotine (i-C_{12}-NN)
4	Dodecanoylnornicotine (n-C_{12}-NN)
5	1′-(3-Hydroxy-10-methylhendecanoyl)nornicotine (i-C_{12}-HONN)
6	1′-(11-Methyldodecanoyl)nornicotine (i-C_{13}-NN)
7	1′-(10-Methyldodecanoyl)nornicotine (a-C_{13}-NN)
8	Nonacosane (n-C_{29})
9	1′-(3-Hydroxydodecanoyl)nornicotine (n-C_{12}-HONN)
10	Tridecanoylnornicotine (n-C_{13}-NN)
11	1′-(3-Hydroxy-11-methyldodecanoyl)nornicotine (i-C_{13}-HONN)
12	3-Methylnonacosane (a-C_{30})
13	1′-(3-Hydroxy-10-methyldodecanoyl)nornicotine (a-C_{13}-HONN)
14	Triacontane (n-C_{30})
15	1′-(3-Hydroxytridecanoyl)nornicotine (n-C_{13}-HONN)
16	2-Methyltriacontane (i-C_{31})
17	1′-(3-Hydroxy-12-methyltridecanoyl)nornicotine (i-C_{14}-HONN)
18	Hentriacontane (n-C_{31})
19	1′-(3-Hydroxytetradecanoyl)nornicotine (n-C_{14}-HONN)
20	1′-(3-Hydroxy-13-methyltetradecanoyl)nornicotine (i-C_{15}-HONN)
21	3-Methylhentriacontane (a-C_{32})
22	1′-(3-Hydroxy-12-methyltetradecanoyl)nornicotine (a-C_{15}-HONN)
23	Dotriacontane (n-C_{32})
24	1′-(3-Hydroxypentadecanoyl)nornicotine (n-C_{15}-HONN)
25	2-Methyldotriacontane (i-C_{33})
26	1′-(3-Hydroxy-14-methylpentadecanoyl)nornicotine (i-C_{16}-HONN)
27	Tritriacontane (n-C_{33})
28	1′-(3-Hydroxyhexadecanoyl)nornicotine (n-C_{16}-HONN)
29	3-Methyltritriacontane (a-C_{34})
30	Tetratriacontane (n-C_{34})
31	2-Methyltetratriacontane (i-C_{35})
32	Pentatriacontane (n-C_{35})

Table II. Efficiency of Cuticular Component Extraction Procedure (N. *stocktonii*)

	% Recovered[a]		
Wash No.	Hydrocarbons	HOacylNN	Duvatriene-diols
1	94.7	99.0	100.0
2	3.7	0.9	--
3	1.6	0.1	--

[a]Based on total amount of material recovered in three sequential methylene chloride extractions. Average of three determinations.

Table III. Comparison of Major Green Leaf Cuticular Component Levels of N. *repanda*, N. *stocktonii*, and N. *nesophila*[a]

	α- + β-Diol[b]	AcylNN[c]	HOAcylNN[d]	Hydrocarbons[e]
	μg/cm²	μg/cm²	μg/cm²	μg/cm²
N. repanda	0.2	1.8	19.4	7.1
N. stocktonii	1.5	<0.1	12.0	9.0
N. nesophila	0.4	0.1	10.8	6.9

[a]Young leaf at onset of flower development, Oxford, NC, 1985.
[b]α- and β-4,8,13-duvatriene-1,4-diol.
[c]C_{12}-C_{13} acylnornicotine, calculated assuming chromatographic response equal to triacontane.
[d]C_{12}-C_{15} hydroxyacylnornicotines, calculated assuming chromatographic response equal to triacontane.
[e]C_{29}-C_{34} hydrocarbons, calculated assuming chromatographic equal to triacontane.

Figure 7. These components are readily characterized by MS and GC-2/MS. The mass spectrum of the major HOacylNN isomer, 1-(3-hydroxy-12-methyltridecanoyl)nornicotine, is shown in Figure 10 and the mass spectrum of its TMS derivative is shown in Figure 11. All HOacylNN homologs and their TMS derivatives yield major fragmentation ions at m/e 147, 175, 189, and 190. Due to the hydroxyl group on the β-carbon of the acid moiety, an intense β-cleavage ion is observed at m/e 219 (291 for TMS). MS data can also be used to determine iso- and anteiso-methyl chain branching from normal isomers. For the iso-methyl branched chain enhanced M-15, M-43 and M-H_2O (or HOTMS), M-18-15, M-18-43, and m/e 43 ions are present. The mass spectrum of anteiso-methyl branched isomers show enhanced ions at M-15, M-29 and M-57, M-HOTMS-15, M-HOTMS-29, M-HOTMS-57, and m/e 57. In contrast, the normal-chain isomers yield mass spectral and group fragmentation typical of n-alkanes. Fragmentation patterns of each of the acylNN compounds clearly indicates the acyl chain length and end branching. The spectra of 1′-dodecanoylNN shows the presence of 12 carbons in the chain and typical straight chain alkane fragmentation. In contrast, the iso-C_{12}-isomer shows the effect of the trimethyl isopropyl group on the intensity of end group fragments of m/e 43 ion (Figure 12). The anteiso-C_{13}-isomer yields fragment ions at M-15, M-29 and M-57 and m/e 57 representative of sec-butyl end group (Figure 13).

The composition of the acylNN-HOacylNN isolates from the Repandae species is given in Table IV. The acylNN constituted about 8% of the N. repanda isolate. All three species had similar acylNN isomer distribution and the C_{14}-HOacylNN was the major component. For the HOacylNN series, some differences in relative distribution of homologs and isomers within each carbon number were observed. Such differences have been observed for two years in field samples grown in Georgia and North Carolina, indicating that the distribution of acids used to form the acylated NN is under genetic control (27).

As shown in Figure 8, the HOacylNN homologs can be separated for other studies by C_{18} reverse-phase chromatography in sufficient quantities to evaluate the effect of carbon number on biological activity. The high level of the iso-C_{14} compound permitted selective isolation of this isomer. Figure 9 shows GC-2 chromatograms of the C_{12}-, C_{13}-, and C_{14}-homolog isomer mixtures from the acylNN-HOacylNN of N. nesophila. In this reverse phase system, the C_{12}-C_{13} acylNN co-elute with the C_{15}-C_{16} HOacylNN. This fact, along with the relative low levels of these compounds, makes their isolation difficult.

Biological Activity of Acylated Nornicotine

Topical applications of 200 μg of a combination of acylNN-HOacylNN isolate from the three species produced 100% mortality to first-instar tobacco hornworm within 48 hrs (Table V). The 50 to 100 μg applications produced 80% or greater mortality after 96 hrs. Even at the low level of 10 μg, 40% mortality was observed after 96 hrs. In contrast, the most toxic of the normal nicotine alkaloids (28), nicotine, when applied at the 500 μg level, produced only 7% mortality after 96 hrs. However, in similar tests with tobacco

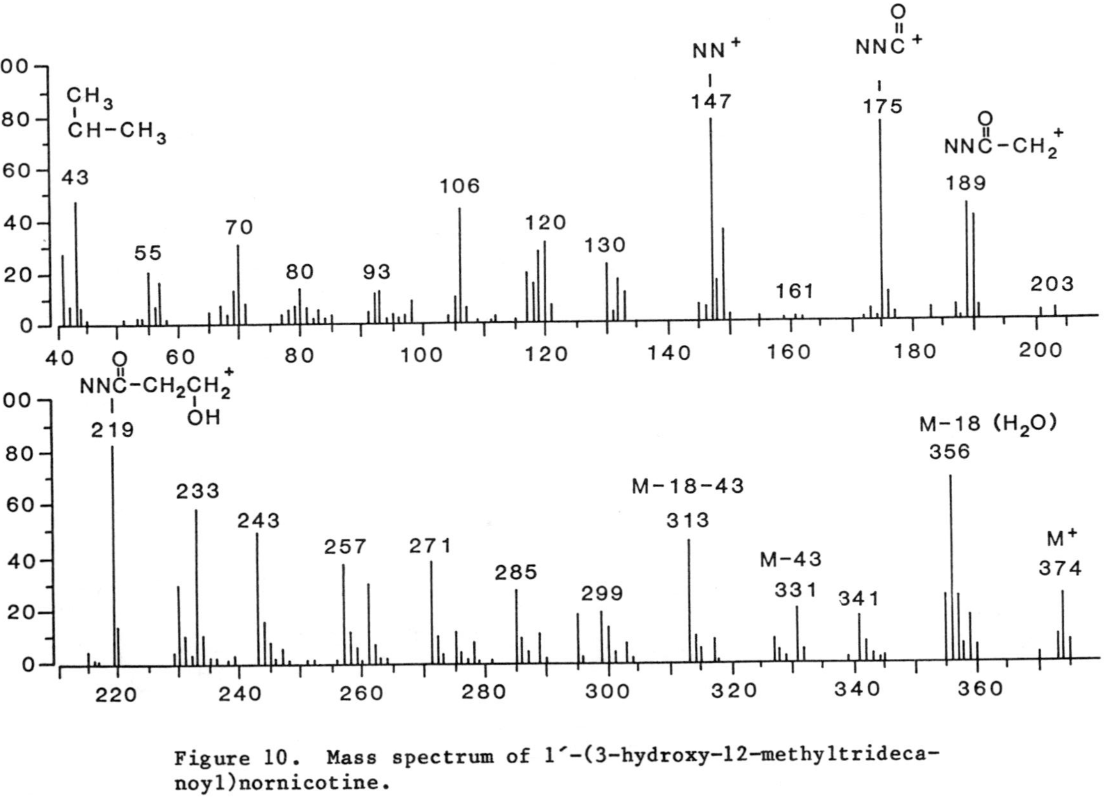

Figure 10. Mass spectrum of 1′-(3-hydroxy-12-methyltrideca-noyl)nornicotine.

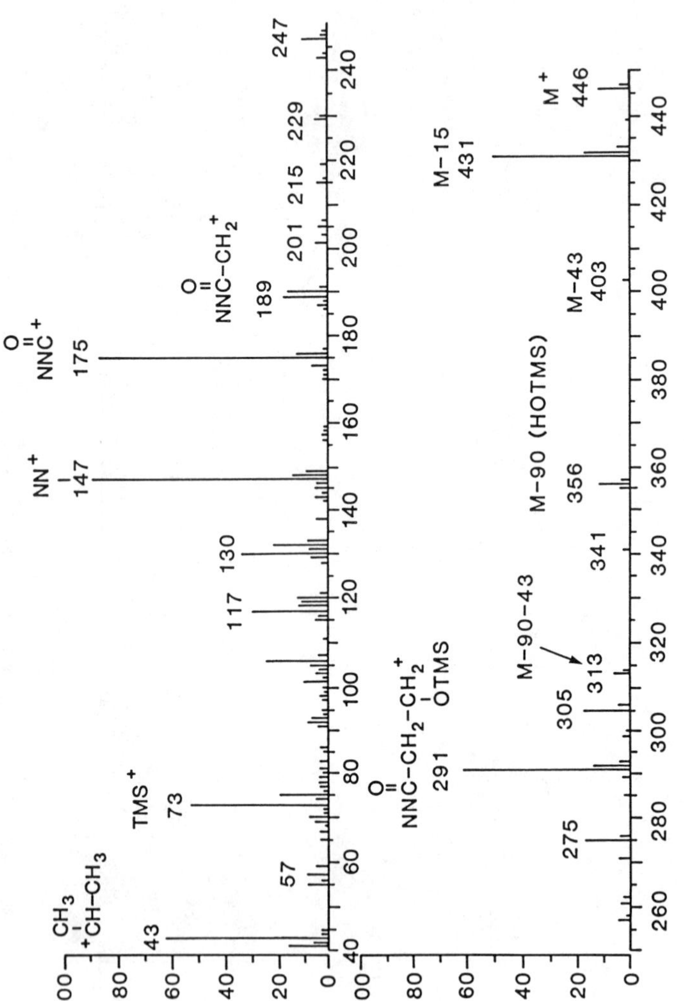

Figure 11. Mass spectrum of 1′-(3-hydroxy-12-methyltrideca-noyl)nornicotine silyl ether.

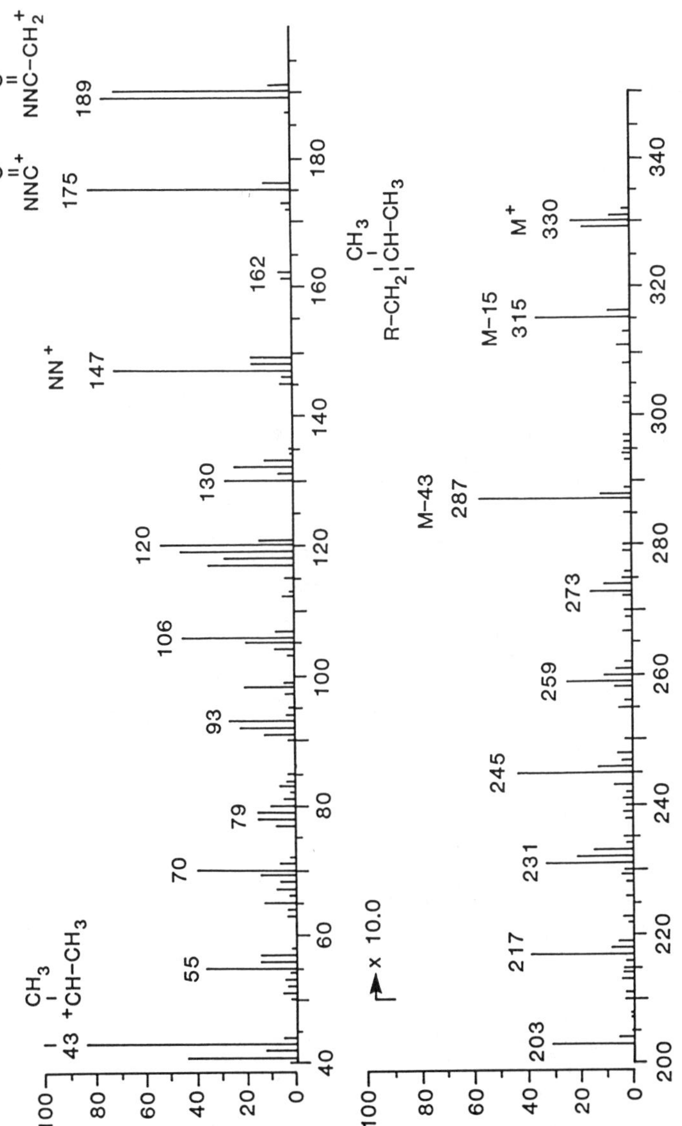

Figure 12. Mass spectrum of 1′-(10-methylundecanoyl)nornicotine.

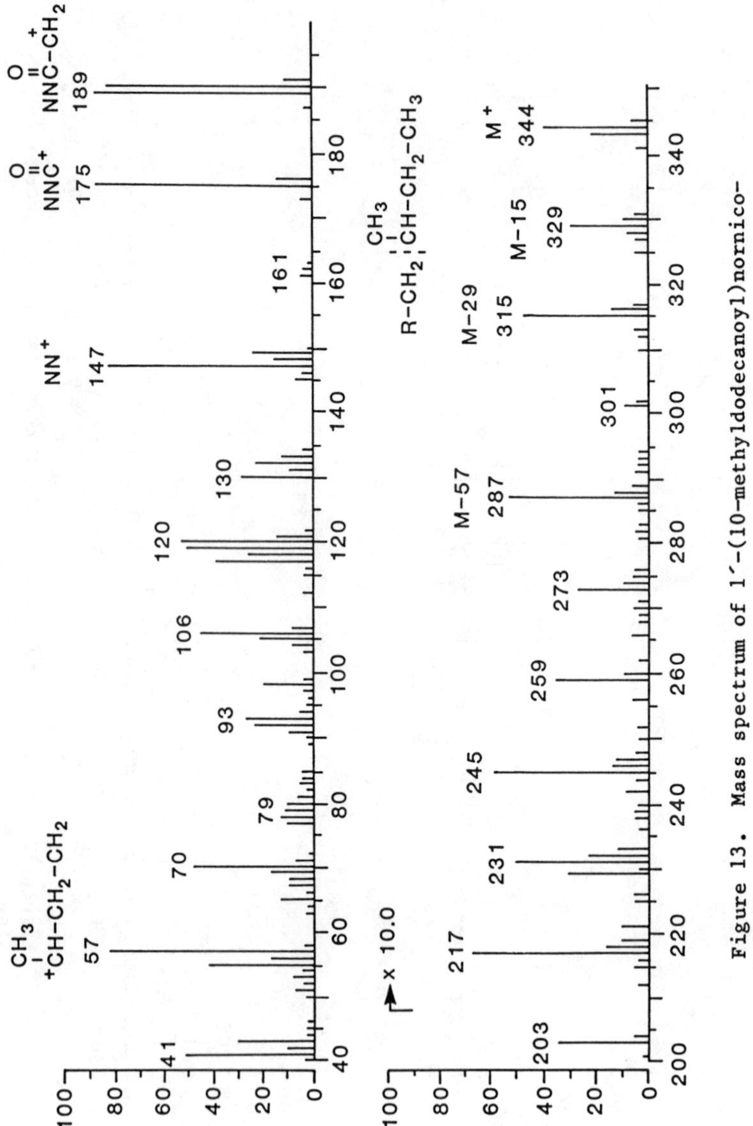

Figure 13. Mass spectrum of 1′-(10-methyldodecanoyl)nornicotine.

Table IV. Composition of Acyl- and Hydroxyacylnornicotine Isolates from *N. repanda*, *N. stocktonii*, and *N. nesophila*

	% Distribution[a]					
	N. repanda		*N. stocktonii*		*N. nesophila*	
C_{12} Acyl	5.7		0.1		0.8	
i		54		52		49
a		--		--		--
n		46		48		51
C_{13} Acyl	2.6		0.1		0.9	
i		5		--		--
a		91		100		100
n		4		--		--
C_{12} HOacyl	2.4		0.4		8.7	
i		62		68		73
a		--		--		--
n		38		32		27
C_{13} HOacyl	7.5		3.8		5.9	
i		28		58		60
a		68		42		40
n		4		--		--
C_{14} HOacyl	76.0		90.5		83.0	
i		87		86		86
a		--		--		--
n		13		14		14
C_{15} HOacyl	5.9		4.8		1.1	
i		6		12		12
a		93		88		88
n		1		--		--
C_{16} HOacyl	--		0.2		0.2	
i		--		65		41
a		--		--		--
n		--		35		56

[a]Calculated assuming unitary chromatographic response.

Table V. Percent Mortality of M. sexta Larvae[a] following Applications of AcylNN-HOacylNN Isolate from N. repanda, C_{14}-HOacylNN, and C_{14}-AcylNN

Treatment/Insect	No. of Larvae	Hours 24	48	72	96
		Cumulative % Mortality[b]			
AcylNN-HOacylNN Isolate[c]					
10 μg	15	0	33	33	40
50 μg	25	8	44	64	80
100 μg	40	60	78	85	93
200 μg	30	90	100	--	--
Control	15	0	0	0	0
EtOH (1 μL)[d]	30	0	0	0	0
Nicotine (500 μg)	15	0	7	7	7
Iso-C_{14}-HOacylNN (1 μm)[e]	7	100	--	--	--
C_{14}-AcylNN (1 μL)[f]	9	89	--	--	--
EtOH (1 μL)	16	0	--	--	--

[a]Larvae were tested late during the first stadium.
[b]Includes moribund larvae which are extensively paralyzed but not yet dead.
[c]Adapted, in part, from Severson et al. (19).
[d]Commercial nicotine in 1 μL EtOH.
[e]98.4% iso and 1.2% normal.
[f]Prepared from nornicotine isolated from N. tabacum cv. TI 1112 and myristyl chloride.

budworm and cabbage looper larvae, no activity was observed. In another bioassay, at the micromole rate, the i-C_{14}-HOacylNN (346 μg) isolate and the C_{14}-acylNN (330 μg) were very toxic to the young larvae. Additional studies are currently underway to determine the contribution of chain length and hydroxylation toward the total toxicity of the molecule for tobacco hornworm larvae.

The antimicrobial properties of the HOacylNN-acylNN homolog isolates and the acetate, benzoate, and 3,5-dinitrobenzoate derivatives of the i-C_{14} HOacylNN are listed in Table VI. All hydroxylated isolates demonstrated some activity against the bacteria tested. No significant differences in activity with carbon numbers was observed for B. subtilis. The i-C_{14} acetate was the most active. B. cereus and M. thermosphactum were susceptible to C_{14}-HOacylNN isolates at all concentrations and were also the most active against the gram negative bacteria, E. cloacae. The benzoate and 3,5-dinitrobenzoate derivatives were completely inactive in all systems.

Similar test data on several synthetic n-acylNNs is given in Table VII. The C_2, C_4, and C_6 compounds were essentially inactive in all systems tested. B. subtilis was moderately susceptible to the C_{10} to C_{14} acylNN and B. cereus was susceptible to C_8 to C_{14} chain lengths. For all gram positive bacteria, the n-C_{14}-acylNN was the most active homolog. Unlike the HOacylNN, the acylNN did not display significant activity against E. cloacae.

Table VIII lists the effects of the HOacylNN and acylNN on the growth of wheat coleoptile. All C_{12}- to C_{14}-HOacylNN and C_8- to C_{14}-acylNN produced 100% growth inhibition at the 10^{-3} M concentration. The i-C_{14}-HOacylNN and its acetate ester and the C_{10}- to C_{14}-acylNN were the most active. The i-C_{14}-HOacylNN benzoates and 3,5-dinitrobenzoate esters were completely inactive. Once again chain length and/or methyl branching were important in relative activity.

Previously we have reported that components in the trichome exudates of N. tabacum, the α- and β-diols, the α- and β-4,8,13-duvatrienols, and the sucrose esters also inhibit the growth of wheat coleoptile. α-Diol and sucrose esters inhibit Bacillus cereus (1,8,9). However, the naturally occurring acylNN and HOacylNN are the most active Nicotiana cuticular components evaluated to date.

One very interesting observation is that the two insects, tobacco budworm and tobacco hornworm, which use tobacco as a host plant, are relatively insensitive to the parent alkaloid, NN, and to other common nicotine alkaloids, in comparison to other insects. However, the HOacylNN and acylNN are selectively insecticidal toward the tobacco hornworm. Studies are currently being initiated to determine the neurophysiological basis for the mechanism of action of this group of alkaloids.

Summary

The leaf trichomes of the species in the Nicotiana section Repandae produces a series of C_{12}- to C_{16}-HOacylNN and C_{12}- to C_{13}-acylNN which can be isolated from other cuticular components by solvent partitioning and Sephadex LH-20 chromatography. N. repanda produces

Table VI. Antibiotic Activity of Various 1′-(3-Hydroxyacyl)nornicotine Isolates and Derivatives

Organism	Amount (μg)	1′-(3-Hydroxyacyl)nornicotine[a] C_{12}-HO[b]	C_{13}-HO[c]	C_{14}-HO[d]	C_{14}-HO[e]	Acetate[f]	Benzoate[f]	Dinitro-benzoate[f]
		Inhibition Zone (dia. mm)[g]						
B. subtilis (+)	50	8	0	0	7	14	0	0
	250	8	8	11	9	18	0	0
	500	11	12	12	8	18	0	0
B. cereus (+)	50	10	10	16	15	11	0	0
	250	13	14	18	18	18	0	0
	500	17	18	20	22	19	7	8
M. thermosphactum (+)	50	8	13	17	17	--	0	0
	250	10	18	17	17	--	0	0
	500	14	21	19	17	--	8	0
E. cloacae (-)	50	0	9	--	9	14	0	0
	250	0	11	--	13	16	0	0
	500	13	12	13	16	16	0	0

[a]Isolated by preparative reverse-phase chromatography.

[b]78.5% iso and 19.5% normal.

[c]29.4% iso, 66.6% anteiso, and 2% normal.

[d]53.6% iso and 45.7% normal.

[e]98.4% iso and 1.2% normal.

[f]98+% i-C_{14} ester; prepared by the treatment of i-C_{14}-HOacylNN with corresponding acetyl, benzoyl, or 3,5-dinitrobenzoyl chloride.

[g]Kirby-Bauer type disc assay; susceptible = greater than 15 mm dia. inhibition zone; moderately susceptible = 10-15 mm dia. inhibition zone; resistant = less than 10 mm dia. inhibition zone.

Table VII. Antibiotic Activity of Various Synthetic 1′-(N-acyl)nornicotines

Organism	Amount (μg)	C_2	C_4	C_6	C_8	C_{10}	C_{12}	C_{14}	C_{16}
		1′-(N-Acyl)nornicotines[a] — Inhibition Zone (dia. mm)[b]							
B. subtilis (+)	50	0	0	0	6	9	8	13	0
	250	0	0	0	12	11	9	14	0
	500	0	0	0	14	15	12	13	0
B. cereus (+)	50	0	0	0	9	13	11	23	8
	250	0	0	0	11	15	14	24	10
	500	0	11	0	15	19	16	25	14
M. thermosphactum (+)	50	0	0	0	0	13	12	20	17
	250	0	0	0	15	20	12	22	20
	500	0	9	0	17	22	12	22	20
E. cloacae (-)	50	0	0	0	0	0	0	0	0
	250	0	8	8	0	0	0	0	0
	500	0	0	0	8	12	0	0	0

[a]Prepared from nornicotine isolated from *N. tabacum* cv TI-1112 and corresponding acyl chloride.

[b]Kirby-Bauer type disc assay; susceptible = greater than 15 mm dia. inhibition zone; moderately susceptible = 10-15 mm dia. inhibition zone; resistant = less than 10 mm dia. inhibition zone.

Table VIII. Effects of 3-Hydroxyacylnornicotine Isolates and Esters and N-Acylnornicotines on the Growth of Wheat Coleoptiles

Compounds	Concentration (Molar) 10^{-3}	10^{-4}	10^{-5}	10^{-6}
	% Inhibition[a]			
HOacylNN Isolates[b]				
i+n-C_{12}	100	54	0	0
i+a+n-C_{12}	100	62	0	0
i+n-C_{14}	100	0	0	0
i-C_{14}	100	99	36	0
HOacylNN Esters[b]				
i-C_{14}-acetate	100	100	34	0
i-C_{14}-benzoate	0	0	0	0
i-C_{14}-3,5-dinitrobenzoate	0	0	0	0
N-acylNN[c]				
C_2	4	0	0	0
C_4	48	0	0	0
C_6	58	5	0	0
C_8	100	46	4	0
C_{10}	100	80	45	3
C_{12}	100	96	4	0
C_{14}	100	100	0	0
C_{16}	77	0	0	0

[a]Relative to control.
[b]See Table VI for composition.
[c]Prepared from nornicotine isolated from *N. tabacum* cv TI-1112 and corresponding acid chloride.

the highest level of these components and the acylNN are about 8% of the mixture. Both types of compounds have a wide range of biological activity, including toxicity to tobacco hornworm larva and plant bacterial growth inhibitors.

Acknowledgment

The investigative reports in this paper (88-7-65) is in connection with a project of the Kentucky Agricultural Experiment Station and is published with the approval of its Director. Funded in part by USDA Specific Cooperative Agreement 58-43YK-5-0039.

Literature Cited

1. Severson, R. F.; Johnson, A. W.; Jackson, D. M. Recent Adv. in Tob. Sci. 1985, 11, 105-74.
2. Severson, R. F.; Cutler, H. G.; Cole, P. D.; Jackson, D. M.; Sisson, V. A.; Johnson, A. W.; Herzog, G. A.; Stephenson, M. G. Proc. 12th Annual Plant Growth Regulating Soc. Mtg., 1985, p. 175.
3. Johnson, A. W.; Severson, R. F. J. Agric. Entomol. 1984, 1, 23-32.
4. Johnson, A. W., Severson, R.F. Tob. Sci. 1982, 26, 98-102.
5. Jackson, D. M.; Severson, R. F.; Johnson, A. W.; Chaplin, J. F.; Stephenson, M. G. Environ. Entomol. 1984, 13, 1023-30.
6. Cruickshank, I. A.; Perrin, D. R.; Mandryk, M. Phytopathol. 1977, 90, 243-49.
7. Cutler, H. G.; Cole, R. J. Plant and Cell Physiol. 1974, 15, 19-28.
8. Cutler, H. G.; Reid, W. W.; Delatang, J. Plant and Cell Physiol. 1977, 18, 711-14.
9. Cutler, H. G.; Severson, R. F.; Cole, P. D.; Jackson, D. M.; Johnson, A. W. ACS Symposium Series 296, Natural Resistance of Plants to Pests, Roles of Alleochemicals 1985, 178-96.
10. Jackson, D. M.; Severson, R. F.; Johnson, A. W.; Herzog, G. A. J. Chem. Ecol. 1986, 12, 1349-59.
11. Goodspeed, T. H. The Genus Nicotiana; Chronica Botanica, Waltham, MA, 1954, p. 102.
12. Severson, R. F.; Arrendale, R. F.; Chortyk, O. T.; Johnson, A. W.; Jackson, D. M.; Gwynn, G. R.; Chaplin, J. F.; Stephenson, M. G. J. Agric. Food Chem. 1984, 32, 566-570.
13. Severson, R. F.; Arrendale, R. F.; Smith, L. B.; McDuffie, K. L.; Sisson, V. A. Abstract of Papers, 38th Tobacco Chemists' Research Conference 1984, 38, 7.
14. Sisson, V. A.; Severson, R. F. Abstract of Papers, 39th Tobacco Chemists' Research Conference 1985, 39, 23.
15. Thurston, R.; Parr, J. C.; Smith, W. T. Proc. 4th Internatl. Tob. Sci. Congr. 1966, p. 424.
16. Jones, D.; Jones, G. A.; Hagery, T.; Creech, E. Exp. Appl. 1985, 38, 157-64.
17. Huesing, J. E.; Jones, D. Phytochem. 1987, 26, 1381-84.
18. Severson, R. F.; Arrendale, R. F.; Snook, M. E.; Sisson, V. A. Ga. J. Sci. 1985, 43, 21.

19. Severson, R. F.; Huesing, J. E.; Jones, D.; Arrendale, R. F.; Sisson, V. A. J. Chem. Ecology 1988, (in press).
20. Zado, E.; Jones, D. Plant Physiol. 1986, 82, 479-84.
21. Arrendale, R. F.; Severson, R. F.; Chortyk, O. T. J. High Resolut. Chromatogr. Chromatogr. Communic. 1983, 6, 436.
22. Arrendale, R.F.; Severson, R. F.; Chortyk, O. T. Anal. Chem. 1984, 56, 1533-37.
23. Cutler, H. G. 11th Proc. Plant Growth Regulator Soc. America 1984, 1.
24. Kurtz, T. E.; Link, R. F.; Tukey, J. W.; Wallace, D. L. Technometric 1965, 7, 95.
25. Severson, R. F.; McDuffie, K. L.; Arrendale, R. F.; Gwynn, G. R.; Chaplin, J. R.; Johnson, A. W. J. Chromatogr. 1981, 211, 111-21.
26. Snook, M. E.; Chortyk, O. T.; Arrendale, R. F. Beitr. Tabakforsch. 1984, 12, 227-43.
27. Gwynn, G. R.; Severson, R. F.; Jackson, D. M.; Stephenson, M. G. Tob. Sci. 1985, 29, 79-81.
28. Yamamoto, I.; Kamimura, H.; Yamashota, R.; Sakai, S.; Moga, M. Agric. Biol. Chem. 1962, 26, 709-16.

RECEIVED September 8, 1988

Chapter 23

Allelochemical Properties of *Nicotiana tabacum* Leaf Surface Compounds

D. R. Lawson[1], D. A. Danehower[1], D. G. Shilling[2], M. L. Menetrez[3], and H. W. Spurr, Jr.[3]

[1]North Carolina State University, Raleigh, NC 27695–7620
[2]University of Florida, Gainesville, FL 32611
[3]Crops Research Laboratory, Agricultural Research Service, U.S. Department of Agriculture, Oxford, NC 27565

The chemical constituents on the leaf surface of green tobacco (Nicotiana tabacum) are diverse in their composition. They include, among other compounds, alkanes, fatty alcohols, wax esters, diterpenes, and sucrose esters. In addition to contributing to the flavor quality of the cured leaf, evidence has been reported that some of these compounds, primarily the diterpenes and the sucrose esters, have biological activity in tobacco-insect and tobacco-microbe interactions. Evidence is less definitive for such activity in tobacco-plant relationships, in that allelopathic potential for these compounds is strictly inferred from *in vitro* bioassays.

In order to further investigate the allelochemical potential of these compounds, preparative chromatographic procedures for isolating the major leaf surface components of Nicotiana tabacum, were developed. Components were then assayed for phytotoxic and microbial activity. The results of this work support previously reported results and suggest additional biological activity which has not been reported to date.

Research over the past 25 years on the allelochemical potential of Nicotiana tabacum, L. leaf surface compounds has identified many of these compounds to be active as insect deterrents and attractants (1-5), antimicrobial agents (4,6-9), and potential allelopathic agents (4,10). The principal leaf surface constituents responsible for most of the reported biological activities are the following secondary metabolites: α- and β-4,8,13-duvatriene-1,3-diols (ADVT and BDVT); α- and β-4,8,13-duvatriene-1-ols (DVT-monols), (12Z)-labda-12,14-diene-8α-ol (Z-AB, Z-abienol) and 6-O-acetyl-2,3,4-tri-O-acyl-α-D-glucopyranosyl-β-D-fructofuranoside (sucrose esters) (Figure 1). The sucrose esters consist of six groups, each group consisting of a mixture of structural isomers

0097–6156/88/0380–0363$06.00/0

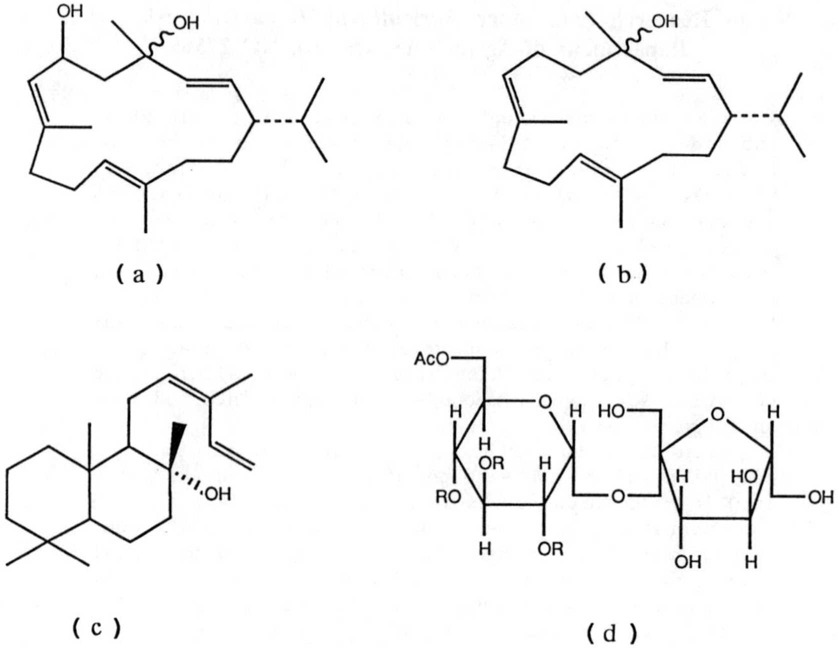

Figure 1. Major N. tabacum leaf surface secondary metabolites: (a) ADVT, BDVT; (b) DVT-monols; (c) Z-AB; (d) Sucrose Esters (R = C_3-C_8 normal and methyl branched fatty acids).

of equal molecular weight. These isomers differ in the chain length and/or position of the short chain fatty acid residues at the 2, 3,or 4 position of the glucose moiety. Each of the six groups of sucrose esters differs from each other by one methylene unit (14 AMU) or multiples thereof. Several of these metabolites are formed by the trichome heads of the glandular hairs found on the leaf surface (14).

The effects of Nicotiana leaf surface constituents on microorganisms and other plants has been investigated less thoroughly than their effects on insects. In 1963, Shepherd and Mandryk (6) attributed the lack of spore (conidia) germination of Peronospora tabacina Adam, the blue mold pathogen, on the leaf surface of N. tabacum L. cv "Virginia Gold" in part to the presence of a water soluble toxin. Much later, Cruickshank et al. (11) isolated two fungitoxic compounds, ADVT and BDVT, from the leaf surface of "Virginia Gold" tobacco, which accounted for most of the inhibitory activity.

Possible inhibitory effects of Nicotiana leaf surface compounds on plant growth have also been identified. Using a wheat (Triticum aestivum L., cv Wakeland) coleoptile segment bioassay, ADVT, BDVT, and the sucrose esters (4,12) have demonstrated potential plant growth inhibitory activity. While this work has piqued interest in the allelopathic potential of these compounds, biological effects on plants have been inferred strictly from in vitro bioassays. No in vivo effects have yet been reported.

The purpose of this study was to further characterize the allelochemical potential of N. tabacum leaf surface compounds on the growth of fungi and plants. Specifically, the effects of this such compounds on P. tabacina spore germination and in vivo biological effects on two common weed species, Echinochloa crus-galli L. Beuvois (barnyard grass) and Sesbania exaltata (Raf) Cory (hemp sesbania) were investigated.

MATERIALS AND METHODS

Isolation of Leaf Surface Components. Whole, green, bud leaves of Nicotiana tabacum (Brazilian domestic variety, Galpao) were washed with gentle agitation 45-60 seconds with HPLC grade methylene chloride in a 25 cm Buchner funnel fitted with Whatman No. 1 filter paper. Galpao was chosen for this work as it contains all of the major cuticular chemicals of interest. The resulting extract was filtered under aspirator vacuum until the plant material was free of excess solvent. The final filtrate from several extractions was concentrated in vacuo at 35°C, and then dried over sodium sulfate. The crude extract was stored in an amber bottle under nitrogen atmosphere at -80°C until needed.

Fractionation by Component Class. Preparative chromatography of the crude extract was carried out on a 2.5 x 68.5 cm bed of Sephadex LH-20 gel (2). Ten ml of a 300 mg/ml solution of the crude extract were loaded onto the column and eluted at 5 ml/min with a 99:1 methylene chloride-methanol mobile phase. One hundred fifty 5 ml fractions were collected and were followed by a single, 500 ml collection. After each run, the column was washed with 1 liter of 97:3 methylene chloride-methanol at 5 ml/min and then re-equilibrated with 99:1 methylene chloride-methanol.

The composition of individual test tube fractions or grouped fractions was determined using thin-layer chromatography (TLC) on silica gel layers and gas chromatography (GC) of silylated derivatives on a DB-5 column (13). Thin layer plates were developed in 9:1 chloroform-methanol. For visualization, TLC plates were sprayed with 10% ethanolic phosphomolybdic acid and then heated at 100°C for 1-3 minutes. Gas chromatographic peak identification was based on retention times of authentic standards and GC/MS data obtained from previous experiments. The Sephadex LH-20 procedure provided an excellent means of fractionating the leaf surface components by chemical class. Elution of the column provided the following fractions (in order of their elution): hydrocarbons, wax esters, Z-AB/DVT-monols, DVT-diols, miscellaneous oxygenated duvanes, and sucrose esters. The secondary metabolite fractions were then subjected to additional preparative chromatographic procedures in order to obtain pure materials.

Isolation of Z-AB/DVT-monols. Preparative separation of the Sephadex LH-20 gel fraction containing these compounds was carried out on a 21.5 x 2.5 cm bed of Merck Silica Gel 60. Two hundred mg of sample were loaded in a 1 ml injection volume onto the column and eluted at 4 ml/min with a 7:3 methylene chloride-hexane mobile phase. Eighty 5 ml test tube fractions were collected. TLC and GC evaluations of individual test tube fractions or combined fraction were carried out using the TLC and GC procedures described above. Thin layer plates were developed with 7:3 methylene chloride-hexane.

Isolation of DVT-diols. Preparative chromatography of the DVT-diol mixture was carried out on a 21.5 x 2.5 cm bed of Merck Silica Gel 60. Two hundred mg of combined Sephadex LH-20 fractions containing the diols were loaded onto the column in a 1 ml injection volume, and the column was eluted at 5 ml/min with 1:1:8 isopropanol-chloroform-hexane. Eighty 5 ml fractions were collected. Chemical composition of individual test tube fractions or grouped fraction collections was determined as described above. Thin layer plates were developed with 1:1:8 isopropanol-chloroform-hexane.

Isolation of Sucrose Esters. Preparative isolation or enrichment of individual sucrose ester groups was carried out on a Whatman Magnum 20 Partisil 10 ODS-3 high-performance preparative column. One hundred twenty-five mg of sample were loaded onto the column in a 1 ml injection volume. The mobile phase was 6:4 acetonitrile-water eluted at 5 ml/min. Eighty 5 ml fractions were collected.

An IBM Instruments 9430 ultra-violet/visible spectrophotometer, set at a wavelength of 225 nm, was used for detection. Absorbance data of the eluate of each test tube was plotted as a histogram. Combinations of fractions were based on histogram profiles.

Peronospora tabacina Conidial Germination Inhibition Assays. Microbial bioassays for the determination of spore germination inhibitory activity of the leaf surface compounds were carried out by the method of Menetrez et al. (23). Stock solutions (2.0 mg/ml) of the various leaf surface components were prepared and 10 fold serial dilutions performed. Fifty microliters of each solution, including methylene chloride and water controls, were then pipetted onto a 1.5% water-agar medium (total volume of agar = 2 ml) contained in a 10 x 35 mm petri dish, and the methylene chloride was allowed to evaporate. Three hundred microliters of a freshly harvested spore suspension containing 1 x 10^5 conidia/ml of water were then pipetted onto the water-agar surface. The petri dishes were covered and placed into a covered tray containing moistened filter paper. The samples were incubated four hours at 18°C and determinations of percent germination were made by counting a minimum of 100 spores and comparing with both methylene chloride and water controls. All experiments were performed in duplicate and were replicated three times.

Plant Growth Inhibition Assays. The phytotoxic activity of the various isolated fractions or purified compounds isolated from Nicotiana tabacum was determined by a previously described bioassay method (15). Some modifications were made for this research as described below. The pure compounds or mixtures were dissolved in either methylene chloride or acetone to establish the desired concentrations. Under a laminar-flow hood, 3.5 ml of each solution were placed into 60 ml glass jars. After evaporation, 0.22 g (60 seeds) of Echinochloa crusgalli L. Beauvois (barnyard grass) or 0.40 g (30 seeds) of Sesbania exaltata (Raf) Cory (hemp sesbania) seed were placed into the jars. Under sterile conditions, 3.5 ml of 15 mM MES [2-morpholinoethanesulfonic acid] buffer (0.1% v/v ethanol), adjusted to pH 6.0, were added to each jar. The jars were then covered, and the seeds were incubated in growth chambers with a 12-h photoperiod and a 28/23°C day/night temperature regime, respectively. The effects of the various compounds were evaluated after 84 h. At this time any qualitative symptoms were noted, and the total plant fresh weight (TFW) was determined. This weight was then used to determine the predicted shoot-plus-root fresh weight (PSRFW) as determined from the following models:

barnyard grass PSRFW = -0.24 + 0.741 (TFW)
hemp sesbania PSRFW = -0.48 + 0.727 (TFW)

The predicted shoot-plus-root fresh weight was used to increase the sensitivity of the bioassay. All experiments were conducted two or three times with three replications.

Data were initially subjected to analysis of variance (16). Single degree of freedom comparisons between the control and fractions eluted from the Sephadex LH-20 column were used to determine significant effects for the initial purification (Table I). Standard deviations were used to statistically evaluate the effects of concentration for each of the purified compounds on the growth of the two bioassay species.

Table I. Phytoxicity of the total leaf-surface extract and partially separated components (fractions) isolated from *Nicotiana tabacum*

Fraction number	Concentration (ppm)	Barnyard grass TFW[1]	Barnyard grass PSRFW[2]	Hemp sesbania TFW	Hemp sesbania PSRFW
		----------- % inhibition[3]	-------------		
24	100	+9.0	+18.7*	3.4	4.7
24	10	9.6	13.2	2.2	3.0
24	1	-	-	4.9	7.6
34	100	+8.9	+18.5*	7.4	3.2
34	10	6.7	9.4	6.7	9.9
39	100	23.1*	34.0*	8.4	13.2
39	10	12.8*	18.2*	1.1	1.2
44	100	4.2	2.8	6.8	10.6
44	10	0	0	+5.1	+8.6
49	100	21.6*	35.0*	11.7	17.6
49	10	30.8*	45.3*	+5.5	+9.6
49	1	-	-	8.0	12.3
64	100	38.1*	61.2*	11.2	17.2
64	10	2.6	7.8	11.2	17.3
64	1	+28.9*	+53.1*	-	-
69	100[4]	43.6*	69.0*	23.2*	36.4*
69	10[4]	38.6*	59.2*	13.5*	21.0*
69	1	+15.7*	+29.1*	-	-
69	0.1	+55.5*	+101.4	-	-
98	100	19.9	32.8*	17.6*	27.0*
98	10	7.7	11.3	+9.0	+14.8
109	100	26.8*	43.3*	18.1*	28.3*
109	10	+9.8	+17.4	8.2	9.3
113	100	+10.2	+18.5	18.2*	27.6*
113	10	+14.4	+24.9*	+6.3	+10.4
113	1	-	-	3.5	5.4
column wash	100[4]	36.2*	57.1*	14.1*	22.1*
column wash	10	15.9*	19.2*	8.3	13.1
column wash	1	+32.3*	+58.9*	-	-
column wash	0.1	+56.7	+103.8*	-	-
crude extract	1000[5]	58.7*	84.9*	27.5*	43.2*
crude extract	100	43.4*	68.1*	19.5*	30.2*
crude extract	10	0.4	0.7	13.7	21.1*
crude extract	1	+14.7	+27.1*	7.6	11.8
crude extract	0.1	+34.6*	+63.4*	0.8	1.2
crude extract	0.01	+42.1	+77.0*	-	-

[1]Total fresh weight.
[2]Predicted shoot-plus-root fresh weight.
[3]Values followed by an asterisk are significantly different from the control at the P=0.05 level. Growth Stimulation indicated by +.
[4]Shoots of barnyard grass were bleached white and root growth of both species was either very inhibited and/or abnormal.
[5]No root growth and very little shoot growth.

RESULTS AND DISCUSSION

Chromatographic Isolations. Preparative chromatography on Sephadex LH-20 gel of Galpao leaf surface extracts yielded enriched diterpene and sucrose ester fractions as indicated in Figure 2. The gel fractions and the crude extract were evaluated for biological activity by determining their effect on fungal spore germination and plant growth. Decisions to further fractionate selected gel fractions were based on these initial bioassay results. Subsequent adsorption and reverse phase preparative chromatographic procedures yielded at least 98% purity of ADVT, BDVT, DVT-monols, and sucrose esters (group V) (Figure 2). Eighty percent purity of Z-AB was obtained. Such purities were routinely obtained from the crude leaf surface extract after just two preparative chromatographic steps. These pure isolates were also evaluated for biological activity.

Peronospora tabacina Spore Germination Inhibition. Spore (conidial) germination assays were conducted on both the component mixtures obtained from Sephadex LH-20 gel chromatography as well as on the pure compounds isolated from the second chromatographic steps. Although all of the major chemical classes represented were tested, our discussion will be limited to those fractions or components which showed significant inhibitory effects. Bioassays of the DVT-monol fractions showed no activity even at the highest concentration tested (50 ppm-data not shown). Z-AB, however, inhibited conidial germination by 37% at 50 ppm. This activity decreased rapidly at lower concentrations (5 ppm), and eventually showed mild stimulatory activity at the lowest concentrations (Figure 3c). The phenomenon of germination inhibition at high concentrations and stimulation at lower concentrations was reproducible and was observed for virtually all the components assayed. Assays of ADVT and BDVT (Figure 3a and b) confirmed the previous observations of Cruickshank et al. (11). Activity was highest at 50 ppm and, once again, decreased rapidly at lower concentrations, eventually leading to slight stimulation of germination. Apparently no significant difference exists between the two isomers with regard to their biological activity versus concentration. This contrasts with the report of Cruickshank et al., who observed slightly higher activity for the β isomer. Bioassays of concentrations between 5 and 50 ppm are planned in order to ascertain the actual ED_{50} (estimated dose required for 50% germination inhibition).

The sucrose ester mixture (Figure 3d) showed very slight inhibitory activity at 50 ppm. Further fractionation of this mixture into Groups I-VI has been accomplished and bioassays are planned in order to ascertain the relative activity of the individual groups.

Plant Growth Inhibition. The plant growth-regulating activities of the compounds isolated from tobacco were evaluated primarily to determine any possible ecological significance. The role of allelopathic compounds in both natural ecosystems and the

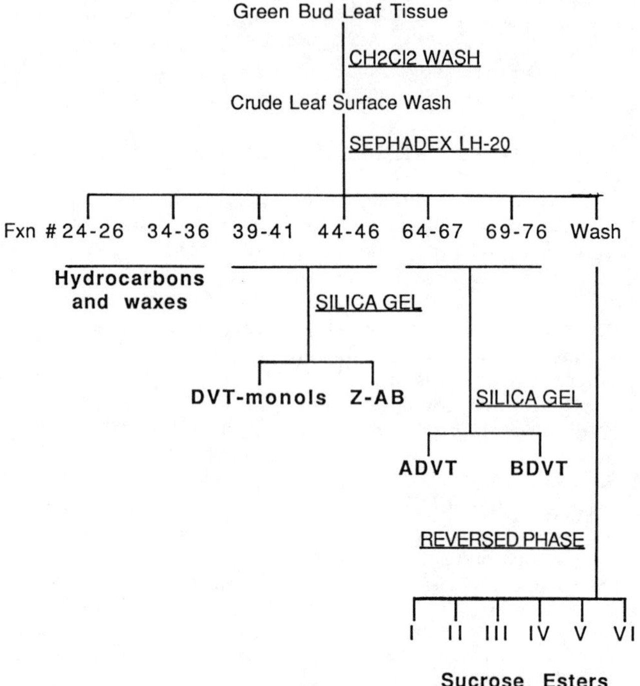

Figure 2. Extraction and separation scheme for the isolation of principle *N. tabacum* leaf surface components.

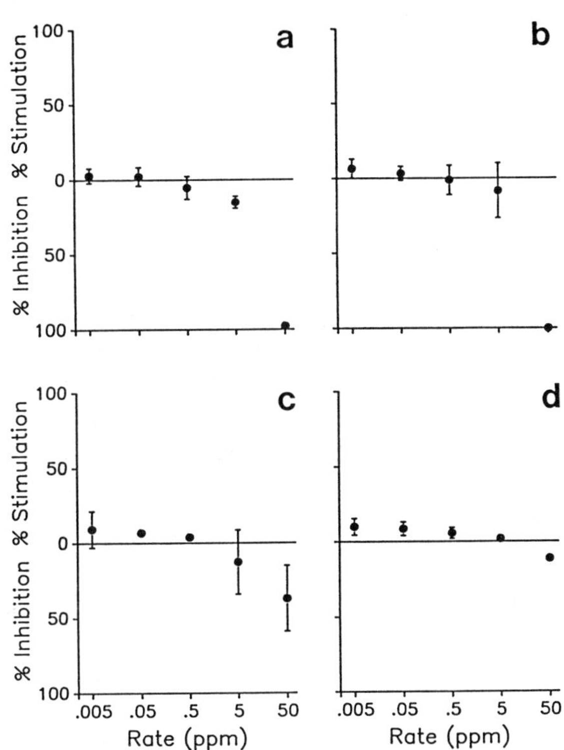

Figure 3. Effects of N. tabacum leaf surface components on P. tabacina germination: (a) ADVT; (b) BDVT; (c) Z-AB (d) Sucrose Esters.

agroecosystem has been well documented (17). Proving a compound phytotoxic is not only important for ecological considerations but can also have technological implications in terms of the development of pesticides that are based on natural product chemistry (18). With these considerations in mind, a whole-plant bioassay was used. This bioassay allowed a compound to be evaluated for its ability to influence the growth of an intact plant. Thus, any observable biological activity has greater potential implications than activity observed using an *in vitro* bioassay. The bioassay used in these studies contrasts with the wheat coleoptile bioassay used by Cutler *et al*. (4) to evaluate similar compounds also isolated from *N. tabacum*.

The initial evaluation of the crude leaf-surface tobacco extract indicated a significant amount of inhibitory and stimulatory activity on both bioassay species (Table I). The quantitative and qualitative activity of the crude extract indicated that compounds were present that had plant growth-regulating activity and warranted further investigation. Therefore, the crude extract was fractionated as previously described in order to determine what compound(s) was causing the biological activity of interest.

Table I shows the effects of the various fractions separated by Sephadex LH-20 gel chromatography on the growth of barnyard grass and hemp sesbania. Fractions 64 and 69 and the column wash (CW) caused the greatest amount of plant growth inhibition. Fraction 69 caused 69% and 36% inhibition of PSRFW at 100 ppm in barnyard grass and hemp sesbania, respectively. The CW not only caused a similar level of inhibition at 100 ppm, but also caused 104% stimulation in barnyard grass PSRFW at 0.1 ppm.

Although the quantitative activity of these fractions was significant, of more interest was the qualitative characteristic of fraction 69 and the CW to cause bleaching (i.e., white shoot tissue) in barnyard grass. This qualitative characteristic not only stressed the importance of using a whole plant bioassay, but also acted as a unique biological marker that was used to follow biological activity of interest during the chemical purification process. By using a bioassay that provided both qualitative and quantitative information during the chemical separation process, the time-consuming task of chemical purification and structural ellucidation was devoted to compounds that caused the biological activity of interest.

Figures 4 and 5 show the plant growth-regulating activity of the purified compounds which were identified as components of the biologically active fractions presented in Table I. Of the terpenes evaluated, BDVT showed the greatest level of plant growth inhibition (Figures 4b and 5b), followed by the α isomer (Figure 4a and 5a). The β isomer caused 30% inhibition and the α isomer caused 5% inhibition at 1 ppm when assayed against barnyard grass. Springer *et al*. (12) also showed that the β isomer was more active than the α isomer in the wheat coleoptile assay system. These compounds also caused bleaching of the shoot tissue in barnyard grass. The other terpenes that were evaluated, DVT-monols, (data not shown) and *Z*-AB (Figure 4c and 5c) also caused significant levels of inhibition of barnyard grass PSRFW. In addition, both of these compounds caused growth stimulation at lower concentrations.

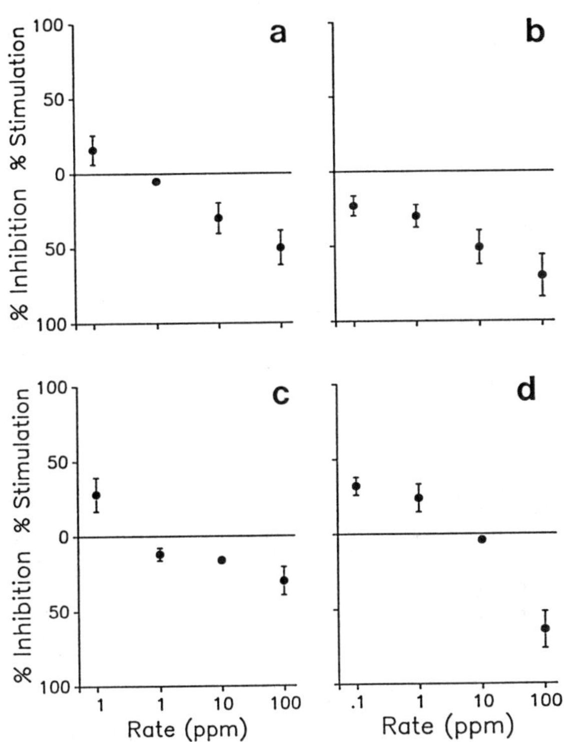

Figure 4. Effects of N. tabacum leaf surface components on Echinochloa crusgalli (barnyard grass) growth: (a) ADVT; (b) BDVT; (c) Z-AB; (d) Sucrose Esters.

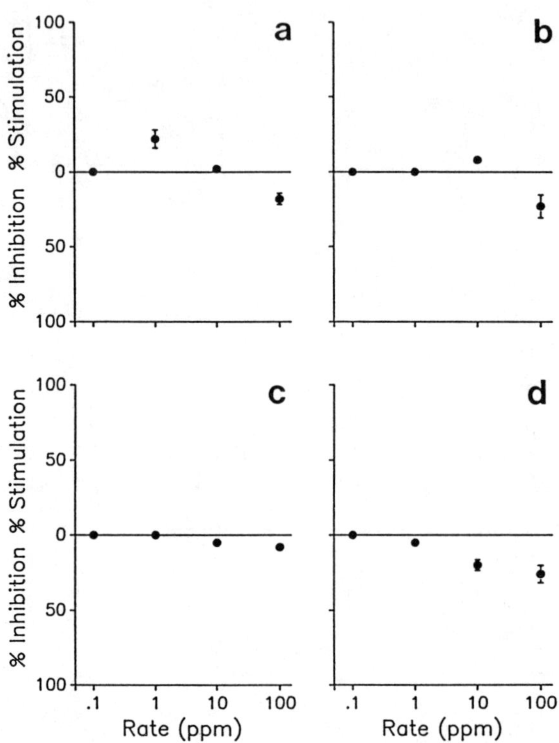

Figure 5. Effects of N. tabacum leaf surface components on Sesbania exaltata (hemp sesbania) growth: (a) ADVT; (b) BDVT; (c) Z-AB; (d) Sucrose Esters.

The quantitative and qualitative activity of the pure compounds accounted for the biological activity associated with the initially separated fractions presented in Table I. Cutler and Cole (10) have previously described the inhibitory properties of these compounds, but not the growth stimulation or the bleaching activity.

The second most active isolates were the sucrose esters (Figure 4d and 5d). The sucrose esters not only inhibited barnyard grass PSRFW comparable to BDVT, but also caused slightly more inhibition of PSRFW (at 10 ppm) of hemp sesbania as compared to BDVT. The inhibitory effects of the sucrose esters of tobacco have been previously demonstrated (4) in a wheat coleoptile assay. In addition, the sucrose esters caused 24% and 32% stimulation of barnyard grass PSRFW at 1 and 0.1 ppm, respectively. The sucrose esters also caused bleaching of barnyard grass shoot tissue. The qualitative and quantitative effects of the purified sucrose esters agreed closely with the activity associated with the CW (Table I).

CONCLUSIONS

The investigations described in this paper confirm and expand upon previous work conducted on the biological activity of *Nicotiana tabacum* leaf surface exudates. The chromatographic procedures allow the isolation of relatively large quantities of these materials, as well as other surface constituents (hydrocarbons, wax esters, etc.), with a minimum number of chromatographic steps, which include simple low pressure chromatographic systems. The Sephadex LH-20 separation is a particularly powerful method. It has high loading capacity, is a very mild separation method based on both partitioning and gel permeation modes, and it results in a clear-cut fractionation into component classes.

Bioassays for *Peronospora tabacina* conidial germination inhibition indicated that, in addition to the known activity of ADVT and BDVT, *Z*-AB also had significant biological activity. The weaker activity displayed by the complex mixture of sucrose esters deserves further work. Assays of individual sucrose ester groups or individual isomers may demonstrate that the biological activity indicated for the sucrose esters as a group is concentrated in a limited number of isomers. For example, Cutler *et al.* (4) tested these compounds against both gram positive and gram negative bacteria. Growth inhibitory activity was indicated against gram positive organisms and activity which was concentrated in the Group V esters in which the 2,3, and 4 hydroxy positions on the glucose moiety are esterified with β-methylvaleric acid.

Plant growth-regulating activity was also demonstrated in the whole plant growth bioassay. Once again, both and BDVT showed considerable activity. Growth inhibition at higher concentrations, and the observation of bleaching activity with these compounds and the sucrose esters is significant. Moreover, the stimulation of plant growth at lower concentrations is indicative of more complex biological properties.

This research confirms that *Nicotiana tabacum* contains plant growth-regulating compounds. The use of the whole plant bioassay allowed both quantitative and qualitative evaluations of isolates during the purification process. Thus, not only were the appropriate compounds identified but time spent separating and identifying chemicals of no biological significance was reduced. Questions still remain concerning the role of these compounds in possible ecological interactions (i.e., allelopathy). In a number of examples, natural products have been shown to leach from plants and elicit an allelopathic response to surrounding vegetation (17, 18). However, the plant growth-regulating activity of the compounds evaluated in this study was demonstrated in glass jars. The extent of their activity in soil has not been studied. In addition, although rainfall has been shown to reduce the levels of these compounds on tobacco leaves, which would indicate that leaching is possible, the concentration that ultimately reaches the soil has not been determined (21). Therefore, the ecological significance of these compounds relative to plants has not yet been ascertained.

Once biological activity has been demonstrated for a natural product, several approaches to their use come to mind. First, plant breeders and geneticists are already actively involved in the introduction of several of these chemical traits into tobacco for the purpose of protecting the plant from insect predation or disease. In the future, biotechnology may supplement this classic genetic approach. These approaches hold promise for the *in vivo* application of *N. tabacum* leaf surface chemicals in the control of pests.

In addition, the sucrose esters in particular, represent an unusual class of natural products that have not been extensively examined. Aside from the work of Severson *et al.* (13), there is only one additional report of the occurrence of these compounds in plants (22). The wide range of biological activity, (apparent) structural simplicity of individual isomers, and potentially inexpensive nature of synthetic starting materials make these compounds an intriguing possibility for lead chemistry in the synthesis of novel pesticides.

Acknowledgments

Paper No. 11504 of the Journal Series of the North Carolina Agricultural Research Service, Raleigh, NC 27695-7601. Florida Agricultural Experiment Station Journal Series No. 8886. The use of trade names in this publication does not imply endorsement by the North Carolina Agricultural Research Service of the products named, nor criticism of similar ones not mentioned.

The authors would like to thank Mr. Clyde W. Adcock, Mr. Henry (Buddy) Quick, Mr. Mark Mossler, Mr. Danai Chow, Mr. Wesley Adcock, Mr. Albert Butler, Mr. John Weeks, and Ms. Marie Hall for their excellent technical assistance. Appreciation is extended to the North Carolina Tobacco Foundation and the R. J. Reynolds Tobacco Company for partial financial support of this work.

Literature Cited

1. Elsey, K. D.; Chaplin, J. F. *J. Econ. Entomol.* 1978, *71*, 723.

2. Jackson, D. M; Severson, R. F.; Johnson, A. W.; Herzog, G. A. J. Chem. Ecol. 1986, 12, 1349.
3. Johnson, A. W.; Severson, R. F. J. Agric. Entomol. 1984, 1, 23.
4. Cutler, H. G.; Severson R. F.; Cole, P. D.; Jackson, D. M.; Johnson, A. W. In Natural Resistance of Plants to Pests; Green, M. B.; Hedin, P. A., Eds.; ACS Symposium Series No. 296; American Chemical Society: Washington, D. C., 1986; p. 178.
5. Huesing, J. R.; Jones, D. Phytochemistry 1987, 26, 1381.
6. Shepherd, C. S.; Mandryk, M. Aust. Soc. Biol. Sci. 1963, 16, 77.
7. Cohen, Y.; Eyal, H.; Goldschmidt, Z.; Sklarz, B. Physiol. Plant Path. 1983, 22, 143.
8. Bailey, J. A.; Burden, R. S.; Vincent, G. G. Phytochemistry 1975, 14, 597.
9. Bailey, J. A.; Vincent, G. G.; Burden, R. S. J. Gen. Microbiol. 1974, 85, 57.
10. Cutler, H. G.; Cole, R. J. Plant and Cell Physiol. 1974, 15, 19.
11. Cruickshank, I. A. M.; Perrin, D. R.; Mandryk,M. Phyto-path. Z. 1977, 90, 243.
12. Springer, J. R.; Clardy, J.; Cox R. H.; Cutler, H. G.; Cole, R. J. Tetrahedron Lett. 1975, 32, 2737.
13. Severson, R. F.; Arrendale, R. F.; Chortyk, O. T.; Johnson, A. W.; Jackson, D. M.; Gwynn, G. R.; Chaplin, J. F.; Stephenson, M. G. J. Agric. Food Chem. 1984, 32, 566.
14. Keene, C. K.; Wagner, G. J. Plant Physiol. 1985, 79, 1026.
15. Shilling, D. G.; Yoshikawa, F. In Allelochemicals: Role in Agriculture and Forestry; G. R. Waller, Ed.; ACS Symposium Series No. 330; American Chemical Society: Washington,D.C., 1987; p. 334.
16. Helwig, J. R.; Council, K. A. (Ed.) SAS User's Guide; SAS Institute, Inc.: Cary, N.C., 1979.
17. Rice, E. L. In Allelopathy; Academic Press: New York, New York, 1984; p. 422.
18. Duke, S. O. In Reviews of Weed Science; Weed Science Society of America: Champaign, IL, 1986; pp. 15-44.
19. Cutler, H. G.; Reid, W. W.; Deletang, J. Plant Cell Physiol. 1977, 18, 711.
20. Davis, R. F. Amer. J. Bot. 1928, 15, 620.
21. Severson, R. F.; Johnson, A. W.; Jackson, D. M. In Recent Advances in Tobacco Science, Vol. 11; 39th Tobacco Chemists Research Conference; Tobacco Abstract Service: North Carolina State University, 1985; p. 105.
22. Burke, B. A.; Goldsby, G.; Mudd, J. B. Phytochemistry 1987, 26, 2567.
23. Menetrez, M. L., Spurr, H. W. Jr., Danehower, D. A., and Lawson, D. R. Phytopathology (in review).

RECEIVED April 20, 1988

Chapter 24

Plant Constituents as Oviposition Deterrents to Lepidopterous Insects

J. A. A. Renwick

Boyce Thompson Institute, Tower Road, Ithaca, NY 14853

The natural resistance of plants to herbivorous insects can often be explained by the presence of specific metabolites that act as toxins or as behavior-modifying agents. Oviposition deterrents can offer a first line of defense against insect pests, and extracts of non-host plants have been effective in protecting otherwise susceptible hosts. A search for active plant constituents that cause avoidance of treated plants for egg laying has led to examination of unacceptable plant species which are closely related to preferred species. The cabbage butterfly, *Pieris rapae* lays eggs on most members of the Cruciferae, but a few species such as *Erysimum cheiranthoides* are avoided. Potent oviposition deterrents have been isolated from this plant and characterized as cardenolides.

Any search for natural products that might be used to control insects inevitably leads us to the natural chemical defense systems that plants have evolved through the ages to protect themselves. The natural resistance of plants against invaders takes many different forms, both physical and chemical, and possible mechanisms have been the subject of many reviews (1, 2, 3, 4, 5). This discussion will be confined to the chemical factors involved.

Allelochemicals that serve to protect plants from insects may function in two ways: (1) by affecting insect development and survival on the plant, or (2) by influencing insect behavior. The behavioral events of most interest here are those involved in the process of host selection. Observations on the avoidance of specific plants by insects can provide the first clues that protective chemicals may be present. More detailed studies on host preferences can provide information on positive as well as negative stimuli affecting host finding and acceptance. If we know what chemicals mediate this behavior, we might have a basis for genetic manipulation of crop plants or for using chemicals directly to protect susceptible plants.

0097–6156/88/0380–0378$06.00/0

Chemical Mediation of Host Selection

Host selection by insects involves one or more behavioral responses that may be mediated by plant constituents. These include orientation, oviposition and feeding (6, 7). However, these phases are not always clearly separated. For example, oviposition generally cannot occur without orientation to the target plant, so analysis of the factors affecting oviposition often requires consideration of the stimuli that elicit landing. The chemical factors mediating landing on and acceptance of potential host plants for oviposition include attractants, repellents, stimulants and deterrents. The importance of understanding the interplay among these factors has been recognized and emphasized by many authors (7, 8). But the idea of utilizing the behavior-modifying chemicals that affect oviposition to combat insect pests has not been widely investigated. The concept of interfering with oviposition is particularly appealing, since it could provide the opportunity to block insect infestations before any damage can occur.

Manipulation of Oviposition

Several different strategies might be employed to manipulate oviposition by insect pests. Firstly, we might interfere with orientation or landing by intercropping with odoriferous plants or through the direct use of repellents. Secondly, oviposition deterrents might be introduced into crop plants as a result of plant breeding or recombinant DNA techniques, or deterrents could be sprayed directly on the plants to be protected. Thirdly, oviposition stimulants might be used to "fool" the insects into laying their eggs on non-host plants which would be unsuitable for larval development. Genetic manipulation might also be possible, in this case, to remove stimulant from the crop plants. The concept of interfering with oviposition as a means of insect control has yet to be seriously tested in the field. Several questions need to be answered before the suggested strategies can be implemented. The use of deterrents may lead to habituation of the target insects. Therefore, the success of this approach may depend on the presence of alternate sites for oviposition. It is likely that combinations of possible strategies will provide the best results, and the idea of using trap crops, or plants treated with stimulants, along with deterrent-treatment of the main crop, has considerable merit.

Despite the widespread interest in oviposition as a target stage for insect control and rapidly expanding research programs in the field, relatively few of the chemicals involved have actually been identified. Probably the best progress in the area has been in the isolation and identification of volatile compounds that attract or stimulate specialist flies such as the onion fly and the carrot fly (9, 10, 11, 12). These compounds function as attractants and/or arrestants so that oviposition is the end result. But the possible involvement of contact stimuli that trigger the final act of egg deposition cannot be excluded. Among the specialist butterflies, the large white cabbage butterfly,

Pieris brassicae, has been widely studied in Europe. This butterfly is stimulated to oviposit by glucosinolates present in its crucifer host plants (13). Another group of butterflies presently being extensively studied are the swallowtails. Flavanone glycosides that stimulate oviposition by Papilio protenor have been identified in Japan (14),

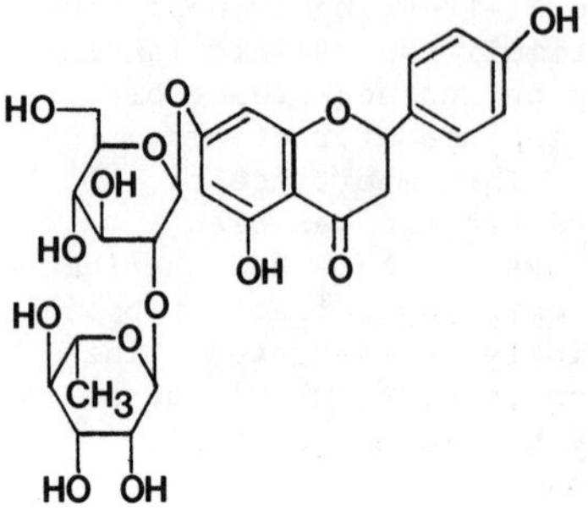

Naringin

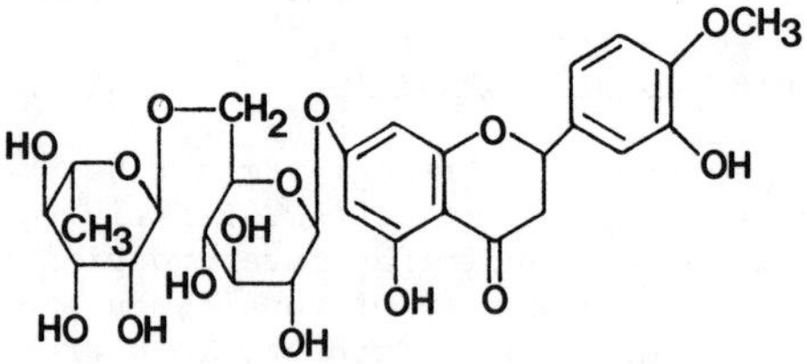

Hesperidin

and Feeny and coworkers recently identified acidic flavonoids that act synergistically with another plant component to stimulate oviposition by the umbellifer-feeding Papilio polyxenes (15). In all the butterflies studied, the critical stimuli are perceived through tarsal receptors (13, 15, 16), so contact with the plant is necessary for host recognition. The factors affecting oviposition by moths are similar to those for butterflies, and the subject has been reviewed by Ramaswamy (17).

Oviposition Deterring Pheromones

The role of inhibitory compounds that discourage oviposition on unsuitable plants has long been recognized, but has only recently been emphasized by Jermy and Szentesi (18). Although we need to know as much as possible about positive as well as negative inputs in plant-insect relationships, from a practical point of view, the idea of blocking oviposition appears to offer the most promise. The potential of oviposition deterrents has recently been highlighted by the discovery of oviposition deterring pheromones. Pioneering work by Prokopy (19) has shown that many species of fruit flies release a marking chemical during and after oviposition. This chemical is smeared onto the fruit surface and serves to discourage further oviposition by other female flies on that fruit. In this way, overcrowding is avoided and the larvae can develop with little or no competition. Similar oviposition deterring pheromones (epideictic pheromones) have been found in a variety of insects where resources are limited and adequate spacing of larvae is essential (20). In most of these cases, physical as well as chemical cues play a role in the spacing mechanism. For example, cowpea weevils achieve almost perfect distribution of their eggs on available seeds by a combination of physical recognition and chemical marking of eggs (21). The cabbage butterfly, *Pieris brassicae*, also uses unknown chemicals as well as visual cues to recognize the presence of egg batches on host plants (22, 23). The identification of oviposition deterring pheromones has proved to be extremely difficult, but the first success has recently been reported by a group in Switzerland. Electrophysiological monitoring was used to isolate the pheromone of the European cherry fruit fly and the active compound has been characterized (24).

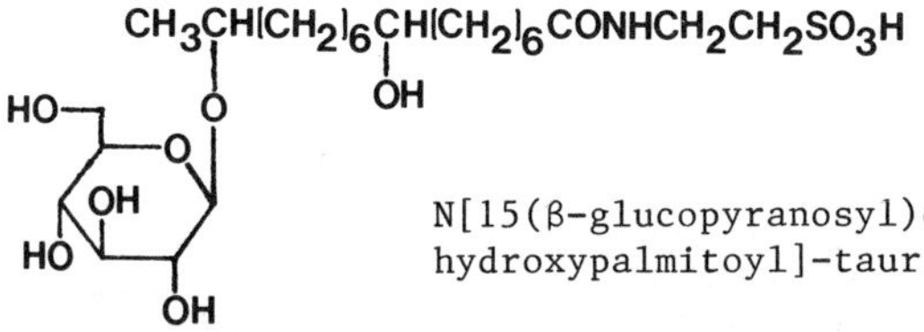

$$CH_3CH(CH_2)_6CH(CH_2)_6CONHCH_2CH_2SO_3H$$

(structure: glucopyranosyl residue with HO, OH, HO, OH groups attached via O at C-15; OH at C-8)

N[15(β-glucopyranosyl)oxy-8-hydroxypalmitoyl]-taurine

Field tests of extracts of this pheromone had previously shown that practical application is a real possibility (25).

Plant Constituents Deterring Oviposition

The success of studies on oviposition deterring pheromones has helped in boosting research on plant-produced deterrents. Several cases have now been reported where insect injury results in the release of plant constituents that may function as oviposition deterrents. One example is the olive fruit fly, which punctures the fruit as it lays its eggs, and the juices flowing from the wound contain chemicals that deter further oviposition (26). Another spacing mechanism was found for the cabbage looper, *Trichoplusia ni*. This moth lays its eggs evenly over available plants and avoids plants that are already occupied by feeding larvae. The larval frass was found to be a source of oviposition deterrent (27), but further studies showed that disruption of plant tissue by the feeding larvae could account for release of active compounds from the plant (28). Oviposition deterrents have now been found in frass of several insects including the European corn borer and species of *Spodoptera* (29, 30, 31), and plant constituents may be responsible for the activity in most of these cases.

It is now clear that several sources of natural oviposition deterrents are available to us. The marking pheromones represent a group of potent species-specific agents for insects which depend on adequate spacing for their survival. Plant constituents offer a more general source of biologically active chemicals. These may be released from damaged host plants, and thus play a role in the distribution of insect populations, or compounds from non-host plants might be more generally involved in protection from a variety of herbivores. Both volatile and non-volatile negative signals may be provided by a plant. The use of volatile repellents to protect a crop plant has often been suggested, and intercropping with herbs or other fragrant plants is an attempt to capitalize on this principle (32, 33, 34). However, most efforts to use oviposition deterrents as protective agents involve non-volatile plant constituents, which are likely to have a more persistent effect. Some early work based on ideas of organic gardeners included the use of tomato homogenates against cabbage butterflies and chestnut extracts against sugar beet moths (35, 36). These studies have been followed by several others aimed at lepidopterous pests of various crop plants (Table I). The usual approach, therefore, has been to test extracts of unrelated, non-host plants as deterrents against a particular insect.

Oviposition Deterrents for *Pieris rapae*

Initial studies on the cabbage butterfly, *Pieris rapae*, included a survey of deterrent activity in a variety of host and non-host plants. In choice bioassays with treated and untreated cabbage plants in greenhouse cages, hexane extracts of non-host plants were all deterrent (44). However, hexane extracts of host plants were also deterrent. These results suggested that many non-specific chemicals in damaged tissues of plants may act as deterrents. But when water soluble material from the same plants was tested, clear differences between hosts and non-hosts were seen (44). The lack of deterrent activity in water extracts of host plants was not surprising since these extracts are known to contain oviposition stimulants (45). As a result of these studies, attention has been focused on crucifers, which we might expect to be attractive to cabbage butterflies, but which are rejected after landing. Two examples of unacceptable crucifers, *Erysimum cheiranthoides* and *Capsella bursa-pastoris*, clearly contain powerful deterrents (44). If these plants also produce stimulants, the deterrents must be particularly potent to outweigh the positive effects of the stimulant. This argument has provided the rationale for our approach of studying plants that are avoided by an insect but which are closely related to preferred hosts.

Erysimum cheiranthoides is a small inconspicuous plant with little foliage for chemical studies. But propagation in the greenhouse is easy, and kilogram quantities were produced over a period of a few months. Separation of polar extracts by partitioning between water and butanol provided a convenient means of showing that both stimulant and deterrent are present in this plant (46).

When butanol extracts of *E. cheiranthoides* were subjected to reversed phase HPLC, with a water-acetonitrile gradient, all the deterrent activity was obtained in a fraction eluting with about 28% acetonitrile. Maximum UV absorption at 219 nm for the major peaks suggested that these compounds may be cardenolides. A positive reaction with Keddes' Reagent confirmed the cardenolide character of the active components (47) and two of the glycosides have been positively identified (48).

Table I. Plant extracts tested as oviposition deterrents against lepidopterous pests

Insect	Plant Extract	Reference
Pieris brassicae *P. rapae* *P. napi*	various	Lundgren (35)
Scrobipalpa ocellatella	*Castanea sativa* (chestnut)	Robert & Blaisinger (36)
Heliothis virescens	various	Tingle & Mitchell (37)
H. virescens	*Sambucus simpsonii* (elderberry)	Tingle & Mitchell (38)
Spodoptera exigua *S. eridania*	*Amaranthus hydridus* (pigweed)	Mitchell & Heath (39)
Spodoptera frugiperda	corn leaves	Williams et al. (30)
Plutella xylostella	various herbs	Dover (40)
Cydia pomonella	various medicinal plants	Adivardi & Benz (41)
Spodoptera litura	Neem kernel	Joshi & Sitaramaiah (42)
Spodoptera frugiperda	Neem kernel	Hellpap & Mercado (43)

The big question that now remains is whether these cardenolides could be effective in protecting cabbage plants in the field. The results of preliminary field tests of crude extracts suggest that some degree of control is possible (49). The concentrations of cardenolides required are likely to be in the microgram per plant range, well below mammalian toxicity levels. However, the toxic reputation of cardenolides may make acceptance of these compounds in an agricultural system difficult, and research efforts are now being focused on other crucifer plants. The approach of isolating deterrents from unacceptable plants that are related to hosts appears to be promising, and this work should serve as a model system for future studies.

Acknowledgments

Research was supported in part by the U.S. Department of Agriculture under Agreement No. 86-CRCR-1-2007.

Literature Cited

1. Painter, R.H. "Insect Resistance in Crop Plants"; University Press of Kansas: Lawrence/London, 1951; 520 pp.
2. Panda, N. "Principles of Host-plant Resistance to Insect Pests"; Allanheld/Universe; New York, 1979; 386 pp.
3. Hedin, P.A., ed. "Plant Resistance to Insects"; Am. Chem. Soc. Symp. Ser. 1983, 208, 375 pp.
4. Green, M.B. & P.A. Hedin, eds. "Natural Resistance of Plants to Pests; Am. Chem. Soc. Symp. Ser. 1986, 296, 243 pp.
5. Robert, P.C. Agronomie 1986, 6, 127-42.
6. Renwick, J.A.A. In: "Plant Resistance to Insects" (P.A. Hedin, ed.); Am. Chem. Soc. Symp. Ser. 1983, 208, 199-213.
7. Miller, J.R.; Strickler, K.L. In: "Chemical Ecology of Insects" (W.J. Bell & R.T. Cardé, eds.); Chapman and Hall: London, 1984; 127-57.
8. Dethier, V.G. Ent. exp. appl. 1982, 31, 49-56.
9. Matsumoto, Y.; Thorsteinson, A.J. Applied Entomol. and Zool. 1968, 3, 5-12.
10. Ishikawa, Y.; Ikeshoji, T.; Matsumoto, Y. Applied Entomol. and Zool. 1978, 13, 115-22.
11. Vernon, R.S.; Pierce, H.D., Jr.; Borden, J.H.; Oehlschlager, A.C. Environ. Entomol. 1978, 7, 728-31.
12. Guerin, P.M., Städler, E.; Buser, H.R. J. Chem. Ecol. 1983, 9, 843-61.
13. Ma, W.C.; Schoonhoven, L.M. Ent. exp. appl. 1973, 16, 343-57.
14. Honda, K. J. Chem. Ecol. 1986, 12, 1999-2010.
15. Feeny, P.P. 1987, personal communication.
16. Traynier, R.M.M. Physiol. Entomol. 1979, 4, 87-96.
17. Ramaswamy, S.B. J. Insect Physiol. (in press).
18. Jermy, T.; Szentesi, Á. Ent. exp. appl. 1978; 24, 258-71.
19. Prokopy, R.J. Environ. Entomol. 1972, 1, 326-32.
20. Prokopy, R.J. In: "Semiochemicals. Their Role in Pest Control" (D.A. Nordlund, R.L. Jones, W.J. Lewis, eds.); Wiley-Interscience: New York, 1981; pp 181-213.
21. Messina, F.J.; Renwick, J.A.A. Ent. exp. appl. 1985, 37, 241-5.
22. Rothschild, M.; Schoonhoven, L.M. Nature 1977, 266, 352-5.

23. Behan, M.; Schoonhoven, L.M. Ent. exp. appl. 1978, 24, 163-79.
24. Hurter, J.; Boller, E.F.; Städler, E.; Buser, H.R.; Bosshard, N.U.; Damm, L.; Kozlowski, M.W.; Schöni, R.; Raschdorf, F.; Dahinden, R.; Schlumpf, E.; Fritz, H.; Richter, W.J.; Schreiber, J. Experientia 1987, 43, 157-64.
25. Katsoyannos, B.I.; Boller, E.F. Environ. Entomol. 1976, 5, 151-2.
26. Cirio, U. Redia 1971, 52, 577-600.
27. Renwick, J.A.A.; Radke, C.D. Environ. Entomol. 1980, 9, 318-20.
28. Renwick, J.A.A.; Radke, C.D. Ent. exp. appl. 1981, 30, 201-4.
29. Dittrick, L.E.; Jones, R.L.; Chiang, H.C. J. Insect Physiol. 1983, 29, 119-21.
30. Williams, A.L.; Mitchell, E.R.; Heath, R.R.; Barfield, C.S. Environ. Entomol. 1986, 15, 327-30.
31. Hilker, M. Naturwissenschaften 1985, 72, 485-6.
32. Perrin, R.M.; Phillips, M.L. Ent. exp. appl. 1978, 24 385-93.
33. Tahvanainen, J.O.; Root, R.B. Oecologia 1972, 321-46.
34. Saxena, K.N.; Basit, A. J. Chem. Ecol. 1982, 8, 329-38.
35. Lundgren, L. Zoologica Scripta 1975, 4, 253-8.
36. Robert, P.C.; Blaisinger, P. Ent. exp. appl. 1978, 24, 632-6.
37. Tingle, F.C.; Mitchell, E.R. J. Chem. Ecol. 1984, 10, 101-13.
38. Tingle, F.C.; Mitchell, E.R. J. Chem. Ecol. 1986, 12, 1523-31.
39. Mitchell, E.R.; Heath, R.R. J. Chem. Ecol. 1985, 11, 609-17.
40. Dover, J.W. Ent. exp. appl. 1985, 39, 177-82.
41. Abivardi, C.; Benz, G. Mitt. Schweiz. Entomol. Gesell 1986, 59, 31-8.
42. Joshi, B.G.; Sitaramaiah, S. Phytoparasitica 1979, 7, 199-202.
43. Hellpap, C.; Mercado, J.C. J. Appl. Ent. 1986, 102, 463-7.
44. Renwick, J.A.A.; Radke, C.D. Ent. exp. appl. 1985, 39, 21-6.
45. Renwick, J.A.A.; Radke, C.D. Environ. Entomol. 1983, 12, 446-50.
46. Renwick, J.A.A.; Radke, C.D. J. Chem. Ecol. 1987, 13, 1771-6.
47. Renwick, J.A.A.; Radke, C.D.; Sachdev, K. in preparation.
48. Sachdev, K.; Renwick, J.A.A. in preparation.
49. Dimock, M.B.; Renwick, J.A.A. in preparation.

RECEIVED April 1, 1988

Chapter 25

Plant Natural Products as Parasitoid Cuing Agents

Douglas W. Whitman[1]

Insect Biology and Population Management Research Laboratory, Agricultural Research Service, U.S. Department of Agriculture, Tifton, GA 31793

Plant natural products play an important role in the orientation of insect parasitoids to their herbivorous insect hosts. Few of these stimulating or attractive plant compounds have been identified, but terpenoids, thiocyanates, and sugars are represented. Another group of synomonal plant natural products is the "green leaf odors," six carbon alcohols, aldehydes, and their derived esters, found in many plants. A better understanding of how plant volatiles and contact stimulants influence the behaviors of insect parasitoids, such as the parasitic Hymenoptera, could aid in the application of biological control.

The herbivorous insects that attack our food and fiber crops are humankind's greatest competitors, causing billions of dollars in damage and control costs annually. Most of these pestiferous species are themselves attacked by microorganisms, predators, and parasitoids, collectively known as natural enemies. One of the larger and more important groups of natural enemies is the parasitic wasps (order Hymenoptera), which number about 100,000 species worldwide. These useful parasitoids make their living by seeking out other insects which serve as food for their young. Usually only one wasp egg is laid on or in the insect host, and the wasp larvae completes its entire development on that particular host, eventually killing it.

The potential value of using natural enemies such as parasitic wasps to control crop pests has long been

[1]Current address: Department of Biology, Illinois State University, Normal, IL 61761

recognized by crop protection specialists. Indeed, the advantages of such "biological control" are manifold. Successful biological control decreases pesticide use, thereby reducing potential environmental contamination as well as chemical disruption of the agroecosystem. The greatest benefit of using natural enemies to control plant pests occurs when the biological control agent becomes permanently established in the environment, thus providing a perennial low-cost means of crop protection.

When it works, biological control can be relatively inexpensive, long-lasting, and environmentally sound. Indeed, there have been many victories in the application of this technique (1), successes which continue to save millions of dollars annually. Unfortunately, however, the majority of attempts to utilize biological control have failed (2, 3). Predators and parasitoids which work well in the laboratory are often dismal failures in the field. The reasons for such mishaps are varied and complex, (4), but often can be attributed to what seems to be an inability or unwillingness of released biological control agents to locate and attack their hosts. Many biological control practitioners, who have spent months carefully culturing large numbers of natural enemies in the laboratory, are frustrated to see the subjects of their research simply fly away following field liberation (5-7). To solve this problem, it is essential to gain an understanding of the factors which initiate and maintain host searching behavior, and a knowledge of the cues used by natural enemies to locate their hosts. Simply put, once we know what turns parasitoids on and how parasitoids find and choose their prey, then we can manipulate these systems for our own advantage.

The factors which govern parasitoid orientation to prey are best understood in the parasitic wasps, where host location and identification is accomplished largely via the chemical modality (Figure 1). Various studies have shown that parasitic Hymenoptera can be stimulated by kairomones originating from such diverse sources as the host's body, frass, scales, hairs, mandibular gland secretions, silk, and even sexual and aggregation pheromones (8-10). It is becoming increasingly clear, however, that plant natural products also play an important role in this process, serving to attract and stimulate Hymenopterous parasitoids as well as other classes of natural enemies (8, 11).

The idea that parasitoids utilize plant volatiles and contact stimulants to locate their prey is not new, but was suggested at least 50 years ago (12, 13). Since then, numerous researchers have documented the attraction of parasitoids to odors of a great variety of plants (Vinson 1981). It is therefore surprising that in only five studies have the actual stimulating plant synomones been identified. Table I lists these studies

Figure 1. Female Microplitis croceipes wasp orienting to volatiles released as its caterpillar host feeds upon a bean leaf.

Table I. Plant Natural Products Which Function as Synomones for Parasitic Insects

Plant Compound	Plant	Herbivore	Parasitoid	Activity	Bioassay	Reference
Sugar sucrose fructose	Fagaceae: *Quercus robur*	Geometridae: *Operophtera brumata*	Tachinidae: *Cyzenis albicans*	Oviposition stimulant	Laboratory: Contact bioassay	14
Thicyanate: allyl isothiocyanate	Brassicaceae: *Brassica oleracea*	Aphididae: *Myzus persicae*	Braconidae: *Diaeretiella rapae*	Walking attraction	Laboratory: Walking olfactometer	15
Terpene: α-pinene	Pinaceae: Various *Pinus* spp.	Scolytidae: *Dendroctonus frontalis*	Pteromalidae: *Heydenia unica*	Flight attraction	Field: Baited trap	16
Terpene: α-cubebene	Ulmaceae: *Ulmus americana*	Scolytidae: *Scolytus multistriatus*	Pteromalidae: *Cheirophachus colon* Eulophidae: *Entedon leucogramma* Braconidae: *Spathius benefactor*, *Dendrosoter protuberans*	Flight attraction	Field: Baited trap	17, 18
Sesquiterpene: α-humulene γ-bisabolene β-caryophyllene oxide spathulenol β-bisabolol gossonorol	Malvaceae: *Gossypium hirsutum*	Noctuidae: *Heliothis virescens*	Ichneumonidae: *Campoletis sonorensis*	Antennation, Ovipositon, Walking attraction	Laboratory: Y-tube Olfactometer	19

and shows the great heterogeneity which characterizes them. It is significant to note that several families of parasitic wasps as well as a tachinid fly are represented. These parasitoids attack a diversity of prey ranging from sap-sucking aphids and leaf-feeding caterpillars to bark-feeding beetles. The stimulating natural products are produced by a diversity of plants representing cultivated and wild taxa, herbs and trees, and deciduous and evergreen species, and include sugars, thiocyanates, and terpenoids (Figure 2), which act as long or short range olfactory attractants, contact arrestants, or oviposition stimulants. Notice also that whereas some compounds were tested in the laboratory with olfactometers, others were examined in the field with baited sticky traps. The diversity of plant, herbivore, and parasitoid taxa, and the variety of active plant components, which are represented in these studies, strongly suggest that plant natural product stimulation of parasitoids is a widespread phenomenon, perhaps a major factor in the orientation of insect parasitoids.

In our laboratory, we recently began a series of studies to elucidate the role of volatile plant natural products in orienting and attracting the braconid wasp *Microplitis croceipes* (Figure 1). Female *M. croceipes* attack and lay their eggs in *Heliothis* caterpillars. These nefarious caterpillars feed upon a great number of important crops including cotton, corn, bean, tomato, and tobacco, causing hundreds of millions of dollars in annual damage worldwide.

We chose to study a series of six carbon alcohols, aldehydes, and their derivitive esters (Figure 3), which are collectively known as "green leaf odors," because of their common occurrence in plants (*20*-*22*). It is these substances, along with plant allelochemicals such as terpenoids and mustard oils, which give each plant group its characteristic odor. The exact function served by the green leaf compounds is unknown. They may have some metabolic purpose, or perhaps deter or harm plant competitors or herbivores. However, data substantiating these roles are scarce. It is known that damaged plants release relatively large quantities of these volatiles (e.g. the odor of a freshly mowed lawn) (*23*, *24*). In fact, some green leaf volatiles appear to be produced only when plant tissue is damaged (*23*).

To test the attractiveness of these substances to *M. cropeipes* wasps, we employed a 75x75x200 cm wind tunnel. Young mated laboratory-reared female wasps were placed in the downwind end of the tunnel and a microcapillary tube containing the test compound was place 100 cm upwind (preliminary experiments had shown that males did not respond to these substances). By using various-sized capillary tubes, we obtained a wide range of release rates for our test compounds. If the substance and the

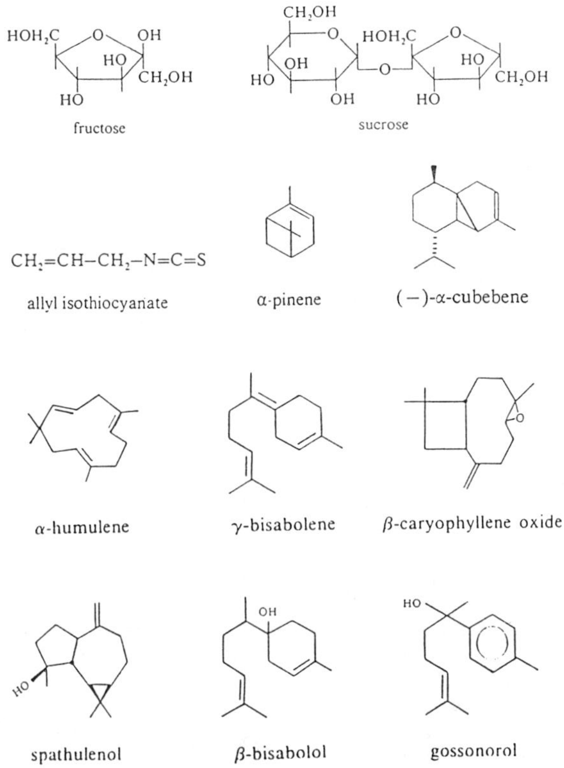

fructose

sucrose

$CH_2=CH-CH_2-N=C=S$

allyl isothiocyanate

α-pinene

(−)-α-cubebene

α-humulene

γ-bisabolene

β-caryophyllene oxide

spathulenol

β-bisabolol

gossonorol

Figure 2. Plant natural products known to stimulate insect parasitoids.

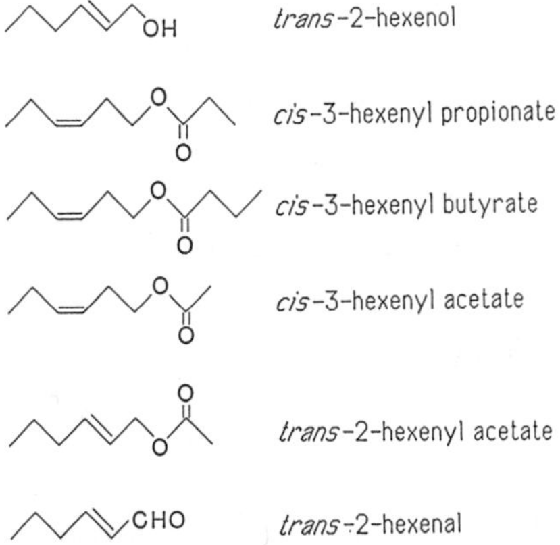

Figure 3. Green leaf volatiles tested for attractiveness with Microplitis croceipes wasps.

dose were attractive to the wasps, they flew upwind and landed on a 1 cm diameter paper target placed near the capillary tube orifice. If the test compound and/or dose was unattractive, wasps did not orient upwind, but instead flew to the ceiling or walls of the wind tunnel. Wasps were either placed directly into the flight chamber (naive females) or first given a preflight stimulatory experience (experienced females). This preflight experience consisted of allowing the wasps to antennate a dried spot of water extract of *Heliothis zea* frass for 30 seconds. Previous work (*25*, *26*) demonstrated that antennal contact with *H. zea* frass or frass extract stimulates and enhances host finding by *M. croceipes*.

After establishing dose-response curves for the various test substances, a release rate of approximately 0.0004 µl/min was chosen for the test. This release rate was selected because it was thought to be within the range that naturally occurs when a caterpillar feeds upon plant tissue, and because this level (for most compounds) elicited a high rate of response from the wasps.

Our results showed that, at the doses tested, the green leaf odors varied widely in their attractiveness to *M. croceipes* females (Table II). Some substances (e.g., *cis*-3-hexenyl butyrate) were extremely attractive to *M. croceipes*, eliciting orientation responses near the highest ever recorded for this insect (*25*, *26*), whereas other substances were relatively unattractive (e.g., *trans*-2-hexenal).

Table II. The Response of Naive and Experienced *Microplitis croceipes* Wasps to Various Green Leaf Volatile Substances Tested in a Wind Tunnel. See text for explanation

Volatile	% Successful Orientation	
	Naive (n=20)	Experienced (n=20)
blank control	0	0
hexane control	0	0
trans-2-hexenal	0	10
trans-2-hexenol	10	55
trans-2-hexenyl acetate	20	55
cis-3-hexenyl acetate	20	70
cis-3-hexenyl propionate	25	65
cis-3-hexenyl butyrate	45	80

As in other studies (*25*, *26*), preflight stimulation had a strong effect on orientation performance. In most cases, wasps were significantly more responsive to the test compounds if they had first been exposed to the dried extract of Heliothis frass. Currently, two hypotheses compete to explain this phenomenon. The sensitization hypothesis holds that the water extract of the Heliothis frass contained host kairomones which stimulated the wasps, perhaps sensitizing them, or increasing their appetitive drive. The second theory involves associative learning. Lewis and Tumlinson (25) recently demonstrated that M. croceipes could be conditioned to respond to novel odors if the novel substances were paired with host kairomones in a classical UCS-CS encounter. A similar type of conditioning may have occurred in our tests when wasps were provided a preflight exposure to frass extract. Perhaps the dried extract of Heliothis frass contained low levels of certain of our green leaf test substances, which became conditioned stimuli when they were associated with the host kairomone acting as an unconditioned stimulus. Thus, increased response to green leaf odors following preflight stimulation may represent learning instead of sensitization. This possibility is currently being explored in our laboratory.

Regardless of how preflight stimulation occurs, the results clearly show that M. croceipes wasps orient to certain of the green leaf odors. Under natural conditions, this chemical-induced anemotactic response would bring wasps to plants where they could obtain food and water (nectar, dew, and honeydew), high humidity, and shelter from wind, rain, sunlight, and predators. However, the fact that only females are attracted by the green leaf volatiles suggests a more female-specific benefit, i.e. host location. When plant tissues are damaged by chewing herbivores, relatively large amounts of these volatile substances are liberated. Perhaps M. croceipes, as well as other parasitic wasps, orient to the source of these emissions, where they find their hosts.

In the field, parasitic wasps confront a diverse and structurally complex habitat in which suitable hosts may be scarce. An ability to locate prey using both long- and short-range cues would be highly adaptive. Likewise, there is undoubtedly selective pressure for plants to evolve mechanisms to attract parasitoids to sites of herbivory. The green leaf odors may represent such a mechanism. In other words, a tritrophic chemical communication may exist between plants and members of the third trophic level, a communication largely unnoticed by humans because of our poorly developed olfactory senses.

Lending support to this hypothesis is the fact that the carnivorous Hymenoptera are thought to have evolved from phytophagous ancestors. Phytophagous insects are well known to respond to green leaf odors (22). It is conceivable that the ancestors of the parasitic Hymenoptera also possessed an ability to orient to plant compounds and retained this sensitivity during their evolution to a carnivorous way of life.

Also of interest is evidence suggesting that undamaged plants release relatively low amounts of green leaf volatiles while damaged plans release relatively large amounts (23, 24). Since herbivores respond to these compounds, it may pay plants to "hide" their odor until they come under attack. At that time, the advantages of releasing parasitoid-attracting synomones could outweigh the disadvantages of attracting additional herbivores. Other studies suggest that the concentration of green leaf volatiles in the tissues of some plants may be quite low; only when the plant is damaged are these volatiles derived enzymatically from precursors (23).

Herbivorous insects might also produce and release green leaf volatiles in their frass as a consequence of their digestion of plant tissue. At this point the relative importance of plant- versus possible insect-produced green leaf odors on host finding by parasitoids is unknown. However, parasitoids would presumably orient to their hosts whether the attractive volatiles arose from the head region (as a consequence of simple mandibular damage to plant tissue) or from the opposite end of the caterpillar.

Significance to Biological Control

The orientation of parasitoids to green leaf odors may have significance for agriculture, particularly in the realm of biological control. Entomologists who mass rear parasitoids in the laboratory for eventual innundative or innoculative release are careful to preserve the vitality, fecundity, and genetic diversity of their colonies. However, under laboratory conditions, where hosts are often presented to parasitoids literally "on a platter," there may be little selection for the ability to respond to long-range plant cues. Indeed, we have observed great differences among our laboratory strains in their response to plant volatiles.

Because plant volatiles differ qualitatively and quantitatively among plant species, and because parasitoids often specialize not on particular hosts, but on a variety of hosts found on specific plant taxa (8), it is likely that parasitoids have evolved innate preferences for certain volatiles. It is important that biological control specialists recognize these features and try to preserve in their insects a sensitivity and a capability to orient to the natural products of their host's foodplant.

The recent finding that parasitoids learn and can be conditioned (25) is also of fundamental importance vis à vis the semiochemical manipulation of natural enemies. Studies in our laboratory indicate that the ability to orient to hosts is strongly related to the quality and quantity of the pre-release semiochemical experience. In fact, without the proper pre-release stimulation, laboratory-reared *M. croceipes* parasitoids show a tendency to disperse and are much less adept at locating their hosts. Such a phenomenon may be partly responsible for the large number of biological control failures alluded to earlier in this chapter. With the proper pre-release chemical conditioning, the field performance of parasitoids might be substantially increased. Perhaps parasitoids might even be "taught" to orient to the semiochemicals characteristic of specific crops, such as cotton or bean.

If parasitoids possess the capability to learn to respond to certain odors, it is possible that they could also become habituated to these same odors. The high concentrations of key host or plant semiochemicals which sometimes permeate insect rearing facilities may thus have a negative influence on subsequent parasitoid performance. Biological control specialists must recognize and compensate for those potential semiochemical effects. Failure to do so may have a profound impact on the success of future biological control release programs.

Another important ramification of our findings relates to biotechnology. It appears that parasitoids orient to volatile plant substances. Perhaps genetic engineering will allow us to create plants which are even more attractive to these natural enemies. Such plants would be designed to release elevated levels of volatiles at sites of herbivore attack, thereby attracting elevated levels of parasitoids. Considering the rapid advancement of biotechnology, the ability to perform such genetic transformations may not be far off.

Above all, our findings suggest that a complex world of semiochemical interactions exists between plants, herbivores, and parasitoids. At present we have only begun to understand these heterogeneous interactions, let alone their applications. However, our initial discoveries suggest that the use of plant natural products to manipulate the enemies of our crop pests may have potential economic and social benefits. It is for these reasons that the highest priorities must be given to the elucidation of these semiochemical relationships.

Acknowledgments

I thank W. J. Lewis for advice and technical support, M. S. Blum, T. Muller, A. Simmons, and K. Smith-Whitman for manuscript review.

Literature Cited

1. van den Bosch, R.; Messenger, P. S. Biological Control; Intext Press: New York, 1973; p. 180.
2. Beirne, B. P. Can. Entomol. 1975, 107, 225-36.
3. Hall, R. W.; Ehler, L. E. Bull. Entomol. Soc. Am. 1979, 25, 280-2.
4. Hoy, M. A. In Biological Control in Agricultural IPM Systems; Hoy, M. A.; Herzog, D. C., Eds.; Academic Press: Orlando, FL; pp. 151-66.
5. Gross, H. R.; Lewis, W. J.; Jones, R. L.; Nordlund, D. A. J. Chem. Ecol. 1975, 1, 431-8.
6. Greany, P. D.; Vinson, S. B.; Lewis, W. J. Bioscience 1974, 34, 690-6.
7. Vinson, S. B. In Biological Plant and Health Protection; Franz. J. M. Ed.; Fortsch. Zool. 32. G. Fischer Verlag: Stuttgart, 1986; pp. 75-87.
8. Vinson, S. B. In Semiochemicals: Their Role in Pest Control; Nordlund, D. A.; Jones, R. L.; Lewis, W. J., Eds.; John Wiley: New York, 1981; Chapter 4.
9. Vinson, S. B. In Insect Communication; Lewis, T., Ed.; Academic: New York, 1984; Chapter 14.
10. Weseloh, R. M. In Semiochemicals: Their Role in Pest Control; Nordlund, D. A.; Jones, R. L.; Lewis, W. J., Eds.; John Wiley: New York, 1981; Chapter 5.
11. Whitman, D. W. In Novel Aspects of Insect-Plant Interactions; Barbosa, P.; Letourneau, D. K., Eds.; John Wiley: New York, 1988; Chapter 1.
12. Bess, H. A. Ann. Entomol. Soc. 1936, 29, 593-613.
13. Thorpe, W. H.; Caudle, H. B. Parasitology 1938, 30, 523-8.
14. Hassell, M. P. J. Anim. Ecol. 1968, 37, 627-39.
15. Read, D. P.; Feeny, P. D.; Root, R. B. Can. Entomol. 1970, 102, 1567-78.
16. Camors, F. B.; Payne, T. L.; Ann. Entomol. Soc. Am. 1972, 65, 31-3.
17. Kennedy, B. H. Bull. Entomol. Soc. Am. 1979, 25, 116-8.
18. Kennedy, B. H. J. Chem. Ecol. 1984, 10, 373-85.
19. Elzen, G. W.; Williams, H. J.; Vinson, S. B. J. Chem. Ecol. 1984, 10, 1251-64.
20. Straten, S. van. Volatile Compounds in Food; Central Institute for Nutrition and Food Research TNO: Zeist, The Netherlands, 1977; p. 300.
21. Visser, J. H.; van Straten, S.; Maarse, H. J. Chem. Ecol. 1979, 5, 13-25.
22. Visser, J. H. Ann. Rev. Entomol. 1986, 31: 121-44.
23. Saijo, R.; Takeo, T. Phytochemistry 1975, 14, 181-2.
24. Wallbank, B. E.; Wheatley, G. A. Phytochemistry 1976, 15, 763-6.
25. Lewis, W. J.; Tumlinson, J. H. Nature 1988, 331, 257-9.
26. Drost, Y. C.; Lewis, W. J.; Zanen, P. O.; Keller, M. A. J. Chem. Ecol. 1986, 12, 1247-62.

RECEIVED May 26, 1988

Chapter 26

Iridoid Glycosides and Aglycones as Chiral Synthons, Bioactive Compounds, and Lepidopteran Defenses

Frank R. Stermitz

Department of Chemistry, Colorado State University, Fort Collins, CO 80523

Iridoid glycosides are relatively widely distributed among families of several higher plant orders and are often present in amounts reaching several percent of the plant dry weight. The high plant concentration of several iridoids has resulted in their use as starting materials for subsequent conversion to substituted cyclopentanes, such as prostaglandin synthons. Some iridoid glycosides have been shown to have antibiotic, antioxidant, plant growth inhibiting or antitumor properties, but these seem to be mainly a property of the aglycone. The bitter taste of iridoids is responsible for their utilization by Lepidoptera as defensive substances and such iridoids have been found in a number of boldly-patterned insect larvae.

Iridoids are higher plant biosynthetic derivatives of **2**, the hemiacetal form of an enol of iridoidal, **1**. Iridoids generally occur in the plant as a glucosides, such as **3**, but are occasionally encountered as glycosides of other sugars. Many iridoids occur as oxidized derivatives of **3**, where the C-4 methyl has been replaced by an aldehyde, carboxylic acid, or carbomethoxy group. The C-4 decarboxylated compounds are also common. Most iridoids are, in addition, further oxidized in the cyclopentane ring. In some plant families, the cyclopentane ring has been cleaved to yield secoiridoids related to secologanin, **4**, which is an important biosynthetic precursor of indole alkaloids and a useful, commercially available synthon. Known iridoids and secoiridoids from plants now number over 300. A recent review (1) with 200 references provided background information as well as a comprehensive discussion of iridoid biosynthesis.

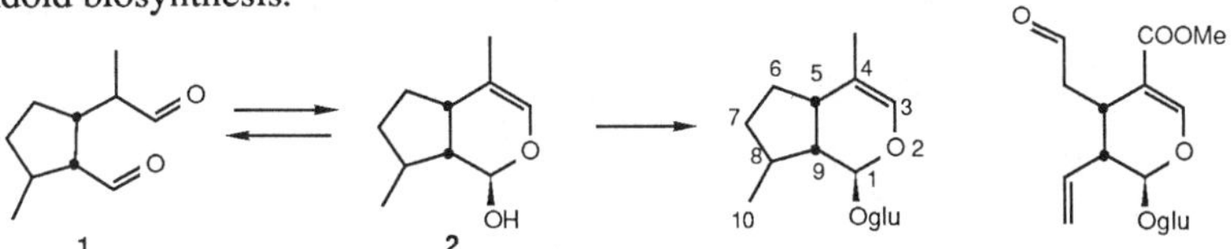

O O OH 6 5 4 3 7 8 9 1 O 2 10 Oglu COOMe O Oglu

1 2 3 4

The present review, which will be illustrative rather than comprehensive, focuses on aspects of recent work showing that iridoids can be exploited as chiral synthons, on recent reports of biological activity of some iridoids and their aglycones, and on some of our own results on iridoid sequestration by aposematic (warningly-patterned or colored) insects.

0097–6156/88/0380–0397$06.00/0

Iridoids as Useful Chiral Synthons

By early 1986 18 papers and 4 patents had appeared in which naturally-occurring iridoids were utilized as chiral starting materials for the preparation of prostanoid synthons or prostaglandins (2). The three iridoids most commonly used have been aucubin, **5**, asperuloside, **6**, and catalpol, **7**. The Italian group has generally used

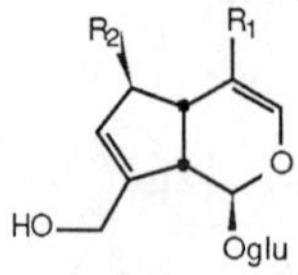

5 R_1 = H; R_2 = OH

12 R_1 = COOMe; R_2 = OH

13 R_1 = COOH; R_2 = H

14 R_1 = COOMe; R_2 = H

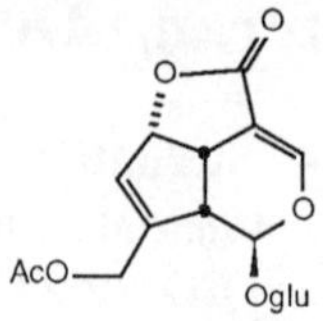

6

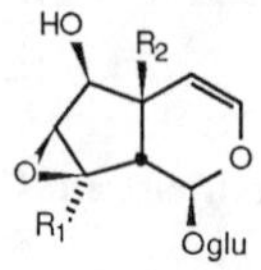

7 R_1 = CH_2OH; R_2 = H

8 R_1 = CH_3; R_2 = OH

5, obtained in 2% yield from the readily available ornamental shrub *Aucuba japonica* (2). In another case (3), **5**, was obtained from *Eucommia ulmoides* (Eucommiaceae). Derivatives of **6** have been obtained from *Asperula odorosa* (Rubiaceae). Weinges and co-workers have generally used **7** and recently reported (4) a 1.6% yield of pure **7** by hydrolysis of the natural catalpol esters from *Picrorhiza kurrooa* (Scrophulariaceae). Thus, 800 g of pure catalpol was obtained from 50 kg of dried plant after ester hydrolysis, activated charcoal absorption, desorption and alumina chromatography. Catalpol (5a) and aucubin (5b) have been converted to (-)-jasmonate or analogs.

Although it has apparently not yet been utilized synthetically, antirrhinoside, **8**, might be valuable, perhaps in the preparation of 11-methylprostaglandins (6). Many common snapdragons (*Antirrhinum* sp.) contain high levels of **8** and might be exploited as iridoid sources. I recently soaked 100 g of fresh plant material of a closely related taxon, *Maurandya antirrhiniflora*, in methanol for two days, evaporated the methanol, extracted the residue with water, washed the water with ether, evaporated the water and triturated the residue with absolute methanol. Evaporation of the methanol yielded 2g of solid **8** which was greater than 90% pure by nmr analysis. Thus, if one is able to work with fresh plant material even higher yields of an exploitable iridoid might be achievable since plant drying has sometimes been found to result in a decrease in extractable iridoids.

Just as secologanin, **4**, is a precursor of indole alkaloids, iridoids are both biosynthetically and synthetically convertible to pyridine monoterpene alkaloids. As a recent example of conversion, we isolated both penstemonoside, **9**, and rhexifoline, **10**, from several *Castilleja* (Scrophulariaceae) species and were able to convert **9** to **10** (7). The pharmacological properties of pyridine monoterpene

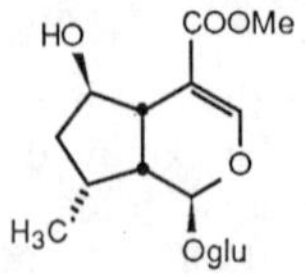

9 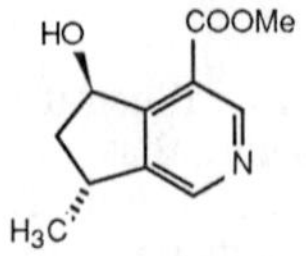 10 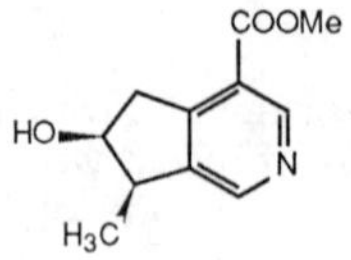 11

alkaloids have been but little explored, although most have been isolated from pharmacologically active plants (8). Many of the pyridine monoterpenes occur as such in the plant, while others, such as cantleyine, **11**, have been found several times in different plant species, but always as artifacts arising from the use of ammonia in the isolation procedure. Other "alkaloids" reported from iridoid-containing plants can also arise as artifacts of the isolation procedure (9).

In general, extensive use of iridoids as chiral synthons will depend on the availability of an easily grown plant species containing the desired iridoid (or iridoid ester) as a major component. When the desired iridoid is not a major component, separation of individual compounds from each other and from contaminating sugars in the extracts may be tedious and time-consuming.

Iridoid and Iridoid Aglycone Biological Activity

As has been pointed out (1), one important factor leading to a doubling of the known iridoid glycosides from 158 in 1980 (10) to over 300 by 1986 has been their common occurrence in folk medicinal plants. This has led to many new isolations and a renewed interest in biological activity studies. Several recent reports have focussed on the activity of the iridoid glucosides relative to the activity of their aglycones.

In only rare instances have iridoid aglycones related to **2** been isolated from plants, although compounds linked at C-1 to alcohols other than sugars, or further transformation products of **2** (such as lactones or cyclic ethers) are known (10). The relative plant proportions of iridoid aglycone vs. glycoside are essentially unknown since normal isolation procedures probably result in the decomposition of the relatively unstable aglycone (11). Recently, however, an isolation technique involving CH_2Cl_2 liquid-liquid extraction of an aqueous iridoid glycoside enzymatic hydrolysis reaction mixture has yielded pure aglycones in high yield (12). In this work, the antimicrobial activity of aucubin aglycone against moulds and bacteria, but not yeasts, was clearly established. Other research groups had previously studied the antimicrobial activity of aucubin and other iridoids (12). It had been shown (13) in a study involving 21 iridoids, that iridoid glycoside antimicrobial activity was only observable after treatment with β-glucosidase. A similar study was reported on antitumor activity. Five isolated and Si gel purified aglycones showed activity against leukemia P388, while none of the glucosides were active (14). The most active were **5** aglycone and (especially) scandoside methyl ester aglycone, **12**. A patent was recently issued (15) for use of **12** as a chiral synthon for preparation of other antitumor compounds.

Several iridoids isolated from *Garrya elliptica*, including geniposidic acid, **13**, and its aglycone were shown to inhibit the growth of wheat embryos (16). **13** has also been shown to be the active antioxidant in a methanol extract of *Plantago*

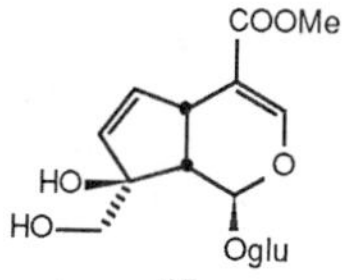

COOMe
HO
HO
O
Oglu
15

asiatica Linne (17). The activity was superior to that of *dl*-α-tocopherol against air oxidation of linoleic acid. Also present in the extract were aucubin, geniposide, **14**, and gardenoside, **15**, but these were much less active than **13**. The aglycone of **13** was not studied.

In each of these cases, the aglycones may not be the active species themselves. Biological activity might instead stem from the aglycone-derived, highly electrophilic oxonium ion, **16**. There are, however, a number of other biologically-active compounds which contain highly substituted 5,6-fused rings and iridoid aglycones could be bioisosters of these. An example is provided by the important indolizidine swainsonine, **17**.

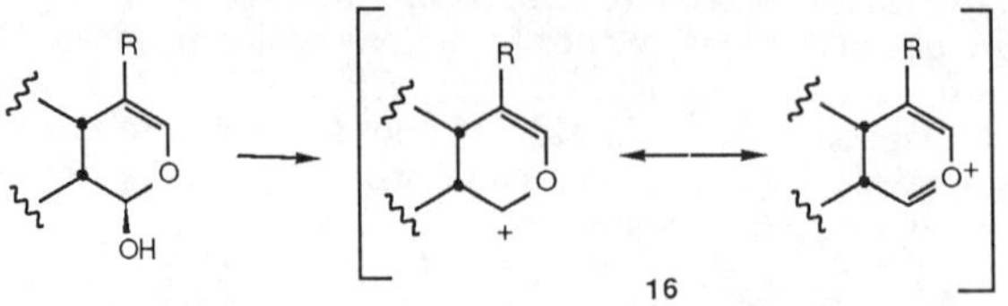

16

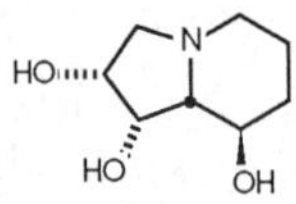

17

Iridoid Sequestration and Aposematic Lepidopteran Larvae

Several recent reports detail the acquisition and sequestration of bitter tasting iridoid glucosides by lepidopteran larvae feeding on iridoid-containing plant species (18, 19, 20, 21). The known cases of sequestration were recently summarized (22). We have discovered additional examples involving some lepidopteran species having particularly boldly-patterned and/or highly colored larval stages, but cryptic imagos.

Poole first described the remarkable similarity between last-instar larval color patterns (black and white longitudinal stripes, with gold-yellow spots) of *Meris alticola* Hulst and *Neoterpes graefiaria* (Hulst), geometrid moths known from Arizona (23). We collected *M. alticola* larvae on *Penstemon virgatus*, raised them to cryptic adults and analyzed adults, empty pupal cases, cocoon silk and emitted meconium for catalpol. A *Neoterpes graefiaria* female adult was collected at a uv light in southern Arizona. After the female had oviposited, eggs were collected, hatched and the larvae raised on *P. barbatus* (23). Larvae, pupal cases, meconium and adults were again analyzed for catalpol. Adult males of both species contained low amounts of catalpol, while meconium and the pupal cases were high in catalpol. Although only one adult female was available (of *Meris alticola*), a high level of catalpol was found in the abdomen, but not in other parts. We presume catalpol to be an important constituent of the eggs. Manipulation of *N. graefiaria* larvae resulted in a reflex bleeding emission, which had a high catalpol concentration. We have suggested that larval Müllerian mimicry, based on the sequestered bitter iridoid catalpol, may be involved with these two geometrids (24).

The vine snapdragon, *Maurandya antirrhiniflora*, a common wildflower in the southwest U.S., is host to another geometrid, *Meris paradoxa* Rindge, and also to a noctuid moth, *Oncocnemis* (Lepopolys) *perscripta*. The strikingly colored and boldly-patterned larvae of these insects have cryptic adult stages. The previously uninvestigated hostplant was found to contain 2% wet weight of antirrhinoside and this iridoid was also found to be present in relatively high concentration in the larvae, but not adults of both species. *M. paradoxa* emits a yellow liquid via reflex bleeding and this also contained antirrhinoside.

Thus the aposematic larva/cryptic adult predator avoidance strategy noted (18) for *Ceratomia catalpae* (Sphingidae), a catalpol sequester, has been extended to several additional Lepidoptera. A number of evolutionary aspects of the "profitability" of such strategies have recently been discussed (25).

Our field work indicates that the iridoid content of the insect larvae and adults is not deterrent to ants or spiders, which can be major predators. It seems likely that vertebrate predators provided the selection pressures resulting in the development of the aposematic patterns. There is evidence that some iridoid-containing lepidoptera are unpalatable to birds (26, 27), but effects on other possible vertebrate predators (such as mice or lizards) are unknown. In wild-collected and lab raised insects, sequestered iridoid content is commonly 2-4% of the insect dry weight and values as high as 10% have been found. In view of some of the other biological activities described above (antioxidant, antimicrobial, for example) it is conceivable that the iridoids could play additional roles in insect survival besides that of a predation defense. If this were the primary selection pressure which resulted in iridoid sequestration, one might, however, expect completely cryptic specialists on iridoid-containing plants to also store these substances. The only case we have studied so far is that of the plume moth *Amblyptilia* (Platyptilia) *pica*, which is a specialist on many iridoid-containing Scrophulariaceae, but which is cryptic in both larval and adult stages. These larvae excrete iridoids in the frass and none can be found in adult insects (28).

Addendum

As a continuation of some earlier work, a very recent report (29) described a new preparation of a prostanoid intermediate starting from the iridoid glucoside loganin. A complete literature review, supplementary to that quoted above (2), is presented on such synthetic uses of iridoids.

Acknowledgment

Preparation of this review was supported by grant CHE-8521382 from the National Science Foundation.

Literature Cited

1. Inouye, H.; Uesato, S. Prog. Chem. Nat. Prod. 1986, 50, 169-236.
2. Bernini, R.; Davini, E.; Iavarone, C.; Trogolo, C. J. Org. Chem. 1986, 51, 4600-4603.
3. Bonini, C.; Iavarone, C.; Trogolo, C.; DiFabio, R. J. Org. Chem. 1985, 50, 958-861.
4. Weinges, K.; Haremsa, S.; Huber-Patz, U.; Jahn, R.; Rodewald, H.; Irngartinger, H.; Jaggy, H.; Melzer, E. Liebigs Ann. Chem. 1986, 46-53.
5. a) Weinges, K.; Gethoeffer, H.; Huber-Patz, U.; Rodewald, H.; Irngartinger, H. Liebigs Ann. Chem. 1987, 361-366. b) Davini, E.; Iavarone, C.; Trogolo, C. Heterocycles 1988, 27, 57-61.
6. Bonaides, F.; Gubbiotti, A.; Bonini, C. Gazz. Chim. Ital. 1985, 115, 45-48.
7. Roby, M.R.; Stermitz, F.R. J. Nat. Prod. 1984, 47, 854-857.
8. Cordell, G.A. in The Alkaloids; Academic Press: New York, 1977; Vol. 16, p. 432.
9. Stermitz, F.R.; Harris, G.H. Tetrahedron Lett. 1985, 26, 5251-5252.
10. El-Naggar, L.J.; Beal, J.L. J. Nat. Prod. 1980, 43, 649-707.
11. Bianco, A.; Guiso, M.; Iavarone, C.; Passacantilli, P.; Trogolo, C. Tetrahedron 1977, 33, 847-850.
12. Davini, E.; Iavarone, C.; Trogolo, C.; Aureli, P.; Pasolini, B. Phytochemistry 1986, 25, 2420-2422.
13. Ishiguro, K.; Yamaki, M.; Takagi, S. Yakugaku Zasshi 1982, 102, 755.
14. Ishiguro, K.; Yamaki, M.; Takagi, S.; Ikeda, Y.; Kawakami, K.; Ito, K.; Nose, T. Chem. Pharm. Bull. 1986, 34, 2375-2379.
15. Isoe, Y.; Takemoto, T.; Inaba, H.; Kan, K. Jpn. Kokai Tokkyo Koho JP 62 53, 982 (CA107, 134198).
16. Cameron, D.W.; Feutrill, G.I.; Perlmutter, P.; Sasse, J.M. Phytochemistry 1984, 23, 533-535.
17. Toda, S.; Miyase, T.; Arichi, H.; Tanizawa, H.; Takino, Y. Chem. Pharm. Bull. 1985, 33, 1270-1273.
18. Bowers, M.D.; Puttick, G.M. J. Chem. Ecol. 1986, 12, 169-178.
19. Stermitz, F.R.; Gardner, D.R.; Odendaal, F.J.; Ehrlich, P.R. J. Chem. Ecol. 1986, 12, 1459-1468.

20. Gardner, D.R.; Stermitz, F.R. J. Chem. Ecol. 1988, 14, in press.
21. Franke, A.; Rimpler, H.; Schneider, D. Phytochemistry 1987, 26, 103-106.
22. Bowers, M.D. in Chemistry and Evolution, K. Spencer, Ed., 1988, in press.
23. Poole, R.W. J. Kansas Ent. Soc. 1970, 43, 292-297.
24. Stermitz, F.R.; Gardner, D.R.; McFarland, N. J. Chem. Ecol. 1988, 14, 435-441.
25. Leimar, O.; Enquist, M.; Sillen-Tullberg, B. Am. Nat. 1987, 128, 469-490.
26. Bowers, M.D. Evolution 1980, 34, 586-600.
27. Bowers, M.D. Evolution 1981, 35, 367-375.
28. Stermitz, F.R.; Harris, G.H.; Wang, J. Biochem. Syst. Ecol. 1986, 14, 499-506.
29. Berkowitz, W.F.; Arafat, A.F. J. Org. Chem. 1988, 53, 1100-1102.

RECEIVED April 5, 1988

Chapter 27

Glycosides: The Interface Between Plant Secondary and Insect Primary Metabolism

Kevin C. Spencer

Department of Medicinal Chemistry and Pharmacognosy, University of Illinois, Chicago, IL 60612

Plant secondary glycosides are proposed to have evolved as specific toxins toward enzymes of herbivore primary digestive metabolism. Insect herbivore digestive enzymes are optimized to process plant foods. An essential component of their digestive effort is the breakdown of sugar-containing plant materials by glycosidases. This essential activity has provided a target for plant toxins, which are delivered as glycosides. Such toxic action against glycosidases is described for cyanogenic glycosides, glucosinolates, iridoid glycosides, phenol glycosides and triterpene glycosides. The toxicity of individual glycosides depends upon their specific interaction with specific glycosidases, wherein they function as inhibitory substrates or release toxic aglycones. This great specificity in delivery and manifestation of toxicity can be measured for given sets of plants and herbivores, and may possibly be exploited in the design of effective, species-specific insecticides.

A plant is, in its essence, a collection of chemicals. A diversity of sensory modalities exist in an insect, collectively constituting a mechanism for mating its digestive specialization with host searching, selection and acceptance (1). The ability of an insect to acquire and internalize appropriate levels and types of nutritional factors is critical to its success (2). These nutritional factors are related to defensive allelochemical regimes in two ways: first, as the effectiveness of the defense impinges upon the value of the nutrients, and second, as the quality of the nutrients may permit the tolerance of secondary chemicals (3).

The quantitative and qualitative variation expressed in allelochemical factors is a powerful force against herbivore specialization (4). The presence of such variability in host-plant populations is evidence for the existence of a strong selection potential promoting specialization of insects upon individual plant chemotypes. Ecological factors are also important, and consideration

0097–6156/88/0380–0403$06.00/0

of the overall physiological efficiency of feeding (5) is a far more useful construct for evaluating insect host-plant interactions than narrower feeding specialization hypotheses. Whatever their evolutionary relationship to plant defensive chemistries, herbivorous insects have been under evolutionary pressure for millenia to optimize their digestive systems for efficient extraction of nutrients from their host plants. Their primary metabolism having thus been adapted to plant nutrient composition, it seems reasonable to hypothesize that insect digestive systems have been forced to adapt to plant secondary compounds as well.

In the following paragraphs the essentials of insect primary metabolism of glucose and its extraction from plant sources through the action of glycosidases is discussed. These and other sugar-processing enzymes are considered in view of their activity upon potentially toxic plant glycosides. Possible interference with the biological activity of glycoside hydrolysis products *in situ* by insect detoxification enzymes is explored, and interference with normal insect glycosidase activity by specialized plant-derived inhibitory substances is found to occur. Finally, the results of experimental tests of the hypothesis that plant glycosides are elaborated primarily as toxins targeted against essential insect glycosidases are given, and the conclusion is reached that this may indeed be the case. If so, the consideration of plant glycoside/insect glycosidase interactions may be fundamentally important in optimizing the design and efficacy of novel and specific pesticides.

Insect Primary Metabolism

The energy needed to drive insect primary metabolism is derived primarily from the uptake of glucose from the diet through catabolic glycosidase activity and from the subsequent glycolysis of alpha-D-glucose (6). An especially efficient glucose transport and transformation ability is required by insects. While their primary metabolism is essentially similar to that of other animals, insects are subject to extreme physiological stresses during metamorphosis, acquisition of cold-hardiness, oviposition and flight. Developmental processes require the rapid and specific depolymerization and repolymerization of alpha-chitin (1,4 linked 2-deoxy-2-acetamido-beta-D-glucose polymer). Acquisition of cold-hardiness requires the rapid production of glycol. Oviposition involves the production of storage glycoproteins and flight depends upon the efficient degradation of trehalose.

Insect Glucose Digestion

Glucose uptake and absorption takes place in the midgut as a very rapid and osmotically-regulated process. Increased sugar or other solute concentrations in the haemolymph decrease the rate of crop-emptying, increased sugar concentrations in the gut increase the rate of crop-emptying, and crop distension regulates the rate of uptake of food. The functional process of glucose transport across membranes is probably a facilitated diffusion which does not depend upon an active carrier mechanism (7).

The uptake of glucose can be potentiated by trehalose formation

(Figure 1). An active transport system involving Malpighian reabsorption has been reported for Locusta migratoria (8) and Calliphora vomitoria (9), which may permit maintenance of an exceptionally high glucose-to-trehalose ratio in the haemolymph of these insects, relative to insects shown to depend upon passive diffusion only (7).

Figure 1. Trehalose.

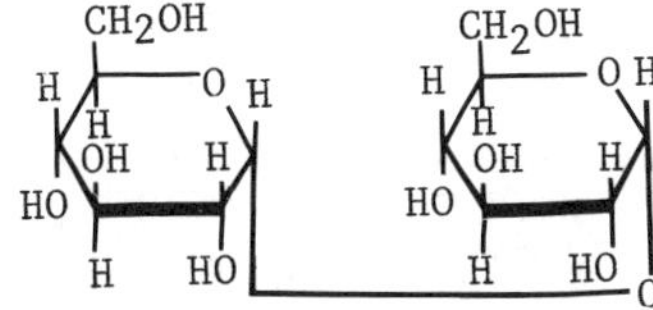

Trehalose (alpha-D-glucopyranosyl-alpha-D-glucopyranoside) is the major form of ready-reserve stored energy in insects, not glycogen as in mammals (10). Concentration of the disaccharide in the haemolymph can reach 2% w/v, more than enough to account for the extreme energy requirements of flight through the direct hydrolysis of trehalose to glucose. The energetics of conversion are sufficient to offset the inefficient insect circulatory system (11). Trehalase, the enzyme catalyzing the hydrolysis of the high-energy storage product, is localized in the foregut- and midgut-epithelia. The latter contains the highest titer of trehalase activity, and is also responsible for the highest rate of absorption of glucose from the diet (in the midgut caeca) (12).

Insect Digestive Enzymes: Glycosidases

Insect beta-glucosidase activity is localized in the foregut- and midgut-epithelia (13). In the cottonseed feeder Dysdercus cingulatus, the highest titers are found in the foregut and increase with continued feeding (14). Phytophagous insects produce digestive enzymes at levels consistent with quantitative stimulation by food in the gut, probably via a protein-sensitive neurosecretory pathway (15,16). All of the secreted midgut enzymes may vary quantitatively together, as was found for 3 carbohydrases, trehalase and trypsin in Locusta migratoria (15). The high pH of many insect digestive system maximizes the activity of alkaline-tolerant enzymes (glycosidases, oxygenases) and permits a high rate of passive uptake of sugars through polar channels (7).

Compartmentalization and regulation of these activities has been observed. Digestive glycosidase activities of Schistocerca gregaria disappear, and are possibly in some way inactivated as they pass to the hindgut (13). It is not known whether they are inactivated and excreted, reabsorbed, digested as part of the inactivation process, or even retained, by an unknown process, proximally above the hindgut.

Insect gut microflora may also be important sources of glycosidases, as they are cellulases (17). These may be especially significant factors in the detoxification and degradation of plant glycosides (18).

Five beta-glycosidases isolated and partially purified from the generalist feeder Locusta migratoria (19) showed hydrolytic activity toward cellobiose, gentiobiose and methyl-beta-glucoside. While a broad range of hydrolytic capability was postulated for this suite of enzymes, it is more probable that each of the five activities represents a partially-separated mixture of more specific activities. Cellobiases are critical in the nutrition of wood-devouring insects. These glycosidases catalyze the terminal degradative step in the digestion of wood cellulose (12). However, the terminal cellulosic activity is dependent upon the presence of an initial cellulase activity: cellobiase alone does not confer the ability to accept cellulose in the diet. Insect carbohydrases, including amylases, transglycosidases, phosphorylases and glucosidases, exhibit specificity of action depending upon substrate linkage conformation (alpha vs beta), conformation (D vs L) and type and number of monosaccharide units (10).

Substrate specificity has been established for plant glycosidases which hydrolyze phenol glycosides, steroidal glycosides, coumarins, flavonoids, cyanogenic glycosides, thioglucosides and oligosaccharides (20-23). Glycosidases have been shown to exhibit substrate specificity dependent upon the structure of the aglycone or the type and number of sugar moieties (20-23). This specificity can be exclusive (24), but is generally more "relaxed" (25). The admission of xenosubstrates to glycosidase sugar binding sites creates a potential for the inhibition of these enzymes.

Insect Detoxification Systems

Before we can postulate an effective inhibitory interaction between plant toxic glycosides and insect glycosidases, we must consider the likelihood that insect detoxification systems will degrade glycosides before they can reach their putative sites of action. The major enzymes utilized by insects in detoxification reactions are the microsomal monooxygenases (mixed function oxidases or MFO). Insect MFO activity is a multienzyme O2/NADPH-requiring system capable of catalyzing oxidation of many substrates (Equation 1). R represents the substrate and X the electron donor.

$$RH + O_2 + H_2X \longrightarrow ROH + X + H_2O \qquad (1)$$

Reactions catalyzed include hydroxylations, dealkylations, oxidations, epoxidations, desulfurations and dehalogenations (26). These varied reactions are carried out by individual hemoproteins, P-450 cytochrome terminal oxidase isozymes (27), which are produced in response to the presence of specific substrates in the gut (28). Specific inducers, inhibitors and synergists of activity are known. The expressed activity of these isozymes is quite variable, and is responsive to many ecological, edaphic, and organismal factors (29). Of the many isozymes postulated, only 3 from insects and 6 from mammals have been characterized. MFO activity can be quantitatively much greater in mammals than in insects on a per unit protein basis (30).

Insects are adept at synthesizing glycosides through the addition of glucose to a substrate via UDP-glucosyltransferase (31), the same process as occurs in plants. Insects are known to synthesize mainly O-glucosides and to a lesser extent S-glucosides, whereas plants also synthesize N-glucosides. Glycosidation is an effective detoxification mechanism (32), especially for phenolics, coumarins, quinones and other conjugated aromatic compounds containing active oxygens. Destructive free-radical formation from these source compounds can be prevented through conjugation of glucose at the active center.

Other detoxification pathways of importance include sulfotransferases and glutathione-S-transferases (33).

Insect glycosidase activity seems to be present earlier in the digestive sequence than MFO activity (foregut/midgut vs midgut), and hydrolysis often has to take place before toxic aglycones are released and act to induce MFO synthesis. Glycosidase activity may be greater than MFO activity as a general consequence of the greater frequency of occurrence of the molecular events of sugar digestion than those of toxin degradation. It seems likely that glycosidase activity will precede MFO activity because of physical location of enzymes, of differences in relative activities, and because it would be unreasonable to expect MFO activity to be needed before toxic aglycones are liberated. A further reason is found in the relaxed specificity shown by some glycosidases. For example, a beta-glucosidase with generalized activity isolated from *Phoracantha semipunctata* (34) increases glucose availability through cellobiose hydrolysis while at the same time releasing toxic aglycones in other hydrolytic reactions (Applebaum, 1985). MFO activity targeted against such toxins would also undoubtedly interfere with glucose assimilation.

Glycosidase Inhibitors

The rate of uptake of glucose, being a molar-diffusive process, is dependent upon the concentration of glucose in the midgut. Inhibition of glycosidase activity will limit the amount of free glucose in the lumen and decrease the rate of uptake of glucose into haemolymph as terhalose (35), which creates conditions of glucose starvation. The insect will respond by emptying the crop faster (36) and eating more (37). In addition, since insect digestive enzymes seem often to be produced in response to the stimulus of protein in food as a prepackaged set of invariant composition (16), defeat of one critical enzyme system will effectively diminish an extended range of digestive activities.

Inhibition of herbivore glycosidase activity by plant glycosides has been found in the case of castanospermine (1,6,7,8-tetrahydroxyoctahydroindolizidine) isolated from *Castanospermum australe* A. Cunn. (Fabaceae) (38). This compound inhibits several insect and mammalian alpha- and beta-glycosidases (39), often competitively (40). Inhibition is efficient at high pH, and disrupts glycoprotein processing (41).

Castanospermine was found to be a feeding deterrent to aphids (42) and *Locusta migratoria* L. but not to *Schistocerca gregaria* Forsk. (43), and to be toxic in a dose-dependent manner to *Callosobruchus maculatus* (44). In this insect and in *Tribolium confusum* toxicity was caused by inhibition of alpha-D-glucosidase, beta-D-glucosidase and beta-D-galactosidase activities in the alimentary tract (45). Castanospermine appears to act as a structural analogue of glucose (43).

A similar compound, swainsonine (an indolizidine triol), was shown to inhibit alpha-mannosidase (46,47). Other pyranose and furanose sugar analogues are known to be potent inhibitors of glycosidases, and in addition, of glucosytransferase (43). One furanose analogue, 1,5-dideoxy-1,5-imino-D-mannitol, has been found to inhibit trehalase (48). These data on sugar analogue action prompted the observation that inhibitors of glycosidases may be an important and common defensive mechanism in plants (43).

Plant Glycosides as Inhibitors of Glycosidases

Plant glycosides which release toxic aglycones upon hydrolysis may have evolved to target the essential process by which insect herbivores digest plant foods and internalize energy. This hypothesis is tested in the experiments described in the following section. A summary of the structural classes of toxic glycosides is given in Table I (after 49-51).

Table I. Structural Classes of Bioactive Plant Glycosides

Terpene glycosides	Glucosinolates
Steroidal glycosides	Lactone glycosides
Saponins	Glycoalkaloids
Phenolic glycosides	Cyanogenic glycosides
Flavonoids	Glycoproteins
Stilbene glycosides	Iridoid glycosides
Xanthone glycosides	Quinone glycosides
Lignan glycosides	Benzoxazine glycosides

Plant glycosides utilized in tests of inhibition were selected to represent the commonly-studied toxic glycosides, and included a thioglucoside, cyanogenic glucoside, iridoid glycoside, phenol glycoside, triterpene glycoside and benzoxazine (Figure 2).

Experimental Tests Of Inhibition

Glycosidase preparations were made by flash-freezing whole starved insect larvae under liquid nitrogen, bisecting the larvae and extracting the frozen midgut under a stereomicroscope, and dipping the extracted tissue into a cold microcell containing 2.0 mL of pH 6.8 phosphate buffer. This initial enzyme mixture was accreted until protein-assay aliquots (micro-Lowry/biuret) showed 10-100 µg total protein. The mixture was then partially purified by passage through a series of small Sephadex columns (G-25, G-100, G-200) in buffer. Beta-glycosidase activity was assayed using 4-nitrophenyl-beta-D-glucoside as a substrate (40). Standard reaction conditions consisted of a 1.0 µg solution of glycosidase in 1.0 mL of buffer to which was added 5.0 umol of nitrophenyl glucoside. The mixture was incubated for 30 min at 37 ^{o}C, then 3.0 mL of pH 10.4 glycine buffer was added to slow the reaction and allow quantitative colorimetric detection of 4-nitrophenol released at 410 nm. The final products of isolation had, on average, only 10 times the activity of the initial enzyme mixture, but were free of proteases and tissue particles, and could be stored in the refrigerator for several weeks.

Tests for inhibition of insect glucosidase activity were carried

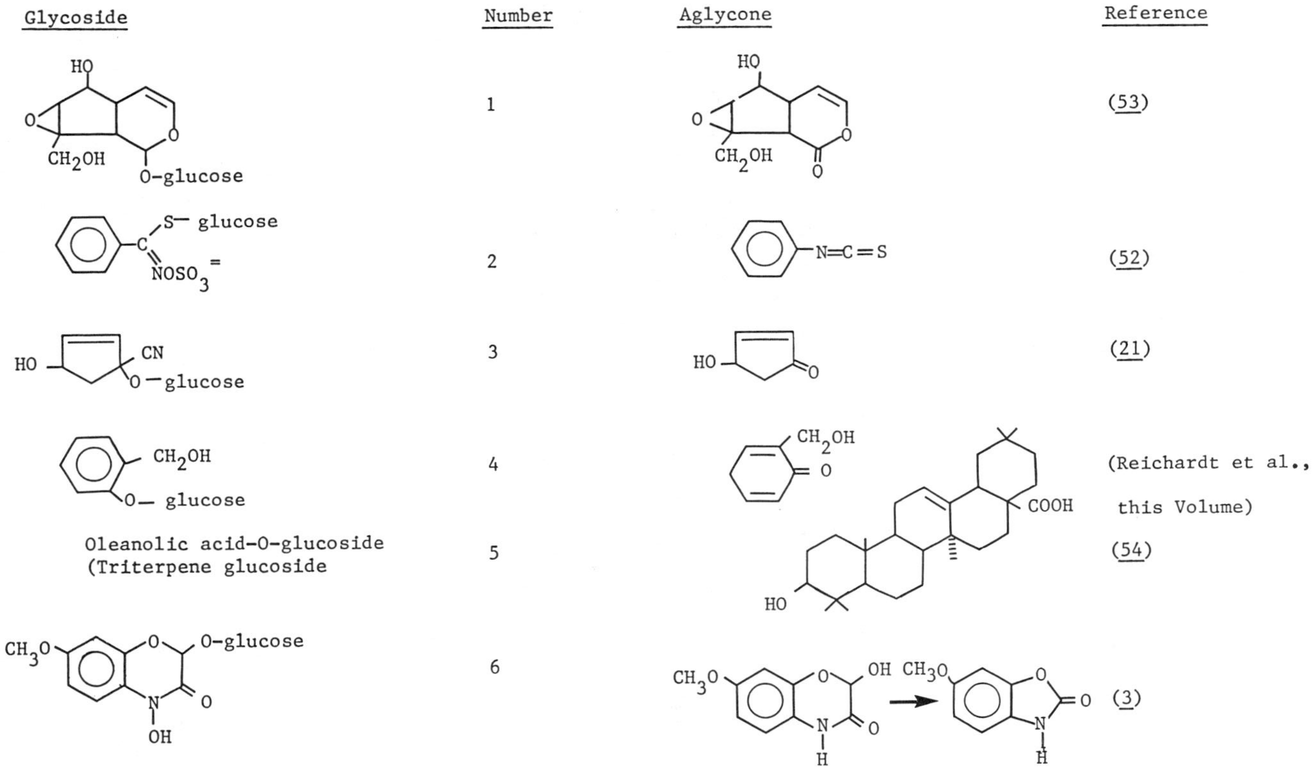

Glycoside	Number	Aglycone	Reference
HO, O, CH_2OH, O, O-glucose	1	HO, O, CH_2OH, O, O	(53)
S— glucose, C, $NOSO_3^=$	2	N=C=S	(52)
HO, CN, O—glucose	3	HO, O	(21)
CH_2OH, O— glucose	4	CH_2OH, O	(Reichardt et al., this Volume)
Oleanolic acid-O-glucoside (Triterpene glucoside	5	COOH, HO	(54)
CH_3O, O, O-glucose, N, O, OH	6	CH_3O, O, OH, N, O, H → CH_3O, O, O, N, H	(3)

Figure 2. Plant glycosides and aglycones tested for inhibitory effects against insect glucosidase fractions.

out by adding 1.0 µg/mL of plant glycoside to the above reaction mixture. When necessary, small amounts of dimethyl sulfoxide or ethanol were added to improve solubility. Control reactions were also run with these solvents. Reactions were conducted as above, and at the end of the incubation period, the amount of 4-nitrophenyl glucoside released was compared with that of the control, and expressed as a percentage.

Separate tests of aglycone activity were also carried out, under the assumption that not all plant glycoside/insect glycosidase interactions proceed to hydrolysis. Aglycones were produced using a specialized microware apparatus depicted in Figure 3. The production chamber is made of a dialysis tubing sac affixed to the distal end of a thistle tube, into which could be added buffer and plant glycoside/plant glycosidase pairs optimal for hydrolysis. A long needle inserted into the tube provides a stream of nitrogen for mixing. The production chamber (1.0 mL volume) was held below the surface of a tube holding 1.0 mL of buffer. This tube contains the insect test glycosidase fraction, and is stirred constantly with a microbar magnetic stirrer. Hydrolysis of the plant glycoside takes place efficiently in the production chamber, and the aglycone produced diffuses efficiently across the membrane into the test solution. During the reaction experiment, sufficient glycoside is introduced into the reaction chamber to yield approximately 1.0 ug/mL of aglycone at equilibrium (usually 2.0 mg of aglycone as glycoside molar equivalent). The pH of the buffer was sometimes lowered to prevent binding of the aglycone to the membrane, and to stabilize ketone products. Samples drawn from the test solution after 30 min at 37 ^{O}C were assayed for relative glucosidase activity as above.

Glucosidase preparations from nine insect species were utilized in the present experiments. These insects were chosen to represent a range of feeding specialists and generalists, including specialists known to be tolerant of certain of the test plant glycosides (Table II).

The purpose of combining individual glycosides with individual

Table II. Test Pairs Of Insects And Their Hostplants Used In Analyses Of Inhibition Of Insect Glucosidases By Plant Glycosides

Insect Species	Host Plant	Interaction
Pieris brassicae L.	Brassicaceae	Specialist
Heliconius erato L.	Passifloraceae	Specialist
Ceratomia catalpa Bois.	*Catalpa speciosa* L.	Specialist
Papilio glaucus canadensis L.	Salicaceae	Generalist Salicin-tolerant
Drosophila mojavensis L.	Cactaceae	Specialist
Heliothis zea L.	*Zea mays* L.	Generalist Specialist *Z. mays*
Spodoptera eridania Cramer		Generalist
Locusta migratoria L.		Generalist
Drosophila melanogaster L.		Generalist

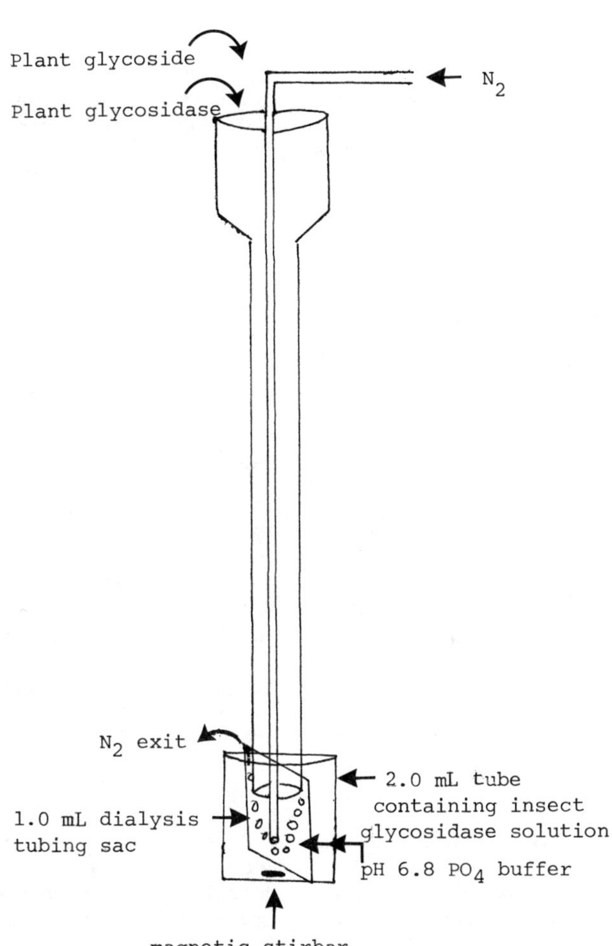

Figure 3. Apparatus for the donation of aglycone hydrolysis products from a plant glucoside/plant glucosidase mixture to a test insect glucosidase solution.

glucosidase fractions in the above reaction system was to determine the following: a) Do plant glycosides inhibit insect glucosidase activities in vitro? b) If such inhibition is detected, does it derive from competitive or noncompetitive behavior? c) If noncompetitive inhibition takes place, what is the mode of action of the inhibitor? d) When inhibition is observed, is there a component of specificity of action which can be related to insect host-plant specialization?

The results of these experiments are summarized in Table III. All plant glycosides tested reduced insect glucosidase activity towards 4-nitrophenyl glucoside to some degree in vitro. Enzyme fractions from specialists were most potently inhibited by glycosides not present in their host plants: H. erato by thioglucoside 2, P. brassicae by cyanogenic glycoside 3, C. catalpae by salicin 4, and P. glaucus by cyanogenic glycoside 3. This strongly indicates that these insects are able to avoid the hydrolysis products corresponding to the compounds found in their preferred host plants, which are seen to be effective inhibitors, by not accepting the glycoside as a substrate for their glycosidases. D. mojavensis is an exception in responding to inhibition by a triterpene glycoside 5 found in its host plant, but this species probably does not encounter this compound in its normal diet as it is first processed to the aglycone by yeasts (21). The triterpene 5 is seen to have little effect in other species. The thioglucoside 2 is not an effective inhibitor of glucosidases, as is expected from its being an S-glucoside, except in H. erato and three generalist species. All of these were found to possess thioglucosidase activity sufficient to produce the aglycone product 2A. This was found to be a noncompetitive inhibitor of glucosidase activity, possibly through alkylation of acidic active sites by thiocyanate (40).

Table III. Inhibitory Effects Of Plant Glycosides And Aglycones (A) Upon Insect Glycosidase Fractions

Source of Glycosidase	Glycoside (Number Refers To Figure 2)											
	1	1A	2	2A	3	3A	4	4A	5	5A	6	6A
	% Activity Relative to Control											
P. brassicae	80	90	90	60	40	40	70	70	100	100	100	100
H. erato	80	70	40	50	90	30	70	60	90	100	90	80
C. catalpae	90	80	90	50	80	30	50	60	90	100	90	70
P. glaucus can.	80	70	90	60	60	40	90	70	90	100	90	80
D. mojavensis	90	90	90	60	90	30	90	70	40	80	100	100
H. zea	80	90	70	60	70	30	80	80	100	100	80	70
S. eridania	80	90	60	60	70	40	80	70	100	100	80	50
L. migratoria	100	100	70	50	80	40	90	70	100	100	90	80
D. melanogaster	100	90	90	50	90	20	100	80	90	90	90	90

The cyclopentenoid aglycone 3A is a potent alkylator and diminished enzyme activity wherever present (21,22). Some degree of noncompetitive inhibition was observed for catalpol 1, salicin 4 and DIMBOA 6, as the recoverable titer of glycoside was detectably smaller than expected after reactions were run. Retention of the corresponding aglycones 1A, 4A, 6A was also observed from reactions run using the aglycone donation apparatus, but quantification was difficult. It appears that both noncompetitive binding and alkylation occur with these compounds, and possibly with the thioglucoside 2 and cyanogenic glucoside 3 as well. Alkylation under certain circumstances seems to be a rational possibility, as can be inferred from consideration of the structures of aglycones shown in Figure 2.

It should be stressed that extrapolation of these results to toxicity *in vivo* should be done with care. We do know that tetraphyllin B is usually toxic in proportion to its alkylating ability to *Heliconius* sp. and to *H. zea* (21,22). Benzylisothiocyanate 2 is a toxin or deterrent to many species (52), and DIMBOA 6 has been shown to reduce fitness in *Spodoptera* sp., but its effect upon *H. zea* is less clear (3). Triterpene glycosides 5 are frequently bioactive, but specificity of action against desert fruit flies has only recently been studied (21). Both catalpol 1 and salicin 4 are such effective feeding deterrents (29,53) that determination of modes of action requires specialized techniques. The oral toxicity of all these compounds toward insects is currently under investigation in this and other laboratories, and it is hoped that these *in vitro* studies will provide some guidance in experimental design. Work is also underway to further purify insect beta-glucosidase fractions, to acquire alpha-glucosidase fractions, and to provide adequate replication for these results. Other test substrates and enzymes are also being studied.

Summary

The primary metabolism of insects depends upon efficient function of glycosidases optimized for extraction of glucose from plant food sources in the diet. The elaboration by plants of substances inhibitory to glycosidases, such as the effective toxin and feeding deterrent castanospermine, suggest that insect digestive enzymes can be targets of plant defensive chemistry. Insect detoxification mechanisms are probably not able to efficiently interfere with glycoside hydrolysis events. Bioactive plant glycosides are ubiquitous in the plant kingdom, and it is proposed that these have arisen in evolution as specific toxins toward insect glycosidases. Preliminary data are presented which support this hypothesis. They demonstrate: a) inhibition of a selection of insect glucosidase fractions by diverse plant glycoside structures, b) the existence of biochemical mechanisms of toxicity including both competitive and noncompetitive modes of inhibition, c) selectivity and specificity of action in inhibitory effect related to insect specialization upon host plants.

Exploitability In Pesticide Design

In the ongoing effort to eliminate major crop pests, special attention has been given to chemical control, including compounds derived from plant sources. Many of these plant-derived natural products are glycosides. We have in the past assumed toxicity for many of these compounds whereas they are in fact toxic only upon hydrolysis by glycosidases. Work in this laboratory and others has shown that these glycosidases are highly specific in action, and may be inhibited either by other glycosidases or glycosides. The determination of toxicity of glycosides will therefore depend upon an assessment of the ability of the enzyme complement of the targeted organism to either potentiate or inhibit hydrolysis and hence toxification.

Literature Cited

1. Miller, J.R.; Strickler, K.L. In Chemical Ecology of Insects; Bell, W.J.; Cardé, R.T., Eds.; Sinauer Associates: Sunderland, MA, 1984; p. 127.
2. Slansky, F., Jr.; Scriber, J.M. In Comprehensive Insect Physiology, Biochemistry and Pharmacology; Kerkut, G.A.; Gilbert, L.I., Eds.; Pergamon Press: New York, 1985; Vol. 4, p. 87.
3. Scriber, J.M. In Chemical Ecology of Insects; Bell, W.J.; Cardé, R.T., Eds.; Sinauer Associates: Sunderland, MA, 1984; p. 159.
4. Cates, R.G.; Redak, R.A. Terpene chemistry of Douglas-fir and its relationship to western spruce budworm success. In Chemical Mediation of Coevolution; Spencer, K.C., Ed.; Academic Press: New York, 1988; in press.
5. Scriber, J.M. In Variable Plants and Herbivores in Natural and Managed Systems; Denno, R.F.; McClure, M.S., Eds.; Academic Press: New York, 1983; p. 373.
6. Candy, D.J. In Comprehensive Insect Physiology, Biochemistry and Pharmacology; Kerkut, G.A.; Gilbert, L.I., Eds.; Pergamon Press: New York, 1985; Vol. 10, p. 1.
7. Turunen, S. In Comprehensive Insect Physiology, Biochemistry and Pharmacology; Kerkut, G.A.; Gilbert, L.I., Eds.; Pergamon Press: New York, 1985; Vol. 4, p. 241.
8. Rafaeli-Bernstein, A.; Mordue, W. J. Insect Physiol. 1979, 25, 241-47.
9. Knowles, G. J. Exp. Biol. 1975, 62, 327-40.
10. Chefurka, W. In The Physiology of Insecta; Rockstein, M., Ed.; Academic Press: New York, 1965; p. 581.
11. Friedman, S. In Comprehensive Insect Physiology, Biochemistry and Pharmacology; Kerkut, G.A.; Gilbert, L.I., Eds; Pergamon Press: New York, 1985; Vol. 10, p. 43.
12. Applebaum, S.W. In Comprehensive Insect Physiology, Biochemistry and Pharmacology; Kerkut, G.A.; Gilbert, L.I., Eds.; Pergamon Press: New York, 1985; Vol. 4, p. 279.
13. Evans, W.A.L.; Payne, D.W. J. Insect Physiol. 1964, 10, 657-74.
14. Sláma, K.; Němec, V. Acta Ent. Bohem. 1981, 78, 1-9.
15. Anstee, J.H.; Charnley, A.K. J. Insect Physiol. 1977, 23, 965-74.
16. Chapman, R.F. In Comprehensive Insect Physiology, Biochemistry and Pharmacology; Kerkut, G.A.; Gilbert, L.I., Eds.; Pergamon Press: New York, 1985; Vol. 4, p. 213.
17. McBee, R.H. Ann. Rev. Ecol. Syst. 1971, 2, 165-76.
18. Scheline, R.R. Mammalian metabolism of Plant Xenobiotics; Academic Press: New York, 1978.
19. Chippendale, G.M. In Biochemistry of Insects; Rockstein, M. Ed.; Academic Press: New York, 1978; p. 1.
20. Hosel, W. In The Biochemistry of Plants; Stumpf, P.K.; Conn, E.E., Eds.; Academic Press: New York, 1981; Vol. 7, p. 725.

21. Spencer, K.C. In Allelochemicals: Role in Agriculture and Forestry; Waller, G.R., Ed.; ACS Symposium Series No. 330, American Chemical Society: Washington, DC, 1987.
22. Spencer, K.C. Chemical Mediation of Coevolution in the Passiflora-Heliconius Interaction. In Chemical Mediation of Coevolution; Spencer, K.C., Ed.; Academic Press: New York, 1988; in press.
23. Nisizawa, K.; Hashimoto, J. In The Carbohydrates; Pigman, W.; Horton, D., Eds.; Academic Press, New York, 1970; 2nd ed., Vol. 2A, p. 241.
24. Hösel, W.; Conn, E.E. Trends in Biochem. Sci. 1982, 7, 219-21.
25. Dale, M.P.; Emsley, H.E.; Kern, K.; Sastry, K.A.R.; Byers, L.D. Biochemistry 1985, 24, 3530-9.
26. Agosin, M. In Comprehensive Insect Physiology, Biochemistry and Pharmacology; Kerkut, G.A.; Gilbert, L.I., Eds.; Pergamon Press: New York, 1985; Vol. 12, p. 647.
27. Nebert, D.W.; Eisen, H.J.; Negishi, M.; Lang, M.A.; Hjelmeland, L.M. Ann. Rev. Pharmacol. Toxicol. 1981, 21, 431-62.
28. Terriere, L.C. Ann. Rev. Entomol. 1984, 29, 71-88.
29. Brattsten, L.B. In Herbivores: Their Interaction with Secondary Plant Metabolites; Rosenthal, G.A.; Janzen, D.H., Eds.; Academic Press: New York, 1979; p. 200.
30. Lindroth, R.L. Adaptations of Mammalian Herbivores to Plant Chemical Defenses. In Chemical Mediation of Coevolution; Spencer, K.C., Ed.; Academic Press: New York, 1988; in press.
31. Smith, J.N. Adv. Comp. Physiol. Biochem. 1968, 3, 173-232.
32. Shono, T.; Unai, T.; Casida, J.E. Pesticide Biochem. Physiol. 1979, 9, 96-106.
33. Dauterman, W.C.; Hodgson, E. In Biochemistry of Insects; Rockstein, M., Ed.; Academic Press: New York, 1978; p. 541.
34. Chararas, C.; Chipoulet, J.M. Comp. Biochem. Physiol. 1982, 72B, 559-64.
35. Treherne, J.E. J. Exp. Biol. 1958, 35, 611-25.
36. Gelperin, A. J. Insect Physiol. 1966, 12, 331-45.
37. Bernays, E. In Comprehensive Insect Phsiology, Biochemistry and Pharmacology; Kerkut, G.A.; Gilbert, L.I., Eds.; Pergamon Press: New York, 1985; Vol. 4, p. 1.
38. Hohenschutz, L.D.; Bell, E.A.; Jewess, P.J.; Leworthy, D.P.; Pryce, R.J.; Arnold, E.; Clardy, J. Phytochemistry 1981, 20, 811-4.
39. Saul, R.; Chambers, J.P.; Molyneux, R.J.; Elbein, A.D. Arch. Biochem. Biophys. 1983, 221, 593-7.
40. Saul, R.; Molyneux, R.J.; Elbein, A.D. Arch. Biochem. Biophys. 1984, 230, 668-75.
41. Pan, Y.T.; Hori, H.; Saul, R.; Sanford, B.A.; Molyneux, R.J.; Elbein, A.D. Biochemistry 1983, 22, 3975-84.
42. Dreyer, D.L.; Jones, K.C.; Molyneux, R.J. J. Chem. Ecol. 1985, 11, 1045-52.
43. Nash, R.J.; Evans, S.V.; Fellows, L.E.; Bell, E.A. In Plant Toxicology; Seawright, A.A.; Hegarty, M.P.; James, L.F.; Keeler, R.F., Eds.; Queensland Poisonous Plants Committee: Yeerongpilly, Queensland, 1985; p. 309.
44. Evans, S.V.; Gatehouse, A.M.R.; Fellows, L.E. Entomol. Exp. Appl. 1985, 37, 257-61.

45. Nash, R.J.; Fenton, K.A.; Gatehouse, A.M.R.; Bell, E.A. Entomol. Exp. Appl. 1986, 42, 71-7.
46. Dorling, P.R.; Huxtable, C.R.; Colegate, S.M. Biochem. J. 1980, 191, 549-56.
47. Colegate, S.M.; Dorling, P.R.; Huxtable, C.R. In Plant Toxicology; Seawright, A.A.; Hegarty, M.P.; James, L.F.; Keeler, R.F., Eds.; Queensland Poisonous Plants Committee: Yeerongpilly, Queensland, 1985; p. 249.
48. Evans, S.V.; Fellows, L.E.; Bell, E.A. Phytochemistry 1983, 22, 770-7.
49. Miller, L.P. In Phytochemistry; Miller, L.P., Ed.; Van Nostrand Reinhold: New York, 1973; Vol. I, p. 297.
50. Bell, E.A.; Charlwood, B.V., Eds. Encyclopedia of Plant Physiology; Springer-Verlag, New York, 1980; Vol. 8.
51. Loewus, F.A.; Tanner, W., Eds. Encyclopedia of Plant Physiology; Springer-Verlag, New York, 1982; Vol. 13A.
52. Van Etten, C.H.; Tookey, H.L. In Herbivores: Their Interaction With Secondary Plant Metabolites; Rosenthal, G.A.; Janzen, D.A., Eds.; Academic Press: New York, 1979; p. 471.
53. El-Naggar, L.J.;Beal, J.L. J. Nat. Prod. 1980, 43, 649-707.
54. Djerassi, C. In Festschr. Arthur Stoll; Birkhauser-Verlag: Berlin, 1957; p. 330.

RECEIVED April 18, 1988

Chapter 28

Chemistry and Biological Activity of Pentatomoid Sex Pheromones

Jeffrey R. Aldrich

Insect and Nematode Hormone Laboratory, Beltsville Agricultural Research Center, Agricultural Research Service, U.S. Department of Agriculture, Beltsville, MD 20705

The Pentatomoidea include devastating pests, as well as beneficial predators. In the pest, *Nezara viridula*, males liberate a long-range attractant pheromone containing a *trans*-epoxide of (*Z*)-α-bisabolene as the major component, and in *Eurygaster integriceps*, vanillin and ethyl acrylate from males attract nearby females after the bugs migrate to wheat fields. Males of an African cotton pest, *Sphaerocoris annulus*, emit a blend of (*Z*)- and (*E*)-4,8-nonadienal, (*Z*)-4-nonenal, and nonanal that is probably an attractant pheromone. In the cottonseed-feeding *Tectacoris diophthalmus*, males that have not found females produce a crystalline deposit of 3,5-dihydroxy-4-pyrone which may be an aphrodisiac and/or an attractant. Pheromones attractive to flying adults have been identified from male predaceous pentatomids (Asopinae) in the genus *Podisus* and artificially mimicked; *R*-(+)-α-terpineol, (*E*)-2-hexenal, and benzyl alcohol for *P. maculiventris*, *S*-(+)-linalool, (*E*)-2-hexenal, and benzyl alcohol for *P. fretus*. Antipodes of chiral monoterpenols are inactive but not inhibitory, and the pheromones are not mutually inhibitory. A complex of parasitoids use these pheromones to locate hosts. Males of 5 other asopine species that prey upon pestiferous beetles yield pheromones comprised of 6,10,13-trimethyltetradecanol and/or the isovalerate ester of this alcohol.

On the night of October 14, 1939, a high school football game in Kansas was invaded by stink bugs: "Spectators were first made aware of their presence by the din caused by the whirring of wings...The band located in the best lighted portion of the stands had to move immediately. The offensive stinkbug smell permeated the neighborhood, especially after spectators and players had crushed the bugs with their shoes and clothing. By the time the game was over...the bugs were from 1 1/2 to 3 1/2 inches deep"(1). This bizarre account of a Thyanta species indicates that pentatomids, alias stink bugs, are not only chemically vile, but also vagile.

The Pentatomidae and allied groups such as shield bugs (Scutelleridae) comprise the superfamily Pentatomoidea. Their stinks are typically mixtures of n-alkanes, C_6, C_8, and C_{10} alk-2-enals, 4-oxo-alkenals and, in adults, alkenyl acetates. These odoriferous secretions are expelled from thoracic glands in adults and abdominal glands in immatures when the insects are attacked (2). Besides their notorious chemical fortification, pentatomoids characteristically possess hypodermic-like mouthparts with which they pierce and suck the sap of developing fruits or, in the case of predatory species, the blood of prey. The most valuable portion of a crop may be severely damaged and exposed to disease, yet the initial injury is often inconspicuous. This subtle assault, combined with the mobility of adult bugs, make pentatomoids difficult to detect and control.

At first researchers in search of sex pheromones for Heteroptera suspected that stink gland esters were attractants (3). However, in the Pentatomoidea (and probably most Heteroptera) this is not so. In fact, the chemical barrage from the stink glands confounds the task of isolating the true sex pheromones. It is now evident that a variety of pheromone glands have evolved independently in male pentatomoids. In some of the predaceous stink bugs (Asopinae), male-specific pheromone glands are extraordinarily large and produce copious amounts of secretion. Though less than a tenth of the over 2500 species of Pentatomidae are predaceous, the extreme reliance of asopines on pheromones has facilitated research on these species. Therefore, the sex pheromone chemistry and biological activity of Asopinae will be discussed first as a framework for the ensuing discussions of the meager data for scutellerid and phytophagous pentatomid sex pheromones. Data on the usurpation of pentatomoid sex pheromones and larval stink gland secretions by parasitoids as host-finding kairomones are also included.

Asopinae

Males in the genera Podisus, Alcaeorrhynchus, and Zicrona have huge pheromone glands opening underneath the wings (4,5). A well-fed Podisus male may contain over a milligram of pheromone in his dorsal abdominal glands (DAGs) and can selectively release the fragrant secretion to call a mate (6). In the two sympatric species of Podisus that have been investigated, P. maculiventris (called the spined soldier bug) and P. fretus, the DAG secretions each contain (E)-2-hexenal (I) and benzyl alcohol (II) as major constituents, but differ in their major monoterpenol component (Figure 1). R-(+)-α-Terpineol (V) is the principal monoterpenol in the pheromone of the spined soldier bug and S-(+)-linalool (III) is the predominant terpene in the P. fretus pheromone but, interestingly, the major monoterpenol of each species occurs as a minor constituent in its sibling species' pheromone (7,8). Other minor components (<2%) include terpinen-4-ol (IV) and trans-piperitol (VI) in the

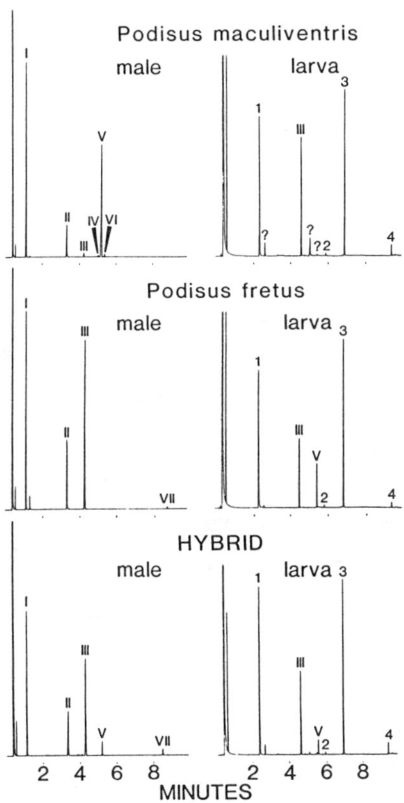

Figure 1. Gas chromatograms of extracts of male pheromone glands and larval exuviae for Podisus spp. and their hybrid (Roman numerals refer to illustrated compounds; 1 = (E)-4-oxo-2-hexenal, 2 = n-dodecane, 3 = n-tridecane, and 4 = tetradecanal).

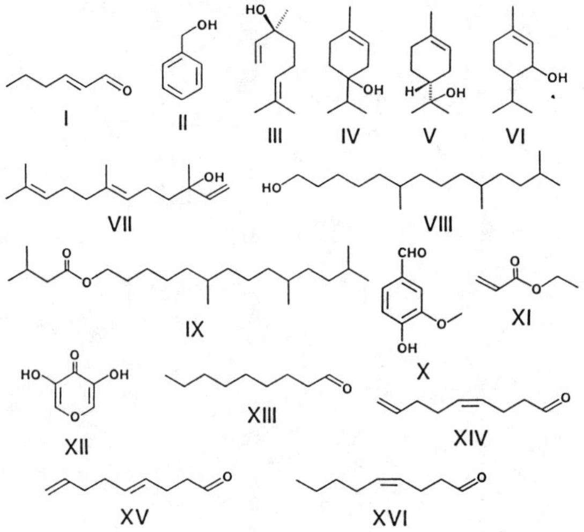

I II III IV V VI

VII VIII

IX X XI

XII XIII XIV

XV XVI

XVII XVIII XIX

pheromone blends of both species and, in the *P. fretus* pheromone, the sesquiterpenoid homologue of linalool, nerolidol (VII). *Podisus fretus* has only been collected in coniferous forests (8). However, the spined soldier bug occurs in deciduous and coniferous forests, and is an important predator in agroecosystems because of its facility for congregating near aggregations of a wide variety of larval crop pests (9-11). In the laboratory, male *P. fretus* paired with female *P. maculiventris* occasionally mated and produced viable hybrids. The DAG secretion of hybrid males contained both α-terpineol and linalool, plus (*E*)-2-hexenal and benzyl alcohol as major constituents (Figure 1).

Early field tests of artificial *P. maculiventris* pheromone containing compounds I-VI showed that *S*-(-)-α-terpineol is inactive but not inhibitory, so pheromones containing a double volume of racemic α-terpineol can be used (6). Field comparisons of partial pheromone formulations to the six component pheromone made with (±)-α-terpineol revealed that only α-terpineol and (*E*)-2-hexenal are required for high attractancy (Table I). Thus, while minor components may play a communicative role during courtship, they are not essential for long-range attraction. Likewise, a formulation of (E)-2-hexenal and (±)-linalool proved highly attractive to *P. fretus* (7). In both *Podisus* spp., males as well as females are attracted. For *P. maculiventris* significantly more males than females were captured in pheromone-baited traps (Table I) whereas for *P. fretus* the reverse was true (7). The response of *Podisus* males to the pheromone of conspecific males may be explained, in part, by the fact that a complex of parasitoids use the pheromones as host-finding kairomones (Table II). Calling males risk discovery by parasitoids, selecting for a counterstrategy whereby "silent" males attracted to the vicinity of calling males try to intercept females without paying the price of parasitization (3). This male counterstrategy may be more pronounced for *P. maculiventris* than *P. fretus* because the spined soldier bug, with its catholic diet, is relatively more abundant and more heavily parasitized than *P. fretus* (Table II). Female spined soldier bugs have small DAGs that produce their own unique blend, including benzaldehyde and nonanal (12,13). Females of the egg parasitoid, *Telonomus calvus* (Hymenoptera: Scelionidae), seem to use both sex-specific spined soldier bug odors as cues; parasitoids are attracted to calling males, but only become phoretic on females (3,14). Another interesting feature of the *Podisus* pheromone system is that an artificial hybrid pheromone attracts both species (Table II), indicating that neither species is inhibited by the major monoterpenol of the other's pheromone.

The tachinid parasitoids attracted to *Podisus* pheromones have previously been recorded from *P. maculiventris* larvae (15), suggesting that these flies use other signals to find larval hosts. When heteropterans molt, the contents of their scent glands are shed with the exuviae. Since tachinids oviposit on the cuticle and these eggs must hatch before the larval host molts, might not scent gland volatiles emanating from the exuviae of a recently molted bug serve as an ideal tachinid kairomone? To test this hypothesis, exuvial extracts of *P. maculiventris*, *P. fretus*, and hybrids were analyzed (Figure 1) and an artificial larval stink gland secretion for *P. maculiventris* was prepared and field tested (Table III).

Table I. Podisus maculiventris caught alive inside traps baited with artificial pheromones[1]

Mean/day/trap			Pheromone blend[3]					
Both sexes[2]	Male	Female	I	II	III	IV	V	VI
6.44 a	4.17	2.28	+	+	-	-	+	-
5.83 ab	3.53	2.31	+	+	-	+	+	+
4.39 ab	2.81	1.58	+	-	-	-	+	-
3.72 ab	2.11	1.61	+	+	+	+	+	+
3.67 b	2.22	1.44	+	+	+	+	+	-
3.42 b	2.11	1.31	+	+	+	-	+	+
2.03 c	1.19	0.83	-	+	+	+	+	+
1.94 c	0.92	1.03	+	-	+	+	+	+
0.17 d	0.17	0.00	-	-	-	-	+	-
0.06 de	0.06	0.00	+	+	+	+	-	+
0.00 e	0.00	0.00	+	-	-	-	-	-
0.00 e	0.00	0.00	-	-	-	-	-	-
Mean =	1.61 a	1.03 b						

[1] Trap design and field placement as described previously (6). Two replicates per treatment were monitored and repositioned daily, and rebaited every three days from April 16 through May 3, 1984.

[2] Data were transformed using log (x+1) and analyzed by Duncan's Multiple Range Test. Means not followed by a common letter are significantly different ($P<0.05$).

[3] The complete pheromone was prepared by blending 1739 µl (E)-2-hexenal (I), 96.8 µl benzyl alcohol (II), 11.8 µl (±)-linalool (III), 9.8 µl 87% ee (+)-terpinen-4-ol (IV), 2123 µl (±)-α-terpineol (V), and 19.6 µl trans-piperitol (VI) (4 ml total), and then mixing this amount of the neat blend with 20 ml of plasticized polyvinyl chloride (PVC). Partial formulations were prepared in the same manner except one or more compounds were omitted and the resulting volume was mixed with 20 ml PVC. Traps were rebaited with ca. 250 mg pheromone-PVC.

Table II. Podisus spp. and parasitoids caught in traps baited with artificial P. maculiventris pheromone, P. fretus pheromone, or hybrid pheromone[1]

Lure[2]	Podisus mac.	Podisus fretus	Parasitoids[3] E. flava	H. aurata	F. crinita	T. calvus
I/II/V	308	0	279	41	18	2
I/II/III	0	83	19	30	0	5
I/II/III-V	151	41	211	19	4	10

[1] Two live-traps and 2 sticky traps per treatment (6), randomized and monitored daily, and rebaited every 3 days from March 27 through May 29, 1985. One P. maculiventris was caught in a control trap.

[2] Lures contained by volume 1 part (E)-2-hexenal (I) and 0.17 part benzyl alcohol (II) with either 2 parts (±)-α-terpineol (V), 2 parts (±)-linalool (III), or 1 part (±)-α-terpineol and 1 part (±)-linalool. Pheromone-PVC (20%) was prepared and used for rebaiting as described in Table I.

[3] Euclytia flava and Hemyda aurata (Diptera: Tachinidae) are endoparasitic on adults and larvae, Forcipomyia crinita (Diptera: Ceratopogonidae) are ectoparasites of adults, and Telenomus calvus (Hymenoptera: Scelionidae) are egg parasitoids.

Each exuvial extract contained (*E*)-4-oxo-2-hexenal, *n*-dodecane, *n*-tridecane, and tetradecanal in equivalent amounts. Surprisingly, linalool is a major constituent of larval *P. maculiventris* DAG secretion (13) and larval *P. fretus* DAG secretion includes both linalool and α-terpineol (Figure 1). The exuvial extract of hybrids is compositionally like that of *P. fretus* except for relatively more linalool (Figure 1). Despite the fact that the DAG secretions of these *Podisus* larvae are species-specific by virtue of their monoterpenol components, there was no indication in the field test that spined soldier bug larvae use the stink gland volatiles as an aggregation pheromone. On the contrary, the only *P. maculiventris* larvae caught during the experiment were attracted to the formulation mimicking the adult pheromone (Table III). On the other hand, both tachinid species were attracted to the blend mimicking the larval *P. maculiventris* DAG secretion, albeit less so than to the artificial adult pheromone, demonstrating that these parasitoids recognize at least two chemical beacons from their *Podisus* hosts.

Asopines in the genera *Perillus* and *Oplomus* have been released (but apparently not established) in Europe and the U.S. for biological control of the Colorado potato beetle and the Mexican bean beetle, respectively (17,18). These predators, and others from the related genera *Stiretrus* and *Mineus*, evidently specialize on certain lepidopteran and coleopteran larvae (9,10). This group of asopines lacks sexually dimorphic DAGs, but instead males possess sternal glands (SGs) conspicuous externally by dense patches of setae on the fourth to sixth abdominal sternites (Figure 2). In these oligophagous predators, starvation (simulating migration?) followed by feeding *ad libitum* stimulates the SGs to secrete 6,10,13-trimethyltetradecanol (VIII) (*S. anchorago* and *O. dichrous*) and/or the isovalerate ester (IX) of this alcohol (*P. bioculatus*, *O. severus*, *O. dichrous*, and *M. strigipes*) (17,19). Although the chirality of the natural products has not been determined, male and female adults of *S. anchorago* flew and larvae walked to field traps baited with racemic VIII, demonstrating that the male SG secretion is an attractant pheromone and that unnatural antipodes are probably not inhibitory (Kochansky, J. P.; Aldrich, J. R.; Lusby, W. R. *J. Chem. Ecol.*, in press). Airborne trapping of volatiles from field-collected *S. anchorago* males verified that VIII evaporates from the SGs of males and is absent from females. In addition, analysis of male-derived airborne extracts revealed that the corresponding aldehyde of VIII is about ten times more concentrated (5%) than in samples prepared by directly extracting SG secretion from the sternal setae. Whether or not inclusion of the minor aldehydic component or only the natural stereoisomer(s) will increase the attractiveness of the synthetic pheromone remains to be determined. The SG secretions of the other asopines containing VIII and/or IX will likely prove to be long-range attractants and possibly close-range mating stimulants. Additional asopine genera bearing pubescent SGs whose secretions have yet to be chemically investigated include *Discocera* (American tropics), *Dorycoris* (Africa and Madagascar), and *Cazira* (Asian tropics) (20).

The asopine genera *Picromerus*, *Dinorhynchus*, *Apateticus*, and *Euthyrhynchus* have neither sexually dimorphic DAGs nor SGs (2). The

Table III. *Podisus maculiventris* and tachinid fly parasitoids caught in traps baited with artificial adult pheromone or artificial larval stink gland secretion[1]

Lure[2]	*P. maculiventris* adults	*P. maculiventris* larvae	*E. flava* female	*E. flava* male	*H. aurata* female	*H. aurata* male
I/II/V	86	6	249	171	42	35
1/III/3	0	0	88	8	11	18
III	0	0	0	0	0	2
Unbaited	0	0	0	0	0	0

[1] Six live-traps and 6 sticky traps per treatment were tied to trees ca. 2 m above ground (*6*). Traps were monitored and rebaited daily from April 25 through June 17, 1986.

[2] (*E*)-2-Hexenal (I)/benzyl alcohol (II)/(±)-α-terpineol (V) were mixed as described in Table II, and (*E*)-4-oxo-2-hexenal (1)/(±)-linalool (III)/*n*-tridecane (3) were mixed in equal volumes. Ten µl of these neat mixtures were applied daily to a rubber septum in the appropriate traps. Unbaited controls were run from April 25 through June 2; for the remainder of the test control traps were baited with 5 µl of (±)-linalool daily. (*E*)-4-Oxo-2-hexenal was synthesized according to Ward and VanDorp (*16*).

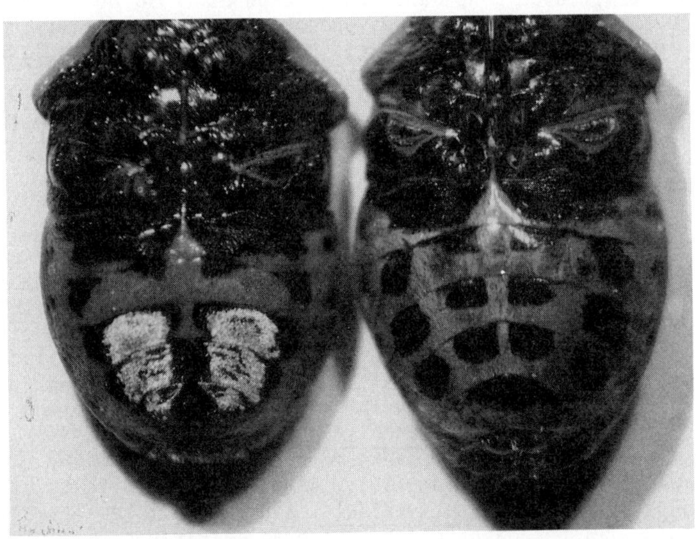

Figure 2. Sterna of *Perillus bioculatus* adults showing the male (left) sternal gland setae (light areas) that are wetted with secretion, and the female (right) where pubescent patches are absent. (Reprinted from ref. 19.)

only hint of pheromone-mediated behavior in these predators is for larvae; E. floridanus immatures are exceptionally gregarious (21) and A. bracteatus larvae hunt singly, yet recongregate to molt (22).

Scutelleridae

Shield bugs are all phytophagous and some are important agricultural pests. Foremost among the pest species is Eurygaster integriceps, a serious wheat pest in the Middle East (18). Adults overwinter in the mountains and migrate to the grain fields in the spring, whereupon males release a scent containing vanillin (X), ethyl acrylate (XI), and presumably other compounds (23,24). While the source of this scent remains unknown, emission is accompanied by a characteristic calling posture and nearby females are attracted to calling males (25). Some scutellerid males have pubescent SGs (also called androconial glands) on sternites 4-6 (20), but in the subfamily Tetyrinae the sternum instead bears stridulatory areas in both sexes (26). The cotton harlequin bug, Tectoris diophthalmus, is an occasional pest of cultivated Malvaceae in Australia and is the only shield bug whose SG secretion has been chemically analyzed; in virgin males the SGs are often loaded with a crystalline deposit of 3,5-dihydroxy-4-pyrone (rubiginol) (XII) suspected to serve as an aphrodisiac (20,27). Nevertheless, XII may also be an attractant because the crystals reportedly sublime quite readily (27). Adults of T. diophthalmus also retain active anterior DAGs that produce nonanal in both sexes (28). In another cotton-feeding shield bug, Hotea gambiae, the anterior DAGs are proportionally larger in males than in larvae, whereas in females they regress in size; in larvae these glands produce mono- and sesquiterpenes (29), but in both sexes of adults (E)-2-hexenol is predominant (30). Finally, males of the tropical West African shield bug, Sphaerocoris annulus, have well developed DAGs that release an odoriferous array of C_9 aliphatic aldehydes (XIII-XVI) that might function as an attractant pheromone (31). Unfortunately, behavioral data as to the raison d'etre for scutellerid male-specific exocrine secretions are wanting.

Phytophagous Pentatomidae

Plant-feeding pentatomids are a large and cosmopolitan assemblage of insects, including many devastating agricultural pests (11,18,32), but as of this writing a synthetic pheromone active in the field has yet to be reported for the group (2). It has been known since 1971 that males of the worldwide pest, Nezara viridula, emit an attractant pheromone (33). In the same year it was reported that the pestiferous stink bug, Euschistus conspersus, mates in aggregations that remain at the same place for days, with individuals coming and going (34). Additional field tests using caged bugs verified that N. viridula males attract females and the parasitoid, Trichopoda pennipes (Diptera: Tachinidae), and that conspecific males and fifth-instar larvae also respond (35). More recently, caged males of a serious fruit pest in Japan, Plautia stali, have been shown to strongly attract females and males of the species, as well as a tachinid parasitoid (36), and in the pentatomoid, *Megacopta punctissimum* (Plataspidae), males initiate mating aggregations similar to those of *E. conspersus* (37,38). Thus, even though males of many phytophagous pentatomoids have glands in the sternal epidermis that are absent or less numerous in females (39), the behavioral data signalling the existence of sex pheromones in these bugs are sparse.

For *N. viridula*, the so-called southern green stink bug or green vegetable bug, airborne volatiles isolated from mature males were attractive to females in an olfactometer (40,41), and to adults, larvae, and *T. pennipes* in the field (Figure 3) (42). *Nezara* males, and males of several other pentatomoids, have areas on the anterior abdominal sternites that are visibly smoother than surrounding cuticle (2). Nonetheless, extracts of the posterior segments of *N. viridula* males were more attractive than extracts of anterior segments (43), suggesting that the modified sternal areas are not involved in pheromone production. Males of a Brazilian strain of *N. viridula* liberate C_{12}, C_{13}, and C_{19} normal hydrocarbons, (*Z*)-α-bisabolene (XVII), and a *trans*-epoxide (XVIII) of the sesquiterpene hydrocarbon (44). The latter compound was the predominant component and the stereochemistry of the natural product was assigned based on the bioassay activities of synthetic diastereomers (44). Southern green stink bug males from the U.S. release XVIII (44%) (and/or the antipode) and the *cis*-isomer XIX (15%) (and/or the antipode), plus the C_{13} (2.3%) and C_{19} (7.4%) alkanes, and 1.4% nerolidol (VII) (42). Male *N. viridula* from France resemble the U.S. strain of the bug in emitting both XVIII and XIX (ca. 2:1 ratio) (44). Therefore, it appears that different pheromone strains of *N. viridula* exist (42).

Concluding Remarks

Males of many predaceous stink bugs have massive, well-defined exocrine glands that secrete attractant pheromones. Some plant-feeding scutellerids also have obvious pheromone-like glands as do the Asopinae (20), but in most phytophagous pentatomoids exocrine gland cells are scattered in the cuticular epidermis and their excess in males is only apparent upon histological examination (39). In species lacking pheromone glands that can be dissected from the bugs, aeration is the best method for pheromone isolation. This approach is fraught with the difficulty of herding bugs into a suitable apparatus without eliciting a defensive chemical discharge. Moreover, pentatomoids frequently hibernate or aestivate as adults before reproducing, and diapausing males of *N. viridula* (and probably males of other species) do not produce pheromone (45). Nevertheless, male-specific blends have been isolated via airborne trapping for economically important *Euschistus*, *Acrosternum*, *Thyanta*, and *Murgantia* species (Aldrich, J. R.; Hoffmann, M. P., unpublished data).

Stink bug damage is often unpredictable because pentatomoids are strong fliers (1,46,47) that haphazardly invade crops. Therefore, pheromone-baited traps for pest species may be useful in detecting incipient infestations. Beyond this now conventional use of pheromones, deciphering the chemical communication systems of

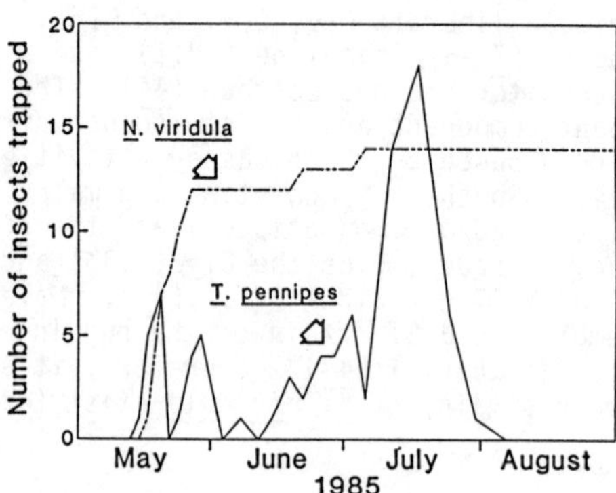

Figure 3. Nezara viridula adults (cumulative) and tachinid parasitoids (daily) caught in traps deployed in Louisiana baited with airborne-trapped volatiles from N. viridula males laboratory-reared in Maryland. (Reprinted from ref. 42.)

harmful and beneficial pentatomoids could lead to improved biological control in a variety of ways. *Nezara viridula* has recently invaded the Sacramento Valley in California (32). The knowledge that at least some parasitoids home in on the *Nezara* attractant pheromone and that this pheromone differs between geographically isolated populations, may temper decisions as to where to seek parasitoids for establishment in California (Hoffmann, M. P., University of California at Davis, personal communication, 1987). A novel use for synthetic pheromones of the southern green stink bug and other exotic pentatomoid pests would be their implementation as artificial kairomones to collect the parasitoids needed for biological control. Similarly, synthetic pheromones of Asopinae might be used to collect exotic predators for biological control programs (17). Native asopines could be augmented and conserved in agroecosystems by the judicious application of artificial pheromones. This approach may be an economically favorable means to tap the vast reservoir of predaceous heteropterans, especially if attractants are discovered for more species from other taxa and found not to be mutually inhibitory. Better biocontrol through chemistry would be a propitious welcome to the next millennium.

Acknowledgments

The author thanks Dr. Larry Douglas, ARS Consulting Statistician, University of Maryland, for statistical analysis, and Susan Wilzer for technical assistance. Mention of commercial products does not constitute an endorsement by the U.S. Department of Agriculture.

Literature Cited

1. Wilbur, D. A. *J. Kans. Entomol. Soc.* 1939, *12*, 77-80.
2. Aldrich, J. R. *Ann. Rev. Entomol.* 1988, *33*, 211-38.
3. Aldrich, J. R. In *Semiochemistry: Flavors and Pheromones*; Acree, T. E.; Soderlund, D. M., Eds.; de Gruyter: Berlin, 1985; pp 95-119.
4. Aldrich, J. R.; Blum, M. S.; Lloyd, H.; Fales, H. M. *J. Chem. Ecol.* 1978, *4*, 161-72.
5. Dupuis, C. *Bull. Soc. Nat. (Paris)* 1952, *7*, 1-4.
6. Aldrich, J. R.; Kochansky, J. P.; Abrams, C. B. *Environ. Entomol.* 1984, *13*, 1031-6.
7. Aldrich, J. R.; Lusby, W. R.; Kochansky, J. P. *Experientia* 1986, *42*, 583-5.
8. Rudmann, A. A.; Aldrich, J. R. *J. Chromatogr.* 1987, *407*, 324-9.
9. Evans, E. W. *Ecology* 1982, *63*, 147-58.
10. McPherson, J. E. *The Pentatomoidea (Hemiptera) of Northeastern North America*; Southern Illinois University Press: Carbondale, 1982.
11. Turnipseed, S.; Kogan, M. In *Natural Enemies of Arthropods in Soybean*; Pitre, H. N., Ed.; South Carolina Agr. Exp. Stn.: Southern Cooperative Series Bull No. 285, 1983; pp 1-6.
12. Aldrich, J. R.; Lusby, W. R.; Kochansky, J. P.; Abrams, C. B. *J. Chem. Ecol.* 1984, *10*, 561-8.

13. Aldrich, J. R.; Kochansky, J. P.; Lusby, W. R.; Sexton, J. D. J. Wash. Acad. Sci. 1984, 74, 39-46.
14. Orr, D. B.; Russin, J. S.; Boethel, D. J. Can. Entomol. 1986, 118, 1063-72.
15. Eger Jr., J. E.; Ables, J. R. Southwestern Entomol. 1981, 6, 28-33.
16. Ward, J. P.; VanDorp, D. A. Recueil 1969, 88, 989-93.
17. Aldrich, J. R.; Lusby, W. R. Comp. Biochem. Physiol. 1986, 85B, 639-42.
18. Clausen, C. P. Introduced Parasites and Predators of Arthropod Pests and Weeds: A World Review; U. S. Department of Agriculture, U. S. Government Print Office: Washington, DC, 1978; pp 23-6.
19. Aldrich, J. R.; Oliver, J. E.; Lusby, W. R.; Kochansky, J. P. Arch. Insect Biochem. Physiol. 1986, 3, 1-12.
20. Carayon, J. Ann. Soc. Entomol. Fr. 1984, 20, 113-34.
21. Oetting, R. D.; Yonke, T. R. Ann. Entomol. Soc. Am. 1975, 68, 659-62.
22. Evans, E. W.; Root, R. B. Ann. Entomol. Soc. Am. 1980, 73, 270-4.
23. Ubik, K.; Vrkoc, J.; Zdarek, J.; Kontev, C. Naturwissenschaften 1975, 62, 348.
24. Vrkoc, J.; Ubik, K.; Zdarek, J.; Kontev, C. Acta Entomol. Bohemoslov. 1977, 74, 205-6.
25. Zdarek, J.; Kontev, C. Acta Entomol. Bohemoslov. 1975, 72, 239-48.
26. Slater, J. A.; Baranowski, R. M. How To Know The True Bugs; Wm. C. Brown Co.: Dubuque, Iowa, 1978; p 43.
27. Knight, D. W.; Staddon, B. W.; Thorne, M. J. Z. Naturforsch. Teil C 1985, 40, 851-3.
28. Staddon, B. W.; Thorne, M. J.; Knight, D. W. Aust. J. Zool. 1987, 35, 227-34.
29. Gough, A. J. E.; Hamilton, J. G. C.; Games, D. E.; Staddon, B. W. J. Chem. Ecol. 1985, 11, 343-52.
30. Hamilton, J. G. C., Gough, A. J. E.; Staddon, B. W.; Games, D. E. J. Chem. Ecol. 1985, 11, 1399-409.
31. Gough, A. J. E.; Games, D. E.; Staddon, B. W.; Knight, D. W.; Olagbemiro, T. O. Z. Naturforsch. Teil C 1986, 41, 93-6.
32. Hoffmann, M. P.; Wilson, L. T.; Zalom, F. G. California Agriculture 1987, 41, 4-6.
33. Mitchell, W. C.; Mau, R. F. L. J. Econ. Entomol. 1971, 64, 856-9.
34. Alcock, J. Psyche 1971, 78, 215-28.
35. Harris, V. E.; Todd, J. W. Entomol. Exp. Appl. 1980, 27, 117-26.
36. Moriya, S.; Shiga, M. Appl. Entomol. Zool. 1984, 19, 317-22.
37. Hibino, Y. J. Ethol. 1985, 3, 123-9.
38. Hibino, Y. J. Ethol. 1986, 4, 91-5.
39. Carayon, J. C. R. Acad. Sci. Ser. III 1981, 292, 867-70.
40. Borges, M.; Jepson, P. C.; Howse, P. E. Entomol. Exp. Appl. 1987, 44, 205-212.
41. Pavis, C.; Malosse, C. C. R. Acad. Sci. Ser. III 1986, 303, 273-6.

42. Aldrich, J. R.; Oliver, J. E.; Lusby, W. R.; Kochansky, J. P.; Lockwood, J. A. J. Exp. Zool. 1987, 244, in press.
43. Pavis, C. Ph.D. Thesis, Universite de Paris-Sud Centre d'Orsay, France, 1986; pp 92-99.
44. Baker, R.; Borges, M.; Cooke, N. G.; Herbert, R. H. Chem. Commun. 1987, 414-16.
45. Brennan, B. M.; Chang, F.; Mitchell, W. C. Environ. Entomol. 1977, 6, 167-73.
46. Hokkanen, H. Ann. Entomol. Fenn. 1986, 52, 28-31.
47. Moriya, S. Appl. Entomol. Zool. 1987, 22, 19-24.

RECEIVED April 1, 1988

Chapter 29

Dithiopolyacetylenes as Potential Pesticides

Eloy Rodriguez

Phytochemical and Toxicology Laboratory, Department of Developmental and Cell Biology, University of California–Irvine, Irvine, CA 92717

Dithiopolyacetylenes (1,2-dithiin cyclohexadiene polyacetylenes) are a group of naturally occurring allelochemicals synthesized primarily in the roots and exhibit a wide-range of antibiotic activities. Some dithiopolynes known to be active in the dark and in the presence of ultraviolet light, are active against fungi and insects. The thiarubrines (1,2-dithiin polyines) occur in a variety of plant species from the family Asteraceae and show promise as antiviral and nematicidal agents. In this brief report, we present our latest in vitro finding on the nematicidal and antiviral action of thiarubrine A.

Nematodes that parasitize root systems are well-recognized pests in agriculture, with growers commonly applying large quantities of synthetic pesticides to control their infestation. Although various synthetic pesticides have been used and banned, studies on the use of naturally derived chemicals as nematicidal agents have been rather limited. Gommers in 1973 (1) established that certain secondary metabolites (polyacetylenes and chromenes) present in the roots of various taxa of the sunflower family (Asteraceae) were effective in controlling nematode infestations. It was found that various species of the Asteraceae controlled and reduced the populations of nematodes in both greenhouse and field conditions. The most effective species belonged to the genera *Rudbeckia*, *Aspilia*, *Ambrosia*, *Chaenactis* and *Gaillardia*. Gommers suggested that it might be possible to eliminate 80-90% of parasitic nematodes by planting a crop of ornamental composites. Although no detailed chemical studies were carried out by Gommers, it was suggested that the possible active agents were thiophene derivatives and dithiopolyacetylenes. The dithiopolyines were discovered by Bohlmann (2) and it was not until 1985 that serious biological tests were initiated. Rodriguez and associates (3) discovered substantial quantities of the thiarubrines in leaves of *Aspilia mossambicensis*

0097–6156/88/0380–0432$06.00/0

and subsequently in various species of Ambrosia. The active constituents were identified as thiarubrine A and B, which were previously found in the roots and tumor cultures of Chaenactis douglasii (4, 8). Thiarubrines A and B (Figure 1) are important medicinal principles present in plant extracts and used in Africa and Canada by native populations for skin infections and intestinal parasites (5). The findings are significant, since previous anthropological studies had suggested that wild apes were swallowing Aspilia leaves in a particular manner for possible therapeutic purposes (3). Detailed chemical investigations proved that the major compounds were red-colored constituents present in high amounts in the young leaves preferred by the wild chimpanzees. The red compound absorbed at 490 nm, 345 nm and 243 nm. High pressure liquid chromatography and mass spectrometry identified the two major dithiopolyacetylenes as 1-(2-methylethyn)-4-(hex-1,3-diyn-4-ene)-2,3-dithiacyclohexa-4,5-diene (thiarubrine A) and 1-(4-methylbut-1,3-diyn)-4-(but-1-yn-3-ene)-2,3-dithiacyclohexa-4,6-diene (Thiarubrine B). The leaves of Aspilia mossambicensis, consumed by the wild apes, were also found to be used by African natives for the treatment of abdominal pains, intestinal worms and skin infections.

Recently, Hudson and associates (7) published preliminary findings on the action of thiarubrine A (in the dark and light) on a variety of bacteria, fungi, viruses and nematodes. As noted in the Experientia paper, thiarubrine A is as effective as the strong photosensitizer α-terthienyl (α-T) which is effective against Candida albicans, Staphylococcus albus, Mycobacterium phlei, Bacillus subtilis, Streptococcus faecalis and E. coli at a concentration of 0.1-1.0 ppm. It is as effective against Candida albicans as fungizone (amphotericin B), at concentrations of 1 ppm in the dark or 0.1 ppm in the light (Figure 2). Thiarubrine A was also evaluated for its antiviral properties in the presence and absence of UV-A radiation (UVA) (Hudson, J., Planta Medica, in press). Two mammalian viruses, murine cytomegalovirus (mCMv) and Sindbis virus (SV), both of which possess membranes, were sensitive to the compound, but only in the presence of UVA radiation. The bacteriophage T4 was slightly affected in the presence of UVA radiation, whereas the bacteriophage M13 was completely unaffected (Tables I and II). These studies suggest that a photoactivated species of thiarubrine A is the active constituent. Although α-terthienyl and thiarubrine A have significant antiviral activity at 10^{-5} ug/ml, an equivalent anticellular effect requires 10^{-2} ug/ml (6).

These data suggest that in addition to lipids some viral proteins may be targets of these compounds, although the mechanism by which such target-directed photoactivity is converted into a block in the viral replication cycle is not understood (Hudson, J., Planta Medica, in press). Indeed, our proposed synthetic and structure-activity studies should provide valuable insights.

In yet unpublished results, investigators at UC Riverside and UC Irvine have found that roots of Aspilia mossambicensis are very resistent to the root nematode, Meloidogyne incognita. Furthermore, petroleum ether extractions of the roots have yielded a substantial

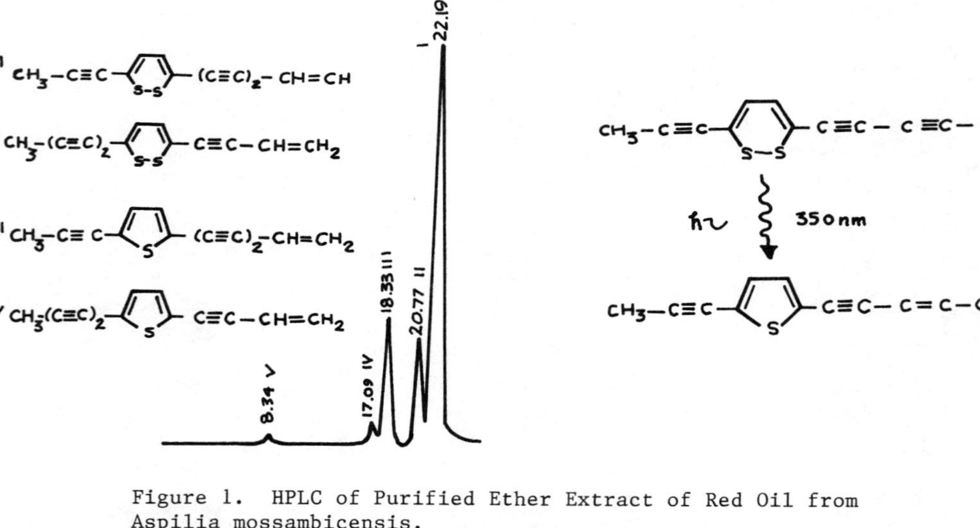

Figure 1. HPLC of Purified Ether Extract of Red Oil from Aspilia mossambicensis.
(Thiarubrine A = I, Thiarubrine B = II, Thiophene A = III, Thiophene B = IV).

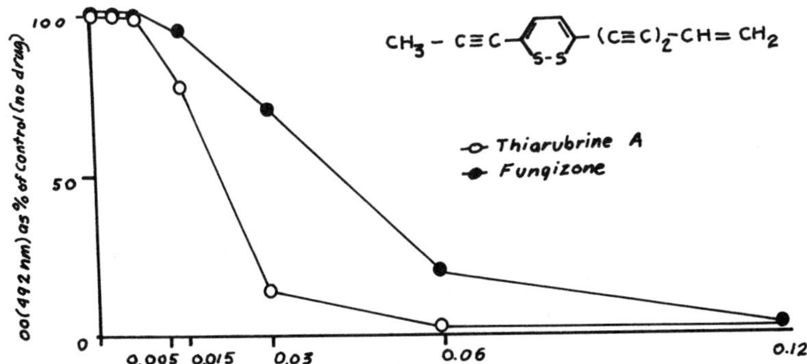

Figure 2. Activity of Thiarubrine A vs Fungizone against <u>Candida</u> <u>albicans</u>. (Data from ref. 5.)

quantity of red oil that contains thiarubrines A and B. Extractions of *Rudbeckia hirta*, *Ambrosia chamissonis* and *A. psilostachya* have yielded good quantities of thiarubrines A and B and some novel dithiopolyines. High pressure liquid chromatography (HPLC) yielded the purified thiarubrine A that exhibited potent *in vitro* activity against the root knot nematode (see Table III). Similar nematicidal activity has been reported for asparagusic acid (*9*). It is also important to note that the activity of thiarubrine A was observed in the absence of light, therefore suggesting that the active nematicidal metabolite does not appear to be a photosensitizer.

Table I. Summary of Antiviral Effects of Thiarubrine A

Virus	Genome	Membrane	Antiviral Effect UV-A	In dark
MCMV	d.s[a] DNA	yes	v. strong	none
Sindbis	s.s. RNA	yes	v. strong	none
T4	d.s. DNA	no	slight	none
M13	s.s. DNA	no	none	none

[a]d.s.= double stranded, s.s.= single stranded.
Hudson *et al.*, (*6*).

Table II. LD_{99} Values for Thiarubrine A and Thiophenes of Natural Origin[b]

Compounds[a]	SV	MCMV
-Terthienyl	30	194
thiarubrine A	1,140	625
Phenylheptatriyne (PHT)	3,168	6,272
Thiophene A[c]	12,000	12,000

[a] Concentration of 0.1 ug/ml
[b] LD_{99} = dose of UV radiation (sec's of exposure x 5 watts/m^2) required to decrease infectivity by 99%
[c] Estimated by comparison with PHT at 10 ug/ml
Hudson *et al.*, (*6*).

Table III. *In vitro* toxicity[1] of thiarubrine A on the second-stage larvae of *Meloidogyne incognita*

Exposure Condition	% Live Nematode[2]					
	D.W.(CK)[3] Control	1% EtoH Control	5ppm	10ppm	15ppm	20ppm
Light, 48 h	97.5	93.3	10.9	2.2	0	0
Dark, 48 h	99.5	93.7	1.7	0	0	0

[1] In 1% ethanol solution
[2] Mean of 8 replicates.
[3] D.W. (CK) = distilled water (check)

Due to the difficulty in obtaining sufficient quantities of thiarubrines for biological testing, we have established root cultures of Aspilia mossambicensis, and tumor cultures of Chaenactis douglasii for obtaining thiarubrines A and B (Norton, R. et al., J. of Plant Physiology, in press) (8). Polyine levels in Aspilia cultures exceeded the levels in roots of intact plants with the percentage composition very similar for cultured and intact roots.

Concluding Remarks

Although only a few dithiopolyacetylenes have been isolated from higher plants, it appears that these red-colored oils have primarily evolved as potent antibiotics that are effective against fungi, viruses and nematodes. It also is not surprising that these compounds are effective against phytophagous insects, since preliminary studies indicate that these compounds affect the hormonal system of larvae. Further chemical and mechanism of action studies are needed to fully understand the pesticidal potential of the dithiopolyacetylenes.

Acknowledgments

ER thanks Elevenia J. Slosson Fund for providing support for the nematicidal studies. We thank Dr. Jan West (UCI), Dr. S. D. Van Gundy (UC Riverside) and Dr. B. Tsai for unpublished data on nematicidal activity of dithiopolyines. We also thank the reviewers for their constructive criticism of the manuscript.

Literature Cited

1. Gommers, F. J. Neth. J. Pl. Path. 1973. 82, 73-77.
2. Bohlmann, F.; Burkhardt, T.; Zdero, C. Naturally Occurring Acetylenes; Academic Press: New York; p. 547.
3. Rodriguez, E.; Aregullin, M.; Nishida, T.; Uhara, S.; Wrangham, R.; Abramowski, Z.; Finlayson, H.; Towers, G.H.N.; Experientia 1985, 41, 419-420.
4. Norton, R. A.; Finlayson, A. J.; Towers, G.H.N.; Phytochemistry 1985, 24, 356-359.
5. Towers, G.H.N.; Abramowski, Z.; Zucconi, A.; Planta Medica 1985 3, 222-229.
6. Hudson, J.; Graham, E. A.; Chan, G.; Finlayson, A. J.; Towers, G.H.N. Planta Medica 1986, 4, 453-457.
7. Hudson, J.; Graham, E. A.; Fong, R.; Finlayson, A. J.; Towers, G.H.N. Planta Medica 1985, 3, 225-230.
8. Cosio, E.; Norton, R.; Finlayson, A.; Towers, G.H.N.; Rodriguez, E. J. of Plant Physiology 1988, 124, 154-164.
9. Takasuga, M.; Yachida, Y.; Anetai, M.; Masumune, T.; Kegasawa, K. Chemistry Letters 1975, 43-44.

RECEIVED May 6, 1988

Chapter 30

Biocidal and Deterrent Activities of Nitrogen Heterocycles Produced by Venomous Myrmicine Ants

Murray S. Blum

Department of Entomology, University of Georgia, Athens, GA 30602

The poison glands of ant species in the genera Solenopsis and Monomorium are outstanding sources of novel nitrogen heterocycles. A large variety of alkaloids belonging to several chemical classes have been identified as venom constituents of these myrmicine species and these compounds have been demonstrated to possess a diversity of biological activities. These alkaloids are undoubtedly of great significance in the chemical ecology of these ants vis-à-vis both prokaryotes and eukaryotes. Comparative analyses of the biocidal activities of these alkaloids against fungi and insects provide strong grounds for regarding them as very promising candidates for reducing pest populations of these organisms. The pronounced insect repellent properties of these nitrogen heterocycles offer additional support for regarding them as of great potential value in man's struggle with his microbial and insect competitors.

In the order Hymenoptera, ants in a few genera in the subfamily Myrmicinae have proven to be singular in producing venoms dominated by alkaloids rather than proteinaceous constituents. Whereas analyses of the venoms of bees, wasps, and most ant species have demonstrated that these secretions are fortified primarily with proteins and polypeptides (1), the venoms of ants in the genera Solenopsis and Monomorium consist mainly of alkaloids (2). These poison gland products include a diversity of piperidines, pyrrolidines, pyrrolines, pyrrolizidines, and indolizidines (3), and in most cases, mixtures of alkaloids characterize each species' venom. Although proteins may constitute minor

0097–6156/88/0380–0438$06.00/0

concomitants of the alkaloids in these venoms (4), the poison glands of these myrmicines are clearly atypical of those of the Hymenoptera in emphasizing small nitrogen heterocycles as their primary products.

The toxinological significance of these alkaloids has become apparent as a consequence of the demonstration that they possess a wide range of pharmacological activities. In particular, the 2,6-dialkylpiperidines produced by *Solenopsis* species (5) have been shown to be powerful cytotoxins capable of producing a variety of biochemical lesions (reviewed in 6). In addition, the pronounced antimicrobial and insecticidal activities of fire ant (*Solenopsis* spp.) venoms (7) have been identified with these alkaloids (8, 9), further documenting their ability to function as toxins for a wide spectrum of organisms. Recent studies on the biological properties of another group of ant-derived alkaloids, 2,5-dialkylpyrrolidines, demonstrate that as in the case of the dialkylpiperidines, they possess pronounced toxicity and repellency.

One of the 2,5-dialkylpyrrolidines characteristic of *Monomorium* spp. (10) has been reported to exhibit considerable insecticidal activity against termite species in the genus *Reticulitermes* (11). Another of these nitrogen heterocycles has been shown to be very active as a repellent against a variety of ant species (12). Significantly, these two activities of the pyrrolidines--insect repellency and toxicity--are readily correlated with the chemical ecology of the ants producing them vis-à-vis other insect species.

In view of the reported biocidal and deterrent activities of a few of these ant-derived alkaloids, it seemed worthwhile to examine the biological properties of a large number of these compounds in some detail. In the present report, an in-depth analysis of the fungicidal, insect deterrent, and insecticidal activities of a variety of the alkaloids synthesized by ants is described. The results of these investigations provide considerable optimism for regarding the poison gland effronteries of myrmicine ants as agents of real potential for the control of man's omnipresent--and ubiquitous--microbial and insect adversaries.

Fungicidal Activities

The *cis*- and *trans*-isomers of the 6-alkyl- or 6-alkylidene-2-methylpiperidines identified as venom products of *Solenopsis* spp. (5) were evaluated as inhibitors of colony growth for a variety of fungal species (13). The alkaloids were prepared by alkylation of the lithuim salt of 2,6-lutidine with the appropriate alkyl bromide or tosylate, followed by reduction of the resulting 2-alkyl-6-methylpyridine (5). The stereoisomers were purified by preparative gas chromatography and determined to be of >98% purity.

These alkaloids included compounds in which the 6-alkyl group was normal C_9, C_{11}, C_{13}, and C_{15}; the 6-alkylidene group was either Z-4-tridecenyl or Z-6-pentadecenyl. In this section, compounds are referred to as cis or trans (relation of substituents at C-2 and C-6) in combination with an abbreviation for the chain length (plus unsaturation) of the group attached to C-6. For example, cis-6-tridecyl-2-methylpiperidine (I) is designated as cis-C_{13} and trans-6-(Z-6'-pentadecenyl)-2-methylpiperidine is designated trans-$C_{15:1}$ (II) (see Figure 1).

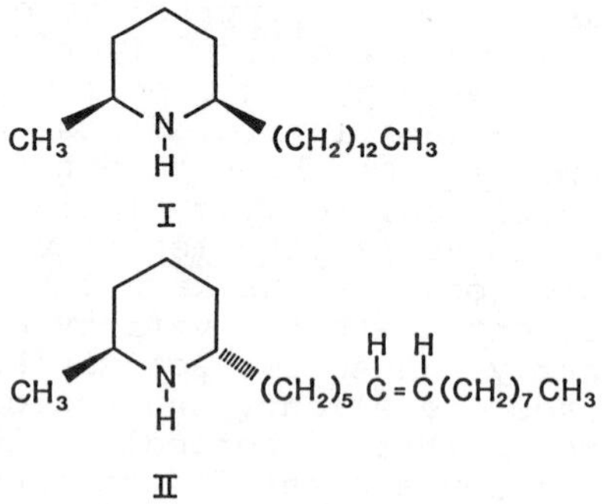

Figure 1. cis-6-tridecyl-2-methylpiperidine (I) (cis-C_{13}) and trans-6-(Z-6'-pentadecenyl)-2-methylpiperidine (II) (trans-$C_{15:1}$).

The fungicidal activities of the alkaloids identified as natural products of fire ants (*Solenopsis* spp.) have been compared to those of two commercial fungicides, 10-undecenoic acid (Un) and griseofulvin (Gris). In addition, the inhibitory activity of 6-undecyl-2-methylpyridine was evaluated in order to compare the fungicidal activity of this alkaloid to its saturated counterpart, the ant-derived compound 6-undecyl-2-methylpiperidine.

Inhibition of colony growth in the presence of candidate compounds was determined for 13 fungal species. Included were three animal pathogens--*Trichophyton rubrum* (TR), *T. mentagrophytes* (TM) and *Microsporum canis* (MC)--and two plant pathogens, *Pythium irregulare* (PI) and *Rhizoctonia solani* (RS), which were obtained as pure isolates from laboratory cultures.

In addition, eight species of fungi were isolated from larvae of two ant species, *Iridomyrmex pruinosus* and the fire ant *Solenopsis invicta*, in order to evaluate the toxicity of ant natural products to fungi that are known to be associated with ants in the colonial milieu. Fungi were transferred from field-collected ant larvae that had been placed on a sterile watch glass supported on water agar. Fungi growing from the larvae were transferred to plates containing potato dextrose agar; pure cultures were obtained by repeated transfers of isolates to fresh medium. *Fusarium oxysporum* (FO), *Cunninghamella echinulata* (CE), *Gliocladium deliquescens* (GD), *Penicillium* species (PS), and *Paecilomyces marquandii* (PM) were isolated from larvae of *I. pruinosus*. Larvae of the fire ant *S. invicta* yielded *Aspergillus zonatus* (AZ), *Zygorhynchus vuilleminii* (ZV), and *Mucor* sp. (MS). No fungi could be isolated from adult ants.

Fungal growth on potato dextrose agar was determined following incorporation of test compounds in warm agar. Inhibitory activity was evaluated in comparison to agar controls lacking test compounds. All fungal species grew vigorously on potato dextrose agar.

Preliminary evaluations demonstrated that the isomers of 6-nonyl-2-methylpiperidine (*cis*-*trans*-C_9) and 6-undecyl-2-methylpiperidine (*cis*- and *trans*-C_{11}) were among the most active inhibitors of fungal growth. As a consequence, the fungicidal activities of these compounds were compared to those of the other alkaloids and commercial fungicides. The results of these investigations are presented in Tables I-III.

Both the *cis* and *trans*-isomers of the 2-methylpiperidines containing 6-undecyl (C_{11}) or 6-tridecyl (C_{13}) substituents were equivalent in activity to 10-undecenoic acid (Un) for all but one species (*Rhizoctonia solani*) of pathogenic fungi (Table I). On the other hand, the alkaloids were considerably more fungicidal than the C_{11} acid for two of the ant-derived fungi, *Gliocladium deliquescens* and *Zygorhynchus vuilleminii*. Although the alkaloids manifested considerable activity as inhibitors of five of the ant-derived fungi (Table I), they were not highly effective against *Cunninghamella echinulata*, *Aspergillus zonatus*, and *Penicillium* species, as was also the case for undecenoic acid. In general, the *cis*- and *trans*-C_{11} and C_{13} isomers were fungicidally equivalent against the 13 species.

A comparison of the activities of the alkaloids with 6-(4'-tridecenyl) ($C_{13:1}$), 6-pentadecyl (C_{15}) and 6-(6'-pentadecenyl) ($C_{15:1}$) side chains with the 6-nonyl isomers (C_9), demonstrates that the C_9 compounds are consistently as, or more active, than the C_{13} or C_{15}

Table I. Percent inhibition of colony growth[a] of fungi by isomers of two *Solenopsis* alkaloids and 10-undecenoic acid (Un)

	Compound[b]				
Fungus[c]	*cis*-C_{11}[c]	*trans*-C_{11}	*cis*-C_{13}	*trans*-C_{13}	Un
TR	100.0	100.0	100.0	100.0	100.0
TM	100.0	100.0	95.0	98.2	100.0
MC	94.5	100.0	87.5	89.9	100.0
PI	100.0	100.0	99.0	99.9	100.0
RS	47.9	56.0	31.3	36.2	100.0
GD	98.9	98.9	93.0	93.7	8.1
ZV	96.9	98.2	76.9	97.3	47.7
MS	89.7	94.8	79.3	83.8	98.4
PM	88.9	91.6	80.3	81.8	88.5
FO	76.5	78.1	69.9	76.6	55.0
CE	48.9	53.3	34.5	33.6	49.3
AZ	20.8	46.3	16.1	14.8	19.7
PS	1.2	16.8	8.9	16.5	9.3

[a] Mean values at maximum incubation period.
[b] 800 ppm at 41 days incubation.
[c] See text for names of fungi and compounds.

Table II. Percent inhibition of colony growth[a] of fungi by four *Solenopsis* alkaloids

Fungus[c]	Compound[b]						
	cis-*trans*-C_9[c,d]	*cis*-$C_{13:1}$	*trans*-$C_{13:1}$	*cis*-C_{15}	*trans*-C_{15}	*cis*-$C_{15:1}$	*trans*-$C_{15:1}$
TR	100.0	100.0	100.0	87.7	92.7	92.3	95.0
TM	100.0	100.0	100.0	84.3	90.6	91.0	94.0
MC	100.0	95.7	96.0	61.4	89.1	90.5	91.0
PI	100.0	99.8	99.8	22.3	38.5	99.0	99.0
RS	98.7	63.6	64.0	29.5	31.8	52.3	52.3
GD	99.6	98.1	98.3	89.6	89.7	97.9	98.0
ZV	100.0	98.1	98.3	53.6	70.6	79.9	97.8
MS	91.9	78.3	81.7	67.9	69.4	67.3	68.8
PM	100.0	92.8	92.1	31.6	34.2	47.8	89.4
FO	98.5	95.3	95.2	34.1	34.1	95.0	94.5
CE	46.3	38.4	38.0	25.2	17.8	21.9	24.8
AZ	35.9	24.0	21.0	22.6	16.1	15.0	16.1
PS	20.0	22.5	28.0	23.7	27.5	16.2	17.5

[a] Mean values at maximum incubation period.
[b] 800 ppm at 53 days incubation.
[c] See text for names of fungi and compounds.
[d] A mixture of 85% *cis* and 15% *trans* isomers.

Table III. Percent inhibition of colony growth[a] of fungi by two *Solenopsis* alkaloids, griseofulvin (Gris), and 10-undecenoic acid (Un)

	Compound[b]			
Fungus[c]	*cis*-*trans*-C_9[c,d]	*trans*-C_{11}	Gris	Un
TR	100.0	100.0	100.0	100.0
TM	100.0	100.0	84.6	100.0
MC	100.0	97.0	100.0	100.0
PI	100.0	98.0	25.7	100.0
RS	53.0	40.6	30.2	100.0
GD	99.6	99.6	78.8	57.0
ZV	100.0	99.8	41.0	46.4
MS	66.0	59.5	33.6	98.7
PM	78.8	91.6	86.6	19.2
FO	78.8	67.2	65.1	38.0
CE	25.2	29.4	52.6	34.6
AZ	29.3	21.5	54.6	21.5
PS	16.3	28.4	28.4	18.5

[a] Mean values at maximum incubation period.
[b] 400 ppm at 59 days incubation.
[c] See text for names of fungi and compounds.
[d] A mixture of 85% *cis* and 15% *trans* isomers.

alkaloids (Table II). The 6-tridecenyl compounds were of equivalent fungicidal activity to the pentadecenyl compounds, whereas the alkaloids with 6-pentadecyl side chains were less inhibitory for several of the fungal species. P. marquandii was singular in being considerably more inhibited by trans-$C_{15:1}$ than cis-$C_{15:1}$. As was the case for the C_{11} and C_{13} alkaloids, three fungal species--C. echinulata, A. zonatus, and Penicillium sp.--were much less sensitive to the C_9, $C_{13:1}$, C_{15}, and $C_{15:1}$ alkaloids than the other species of fungi (Table II).

A comparison of the fungicidal activities of the C_9 and C_{11} alkaloidal isomers with two commercial fungicides, 10-undecenoic acid and griseofulvin at 400 ppm, demonstrates a great deal of variability in the growth responses of the different fungal species. For example, whereas all compounds were of equivalent activities against the animal pathogens--T. rubrum, T. mentagrophytes, and M. canis-- griseofulvin was not very inhibitory to one of the plant pathogens, P. irregulare (Table III). On the other hand, undecenoic acid was considerably more fungicidal with one of the plant pathogens, R. solani, and one of the ant-derived fungi, Mucor species. Conversely, the alkenoic acid was less inhibitory than the other compounds against three of the other ant-derived fungi, G. deliquescens, P. marquandii, and F. oxysporum (Table III). The alkaloids exhibited more fungicidal activity against G. deliquescens and Z. vuilleminii than the commercial products. As was found with the other alkaloids, none of the compounds were highly inhibitory against C. echinulata, A. zonatus, and Penicillium species.

Overall, the C_9 and C_{11} alkaloids compare favorably to 10-undecenoic acid and griseofulvin as fungicides. In general, fungicidal activity was similar in the range C_9-C_{13}, with a decrease in efficacy for some species occurring with the C_{15} compounds. The alkaloids with 6-alkylidene side chains were generally more inhibitory than their saturated counterparts. No significant differences were observed between the cis- and trans-isomers.

6-Undecyl-2-methylpyridine exhibited significantly less fungicidal activity than the other alkaloids for eight of 14 species. This unsaturated heterocycle was about as active as its saturated counterparts, cis- and trans-6-undecyl-2-methylpiperidine, against R. solani and three of the ant-derived fungi, C. echinulata, A. zonatus, and Penicillium species.

The fungicidal activities of a series of 2,5-dialkylpyrrolidines and -pyrrolines have been studied recently (14). Both the pyrrolidines and pyrrolines exhibit considerable inhibitory activity against a variety of fungal species, but as in the case of the

2,6-dialkylpiperidines, fungicidal activity varies depending on the nature of the alkyl groups.

Insecticidal Activity

The venom of the fire ant *Solenopsis invicta* possesses considerable *topical* insecticidal activity, a property not generally identified with proteinaceous venoms. The contact toxicity of the venom is clearly due to the presence of the 2,6-dialkylpiperidines, which have recently been studied as termiticides against a species of *Reticulitermes* (15). Both *cis*- and *trans*-6-undecyl-2-methylpiperidine possess an LD_{50} (μg/mg termite) of about 0.6, which is the approximate LD_{50} of *cis*- and *trans*-6-tridecyl-2-methylpiperidine. *cis*-6-Pentadecyl-2-methylpiperidine is somewhat less toxic (1.14 ± 0.08 μg), whereas the *trans*-isomer is similar in toxicity to the C_{13} compounds. The alkaloids with the 6-alkylidene moities are as toxic or more toxic than their saturated counterparts. In general, the toxicities of the dialkylpiperidines are similar to that of nicotine (0.5 ± 0.02 μg) when evaluated against termite workers.

Similar results have been obtained in a study of the topical activity of a 2,5-dialkylpyrrolidine against workers of three European species of *Reticulitermes* (11). Likewise, both 2,5-dialkylpyrrolidines and -pyrrolines exhibit considerable contact toxicity when evaluated against workers of North American species of termites (*Reticulitermes* spp.) (15). These compounds are approximately as toxic as the 2,6-dialkylpiperidines to these termite species.

In addition to these nitrogen heterocycles, several dialkylindolizidines and dialkylpyrrolizidines have been evaluated as termiticides against termite workers. 3-Hexyl-5-methylindolizidine is about as toxic (0.74 ± 0.05 μg) as the dialkylpiperidines and dialkylpyrrolidines, whereas 3-ethyl-5-methylindolizidine is considerably less active (14). The 3,5-dialkylpyrrolizidines were among the most toxic compounds tested, indicating that bicyclic alkaloids such as 3-methyl-5-(8'-nonenyl)pyrrolizidine may be of considerable importance in the chemical ecology of ants vis-à-vis their prey insects.

Repellent Activities

The demonstration that *trans*-2-butyl-5-heptylpyrrolidine, the only alkaloid produced in the venom of a thief ant (*Solenopsis* sp.), functioned as an outstanding repellent for other ant species (12), raised the possibility that these venom alkaloids may also be utilized as effective insect deterrents. Indeed, results of a recent investigation (17) document the efficacy of these nitrogen heterocycles as repellents for several aggressive ant species.

Ten species of ants which included members of both the subfamilies Myrmicinae and Dolichoderinae were given access to liquid food that had been fortified with 1 or 2 µg of a diversity of synthetic alkaloids previously identified as venomous products of ants. The repellency of these heterocycles to species that either produce or do not produce alkaloids was analyzed and it was demonstrated that some of these compounds are powerful repellents for hungry ant workers. Furthermore, there was a tendency for alkaloid-producing species to be less deterred by these compounds than species that are not known to produce alkaloidal venoms (17).

Although each ant species exhibited an idiosyncratic response to the spectrum of alkaloids, certain compounds were significantly more repellent for all species than others. The 2,6-dialkylpiperidines were among the most effective repellents, especially those with shorter 6-alkyl or 6-alkylidene groups. In addition, dialkylpiperidines with the *cis* configuration were more deterrent than the *trans*-isomers (17). 3,5-Dialkylindolizidines were generally less repellent than the dialkylpiperidines whereas some of the 2,5-dialkylpyrrolidines were of equivalent deterrency to the piperidines.

In general, a variety of these nitrogen heterocycles were effective repellents at concentrations similar to those found in the venoms that ant workers can secrete from their poison gland reservoirs. These results are certainly consistent with the conclusion that ant-derived alkaloids can be utilized as highly effective agents of deterrence for competitive species.

Conclusions

The venoms of many species of myrmicine ants are fortified with a large diversity of novel alkaloids that include a variety of mono- and bicyclic compounds. These compounds possess a wide range of biological activities that include pronounced fungicidal, insecticidal, and repellent properties. In view of the liklihood that these compounds have been evolved to counter the ubiquitous and omnipresent pathogens and invertebrate adversaries of these ants, these heterocycles must be regarded as a potential treasure trove of biologically active natural products for utilization by humankind. The need for new classes of fungicides is manifest (18) and the same can be said of insecticides whose use is severely limited by either environmental restrictions and/or insect resistance. The ant-derived alkaloids, whose biocidal activities are relatively pronounced, could be further exploited via development of synthetic analogues for structure-activity investigations. Stereoselective syntheses of the 2,6-dialkylpiperidines are available (5, 19) and enantioselective synthesis of these compounds has recently been reported (20). Similarly, several syntheses for the 2,5-dialkylpyrrolidines have been described (21, 22) which include both stereoselective (23) and chirospecific (24) routes for obtaining these compounds.

The emergence of ants as an incredibly successful group of animals must surely reflect, among other things, the utilization of their alkaloidal venom arsenals as effective offensive and defensive weapons. There is no reason why humankind should not appropriate these weapons for its own use. And there is no time like the present!

Literature Cited

1. Schmidt, J. P. Ann. Rev. Ent. 1982, 27, 339.
2. Jones, T. H.; Blum, M. S.; Fales, H. M. Tetrahedron 1982, 38, 1949.
3. Jones, T. H.; Blum, M. S. In "Alkaloids: Chemical and Biological Perspectives"; Pelletier, S. W., Ed.; John Wiley and Sons, Inc., 1983; Vol. 1, pp. 33-84.
4. Baer, H.; Liu, T. Y.; Anderson, M. C.; Blum, M. S.; Schmid, W. H.; James, F. J. Toxicon 1979, 17, 397.
5. MacConnell, J. G.; Blum, M. S.; Fales, H. M. Tetrahedron 1971, 26, 1129.
6. Blum, M. S. In "Bioregulators for Pest Control"; Hedin, P. A., Ed.; ACS Symp. Ser. No. 276, 1985, pp. 393-408.
7. Blum, M. S.; Walker, J. R.; Callahan, P. S.; Novak, A. F. Science 1958, 128, 306.
8. Blum, M. S. In "Insect Poisons, Allergins, and Other Invertebrate Venoms"; Tu, A. T., Ed.; Maurice Dekker, Inc., 1984; pp. 225-242.
9. Jouvenez, D. P.; Blum, M. S.; MacConnell, J. G. Antimicrob. Ag. Chemother. 1972, 2, 291.
10. Jones, T. H.; Blum, M. S.; Howard, R. W.; McDaniel, C. A.; Fales, H. M.; Dubois, M. B.; Torres, J. J. Chem. Ecol. 1982, 8, 285.
11. Clément, J.-L.; Lemaire, M.; Lange, C. C. R. Acad. Sci. Paris 1986, 303, 669.
12. Blum, M. S.; Jones, T. H.; Hölldobler, B.; Fales, H. M.; Jaouni, T. Naturwissenschaften 1980, 67, 144.
13. Cole, L. K.; Blum, M. S. Unpublished results, 1987.
14. Blum, M. S.; Jones, T. H.; Cutler, H. G. Unpublished results, 1987.
15. Blum, M. S.; Jones, T. H.; Escoubas, P. Unpublished results, 1987.
16. Jones, T. H.; Blum, M. S.; Andersen, A. N.; Fales, H. M.; Escoubas, P. J. Chem. Ecol. 1988, 14, 35.

17. Tomalski, M. D.; Jones, T. H.; Blum, M. S.; Everett, D. M. Unpublished results, 1987.
18. Ragsdale, N.N.; Siegel, M.R. In "Bioregulators for Pest Control"; Hedin, P.A., Ed.; ACS Symp. Ser. No. 276, 1985, pp. 35-45.
19. Moriyama, T.; Doan-Huynh, D; Monneret, C.; Khuong-Huu, Q. Tetrahedron Lett. 1977, 825.
20. Taber, D.F.; Deker, B.; Fales, H.M.; Jones, T.H.; Lloyd, H.A. J. Org. Chem., in press, 1988.
21. Jones, T.H.; Franko, J.B.; Blum, M.S.; Fales, H.M. Tetrahedron Lett. 1980, 21, 789.
22. Ritter, F.J.; Stein, F. U.S. Patent 4,075,320, Feb. 21 (1978).
23. MacDonald, T.L. J. Org. Chem. 1980, 45, 193.
24. Shiosaki, K.; Rapoport, H. J. Org. Chem. 1985, 50, 1229.

RECEIVED April 26, 1988

Chapter 31

Protecting Crops and Wildlife with Chitin and Chitosan

M. L. Bade[1] and R. L. Wick[2]

[1]Department of Biology, Boston College, Chestnut Hill, MA 02167
[2]Suburban Experiment Station, University of Massachusetts, Waltham, MA 02154

Chitin and its deacetylation derivative chitosan show inhibitory activity against crop-damaging fungi and nematodes. A novel method for isolating chitin in forms that retain the high degree of native structure has been developed; its future large-scale employment promises formulation of less toxic pesticides to replace some of the more dangerous ones now in use. The ability of high grade chitosans to be cold cast into high tensile strength plastics has been demonstrated. Plastics based on petrochemicals are not degraded in the environment within a reasonable time span and damage wildlife and fisheries. Chitosan-based plastics have desirable properties and are degradable; their degradation could become smoothly integrated into existing biogeochemical cycles.

Chitin and chitosan are not widely regarded as biologically active; chitin, especially, is considered to be chemically inert and difficult to hydrolyze even by means of specific enzymes. It has now been shown (1) that lack of reactivity and lack of predictability of performance in subsequent reactions by purified chitin and chitosan is an artefact of isolation and can be avoided. Large-scale agricultural and other applications thus become feasible. Among these are formulation of fungicides and nematocides of low toxicity, and manufacture of plastic that combines high tensile strength and good shelf life with total biodegradability after discard.

Chitin and Chitosan.

The natural product chitin consists predominantly of unbranched chains of β-(1,4)-linked N-acetylglucosamine residues (2). It seems to be nearly ubiquitously distributed in invertebrate exoskeletons (3). Annual synthesis in the biosphere is estimated at 100 billion tons (4). An important derivative, chitosan, results from deacetylation of sufficient amine groups to convert very insoluble chitin to a material soluble in dilute weak acid such as acetic.

0097–6156/88/0380–0450$06.00/0

Since solubility is the chief criterion for distinguishing chitosans from parent chitin, a number of substances differing in degree and internal distribution of free amine groups are described by the term chitosan.

Natural Sources. Typically, fibrous chitin in animal exoskeletons is embedded in a protein matrix; it may additionally be associated with other materials. In the most common source of chitin, shellfish waste, chitin and its protein matrix are invested with crystalline calcium salt deposits and lipoid coloring materials. Table I shows the makeup of Red crab waste from a commercial cannery operation in Danvers, MA.

Table I. Composition of Red Crab Waste

Component	w/w%
Water	45
Edible protein	20
Ca salts	20
Carotenes	trace
Structural protein	5
Chitin	10

It is important to note that chitin *in situ* exhibits a high degree of internal order prior to possible damage inflicted during isolation. This is evident from Figure 1 which shows an electron micrograph of decalcified crab claw tendon in which the protein has not been disturbed and the chitin is largely unmodified. Fibrils of <0.1 μ diameter are visible on the right side of the figure; the left shows similar fibrils in slightly angled cross section.

Chitin/Chitosan Isolation: Current Commercial Methods. Extraneous material is removed in the purification of chitin and/or its conversion to chitosan. Processing to desired products typically involves the following sequence of operations (c.f. 6, 7) (Table II):

Table II. Typical Processing Sequence in Chitin/Chitosan Production

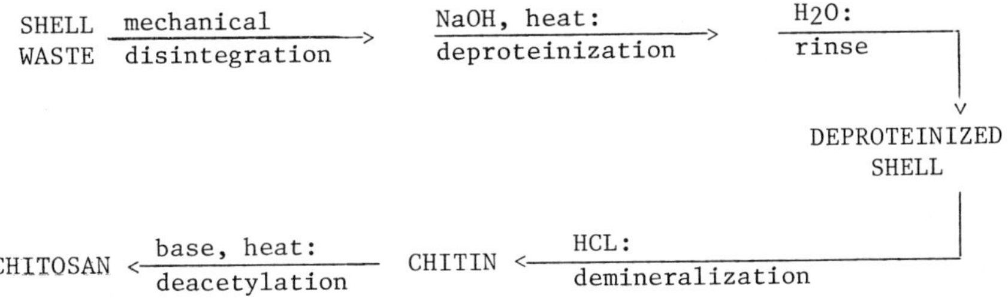

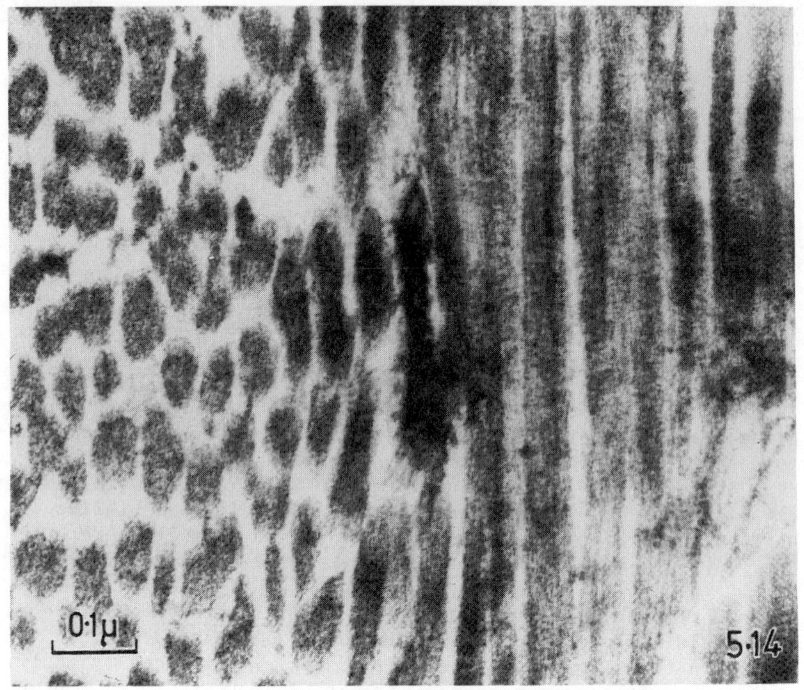

Figure 1. Orderly Array of Chitin-Protein Complex *in situ*. Transmission electron micrograph of decalcified crab claw tendon; 120,000 x magnification. Unstained fibers and fibrils are chitin; they are seen to be intermingled with electron dense protein matrix. (Reproduced with permission from Ref. 6. Copyright 1975, Springer Press.)

The bulk of the commercial chitin/chitosan presently on the market in the U.S.A. is prepared in this manner. However, this operational sequence inevitably leads to extensive collapse of native chitin fine structure (see Figure 2). Such collapse of tertiary structure is irreversible; even if fibers are secondarily formed from intermittently collapsed chitin, the native ultrastructure is not regained (Bade & Stinson, unpublished). In addition, chitin is often bleached in preparation (8, 9, 10) and this results in extensive chain breakage and unpredictable chemical alteration of component sugars (Bade & Stinson, unpublished). Harsh chemical deacetylation conditions inflict further damage, to the point where some commercial chitosan preparations cannot act as inducers of microbial chitosanases (11).

As a result of such maltreatment, "purified" chitins and chitosans have long been known to react unpredictably in practical applications. Since the chemistry of damage was not understood, attempts at preparing chitin with predictable properties have frequently utilized collapsed and otherwise damaged chitin/chitosan as the starting materials (12-18). In other instances, chitins have retained their structure at the expense of gross contamination with proteins (19, 20).

Chitin Isolation: Improved Method. The senior author has developed a method for isolating pure animal chitins with intact, covalently stabilized fine structure (1; pat. pend.). In an important change from existing usage, the new procedure specifies demineralization as the obligatory first step in chitin purification; only then is the shell residue deproteinized. The decalcified, deproteinized "protochitin" thus obtained retains its native fibrous structure in compacted form. Something close to the original fibrous structure can be reconstituted by suspending protochitin in chilled dilute ester-forming acid. Under appropriately controlled conditions, a colloidal suspension of high viscosity is formed from which elongated particles can be precipitated and harvested. The procedure preserves the original fibrous structure of the chitin throughout the isolation procedure and restores the spaces formerly occupied by the protein matrix. This is achieved through insertion of sparse internal ester cross links within the particles (D.S. $<<0.05$). These cross links functionally replace the stabilizing structural proteins previously removed under conditions that minimize damage to native fine structure. The product is > 98% pure chitin that is fibrous in the light and scanning electron microscope. Figure 3A shows sulfate-stabilized insect chitin at 100 x magnification. Sulfate-stabilized Red crab chitin is shown at 3600 x magnification in Figure 3B. The smooth exterior of the fibrous particles is in striking contrast to the rough, disordered picture given by collapsed chitin; orderly appearance is mirrored in interior order which is expressed e.g. in rapid, specific and predictable reactions of ester-stabilized chitins. They can therefore be derivatized in predictable fashion (1, 20). Uniform structure and accessiblity of relevant bonds in ester-stabilized chitin inter alia permit derivatization to chitosan of unusually high quality by very mild chemical techniques (Bade, unpublished).

Agricultural Applications. Since chitin is nearly as abundant in the

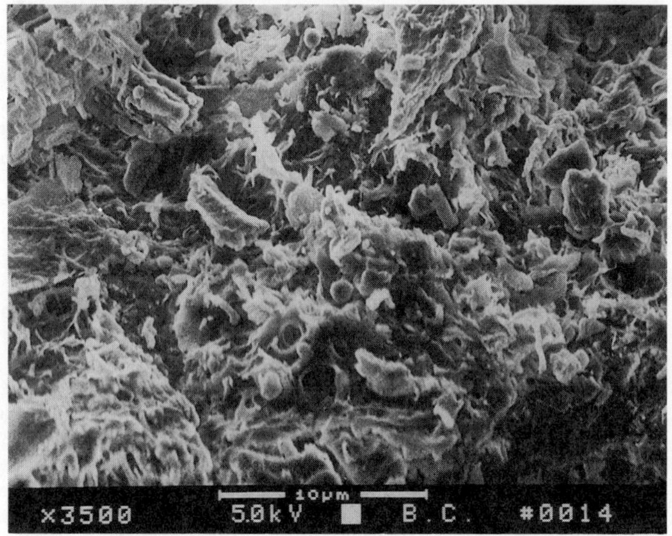

Figure 2. Collapsed Chitin.
Chitin resulting from improper processing of crustacean carapace, i.e. deproteinization followed by HCl decalcification, and showing extensive loss of native structure. 5000 x magnification, scanning electron microscopy. Courtesy Rev. F. Venuta, S.J., Boston College.

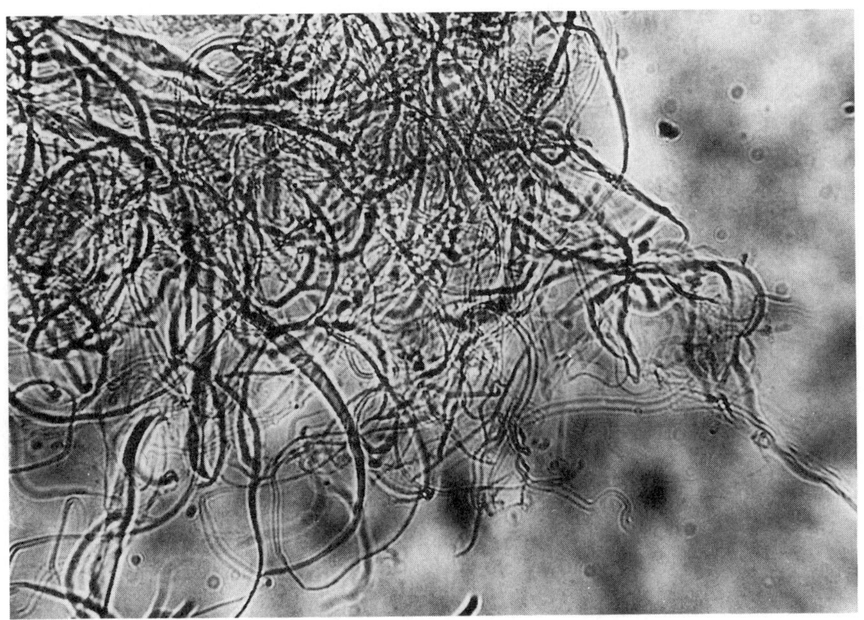

Figure 3A. Fibrous Ester-Stabilized Insect Chitin. Sulfate-stabilized insect chitin prepared in Bade's laboratory from fresh insect integument. 100 x magnification, light microscopy.

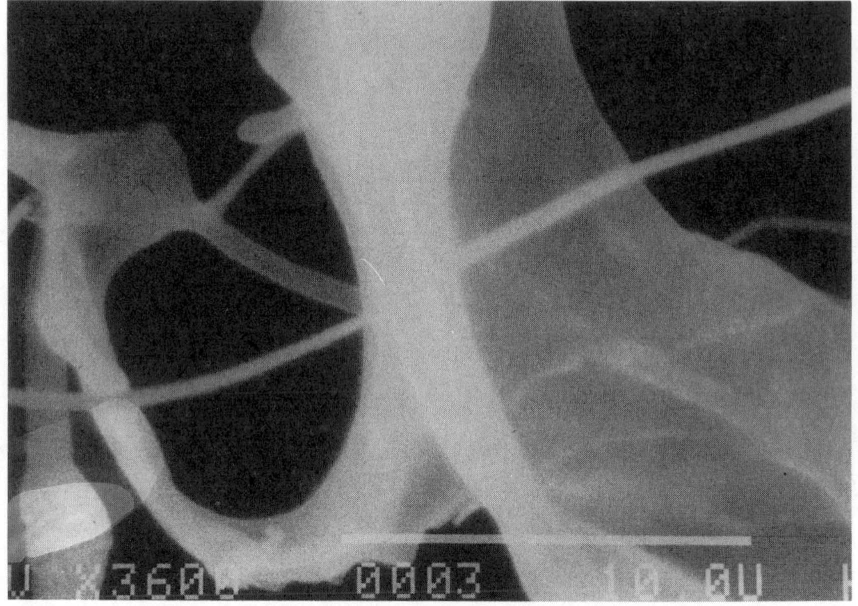

Figure 3B. Fibrous Ester-Stabilized Crab Chitin. Sulfate-stabilized chitin prepared in Bade's laboratory from fresh crab carapace. 3600 x magnification, scanning electron micrograph. SEM preparation courtesy of Hercules, Inc.

biosphere as cellulose, availability of pure chitins with predictable performance in chemical and enzymatic reactions makes practicable large scale technological uses for chitin and chitosan. Two applications of interest to agriculture are:

1) formulation of safe biodegradable pesticides, and
2) manufacture of high tensile strength biodegradable plastics.

Plant Protection.

Against Fungi. Chitin alone or in combination with other organic materials has repeatedly been found to lessen the severity of symptoms of fungal infection in plants (22, 23; see these references also for citations of prior work). This benefit was initially ascribed to a change in soil microflora so as to "encourage ... organisms antagonistic to the pathogen" (22), increases being noted specifically in actinomycetes (*ibid*) and more generally in mycolytic bacteria (24). Induced or cross resistance achieved by prior inoculation of plants with a fungal race pathogenic on another plant species has also been described (25). Many similar observations assign an active role to the plant in its own defense. Plants have been shown to protect themselves in response to injury or stress by producing phytoalexins, defined as "substances with antibiotic activity that function as growth inhibitors of phytopathogenic organisms, chiefly fungi" (26). A widely held view is that an "elicitor" becomes attached to and may enter the plant cell where it stimulates the rise of phytoalexins and hydrolytic enzymes such as chitinases (27). This effect may be mediated through the plant hormone ethylene (28) or it may be more direct: Using immature pea pods, Hadwiger and co-workers (29) have shown that chitosan oligomers (D.P. approx. seven) elicit the rise both of the phytoalexin pisatin and of a large number of proteins among which are endochitinase and endo-β-1,3-glucanase. Hadwiger *et al*. have also observed a reduction in RNA synthesis in the eliciting fungus; this is probably the result of entry of the phytoalexin into the interior of the fungus (30).

Contact of chitosan with plant cells allows chitosan to enter and induce disease resistance responses biochemically identical to those induced by exposure of cells to certain *Fusarium* spores (31). Results have been reported of both seed and foliar treatments of field crops with commerical chitosan (32). Foliar treatments on wheat showed inconsistent and sometimes phytotoxic effects. Seed treatments ranging from 60μg - 1000μg chitosan per gram of seed were carried out over five years on winter and spring wheat, peas, and lentils. Yield increases of 10-30 percent were obtained; reduction in damp-off, logging, and other symptoms of fungal infection were seen and no case of yield decrease due to chitosan treatment occurred during the test period. Chitosan seed treatments are offered commercially.

Against Nematodes. Nematodes are microscopic or macroscopic roundworms which are among the most abundant animals on Earth. A number of genera have adopted parasitic lifestyles and they cause important diseases of humans and other animals. Phytoparasitic nematodes inflict extensive damage on field and greenhouse crops and ornamentals like turf (33).

Following embryonation, each nematode hatches from an egg which is surrounded by several protective layers; one of these is composed of chitin. Hatching is made possible by a chitinase and other enzymes which are released from the egg in response to stimuli signaling an environment favorable to development of the parasite(ibid.).

Although suppression of nematode damage to plants by chitin application to soil has repeatedly been observed, the inhibitory effect of chitin and derivatives on plant parasitic nematodes has been studied in less detail than has fungicidal activity. Results from representative publications are summarized in Table III. Common observations are:

1) reduction or prevention of effects of nematode infestation such as root gall or root knot formation;
2) increase in soil microorganisms including free living nematodes seen as beneficial to plants;
3) a fine line between application rates that have nematode-suppressant effects and rates that are phytotoxic;
4) marked differences between plant genera and families in sensitivity to phytotoxicity;
5) marked differences in effects on both plants and nematodes between types of chitin preparation;
6) marked differences in phytotoxic effects seen depending on whether application occurred in the greenhouse or the field, diminution of phytotoxicity generally being observed in field application.

Application of chitin to agricultural soils is beset by problems which reduce the probability for practical use at present. There are very few scientific studies demonstrating the mechanism(s) by which chitin functions in suppression of nematodes. One theory holds that a transient high spike of NH_3 is formed in the soil which is harmful to the parasitic roundworms (30). Another theory is that "beneficial" microorganisms are favored when chitin is available as a substrate, and that reduction of parasitic nematodes occurs secondarily (34). Both mechanisms are probably operative either simultaneously or sequentially (40).

Additional major problems discourage field applications in the immediate future. Among these are low bulk density (30 to 50 lbs. per cubic foot), the quantities required (2,000 - 10,000 lbs. per acre), together with the need for incorporation into the soil. Incorporation would be difficult at best with established perennial crops such as turf. More important, nematocidal levels of chitin may also be toxic to plants. The hypothetical NH_3 or NO_2 spike has been blamed for phytotoxicity as well as nematode suppression (37). The variety of chitin preparations used by various investigators under non-uniform experimental conditions adds to the confusion.

Feasible Experiments and Applications. Chitin and chitosans of known and reproducible properties can now be prepared in bulk, and can be further derivatized to give materials with predictable performance. Bade's laboratory has established the following: Ester-stabilied chitins can be depolymerized by chitin endochitinases and then derivatized to chitosans or other materials; enzymatic as well as chemical deacetylation of stabilized chitins to high quality

Table III. Summary of Soil Amendment by Chitin

Chitin	Level(w/w)	Location	Nematodes	Plants	Ref.
"Crude, refined."	0.1 - 1.0%	Greenhouse	Free living ↑; parasitic eliminated.	Phytotoxic (tomatoes).	34
Same	Same	Field	Tomato root gall formation↓.	No toxicity observed.	34
"Raw"	0.1 - 0.8%	Greenhouse	2 parasitic species .	No toxicity observed.	35
Crustacean	0 and 2%	Greenhouse	Root knot formation prevented.	Killed squash; <toxic, tomatoes.	36, 37
Flakes	0.5 -4.0%	Field	1% up: No root knot formation.	Heaviest plants 1-1.5%; 2% up phytotoxic.	38
"ClandoSan"	0.05 - 0.3%	Greenhouse	At 0.2% parasitic↓; root galls down 50%.	Toxic to tomatoes, beans; corn ok.	39
Chitin-protein	1%-5%	Randomly chosen soils (petri dishes)	Free living ↓; nematophagous fungus observed ↑. 5% most effective.	Not tested.	20

chitosans is feasible, and labeled chitins and chitosans can be prepared (41) and used to trace the fate of such materials following application to soil or plants.

Experiments can now be performed utilizing highly purified chitin of known composition and structure on a sufficiently large scale to answer questions such as: What is/are mechanism(s) of plant protection against nematode-incited disease? Do some plant species respond differently from others? Can effective protection be provided by chitooligomers which could be made soluble so as to make application by drenching feasible? Can plant protection be achieved reliably without the risk of phytotoxicity?

Despite the present difficulties with field application, the picture is bright with respect to developing useful, protective chitin amendments for greenhouse container soils. Another attractive possibility is that of developing a formulation of chitin/chitosan or derivatives that would combine solubility in water and biodegradability with hatching inhibition of parasitic nematode eggs. To this end, the egg chitinase needs to be studied. Bade's laboratory has shown that chitins whose structure is sufficiently disturbed are effective inhibitors of chitinases from microorganisms and insects; a similar effect on nematode egg chitinase can be expected. Since parasitic namatodes hatch in response to favorable environmental stimuli, seasonal application of a hatching inhibitor for maximum effectiveness could presumably be planned. Such an inhibitor might also find routine uses in veterinary practice.

Plastics.

The reference in the title to "Protecting ... Wildlife..." is only in part designed to call attention to the need for replacing dangerous pesticides with more benign versions. Of at least equal importance is the necessity to stem the potential for destruction inherent in the use of non-biodegradable plastic.

Plastic damage. If long-lived plastic bags are buried in landfills or are discarded, their contents are effectively cut off from mineralization, while buoyancy and resistance to weathering of plastic serves to concentrate and accumulate it in the uppermost surface layers of oceans and lakes. Amounts and types of plastic debris in the North Atlantic system are being examined in an ongoing study(42). Results to date indicate that plastic accidentally lost or indifferently dumped at sea heavily litters the shores even of remote islands far removed from significant local sources; Bermuda and the Bahamas are examples. On Bonaire beaches, 150,000 plastic particles per square meter have been counted. For the upper 3 cm of sand in a width of 1-2 m, volumetrically more plastic than sand was found; 90% of the plastic consisted of polyethylene pellets of which many bore teeth marks from fish (J. Wilber, Woods Hole Oceanographic Institute, personal communication, 1988). This material does more than esthetic damage: "The Marine Sciences Research Center at Stony Brook, L.I., has estimated that 30 percent of the fish in the world's oceans have tiny pieces of plastic in their stomachs that interfere with digestion" (43).

Wilber's study (42) has further established that shore waters of some sites including Cape Cod and the Florida Keys act as "sieves" for freshly discarded plastic trash and lost gear from fishing boats. Such material maims and kills large marine animals including whales seals, and sea turtles that become entangled in it. Because such damage is widespread rather than concentrated as are oilspills, it is more difficult to quantitate. A survey begun in 1976 of one population of mammals, the northern fur seals of the Pribilof Islands, led to the conclusion that plastic entanglement was killing up to 40,000 fur seals a year, or about 5% of the herd (44). Similar mechanisms for damage to the ones described are assumed to be operating in other ocean systems. Figure 4 shows a yearling California sea lion wearing a six-pack strap that probably slipped over his head when he played with it as a pup. Figure 5 depicts another California pinniped entangled in a piece of lost gill netting; such "ghost nets" go on "fishing" for years. Birds are not exempt. They are trapped by floating produce bags and six-pack straps (Figure 6). Birds also accumulate ultimately lethal plastic loads from the fish they eat and digest (ibid).

Chitin/Chitosan-Based Plastics. Environmental damage done by plastics is caused by precisely those characteristics for which they were developed: They outlast and outperform products made from natural materials (44). In replacement, materials are needed that combine lightness, high tensile strength, and durability during use with a rapid rate of degradation after discard. Here, too, chitin, or chitin derivatized into chitosan, can be employed. Chitosan is easily cast into films and filaments of high tensile strength provided its structural integrity is not compromised during preparation. The following abstract provides details:

"The formation of films and fibers from chitosan is dependent on the structure of the bulk chitosan from which it is cast. The structure of the bulk chitosan is related, in turn, to the processing steps in preparing chitin and chitosan from the shell, and it may be influenced ... by the species of crustacea used as the starting material ... The film-forming qualities appear to correlate well with the structure as defined by x-ray diffraction which is indicative of the molecular structure of the polymer.

Tough flexible films, with a tensile strength of 20,000 psi and an elongation of 6 percent, have been cast from dilute acetic or formic acid solutions. These films are virtually impervious to air and water" (45).

Chitin and chitosan are internally cross linked to a much greater degree than is cellulose; each N-acetylglucosamine residue in chitin is linked by an estimated eight hydrogen bonds to residues in surrounding chains which gives this material exceptionally high tensile strength in three dimensions compared to cellulose or starch. Indeed, a lobster shell may be looked upon as a piece of tough, flexible chitin plastic rendered hard and impact-resistant by infiltration of an inorganic filler, calcium carbonate, and lasting as long as needed by its living occupant. A flexible, transparent bag cast in 1980 out of chitosan prepared from Virginia Blue Crab chitin shows no visible change in properties to date. Differences in structure and performance of chitins based on biological origin have been confirmed in Bade's laboratory by several lines of evidence; since distinct properties are reproducible, they can be allowed for in future technological applications.

Figure 4. California Sea Lion Wearing Six-pack Strap. Yearling sea lion with six-pack strap embedded in his neck tissue. Photograph by George Antonelis, National Marine Fisheries Service. (Reproduced with permission from Ref. 54. Copyright 1987 Defenders of Wildlife.)

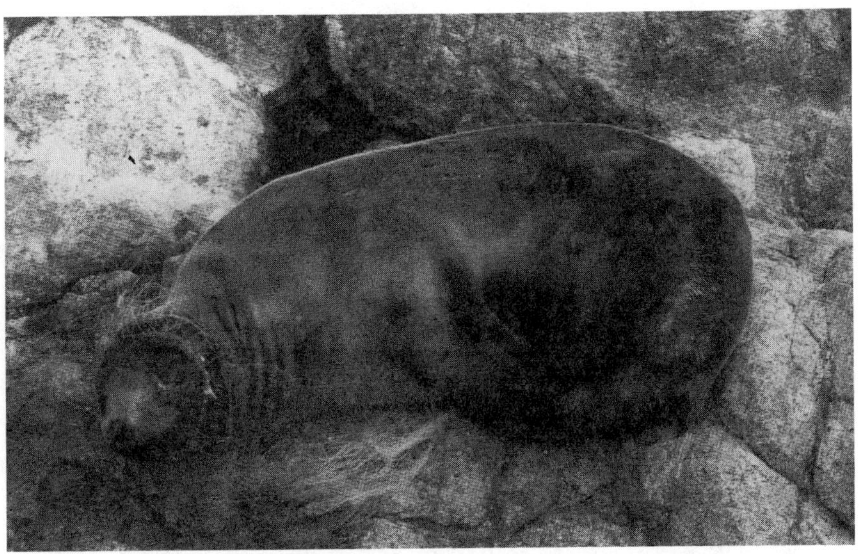

Figure 5. California Pinniped Wearing "Ghost Net." Photograph by Jack D. Swenson. (Reproduced with permission from Ref. 54. Copyright 1987 Defenders of Wildlife.)

Figure 6. Dead Juvenile Gull.
Gull perished from interference by six-pack strap. Photograph courtesy of Michigan Dept. of Natural Resources. (Reproduced with permission from Ref. 54. Copyright 1987 Defenders of Wildlife.)

Biodegradability. Microorganisms that degrade chitosan are abundant in all soil types examined (46, 47) and in some, may contribute a major fraction of microbial biomass (11). Chitosan films disappeared completely within three to four weeks in North Atlantic waters off the coast of Massachusetts (B. Averback, Mass. Inst. Technology, personal communication, 1987). Recognizable pieces of chitinous shells do not accumulate in estuarine environments in the Gulf of Mexico (48); their mineralization rate in Dutch tidal flats is estimated at 90 gm x cm^{-2} x yr^{-1} (49). Chitinases from vertebrate digestive tracts have been described (50-52) and they are inducible in many gram-negative bacteria and Actinomycetes (53). Inducible extracellular exo- and endochitinases from Streptomyces plicatus have recently been separated in the senior author's laboratory. The endochitinase, which rapidly reduces the size of macromolecules, has the low temperature optimum of 15°C, and retains good activity at 4°C where liquid water overwinters.

Because chitin and chitosan do not accumulate in the environment, chitinases and chitosanases must already play a significant role in biogeochemical cycling of limiting elements including nitrogen. Chitin/chitosan are natural products that would only temporarily be diverted from natural breakdown pattern if their use in plastics were to become widespread. Discarding plastic made from such an abundant natural resource should therefore have minimal environmental impact and it may be predicted that degradation will easily become integrated into existing biogeochemical cycles. Contrary to the present situation, such integration should prove life-enhancing; Leaf- and garbage bags formed from fibrous chitin/chitosan will inject needed fixed nitrogen into leaf and other compost deficient in this element. Plastic mulches would break down even in northern environments where the growing season is short but degradative processes continue year-round. In aquatic environments, where damage to wildlife by plastic is particularly acute, cyclic ecosystems along the model of mangrove swamps might well develop with the use of chitin/chitosan-based plastics.

This contrasts with plastics presently in use. Petrochemical-based plastics are perishable only over very long periods; hundreds of years are usually estimated (43). "Partially biodegradable" materials do not offer a viable alternative. Starch-based biodegradables cannot contain more than 5% of starch before tensile strength is adversely affected (A. Andrady, Research Triangle Institute, personal communication, 1988). UV-degradable plastic stops breaking down just when such an effect is most needed: In a field under soil crumbs and plant material, on the forest floor, half-buried in salt marshes. It seems doubtful that any modification to existing plastics will render them as readily biodegradable while leaving them as stable during use as chitin and chitosan naturally are.

Conclusions.

Chitin and chitosans constitute an alternative biomass of considerable promise. In the pesticidal applications, it will be advantageous to understand more about affected systems to permit constructive interference in given life-cycles with minimal harm to other species, but the work remaining there, and with respect to plastics manufacture, represents no more than application of established principles to systems studied incompletely in the past. Success in the ventures here outlined can be predicted with confidence.

Literature Cited:

1. Bade, M. L.; Stinson, A.; Moneam, N. A.-M. Conn. Tiss. Res. 1988, 17, 137-151.
2. Foster, A. B.; Webber, J. M. Adv. Carbohydr. Chem. 1960, 15, 371-390.
3. Jeuniaux, Ch. Chitine et Chitinolyse; Masson & Cie.: Paris, 1963; p. 25.
4. Berkeley, R. C. W. In Microbial Polysaccharides and Polysaccharases; Berkeley, R. C. W.; Gooday, G. W.: Elwood, D. C. Eds.; Academic: London, 1979; Chapter 9.
5. Richards, A. Glenn. The Integument of Arthropods; Univ. Minesota Press: Minneapolis, 1951; p. 26.
6. Neville, A. C. Biology of the Arthropod Cuticle; Springer: New York, 1975; p. 174.
7. Brine, C. J. In Chitin, Chitosan, and Related Enzymes; Zikakis, J. P. Ed.; Academic: New York, 1984; p. xviii.
8. Muzzarelli, R. A. A. Chitin; Pergamon Press: New York, 1977; pp. 89-102.
9. Jeuniaux, Ch. Methods Enzymol. 1967, 8, 644-650.
10. Leuba, J.; Stossel, P. In Chitin in Nature and Technology; Muzzarelli, R. A. A.; Jeuniaux, Ch.; Gooday, G. W., Eds.; Plenum: New York, 1985; 209-222.
11. Davis, B.; Eveleigh, D. E. In Chitin, Chitosan, and Related Enzymes; Zikakis, J. P., Ed.; Academic: New York, 1984; 161-179.
12. Molano, J.; Duran, A.; Cabib, E. Anal. Biochem. 1977, 83, 648-656.
13. Ohtakara, A.; Mitsutomi, M.; Uchida, Y. Jour. Ferment. Technol. 1979, 57, 169-177.
14. Dunn, H. J.; Farr, M. P. 1977, U. S. Patent # 4,034,121.
15. Austin, P. R. 1975; U. S. Patent # 3,879,377.
16. Austin, P. R. 1975; U. S. Patent # 3,892,731.
17. Austin, P. R. 1977; U. S. Patent # 4,059,457.
18. Austin, P. R.; Brine, C. J. 1977; U. S. Patent # 4,027,727.
19. Muralidhara, H. S. 1981; U. S. Patent # 4,293,098.
20. McCandliss, R. J.; Eastwood, B. J.; Milch, R. A. 1985; U. S. Patent # 4,536,207.
21. Bade, M. L.; Stinson, A.; Moneam, M. A.-M. 1985. In Glycoconjugates: Proceed. VIIIth Internat. Sympos.; Davidson, E. A.; Williams, J. C.; Di Ferrante, N. M., Eds.; Praeger: New York, 1985; 2, 480-481.
22. Buxton, E. W.; Khalifa, O.; Ward, V. Ann. appl. Biol. 1965, 55, 83-88.
23. Okafor, N. Trans. Brit. mycol. Soc. 1970, 55, 483-485.
24. Mitchell, R.; Alexander, M. Nature (Lond.) 1961, 190, 109-110.
25. Cook, R. J.; Baker, K. F. The Nature and Practice of Biological Control of Plant Pathogens. St. Paul: Amer. Phytopathol. Soc. 1983, p. 127-128.
26. Brewer, M.; Scott, T. (Eds.) Concise Encyclopedia of Biochemistry; Walter de Gruyter: Berlin, 1983; p. 354.

27. Boller, T. In Chitin in Nature and Technology; Mazzarelli, R. A. A.; Jeuniaux, Ch.; Gooday, G. W. Eds.; Plenum: New York, 1985; 223-230.
28. Hislop, E. C.; Hoad, G. V.; Archer, S. A. In Fungal Pathogenicity and the Plant's Response; Byrde, R. J. W.; Cutting C. V., Eds. London: Academic 1973; p. 104.
29. Kendra, D. F.; Mauch, F.; Riggleman, R. C.; Walker-Simmons, M.; Hadwiger, L. A.; Exper. Mycol. 1984, 8, 276-281.
30. Hadwiger, L. A.; Kendra, D. F.; Fristensky, B.; Wagoner, W. In Chitin in Nature and Technology; Muzzarelli, R. A. A.; Jeuniaux, Ch.; Gooday, G. W., Eds.; Plenum: New York, 1985; 209-222.
31. Hadwiger, L. A.; Beckman, J. M. Plant Physiol. 1980, 66, 205-211.
32. Hadwiger, L. A.; Fristensky, B.; Riggleman, R. C. In Chitin, Chitosan, and Related Enzymes; Zikakis, J. P. Ed.; New York: Academic 1984, 291-302.
33. Schmidt, G. D.; Roberts, L. Foundations of Parasitology (2nd Ed.); C. V. Mosby: St. Louis, 1981; ch. 23, 24.
34. Mankau, R.; Das, R. Jour. Nematol. 1969, 1, 15-16.
35. Miller, P. M., Sands, D. C.; Rich, S. Plant Disease Reporter 1973, 57, 438-442.
36. Mian, I. H.; Godoy, G.; Shelby, R. A.; Rodriguez-Kabana, R.; Morgan-Jones, G. Nematropica 1982, 12, 71-84.
37. Culbreath, A. K.; Rodriguez-Kabana, R.; Morgan-Jones, G. Nematropica 1985, 15, 49-75.
38. Godoy, G.; Rodriguez-Kabana, R.; Shelby, R. A.; Morgan-Jones, G. Nematropica 1983, 13, 63-74.
39. Spiegel, Y.; Cohn, E.; Chet, I. Plant and Soil 1986, 95, 87-95.
40. Spiegel, Y.; Chet, I.; Cohn, E. Plant and Soil 1987, 98, 337-345.
41. Bade, M. L. Biochem. Biophys. Acta 1974, 372, 474-477.
42. Wilber, R. J. Oceanus 1987, 30, 61-68.
43. Browne, M. Plastic Trash Problem Defies World Technology; New York Times 1987, Sept. 6, VI.
44. Weisskopf, M. Smithsonian 1988, 18, 58-66.
45. Averbach, B. L. In Proceed. 1st Internat. Confer. Chitin/Chitosan; Muzzarelli, R. A. A.; Pariser, E. R., Eds.; Mass. Inst. Technol.: Cambridge 1978, MITSG-78-7; 199-209.
46. Fenton, D.; Davis, B.; Rotgers, C.; Eveleigh, D. E. In 1st Internat. Confer. Chitin/Chitosan; Muzzarelli, R. A. A.; Pariser, E. R., Eds.; Mass. Inst. Technol.: Cambridge, 1978, MITSG 78-7; 525-541.
47. Poulicek, M.; Goffinet, G.; Voss-Foucart, M. F.; Bussers, J. C.; Jaspar-Versali, M. F.; Toussaint, C. In Chitin in Nature and Technology; Muzzarelli, R. A. A.; Jeuniaux, Ch.; Gooday, G. W., Eds; Plenum: New York, 1985; 547-550.
48. Becker, R. Ann. Report, 1971-73; Center for Wetland Resources, Louisiana State University: Baton Rouge, LA, 1975, LSU-SG-75-01; 52 pp.
49. Pel, R.; Gottschal, J. C. In Chitin in Nature and Technology; Muzzarelli, R. A. A.; Jeuniaux, Ch.; Gooday, G. W., Eds.; Plenum: New York, 1985; 539-546.
50. Jeuniaux, Ch.; Cornelius, C. In Proceed. 1st Internat. Confer. Chitin/ Chitosan; Muzzarelli, R. A. A.; Pariser, E. R., Eds.; Mass. Inst. Technol.: Cambridge, 1978, MITSG-78-7; 542-549.
51. Dandrifosse, G. Jour. Physiol. 1974, 68, 387-384.

52. Okutani, K. In _Proceed. 1st Internat. Confer. Chitin/Chitosan_; Muzzarelli, R. A. A.; Pariser, E. R., Eds.; Mass. Inst. Technol.: Cambridge, 1978, MITSG-78-7; 582-586.
53. Gooday, G. W. In _Microbial Polysaccharides and Polysaccharases_; Berkeley, R. C. W.; Gooday, G. W.; Ellwood, D. C., Eds.; London: Academic 1979; Ch. 19.
54. MacKenzie, W. H. _The Trashy Sea Around Us_; Defenders 1987, May/June, 30-37.

RECEIVED May 23, 1988

INDEXES

Author Index

Affiliation Index

Subject Index

B

Production by Rebecca A. Hunsicker
Indexing by Deborah H. Steiner

Elements typeset by Hot Type Ltd., Washington, DC
Printed and bound by Maple Press, York, PA